# Springer Climate

Springer Climate is an interdisciplinary book series dedicated to climate research. This includes climatology, climate change impacts, climate change management, climate change policy, regional climate studies, climate monitoring and modeling, palaeoclimatology etc. The series publishes high quality research for scientists, researchers, students and policy makers. An author/editor questionnaire, instructions for authors and a book proposal form can be obtained from the Publishing Editor.

**Now indexed in Scopus® !**

Anindita Sarkar · Nairwita Bandyopadhyay
Editors

# Gender-Transformative Approaches for Climate Change Adaptation

## Policies and Practices from the Global South

Springer

*Editors*
Anindita Sarkar
Center for Development Research (ZEF)
University of Bonn
Bonn, Germany

Department of Geography
Miranda House
University of Delhi
Delhi, India

Nairwita Bandyopadhyay
Department of Geography
Haringhata Mahavidyalaya
University of Kalyani
Nadia, West Bengal, India

ISSN 2352-0698 ISSN 2352-0701 (electronic)
Springer Climate
ISBN 978-3-031-90235-2 ISBN 978-3-031-90236-9 (eBook)
https://doi.org/10.1007/978-3-031-90236-9



This Springer imprint is published by the registered company Springer Nature Switzerland AG
The registered company address is: Gewerbestrasse 11, 6330 Cham, Switzerland

# Foreword

With the impacts of climate change looming large, the world climate community is, raising urgent calls for more effective adaptation. However, climate change adaptation is most often complex and embedded in social, economic, and political structures that can exacerbate existing inequalities. Nowhere is this more evident than in the Global South, where both climate impacts and sometimes adaptation strategies have disproportionately affected socially excluded groups, particularly women. Colonial legacies of economic and political inequalities and deep-rooted social and cultural norms compound these inequities. So, addressing these disparities requires a nuanced understanding of the structural barriers and the need to find context-specific and localized solutions.

Mainstream literature on climate change perpetuates the concept of women as victims and as disadvantaged obscuring the narratives of women as leaders, innovators, and agents of change. Adaptation strategies need to go beyond mere inclusion to actively tackle the structural inequalities so that we are all responsible for finding solutions and can deal with climate change head-on. Gender-transformative approaches—ones that not only address gender disparities but reshape power dynamics, foster agency, and ensure equitable participation in decision-making—are vital parts and can be a solution to this debate.

*Gender-Transformative Approaches for Climate Change Adaptation: Policies and Practices from the Global South* is thus a timely and essential contribution to the discourse on climate justice. This book brings together diverse perspectives, case studies, and policy frameworks that highlight the intersection of gender, climate adaptation, and sustainable development. It underscores the urgent need to integrate gender-transformative strategies into climate policies, ensuring that adaptation efforts are not just inclusive but actively contribute to reshaping systems of inequality.

The insights captured in these pages underscore an essential truth: climate resilience for all cannot be achieved without gender equity. As we stand at the crossroads of a climate crisis, this book serves as both a call to action and a roadmap for building a more just and sustainable future. I compliment the two editors of this book for representing a tapestry of experiences, cultures, and backgrounds, united by their commitment to amplifying the voices and perspectives, often unheard, in the broader discourse on climate adaptation. I hope that this work will inspire and inform transformative policies and practices that make climate adaptation not just sustainable, but just.

Dr. Nicoline de Haan
Director
CGIAR GENDER Platform
Nairobi, Kenya

# Acknowledgments

This book *Gender-Transformative Approaches for Climate change Adaptation: Policies and Practices from the Global South* discusses the importance of gender-transformative approaches in climate change adaptation, highlighting the need to move beyond gender-sensitive and gender-responsive policies to address the root causes of gender inequality.

The completion of this edited volume on gender-transformative approaches would not have been possible without the contributions and support of numerous individuals and organizations.

First and foremost, we express our deepest gratitude to the contributing authors whose insightful chapters form the core of this book. Their dedication, expertise, and passion for advancing gender equality have been truly inspiring.

We are also profoundly thankful to our colleagues and reviewers who provided invaluable feedback throughout the editing process. Their constructive critiques and suggestions have significantly enhanced the quality of this work.

Our sincere thanks go to the team at Springer, particularly to Springer Climate Series Editor, John Dodson; Publishing editor, Robert Doe; and Production editor, Amudha Vijayarangan for unwavering support, guidance, and professionalism in bringing this book to publication. Their commitment to fostering academic discourse on critical social issues is commendable.

We are also indebted to the various institutions and organizations that supported this endeavour, either through funding, research collaboration, or providing resources that facilitated the development of this book.

Lastly, we would like to acknowledge our families and friends for their understanding and encouragement during the many hours spent working on this project. Their support has been a constant source of strength and motivation.

To all who have contributed to this work, directly or indirectly, we extend our heartfelt thanks.

Prof. Dr. Anindita Sarkar
Dr. Nairwita Bandyopadhyay

# Contents

**Part I Concepts and Overview of Gender and Climate Change**

1 **Contextualizing Gender-Transformative Approaches for Climate Change Adaptation** ........ 3
Anindita Sarkar and Nairwita Bandyopadhyay

2 **Exploring Gendered Dimensions of Climate-smart Agriculture for Food Security in the Global South: A Systematic Literature Review** ........ 21
Neelam Singh, Dibakar Sahoo, and Prasanta Moharaj

3 **The Intersection of Climate Change and Gender-Based Violence in Sub-Saharan Africa: A Synthesis** ........ 45
Douglas Nyathi, Joram Ndlovu, Stanley Maphosa, Thembelihle Nyathi, and Nkosilamandla Kunene

4 **Climate-Related Disasters and Gender Inequalities: Unveiling Key Elements Through Disaggregated Data Tool** ........ 65
Gabriela Azevedo Couto, Maurício Trentin, Miguel Angel Trejo-Rangel, Debora Pignatari Drucker, and Carlos A. Nobre

**Part II Vulnerability to Climate Change and Disproportionate Burden on Women**

5 **Climate Extremes and Disproportionate Burden on Women in the Global South: A Way Forward** ........ 87
Mumtaz Alam and Kunal Avishek Gounder

6 **Understanding 'Gendered' Impacts of Climate Change and the Role of Women in Adaptation: Evidence from the Indian Sundarbans** ........ 105
Pratyay Raha, Amrita Sen, and Debarchana Biswas

**7 The Gender Divide: Factors Amplifying Women's Vulnerability During Extreme Events in Bangladesh** .............. 123
Iffat Ara Jasmin, Atia Sanjida Sushoma, Imran Hosen, and Ananda Kumar Biswas

**8 Climate Change and Food Security in Sub-Saharan Africa: An Analysis from Gendered Lens** .............................. 141
Chandani Tiwari, Sakshi Bahuguna, and Aparna P. Krishnan

**9 Climate Change Migration and Gender Vulnerability: Challenges and Solutions** ...................................... 161
Junaid ul Shafi

**10 Assessment of Women Farmers' Drudgery in Rice Farming with Climate Change: A Case Study from South India** ............ 183
Dhanya Punnoli, Jayarajan Kunnampalli, and Sreeraj Punnoli

**11 Hope and Despair: Towards a Gendered Emotional Geography of Water in a Changing Climate in the District of Bankura, West Bengal** ................................. 201
Tanusree Paul and Suranjana Karmakar

**Part III Women as Change-Makers: Gender Transformative Climate Adaptation**

**12 Gender-Transformative Adaptation for Food Security, Rural Livelihoods, and Agriculture: The Case Study of Mal Pahariya Women in Eastern Coalfield Regions of India** ........... 227
Aparajita Banerjee

**13 Women Leading the Way: Female-Led Adaptation Initiatives in Western Himalayas** ................................ 247
Katyayini Sood, Abhishek Kumar, and Vipasha Negi

**14 Can Women's Empowerment Lead to More Effective Climate Action? A Transformative Experience from Rajasthan, India** ...... 269
Kritika Mishra and Farhat Naz

**15 Uplifting Women Farmers by Adopting Climate-Smart Practices with Climate Finance: Evidence from Central and Eastern States of India** ................................. 291
Tarun Katoch and Parul Gurjar

**16 Women on the Frontline: Agriculture and Climate Change in Uttar Pradesh, India** ................................ 317
Surabhi Singh and Farhat Naz

**Part IV Addressing Gender In Climate Change Adaptation: Policies And Programmes**

**17 Gender Integration in Climate-Resilient Agriculture: Exploring Policies, Programmes, and Transformative Pathways** ........ 339
Mamta Mehar and Twinkle Jaspal

**18 Addressing the Crisis of Climate-Induced Internally Displaced Women in Bangladesh** ........ 355
Zelina Sultana

**19 Women's Resilience in Adapting to Climate Change on the North Coast of Java Island, Indonesia** ........ 377
Hari Harjanto Setiawan, Badrun Susantyo, Nyi Raden Irmayani, Habibullah, Mery Ganti, Abdullah, Luthfi Setiadiwibawa, Euis Setiawati, Ineu Maryani, Suhandy Siswoyo, Djaenudin, Mahdi Ibrahim Tanjung, Karno, Ruslan Halifu, and Rizal Akbar Aldyan

**20 Climate Crisis and "*City Citizens*": A Look into the Gender-Inclusive Disaster Management Policies of Indian Smart Cities** ........ 397
Rituparna Bhattacharya and Sramana Mukherjee

**Author Index** ........ 425

**Subject Index** ........ 427

# Contributors

**Abdullah** Telkom University, Bandung, Indonesia

**Mumtaz Alam** Department of Social Sciences, College of Humanities and Education, Fiji National University, Suva, Fiji

**Rizal Akbar Aldyan** National Research and Innovation Agency (BRIN), Jakarta, Indonesia

**Sakshi Bahuguna** Abu Dhabi University, Abu Dhabi, United Arab Emirates

**Nairwita Bandyopadhyay** Department of Geography, Haringhata Mahavidyalaya, University of Kalyani, Nadia, West Bengal, India

**Aparajita Banerjee** German Institute of Development and Sustainability (IDOS), Bonn, Germany

**Rituparna Bhattacharya** Institut für Geographie, Friedrich-Alexander-Universität Erlangen-Nürnberg, Erlangen, Germany

**Ananda Kumar Biswas** Development Studies Discipline, Khulna University, Khulna, Bangladesh

**Debarchana Biswas** Department of Humanities and Social Sciences, Indian Institute of Technology, Guwahati, India

**Gabriela Azevedo Couto** National Institute for Space Research, São José Dos Campos, Brazil

**Djaenudin** National Research and Innovation Agency (BRIN), Jakarta, Indonesia

**Debora Pignatari Drucker** Embrapa Digital Agriculture, EMBRAPA, Campinas, Brazil

**Mery Ganti** National Research and Innovation Agency (BRIN), Jakarta, Indonesia

**Kunal Avishek Gounder** College of Humanities and Education, Fiji National University, Suva, Fiji

**Parul Gurjar** Bankers Institute of Rural Development, Lucknow, Uttar Pradesh, India

**Habibullah** National Research and Innovation Agency (BRIN), Jakarta, Indonesia

**Ruslan Halifu** National Research and Innovation Agency (BRIN), Jakarta, Indonesia

**Imran Hosen** Department of International Relations, University of Dhaka, Dhaka, Bangladesh

**Nyi Raden Irmayani** National Research and Innovation Agency (BRIN), Jakarta, Indonesia

**Iffat Ara Jasmin** Department of International Relations, University of Dhaka, Dhaka, Bangladesh

**Twinkle Jaspal** Assistant Manager, MicroSave Consulting, Delhi, India

**Suranjana Karmakar** Women's Studies Centre, Institute of Humanities and Social Sciences, Visva-Bharati, Santiniketan, India

**Karno** Institute of Home Affairs Government, Jakarta, Indonesia

**Tarun Katoch** Dvara E-Registry, Hyderabad, Telangana, India

**Aparna P. Krishnan** Symbiosis School of Economics, Symbiosis International University, Pune, India

**Abhishek Kumar** People for Integrated Planning and Adaptable Living (PIPAL) Foundation, New Delhi, India

**Nkosilamandla Kunene** Department of Development Studies, Lupane State University, Lupane, Zimbabwe

**Jayarajan Kunnampalli** Department of Geography, Government College Chittur, Palakkad, Kerala, India

**Stanley Maphosa** Institute of Pan African Thought, University of Johannesburg, Guateng, South Africa

**Ineu Maryani** Guidance and Counseling Study Program, Faculty of Educational Sciences, Indonesia University of Education, Bandung, Indonesia

**Mamta Mehar** Evidence Synthesis Consultant, Climate Change Adaptation and Mitigation Impact Action Platform of the CGIAR, Delhi, India

**Kritika Mishra** PhD Research Scholar, School of Liberal Arts, Indian Institute of Technology, Jodhpur, Rajasthan, India

**Prasanta Moharaj** School of Law, Dayananda Sagar University, Bangalore, India

**Sramana Mukherjee** Department of Humanities and Social Sciences, National Institute of Technology Rourkela, Rourkela, Odisha, India

**Farhat Naz** Associate Professor of Sociology, School of Liberal Arts, Indian Institute of Technology, Jodhpur, Rajasthan, India

**Joram Ndlovu** School Social Sciences, Howard College, University of KwaZulu-Natal, Glenwood, South Africa

**Vipasha Negi** Punjab State Council for Science and Technology, Chandigarh, India

**Carlos A. Nobre** Institute of Advanced Studies, University of São Paulo, São Paulo, SP, Brazil

**Douglas Nyathi** School of Public Management, Governance and Public Policy, University of Johannesburg, Gauteng, South Africa

**Thembelihle Nyathi** Department of Development Studies, Lupane State University, Lupane, Zimbabwe

**Tanusree Paul** Women's Studies Centre, Institute of Humanities and Social Sciences, Visva-Bharati, Santiniketan, India

**Dhanya Punnoli** Department of Geography, Government College Chittur, Palakkad, Kerala, India

**Sreeraj Punnoli** Senior Software Developer, Ubisoft, Halifax, Canada

**Pratyay Raha** School of Art, College of Design and Social Context, RMIT University, Melbourne, Australia

**Dibakar Sahoo** Department of Economics, Faculty of Social Sciences, Banaras Hindu University, Varanasi, Uttar Pradesh, India

**Anindita Sarkar** Centre for Development Research (ZEF), University of Bonn, Bonn, Germany;
Department of Geography, Miranda House, University of Delhi, Delhi, India

**Amrita Sen** Department of Humanities and Social Sciences, Indian Institute of Technology, Kharagpur, India

**Luthfi Setiadiwibawa** Winaya Mukti University, Bandung, Indonesia

**Hari Harjanto Setiawan** National Research and Innovation Agency (BRIN), Jakarta, Indonesia

**Euis Setiawati** Religious Training Center Bandung, Indonesia's Ministry of Religious Affairs (MORA), Bandung, Indonesia

**Junaid ul Shafi** School of Law, BML Munjal University, Gurgaon, India

**Neelam Singh** Department of Economics, Faculty of Social Sciences, Banaras Hindu University, Varanasi, Uttar Pradesh, India

**Surabhi Singh** Ph.D. Scholar, School of Liberal Arts, Indian Institute of Technology, Jodhpur, India

**Suhandy Siswoyo** Department of Architectural Engineering Education, Indonesia University of Education, Bandung, Indonesia

**Katyayini Sood** Department of Geography, University of Bonn, Bonn, Germany

**Zelina Sultana** Faculty of Business and Law, Anglia Ruskin University, Cambridge, UK;
Department of Law, Jagannath University, Dhaka, Bangladesh

**Badrun Susantyo** National Research and Innovation Agency (BRIN), Jakarta, Indonesia

**Atia Sanjida Sushoma** Department of Women and Gender Studies, University of Dhaka, Dhaka, Bangladesh

**Mahdi Ibrahim Tanjung** Ministry of Public Works, Bandung, Indonesia

**Chandani Tiwari** Symbiosis School of Economics, Symbiosis International University, Pune, India

**Miguel Angel Trejo-Rangel** School of Architecture, Planning and Environmental Policy, University College Dublin, Dublin, Ireland

**Maurício Trentin** 2MONOPÓLIO, INCODE Tech, TIMESAFE S.a.a.S, BITBOT, São Paulo, Brazil

# Abbreviations

| | |
|---|---|
| AAP | Agrometeorological Advisory Programme |
| ABD | Area-Based Development |
| ACIS | Agroclimatic Information in Southeast Asia |
| AI | Artificial Intelligence |
| AMRUT | Atal Mission for Rejuvenation and Urban Transformation |
| ATFP | Agricultural Total Factor Productivity |
| BCT | Barefoot College in Tilonia |
| BPM | Block Project Manager |
| BRAC | Bangladesh Rural Advancement Committee |
| CCAFS | CGIAR Research Program on Climate Change, Agriculture and Food Security |
| CGIAR | The Consultative Group on International Agricultural Research |
| CGWB | Central Ground Water Board |
| CICERO | Centre for International Climate and Environmental Research |
| CIDP | Cashew Infrastructure Development Project |
| CIID | Climate-Induced Internal Displacement |
| CIIDW | Climate-Induced Internally Displaced Women |
| CIMMYT | International Maize and Wheat Improvement Center, CGIAR Research Center |
| CJ | Climate Justice |
| COP | Conference of Parties |
| CSA | Climate-Smart Agriculture |
| CSV | Climate-Smart Village |
| CWR | Clear Water Reservoirs |
| DRC | Danish Refugee Council |
| DRR | Disaster Risk Reduction |
| DSAP | Development of Sustainable Agriculture in the Pacific Programme |
| DSR | Direct Seeded Rice |
| EBA | Ecosystem-Based Adaptation |
| FAO | Food and Agriculture Organization |
| FGD | Focus Group Discussion |

| | |
|---|---|
| FHTC | Functional Household Tap Connection |
| FPE | Feminist Political Ecology |
| FPO | Farmer Producer Organizations |
| GAD | Gender and Development |
| GBV | Gender-Based Violence |
| GCCASP | Gender Climate Change and Agricultural Support Programme |
| GDI | Gender Development Index |
| GHG | Green House Gases |
| GHI | Global Hunger Index |
| GII | Gender Inequality Index |
| GLS | Generalized Least Squares |
| GOI | Government of India |
| GSI | Geological Survey of India |
| GTA | Gender-Transformative Adaptation |
| GTR | Gender-Transformative Research |
| HDI | Human Development Index |
| HKH | Hindu Kush Himalayas |
| HYV | High Yielding Variety |
| IBGE | Instituto Brasileiro de Geografia e Estatística |
| ICAR | Indian Council of Agricultural Research |
| ICIMOD | International Centre for Integrated Mountain Development |
| ICRISAT | International Crops Research Institute for the Semi-Arid Tropics |
| ICSSR | Indian Council of Social Science research |
| ICT | In-Laid-Digital Technology |
| IDI | In-Depth-Interviews |
| IDP | internally Displaced Persons |
| IFC | International Finance Corporation |
| IFRC | International Federation of Red Cross |
| IGWG | Inter-Agency Gender Working Group |
| IIED | International Institute for Environment and Development |
| ILO | International Labour Organization |
| IMD | Indian Meteorological Department |
| IOM | International Organization for Migration |
| IPCC | Intergovernmental Panel on Climate Change |
| IPV | Intimate Partner Violence |
| IRRI | International Rice Research Institute |
| ISD | Indian Sundarbans Delta |
| ITEC | Indian Technical and Economic Cooperation |
| JJM | Jal Jeevan Mission |
| KII | Key Informant Interviews |
| KSDMA | Kerala State Disaster Management Authority |
| KSHEC | Kerala State Higher Education Council |
| LAC | Latin American and Caribbean Countries |
| LGBTQ | Lesbian, Gay, Bisexual, and Transgender |
| MFIN | Microfinance Institutions Network |

| | |
|---|---|
| MGNREGS | Mahatma Gandhi National Rural Employment Guarantee Mission |
| MTR | Machine Transplanted Rice |
| NABARD | National Bank for Agriculture and Rural Development |
| NAP | National Adaptation Plan |
| NAPCC | National Action Plan on Climate Change |
| NDC | Nationally Determined Contributions |
| NGO | Non-Governmental Organization |
| NTFP | Non-Timber Forest Product |
| NUA | New Urban Agenda |
| OBC | Other Backward Class |
| OHR | Overhead Reservoir |
| OLS | Ordinary Least Squares |
| OXFAM | Oxford Committee for Famine Relief |
| PCI | Problem Confrontation Index |
| PHE | Public Health Engineering Department |
| PIM | Participatory Irrigation Management System |
| PSP | Participatory Scenario Planning |
| PTG | Primitive Tribal Group |
| PTSD | Post-Traumatic Stress Disorder |
| PVTG | Particularly Vulnerable Tribal Group |
| RLC | Radio Listeners' Clubs |
| RS | Remote Sensing |
| RWS | Rural Water Supply |
| SA | South Asia |
| SAI | Sustainable Agricultural Intensification |
| SAPCC | State Action Plans on Climate Change |
| SC | Scheduled Castes |
| SCALA | Scaling up Climate Action in Land Use and Agriculture Sectors |
| SCM | Smart Cities Mission |
| SDG | Sustainable Development Goals |
| SEA | South East Asia |
| SGBV | Sexual and Gender-Based Violence |
| SHG | Self Help Group |
| SME | Small and Medium Enterprises |
| SOGIESC | Sexual Orientation, Gender Identity and Expression, and Sex Characteristics |
| SSA | Sub-Saharan Africa |
| SSN | Social Safety Net |
| ST | Scheduled Tribes |
| STI | Sexually Transmitted Infections |
| SWRC | Social Work and Research Centre |
| TFP | Total Factor Productivity |
| UN | United Nations |
| UNDP | United Nations Development Programme |
| UNDRR | United Nations Office for Disaster Risk Reduction |

| | |
|---|---|
| UNFCCC | United Nations Framework Convention on Climate Change |
| UNFPA | United Nations Population Fund |
| UNICEF | United Nations International Children's Emergency Fund |
| USAID | United States Agency for International Development |
| UTI | Urinary Tract Infection |
| VAWG | Violence Against Women and Girls |
| VSLA | Village Savings and Loan Associations |
| VWSC | Village Water Sanitation Committee |
| WBADMIP | West Bengal Accelerated Development of Minor Irrigation Project |
| WDS | Water Detention Structures |
| WHO | World Health Organization |
| WID | Women in Development |
| WIRE | Women in Rwandan Energy |
| WMO | World Meteorological Organization |

# List of Figures


Fig. 1.1 Three-step framework for gender-transformative approaches to climate change adaptation. *Source* Prepared by authors ..... 10
Fig. 2.1 Preferred reporting items for systematic reviews and meta-analysis (PRISMA) flow diagram. *Source* Prepared by the authors ..... 27
Fig. 2.2 Trend of published articles included in the sample. *Source* Prepared by the authors ..... 28
Fig. 2.3 Most relevant sources based on the number of documents. *Source* Author's compilation using Biblioshiny ..... 29
Fig. 2.4 Trend of topics covered under the CSA studies. *Source* Authors' compilation using Biblioshiny ..... 30
Fig. 2.5 Thematic evolution map of the keywords included in the study. *Source* Authors' compilation using Biblioshiny ..... 31
Fig. 2.6 Conceptual framework. *Source* Prepared by authors ..... 35
Fig. 4.1 Images produced for Sobrevidas© platform and social networks. *Source* Elaborated by the author (Couto 2023) ..... 73
Fig. 4.2 Sobrevidas© project. 2024 *Source* Elaborated by the author Mauricio Tretin ..... 74
Fig. 4.3 Sobrevidas© project: Survivor context. 2024 *Source* Elaborated by the author Mauricio Treitin ..... 78
Fig. 6.1 Voices from the margins, Beguakhali. *Source* Field 2023 ..... 111
Fig. 6.2 Resilient roots: Mangrove plantation done by women villagers. *Source* Field 2023 ..... 115
Fig. 7.1 Sub-national risk index of Bangladesh, 2022. 2022 *Source* Bangladesh Inform Sub-National Risk Index 2022 (Lewis and Hasan) ..... 129
Fig. 8.1 Agricultural total factor productivity index of SSA (as a whole), 2000–2022. *Source* Author's compilation ..... 147

Fig. 8.2 Aggregate gender development index of SSA, 2000–2022. *Source* Author's compilation ..... 152
Fig. 8.3 Population above 25 years of age with some kind of secondary education in SSA, 2000–2022. *Source* Author's compilation ..... 152
Fig. 8.4 Gender inequality index of SSA, 2000–2022. *Source* Author's compilation ..... 153
Fig. 8.5 Labor force participation rate in SSA, 2000–2022. *Source* Author's compilation ..... 153
Fig. 9.1 Global trends of forced displacement 2014–2023. 2023 *Source* UNHCR, Global Trends forced Displacement,Copenhagen ..... 164
Fig. 10.1 Location of study area ..... 186
Fig. 10.2 Block-wise rice crop concentration index in the study area. *Data Source*: Department of Economics and Statistics, Palakkad for the year 2016 ..... 188
Fig. 10.3 Pictures of the survey area in Bharathapuzha Basin. *Source* Taken by authors during field survey ..... 190
Fig. 10.4 Temporal range in day and nighttime mean highest temperatures in the study area, 1991–2023. *Source* Author's compilation from field data ..... 192
Fig. 10.5 Rainfall recorded during the crop growing season in the study area, 2023. *Source* Author's compilation from field data ..... 192
Fig. 10.6 Severity of the climate and weather faced by surveyed farmers. *Source* Author's compilation from field data ..... 193
Fig. 10.7 Problem confrontation index among the surveyed farmers. *Source* Author's compilation from field data ..... 194
Fig. 11.1 Location of study sites. *Source* Prepared by authors ..... 206
Fig. 11.2 Methodology tree. *Source* Prepared by the authors ..... 208
Fig. 11.3 Non-participant observation at village Water Sanitation Committee (VWSC) meeting in Sonamukhi Block, Bankura. *Source* Fieldwork in Bankura, 2024 ..... 209
Fig. 11.4 Key informant interview with the officials in Gram Panchayat. *Source* Fieldwork in Bankura, 2024 ..... 209
Fig. 11.5 Resource maps drawn by female participants. *Source* Fieldwork in Bankura, 2024 ..... 213
Fig. 11.6 Resource maps drawn by male participants. *Source* Fieldwork in Bankura, 2024 ..... 214
Fig. 12.1 Location of the study area. *Source* Prepared by Adrija Roy ..... 232
Fig. 12.2 Two Mal Pahariya women overlooking coal mines near their homes. *Source* Photo by the author ..... 233
Fig. 12.3 Focus group discussion. *Source* Photo by the author ..... 233
Fig. 13.1 Impact web developed during interviews. *Source* In-depth interviews conducted during the fieldwork ..... 254

Fig. 13.2 Pictures from the fieldwork .......... 257
Fig. 14.1 Showing gender equality at the core. *Source* Created by authors .......... 272
Fig. 14.2 Location of Barefoot College in Tilonia village. *Source* Created by authors .......... 273
Fig. 14.3 Images captured from a wall in the Barefoot College campus with inspiring words that reflect the institution's commitment to empowerment and positive change. *Source* Images captured by the Authors during the field visits conducted in 2022–2023 .......... 274
Fig. 14.4 Seven core principles on which the various programmes are implemented at BCT. *Source* Created by authors .......... 277
Fig. 14.5 Triple-win with the solar programme at BCT. *Source* Created by authors .......... 278
Fig. 14.6 Solar energy generation and training centre at BCT. *Source* Images captured by the authors during the field visits conducted in 2022–2023 .......... 280
Fig. 14.7 Impacts of BCT on water conservation and climate change adaptation. *Source* Barefoot College Social Work and Research Centre (2024) .......... 283
Fig. 14.8 Values of Barefoot College. *Source* Barefoot College Social Work and Research Centre (2024) .......... 286
Fig. 14.9 Linking programmes at BCT for promoting SDGs. *Source* Created by authors .......... 288
Fig. 15.1 Map showing Indian states selected for study. *Source* Created by Authors, 2025 .......... 299
Fig. 15.2 Conceptual analysis of study. *Source* Author's conceptualization, 2025 .......... 300
Fig. 15.3 Membership mobilization in FPOs. *Source* Field photograph by Dvara E-Registry, 2024 .......... 300
Fig. 15.4 Interface of Krishak App (Farmer App), FPO Dashboard, and Krishak Sathi App (Farmer's friend App). *Source* Screenshot of FPO dashboard and apps by Dvara E-Registry, 2025 .......... 301
Fig. 15.5 KhetScore parameters and sample score report. *Source* Screenshot by Dvara E-Registry, 2025 .......... 302
Fig. 15.6 Value-added products. *Source* Field photograph by Dvara E-Registry, 2025 .......... 306
Fig. 15.7 Agricultural input Shop of FPO. *Source* Field photograph by Dvara E-Registry, 2025 .......... 307
Fig. 15.8 Seed distribution at Farmer's doorstep. *Source* Field photograph by Dvara E-Registry, 2025 .......... 307
Fig. 15.9 Soybean Seed bags produced by FPO. *Source* Field photograph by Dvara E-Registry, 2025 .......... 308

Fig. 15.10 The "Seed Cafeteria", where 21 Varieties of rice are grown on 1 acre of farmland. *Source* Field photograph by Dvara E-Registry, 2025 ... 309
Fig. 15.11 Demonstrating bio-agricultural inputs for plant root treatment. *Source* Field photograph by Dvara E-Registry, 2023 ... 310
Fig. 15.12 A digital training session for women farmers. *Source* Field photograph by Dvara E-Registry, 2025 ... 311
Fig. 16.1 Location map of Saharanpur district in Uttar Pradesh. *Source* Prepared by Author ... 322
Fig. 16.2 Sampling design of study. *Source* Sampling plan by authors ... 322
Fig. 16.3 Indicators for analysing gender inequality in ggriculture. *Source* Conceptual diagram prepared by authors to understand gender inequality in agriculture ... 323
Fig. 16.4 Women working in sugarcane field. *Source* Author's image from the field ... 325
Fig. 16.5 Shift in sources of irrigation after green revolution in Saharanpur district. *Source* Prepared by authors based on data accumulated from different reports of Land Use Statistics from the Directorate of Economics and Statistics, Department of Agriculture & Farmers Welfare, Ministry of Agriculture and Farmers Welfare, Government of India ... 329
Fig. 16.6 Climate resilient practices in study area. *Source* Author's images from the field ... 332
Fig. 19.1 Framework for Women's Resilience in Adapting. *Source* Adapted from Rej et al. (2023) ... 382
Fig. 19.2 Conceptual framework of data and methodology used. *Source* Created by authors ... 383
Fig. 19.3 Research results framework for women in adapting to climate change. *Source* Created by authors ... 384
Fig. 19.4 A woman is working looking for shrimp and crabs on the beach. *Source* By the authors ... 386
Fig. 20.1 Number of urban areas with population over 7,50,000 affected by disasters. *Source* Gencer and UNDRR (2017) ... 399
Fig. 20.2 Spatio-temporal depiction of risk reduction strategies. *Source* United Nations Office for Disaster Risk Reduction (2019) ... 400
Fig. 20.3 Map depicting the hundred smart cities of India. *Source* UN-Habitat (2023) ... 405
Fig. 20.4 Conceptual framework. *Source* Prepared by the authors ... 406

Fig. 20.5 ICT interventions for climate resilience in smart cities. *Source* UN-Habitat (2023) .............................. 412
Fig. 20.6 Type of disaster risk reduction strategies in different smart cities. *Source* Authors findings ..................... 416

# List of Tables


| | | |
|---|---|---|
| Table 2.1 | Search strings used in the database search | 27 |
| Table 2.2 | Inclusion and exclusion criteria for the documents undertaken in the study | 28 |
| Table 2.3 | Most cited documents | 30 |
| Table 8.1 | Variables and data sources | 149 |
| Table 8.2 | Cross-sectional dependency test | 150 |
| Table 8.3 | GLS estimation results | 150 |
| Table 8.4 | Gender-responsive climate change actions in a few countries of SSA | 155 |
| Table 10.1 | Socioeconomic characteristics of the respondents | 189 |
| Table 11.1 | Study sites | 207 |
| Table 11.2 | Socio-economic profile of FGD participants | 210 |
| Table 18.1 | Reviewed legal documents with provisions for climate change migrant/displaced population | 359 |
| Table 18.2 | Distinguishing features of climate-induced displacement and climate-induced migration | 360 |
| Table 18.3 | Pre- and post-displacement issues of women | 361 |
| Table 18.4 | Findings and suggestions for CIIDW | 371 |
| Table 20.1 | Identification and management of disaster risks in different smart cities | 410 |
| Table 20.2 | Types of disaster risk reduction strategies in different smart cities | 413 |
| Table 20.3 | Smart cities with gender-inclusive disaster resilience projects | 414 |
| Table 20.4 | Key case studies of gender-inclusive disaster resilience projects | 415 |

# Part I
# Concepts and Overview of Gender and Climate Change

# Chapter 1
# Contextualizing Gender-Transformative Approaches for Climate Change Adaptation

**Anindita Sarkar and Nairwita Bandyopadhyay**

**Abstract** The topic of this chapter is to contextualize and unpack gender-transformative approaches in climate change adaptation pathways. The conceptual understanding of transformative approaches is motivated primarily by three questions. First, what does the gender-transformative approach mean in the context of climate change adaptation, and how is it different from gender-sensitive and gender-responsive approaches? Second, why do we need gender-transformative approaches, and how can climate change adaptation pathways provide opportunities for social transformations associated with sustainable development? Third, how can these social transformations take place or what can be the framework for Gender-transformative approaches? Synthesizing existing literature and knowledge on climate change adaptation and feminist literature, the study finds that transformative changes are inevitable when communities and societies change their ways of livelihood and lives to adapt to climate change. These pathways are bound to change the gendered roles and traditional socio-cultural ways of life both negatively and positively, leading to social transformations. We argue that a combination of Gender-transformative approaches in research and appropriate adaptation programmes have the potential to shift "ways of life" from "business-as-usual." While adaptation policies can place vulnerable women at the centre of adaptation efforts, Gender-transformative approaches in research can re-imagine unique methods that lead to co-creation and co-learning of new roles and values giving women the agency to become the "new change-makers." This can be done by innovative pathways that question and try to change the traditional societal narratives to new stories of change.

---

A. Sarkar (✉)
Center for Development Research (ZEF), University of Bonn, Bonn, Germany
e-mail: anindita@uni-bonn.de

Department of Geography, Miranda House, University of Delhi, Delhi, India

N. Bandyopadhyay
Department of Geography, Haringhata Mahavidyalaya, University of Kalyani, Nadia, West Bengal, India
e-mail: nairwita@gmail.com



A. Sarkar and N. Bandyopadhyay (eds.), *Gender-Transformative Approaches for Climate Change Adaptation*, Springer Climate,
https://doi.org/10.1007/978-3-031-90236-9_1

**Keywords** Gender-transformative approaches · Gender-transformative research methodologies · Climate change adaptation · Gender · Women · Vulnerability

## Introduction

Climate change adaptation is a normative outcome. The adaptation pathways offer opportunities for change because they are designed as a sequence of actions, that can be implemented progressively and developed to meet adaptation needs depending on future dynamics (Werners et al. 2021). Gender inequalities and power dynamics make women more vulnerable and marginalized due to reasons like limited access to resources, opportunities and assets. Such constraints put formidable barriers for women to adapt to climate change impacts (Resurrección et al. 2019). Gendered roles and cultural norms further abstain women from making decisions and choosing the way they want to live. Without addressing these inequalities, any adaptation measure implemented on the ground will have a greater possibility of exacerbating social injustice and escalating inequalities leading to maladaptation (Berkhout et al. 2024). However, since adaptation pathways are continuous processes (Fazey et al. 2016), they have the potential to promote collaborative learning and greater scope for enabling social transformations.

Widening gender inequalities in global development is considerable and now well-established (Mullinax et al. 2018). The Sustainable Development Goals (SDGs) that propagate the critical objective of "leaving no one behind" need to also start focusing on a renewed goal for equitable social transformation, to address the root causes of gender inequality. Adaptation plans will likely be gendered undermining social relations, practices, and processes because structural inequalities are complex, unique, context-specific, and place-based (Eriksen et al. 2021). Moreover, since climate change impacts cut across the intersections of age, class, ethnicity, gender, and space; adaptation plans also require different strategies to address that (Erwin et al. 2021). Thus, the solutions need to be in place and context-based and consider the nuances of gender norms, deep-seated power relations, and individual and collective agency for transformation.

There is both a knowledge and policy gap on what extent and how climate change interacts with social and spatial inequality and how one can estimate and measure gender-transformative change within climate change adaptation plans and interventions. Though gender-transformative approaches are being progressively recognized as an important aspect of persistent progress toward gender equality, there remain important methodological gaps. It is time that new paths of adaptations are re-imagined, and new research methodologies redesigned.

The objective of this chapter is to provide a background to the book "*Gender-Transformative Approaches for Climate Change Adaptation: Policies and Practices from the* Global South." This book is an attempt to capture the relationship between vulnerabilities created by structural inequalities and current adaptation strategies in a varied mix of ecological, social, economic, and geographical contexts concerning the Global South. This introductory chapter is divided into three core sections. First, it unpacks the concept of Gender-Transformative Adaptation (GTA) and explains how it is different from the concept of Gender-Sensitive Adaptation and Gender-Responsive Adaptation. Second, it discusses why we need to think about GTA and how climate change adaptation can be used as potential for social transformation. Third, it talks about how these pathways for social transformation can be chartered through policy interventions and gender-transformative research methodologies. In the concluding section, the chapter will add a prologue to the book analysing how the constellation of studies in this volume contributes to new knowledge and methodologies towards this direction.

## Concept of Gender-Transformative Adaptation

GTA is about transforming social systems, norms, and cultures that perpetuate inequality. It is a way by which individual gender-equitable behaviour is addressed on one hand and, on the other, attempts to change the power dynamics and unequal social structures that serve to perpetuate inequalities (Hillenbrand et al. 2015; Jerneck 2018). For example, programmes that address women's self-improvement, adopting "women only" policies, or approaches that question men's behaviour and attitudes can be called transformative because they compel to change the gendered roles and socio-cultural values that position women at the bottom of the power structure.

Both evidence and theories verify that social and cultural values change with changing times. For example, Greenfield (2016), in his theory of social change and human development, discusses how cultural values and social norms (socialization processes) and human behaviour shift in anticipated ways to adapt to new situations (learning environments) (Greenfield 2016). To adapt to new changing conditions, socio-cultural values for social relationships shift from hierarchical to egalitarian gender relations, manifesting in changing gendered roles and expected norms (Uhls and Greenfield 2011). In this context, we can argue that climate change adaptation pathways have the potential to bring about such social transformations as the Industrial Revolution and globalization did in the past.

However, GTA is based on the three proven bases of social transformation. First, it is based on the assumption that women's adaptation strategies will vary from men depending on their socio-economic background and cultural contexts. Second, adaptation policies following positive affirmation for targeted beneficiaries have the power to influence and transform broader structural inequalities concerning not only gender but also its intersectional dimensions. Third, it can provide agency to the adaptation strategies and equip individuals or even communities to change the existing vulnerable conditions (Hillenbrand et al. 2015).

Since mid-2000, gender dimensions of climate change in research have initiated advocacy for formulating gender-responsive climate adaptation policies and programmes particularly tailored to advance gender equality aligning with the Paris Agreement (UNFCCC 2015). At present, many national climate policies recommend fair distribution of opportunities and risks so that the burden of the transition does not disproportionately disadvantage any particular gender, race, ethnicity or class (Chan et al. 2024). Unfortunately, most national policies that boast of gender sensitivity merely use gender-neutral language or at the most evaluate gender vulnerability by using disaggregated data (Holvoet and Inberg 2014; Nyasimi et al. 2018; Mersha and van Laerhoven 2019; Singh et al. 2021). Such policies neither evaluate nor address the complex ways in which marginalization and oppression work (Djoudi et al. 2016; Rao 2019; Lau et al. 2021; Singh et al. 2021). Failing to recognize how gender vulnerabilities get augmented in societies with the interaction of several social factors, such climate adaptation policies inadvertently cannot be called gender-sensitive in the real sense in its implementation.

Many Gender-Responsive policies follow a Gender-Sensitive or Gender-Accommodating approach that considers the inequitable gender norms, roles, and relationships. For example, Gender-Sensitive policies recommend labour-saving technologies in climate-smart agriculture to reduce the disproportionate burden of labour that falls on women. Though this can help in freeing up women's time (Grassi 2015), it does not effectively reduce their burden of gendered roles nor alter the division of labour. In other words, the Gender-Sensitive policies though acknowledge women's vulnerability due to prevailing social structures, they do not address the drivers of gender inequality that perpetuate and legitimize women's disproportionate care-work burdens and undervalue their contributions. They keep the status quo and do nothing to challenge the established socio-cultural norms and gender power relations.

These limitations of Gender-Responsive and Gender-Sensitive policies have given rise to the new paradigm of Gender-Transformative approaches in climate adaptation interventions. The gender-transformative approach is a multi-dimensional yet holistic approach that is based on the idea that interventions can enable women "to understand and challenge the social norms that create these inequalities between men and women" (Njuki et al. 2016; Godfrey-Wood and Otto-Naess 2016). In other words, Gender-Transformative approaches call for transforming systems that continue to maintain gender inequality by transforming individual gender-equitable behaviour on one hand and on the other address the power dynamics and unequal social norms that continue to reinforce gendered inequalities (Hillenbrand et al. 2015; Jerneck 2018). Some examples can be programmes that address women's self-improvement like targeted training and skill development for women or adopting "women only" approaches like special agricultural loans only targeting women farmers. Such interventions can challenge social norms and change cultural values encouraging (or maybe forcing) men to change their behaviour and attitudes.

## Significance of Gender-Transformative Adaptation

Although the negative impacts of climate change vary across regions, evidence shows women to be disproportionately disadvantaged than men (Thomas et al. 2019). Women experience multi-dimensional vulnerabilities that are associated with deep-seated structural inequalities. For example, in several parts of the world, informal labour markets mostly employ women; men usually earn higher wages than women, and women have to work part-time to adjust to the caregiving and domestic roles in their families (Ravera et al. 2016). Lower-income makes them less capable of livelihood diversification and limits their investment capacities to cope with production volatility and shocks (Rao 2019). Rural women living in the Global South are most vulnerable to the negative impacts of climate change because most of them depend on livelihoods based on natural resources that are informal with unstructured labour rubrics (agriculture, fisheries, and forestry) (UN Women 2020; Chanana-Nag and Aggarwal 2020; Deininger et al. 2023). They are also vulnerable due to the biased laws and regulations of some countries (Erman et al. 2021; UNFCCC 2022), with limited legal ownership land rights (FAO 2023) hindering access to other associated resources like loans, credits, and infrastructure. Thus, climate change exacerbates the vicious cycle of vulnerability of women leading to increased unemployment, loss of income, instability, and casualization in the labour market.

Lack of income opportunities, deprivations, and poverty have many subsequent repercussions. It can create conditions restricting women from availing reproductive and maternity health services (Plan International 2019; van Daalen et al. 2020; Women Deliver 2021), drop-out from schools and/ or early marriages to alleviate financial hardship (UNFPA 2022; Women Deliver 2021) and even increase exposure to gender-based violence, domestic abuse, transactional sex, sexual abuse, and exploitation. There is evidence to show how extreme weather events, environmental stress, and climate-induced conflicts lead to an increase in trafficking and even female genital mutilation (Le Masson et al. 2019; Memon 2020; Women Deliver 2021; Esho et al. 2021; Desai and Mandal 2021; Rodrigues 2022).

The differential gender roles and expectations also impact women's freedom, choices, and opportunities to acquire skills and capacity to cope with climate shocks and participation in decision-making.[1] The manifestation of these roles becomes one of the greatest impediments to coping with climate risks and adapting to climate change. They lack capacity due to a lack of opportunities for education, skills, training, and income (Huyer et al. 2015). For example, studies have found more women die from floods because either they did not get an opportunity to learn to swim (Alam and Collins (2010) or they were not socialized to go out of their houses on their own (Bradshaw and Fordham 2013). Moreover, other factors like prohibitive gender norms, discriminatory social practices, and the gendered burden of responsibilities within the households limit their capacity to draw benefits and even participate in climate programmes (Seager et al. 2016; IUCN 2020). For example, some reports

[1] Momsen (2006) refer this as the visible and invisible social and cultural roles that men and women engage in societies due to the intrinsic power relations between them.

note male leaders and members often hold negative perceptions and gendered biases on skills and knowledge disallowing women's participation and their voices to be heard (Staples and Natcher 2015).

Women also lack access to assets both tangible like land, and technology and intangible like climate and agricultural information, insurance, and credit (Tuana 2013; Jost et al. 2016; Rao 2019; Goh 2020). So, the climate stresses further weaken women's social, financial, and economic status in society (Call and Sellers 2019; Glazebrook et al. 2020; Erman et al. 2021; UNFCCC 2022).

There are several pieces of evidence to show how climate change has recreated and added to the gendered roles and has ensured unequal social norms are entrenched. For example, women need to walk longer distances to gather water and fuelwood to cope with the impacts of climate change, escalating the time burden (Ayales et al. 2019; Zaman 2020; Castañeda et al. 2020). Some even face stalking, verbal threats, and harassment doing this task (Zaman 2020). In all cases of gender-selective labour-based migration caused by climate stress (Rao et al. 2019), women who are left behind are forced to cope with seasonal and extreme weather events. They are forced to take over additional caregiving, agricultural, and household duties with fewer resources such as land, inputs, and other services (Gaard 2015; Rigaud et al. 2018; FAO 2023). They are also found to be at greater risk of assault, harassment, and sexual abuse outside of their domestic spaces (Anderson and Overby 2021).

Considering the heightened nature of vulnerability, GTA is not only significant and urgent, but it can also be used as a win–win situation. On one hand, it has the potential to address the inherent drivers of gendered vulnerabilities to address and cope with the negative impacts of climate change. On the other hand, it can make climate change adaptation an opportunity to plan adaptation pathways catered to provide opportunities to women for social transformations associated with sustainable development. Studies have shown how women can be important agents of change in unique ways despite experiencing multi-dimensional vulnerabilities (Huyer 2016; MacGregor 2010). However, climate adaptation research has put far greater attention and emphasis on women's vulnerability, overshadowing the efforts that women make to address and participate in adaptation. Unfortunately, the same narrative is echoed within the international climate policy frameworks (Huyer 2016). These discourses completely ignore the fact that the most vulnerable will have to proactively adapt to survive.

Women are successfully involved in many different types of adaptation strategies like the adoption of climate-smart practices, green jobs, and livelihood changes, as that often go unacknowledged (Shaw and Kristjanson 2013; Aregu et al. 2018). These GTA strategies have contributed positively to both households and communities in the form of improved food security, increased capacity, and leadership (Ouédraogo et al. 2018). It is estimated that the transition to low-carbon economies will offer new job opportunities and green business innovations that will have the potential to address these entrenched disadvantages and empower women.

## Pathways for Social Transformations Through Climate Change Adaptation

The pathways for social transformations can be linked to pathways for climate change adaptation. For correcting root causes for unequal gender roles and drivers of vulnerability among women, particularly for women belonging to poor and marginal social groups, adaptation strategies need to be well planned. The adaptation plans can follow a "three-step framework" to accommodate GTAs (Fig. 1.1). The local adaptation plans need to be context-specific and localized for effective outcomes. The first step is to generate, synthesize, and analyse evidence to understand the local context concerning local climate events, intersectionality, and power dynamics within households and communities. These important insights can come from smart interventions to address structural inequalities at the household and community level. For example, local adaptation plans can have components for compulsory space to engage women's organizations, marginal groups, and civil society in the formation of plans and modes of implementation. For financing of plans, positive affirmation can be applied for women, where a certain quota for agricultural loans, or credit for livelihood diversification is reserved for women. Such policies can give agency to women, forcing society to adjust to new norms and breaking barriers. With local success, such plans can then be scaled up and out.

The second step is to start legislating laws that give women rights to own land (correct inheritance laws, rebates on land registration tax for land owned by women), access resources, and even form strict laws to safeguard women from gender-based violence, child marriages, and trafficking. Policies can also address equal wages for men and women, special skills and training for women, and most importantly advocate for women leaders and change-makers to motivate and encourage change in others. Sectors that are projected to have growth in green jobs are traditionally male-dominated, like the energy, manufacturing, construction, and transport sectors. At present, women's participation is limited in these sectors due to biased hiring practices crippled with discriminatory laws and social norms; inconducive and inflexible childcare policies; unsafe workspaces without preventative measures against sexual harassment and most importantly, lack of access to mentoring, networking, and training opportunities. This situation is unlikely to change without immediate interventions like removing legal barriers and forming internal company policies that enable women's employment in male-dominated technical jobs (UN Women 2021). In the long term, interventions are needed to challenge social norms.

The third step is financing capacity and skill development, focusing on local risks and drivers. For long-term return, the funding strategies for adaptation plans for local communities must be equipped to give access to both tangible assets like land and technology and intangible assets like information and skills so that communities can readjust and recover from the negative impacts. It is estimated that 213 million cumulative jobs could be generated by climate adaptation interventions worldwide between 2020 and 2030 (IFC 2021). The global average workforce participation rate for women is less than 47 per cent as compared to 72 per cent for men (ILO 2021).

| Step 1 | | | Step 2 | | | Step 3 | | |
|---|---|---|---|---|---|---|---|---|
| **DATA AND EVIDENCE** | | | **POLICIES AND PLANS** | | | **FINANCE AND CAPACITY** | | |
| **Generate evidence** Collect gender-disaggregated data, evidence synthesis, and analysis to understand the local context concerning local climate events, intersectionality, and power dynamics within households and communities. | | | **Form policies and strategies for adaptation** Safeguard gender-responsive laws, policies, and regulations that nullify discriminatory practices to facilitate just transition in local contexts. | | | **Implement policies and execute adaptation plans.** Advanced gender-smart and just climate adaptation plans have an inbuilt mechanism for training to enhance knowledge, skills, capabilities, and opportunities. | | |
| **APPROACHES** | | | | | | | | |
| Address Structural inequalities from the regional to the household level | | | Provide access to decent work and opportunities while limiting over-work burdens | | | Access to and control over tangible and intangible assets | | |
| Promote leadership among women to lead climate processes and implementation | Engage women's organizations, marginal groups, and civil society | Change the rules and challenge social norms that create barriers | Ensure women get decent jobs and opportunities for self-employment | Enhance education and skills for women workers to diversify if needed | Champion women innovators and entrepreneurs and support in decision-making on investments | Ensure women have access to and control over natural and communal assets | Ensure women have access to and control over technology | Ensure women can access assets and technologies with skill and confidence |
| National and regional level ⟷ | Regional and community-level ⟷ | Community and household level ⟷ | | | | | | |

Source: Prepared by authors

**Fig. 1.1** Three-step framework for gender-transformative approaches to climate change adaptation. *Source* Prepared by authors

This suggests there is enough room for increasing women's workforce participation and gains from the Paris Agreement alignment to generate green jobs with appropriate policy interventions and social change. To take advantage of these labour market demands, funds need to be channelized for appropriate capacity and skill development for women. The nonlinear, complex processes of change need to be recognized and navigated well to make the best contributions through development programming.

While policies can give an agency to women for social transformation, research methodologies can also foster and nurture positive transformation. Unfortunately, policy implementation is often inadequate due to the deeply ingrained nature of gender inequality both at the social and political level, which further gets complex due to the intersectionality of discrimination. Evidence suggests that several efforts to empower women often fail due to a lack of consideration of the inherent structural inequality in society, gendered roles, and cultural barriers. We believe that one of the solutions to this nexus is finding ways to inculcate Gender-Transformative Research (GTR). GTR is about investigating, understanding and addressing the structural processes and relationships of power dynamics between and among groups of men and women. It also explores the intersectionalities of gender and how gender overlaps with other social identities like class, race, nationality, socio-economic status, ethnicity, and educational levels to create multi-dimensional inequalities leading to multiple levels of disadvantages (Hopkins 2019).

The GENDER Impact Platform of CGIAR defines GTR Approaches as "A method for tackling the root causes of gender-based inequalities, in particular, constraints in gender norms, and thereby kick-starting a process toward greater gender equality" (CGIAR Gender Impact Platform 2023, online resource). Mullinax et al. (2018: 4) define GTR as a means "to promote gender equality and to empower women and girls not just for individual self-improvement, but to transform gender power dynamics and structures at the community and societal levels.". It is "participatory, reflexive, critical, action-oriented," and it "involves collaborative integration of research and practice" as well as a "commitment to decolonize approaches to development and research" (Lopez et al. 2023: 3).

The GTR methodologies have three basic ground rules about where and how they conduct research. First, gender is the focus of fieldwork. The research incorporates methods to collect gender-disaggregated data by interviewing both men and women. Second, it follows participatory research to build connection and empathy with the participants to understand the root causes of structural inequality in local communities. Third, and probably the most important is that the researchers use their research interviews to ask questions to inform participants and initiate a discussion to transform discourses and popular narratives in the communities. In other words, research participants, including both men and women, are engaged to jointly discuss issues of power and gender inequality in households and communities in focus group discussions to change perspectives, raise awareness, and open avenues for transformation. The final goal is to co-learn and co-create adaptation plans to steer communities to re-imagine gender roles and responsibilities for social transformations. The tangible research outputs are scientifically documented knowledge products that can be fed into policy dialogues, tools for monitoring and evaluation, and even new strategies and plans on the ground.

## Introducing the Book to the Readers

It is increasingly being understood that both mitigation and adaptation to climate change will require local and global societies to change the way of their lives. Human adaptations to the negative impacts of climate change are bound to lead to transformations. Gender relations will be an integral feature of such transformations because women are most exposed to the effects of climate change, and they will need to devise their ways to survive and cope with the negative effects even if not supported and assisted. On one hand, climate change has made women and men differentially vulnerable, on the other it has shown how women can be differently abled due to their socially constructed roles and behaviour. Scholars have theorized and shown the gendered nature of climate change impact (Pearse 2017). The (gendered) adaptation practices now need to be contextualized with local socio-economic and cultural contexts rather than juxtaposing the women's intrinsic vulnerability. Such adaptation pathways may undermine the hidden opportunities for social transformation. We are yet to communicate how our attempts to adapt to climate change can become responsive to these differences that are complex, place-based, and context-specific yet exist across all countries and regions of the world.

So, the question arises, will adaptation pathways lead to greater gender inequality, or will they provide new opportunities for change? Do women have better insights about nature and, hence, possess inherent adaptive capacity that men naturally do not possess? And if so, how should the adaptation pathways be designed? This book adds to this debate where studies on gender and climate change adaptation portray both the facts—"women as the vulnerable" and "women as the change maker."

Research on the relationship between gender and climate change has become crucial, especially in the Global South, where the effects of climate change are most noticeable. Diverse and insightful perspectives on the intersection of gender and climate change impacts in the Global South can be found in this volume. It offers important insights for scholars, practitioners, and policymakers exploring critical themes such as gendered vulnerabilities to climate-induced disasters, women's leadership in climate-smart agriculture, and the gendered dimensions of migration and resource conflicts. The chapters highlight both the disproportionate challenges faced by women and marginalized communities, as well as their vital roles in building resilience and driving sustainable solutions through case studies and empirical research. These contributions unitedly provide a comprehensive understanding of the gendered impacts of climate change and advocate for inclusive, intersectional approaches to climate adaptation and policymaking.

The book is divided into four major sections, each addressing crucial aspects of the relationship between gender and climate. The first section offers a conceptual framework by examining the ways in which gender interacts with the effects of climate change and adaptation measures. Sarkar and Bandopadhyay present the concepts of gender-transformative climate change research and represent innovative methodological tools to combine gender-transformative research approaches in adaptation pathways to address and transform underlying unequal societal structures and entrenched gendered roles. By examining the gendered aspects of climate-smart agriculture (CSA), Singh et al. demonstrate a critical need for gender-sensitive CSA policies to foster inclusive climate adaptation, directly contributing to sustainable development goals (SDGs). In their analysis of the relationship between gender-based violence and climate change, Nyathi et al. draw on case studies from Sub-Saharan Africa to investigate how environmental stresses due to climate change, like droughts, floods, and resource scarcity, worsen existing gender inequities and trigger violence. Curoto et al. emphasize the necessity of gender-sensitive disaster response policies and show how unique tools can be used to collect real-time disaggregated data to capture gendered vulnerabilities during extreme events.

The second section of the book explores how climate change puts women of the Global South at greater risk, making them more vulnerable than men. Alam and Gounder examine how women's burdens are made worse by climate extremes, which restrict their access to resources and means of subsistence. Junaid analyses the gendered impacts of climate change on migrants and refugees, highlighting the vulnerabilities faced by women and sexual minorities. The gendered effects of climate-related disasters and relocation are also demonstrated by regional case studies of coastal regions of Bangladesh by Jasin et al. and the Indian Sundarbans by Raha et al. In their gendered analysis of food security in Sub-Saharan Africa, Tiwari et al. highlight the vital role that women play in food systems in spite of systemic obstacles. Furthermore, case studies from the southern and eastern states of India by Punnoli et al. and Paul and Karmakar show how the negative effects of climate change disproportionately and sometimes differently affect women farmers and women in water-insecure communities. Women are both physically and psychologically more affected than their male counterparts.

Women as leaders in climate adaptation are highlighted in the third section of the book, which moves the emphasis from vulnerability to agency. Banerjee discusses the example of Mal Pahariya women in India, whose livelihoods and food security are improved by climate-smart farming methods. Women empowerment is noticed among the marginalized communities of coal mining areas of eastern India that are sowing seeds of transformation. Sood et al. show how grassroots leadership can promote resilience by highlighting adaptation projects conducted by women in the Western Himalayas. Mishra and Naz use examples from Rajasthan, India, to examine how women's empowerment leads to more successful climate action through the promotion of green jobs at the community level. Katoch and Gurjar highlight the role of external agencies in promoting CSA among rural women for livelihood and income by facilitating them with financial, technical, and market support. Singh and Naz, in their study in Uttar Pradesh, highlight the vulnerability and potential of women in areas affected by uncertain rainfall and groundwater depletion.

The final section of the book addresses policy responses to gendered climate impacts, emphasizing the need for inclusive and transformative approaches. Mehar and Jaspal emphasize effective initiatives and transformational paths in their support of including gender in climate-resilient agricultural strategies. Sultana calls for protections for internally displaced women in Bangladesh and draws attention to displacement brought on by climate change. Women's adaptability to Indonesia's coastal climate difficulties is documented by Setiawan et al. Bhattacharya and Mukherjee conclude by evaluating gender-inclusive disaster management strategies in Indian smart cities and advocating for frameworks of urban resilience that prioritize the experiences of women.

The various contributions in this book collectively highlight how vital it is to address gender disparities in climate action. The book captures relationality, scale, and local knowledge of adaptation strategies to give local and as well global policy direction. Moreover, since gender intersects with multiple forms of discrimination based on ethnicity, class, race, sexuality, age, and caste, it incorporates these nuances in the form of case studies to mainstream policy decisions. It unpacks how societies deal with and adapt to the impacts of climate change and how gender will affect and be affected by impacts and responses to these adaptations. The book celebrates women's agency in promoting inclusive and sustainable adaptation measures while also drawing attention to the disproportionate costs they bear. These contributions provide insightful information for scholars, practitioners, and politicians who want to promote gender-equitable climate solutions by bridging the gap between research and practice.

## Conclusion

Innovative policies and actions are capable of bestowing agency of individuals and collective groups that can promote gender-transformative adaptation. It is crucial to use a gender-transformative perspective in adaptation and mitigation efforts because gender disparities influence vulnerabilities, experiences, and reactions to climate change. The chapters in this book explore how gender is an important way to understand inequalities in emerging adaptation efforts and programmes in key systems and how gender-transformative approaches can enable gender-transformative adaptation.

While recognizing the existing structural inequalities and complex multifaceted, and intersecting forms of discrimination, our book highlights the importance of considering gendered roles and social norms in climate change adaptation efforts, as neglecting these factors can exacerbate social injustice and inequalities, leading to maladaptation.

While there has been some progress in integrating gender considerations in climate change adaptation, there are still significant gaps in achieving gender equality and women's empowerment. So, it is time that GTAs are applied to climate change adaptation research that moves beyond a diagnostic focus to consider how complex dynamics of power and institutional structures both shape gender relations and get shaped by climate-related stressors. The methods and tools used in such research can provide insights into root causes and drivers of women's vulnerability as well as their localized groundbreaking adaptive strategies to deal with climate change impacts. A combination of policies and GTR can eventually lead to necessary system change manifesting in social transformation.

We believe future work should focus on applying gender-transformative adaptation to climate change research and focus on how the most vulnerable communities to climate change can become less vulnerable. The strategies of adaptation can no longer follow business-as-usual and rather should now aim to change the basic attributes of the systemic structure in response to climate and its effects. The question we should ask is, "What kinds of institutional, organizational and structural changes can reduce their vulnerabilities? What kind of pathways can be adapted to respond to women's greater vulnerability? More work is necessary to critically examine the intersectionalities based on class, race, ethnicity, and caste and not consider "women" as a homogenous group in adaptation programmes. Further research is required to comprehend the role of women at the local level in managing natural resources leading to adaptation to climate change impacts. The most important question of practical importance is to see how adaptation policies are received by various groups of populations.

Over the past decade, international and national climate policies have become more gender-sensitive in their framings, but they have failed to reduce gendered vulnerability to climate change (Huyer et al. 2020; Acosta et al. 2015). The common narrative of the policies is to focus on women's vulnerability to climate change, which is also well-established in the literature. However, this discourse has led to policies targeting the "symptoms" of vulnerability rather than addressing the root cause. This is a prevalent practice as, most often, the implementors do not want to challenge the existing and deeply rooted social and cultural norms. It may also be because, in most cases, the implementors themselves belong to the same social structure and prefer to keep the status quo and not risk social tension. In such a situation, GTR may intimidate project teams because it requires questioning numerous complex social, cultural and ethical issues. Thinking about culture is still limited in most adaptation projects (Cannon 2015), and ethical justice and negative cultural issues and norms around gender are not at the forefront. There is no denying the fact that integrating gender transformation as a goal into projects is extremely challenging and will require special training, expertise and additional effort.

No longer can we deny the need for justice and equality at the heart of climate change adaptation decision-making processes. Transformative pathways are possible only when we include the voice of the disenfranchised in this process (Schipper et al. 2020; MacGregor 2020). These voices should not get muddled in the discourse of "victimhood narrative"; rather, new evidence should counter this claim, positioning women as "climate change-makers". Encouraging and acknowledging women's leadership and empowerment is critical and urgent for two reasons. First, it will give women the legitimacy to represent their voices in formal policy dialogues. Second, it can be like sowing the seeds of transformation. However, one should be careful not to position women and girls as environmental stewards based on the existing social norms as a caregiver. Local adaptation and mitigation plans should give space to their decision-making and not add the burden of work and labour. Time will only tell what can happen if adaptation pathways are made gender transformative.

## References

Acosta M, Ampaire E, Okolo W, Twyman J (2015) Gender and climate change in Uganda: effects of policy and institutional frameworks. CCAFS Info Note. Copenhagen, Denmark. CGIAR Research Program on Climate Change, Agriculture and Food Security (CCAFS)

Alam E, Collins AE (2010) Cyclone disaster vulnerability and response experiences in coastal Bangladesh. Disasters 34(4):931–954

Anderson GD, Overby R (2021) The impact of rape myths and current events on the well-being of sexual violence survivors. Viol against Women 27(9):1379–1401

Aregu L, Choudhury A, Rajaratnam S, Locke C, McDougall C (2018) Gender norms and agricultural innovation: Insights from six villages in Bangladesh. J Sustain Dev 11(4):270–287

Ayales I, Blomstrom E, Rivera VS, Pedraza D, Briceño PMP (2019) Climate migration in the Dry Corridor of Central America: integrating a gender perspective. Christian Aid. https://wedo.org/climate-migration-in-the-dry-corridor-of-central-america-integrating-a-gender-perspective/
Berkhout F, Dow K, Thomas A (2024) Delayed, abrupt and unjust: an institutionalist perspective on limits to climate change adaptation. Clim Risk Manag 44:100611
Bradshaw S, Fordham M (2013) Women and girls in disasters. Department for International Development, London
Call M, Sellers S (2019) How does gendered vulnerability shape the adoption and impact of sustainable livelihood interventions in an era of global climate change? Environ Res Lett 14(8):083005
Cannon T (2015) Disasters, climate change and the significance of 'culture'. In: Cultures and disasters. Routledge, pp 88–106
Castaneda Carney I et al (2020) Gender-based violence and environment linkages. IUCN, Gland, Switzerland
CGIAR Gender Impact Platform (2023) Gender transformative approaches to overcome constraining gender norms. https://gender.cgiar.org/toolsmethods-manuals/gender-transformative-approaches-overcomeconstraining-gender-norms
Chan T, Wang JA, Higham C (2024) Mapping justice in national climate action: a global overview of just transition policies. Grantham Research Institute on Climate Change and the Environment
Chanana-Nag N, Aggarwal PK (2020) Woman in agriculture, and climate risks: hotspots for development. Clim Change 158(1):13–27
Deininger F, Woodhouse A, Kuriakose AT, Gren A, Liaqat S (2023) Placing gender equality at the center of climate action. World Bank Group Gender Thematic Policy Notes Series. http://hdl.handle.net/10986/39436
Deliver W (2021) The link between climate change and sexual and reproductive health and rights: an evidence review. Climate-Change-Report
Desai BH, Mandal M (2021) Role of climate change in exacerbating sexual and gender-based violence against women: a new challenge for international law. Environ Policy Law 51(3):137–157
Djoudi H et al (2016) Beyond dichotomies: gender and intersecting inequalities in climate change studies. Ambio 45(S3):248–262
Eriksen S et al (2021) Adaptation interventions and their effect on vulnerability in developing countries: help, hindrance or irrelevance? World Dev 141:105383
Erman E, De Vries S, Thies S, Kayenat K, Mirai M (2021) Gender dimensions of disaster risk and resilience: existing evidence. World Bank Group, Washington DC
Erwin A et al (2021) Intersectionality shapes adaptation to social-ecological change. World Dev 138:105282
Esho T, Komba E, Richard F, Shell-Duncan B (2021) Intersections between climate change and female genital mutilation among the Maasai of Kajiado County, Kenya. J Global Health 11
FAO (2023) The status of women in agrifood systems. Food and Agriculture Organization of the United Nations. https://www.fao.org/documents/card/en?details=cc5343en
Fazey I, Wise RM, Lyon C, Câmpeanu C, Moug P, Davies TE (2016) Past and future adaptation pathways. Clim Dev 8(1):26–44
Gaard G (2015) Ecofeminism and climate change. In: Women's studies international forum, vol 49. Pergamon, pp 20–33
Glazebrook T, Noll S, Opoku E (2020) Gender matters: climate change, gender bias, and women's farming in the global South and North. Agriculture 10(7):267
Godfrey-Wood R, Naess LO (2016) Adapting to climate change: transforming development. IDS Bull 47(2):49–62
Goh K (2020) Flows in formation: the global-urban networks of climate change adaptation. Urban Studies 57:2222–2240
Grassi F (2015) Running out of time: the reduction of women's work burden in agricultural production. FAO. https://openknowledge.fao.org/handle/20.500.14283/I4741E

Greenfield PM (2016) Social change, cultural evolution, and human development. Curr Opin Psychol 8:84–92
Hillenbrand E, Karim N, Mohanraj P, Wu D (2015) Measuring gender-transformative change: a review of literature and promising practices. CARE Publications USA, Atlanta
Holvoet N, Inberg L (2014) Gender sensitivity of Sub-Saharan Africa national adaptation programmes of action: findings from a desk review of 31 countries. Clim Dev 6(3):266–276
Hopkins P (2019) Social geography I: intersectionality. Prog Hum Geogr 43(5):937–947
Huyer S (2016) Gender and international climate policy: an analysis of progress in gender equality at COP21. CCAFS Info Note. Copenhagen, Denmark: CGIAR Research Program on Climate Change, Agriculture and Food Security
Huyer S, Acosta M, Gumucio T, Ilham JIJ (2020) Can we turn the tide? Confronting gender inequality in climate policy. Gend Dev 28(3):571–591
Huyer S, Twyman J, Koningstein M, Ashby J, Vermeulen SJ (2015) Supporting women farmers in a changing climate: Five policy lessons. Copenhagen, Denmark. https://cgspace.cgiar.org/rest/bitstreams/60479/retrieve
IFC (2021) Gender X climate: pursuing gender-inclusive climate investments. https://www.ifc.org/content/dam/ifc/doclink/2024/pursuing-gender-inclusive-climate-investments.pdf
International Labour Organization (ILO) (2021) User's manual to the ILO's guidelines for a just transition towards environmentally sustainable economies and societies for all. ILO
IUCN (2020) Gender and the environment: what are the barriers to gender equality in sustainable ecosystem management? https://www.iucn.org/news/gender/202001/gender-and-environment-what-are-barriers-gender-equality-sustainable-ecosystem-management
Jerneck A (2018) Taking gender seriously in climate change adaptation and sustainability science research: views from feminist debates and Sub-Saharan small-scale agriculture. Sustain Sci 13(2):403–416. https://doi.org/10.1007/s11625-017-0464-y
Jost C et al (2016) Understanding gender dimensions of agriculture and climate change in smallholder farming communities. Clim Dev 8(2):133–144
Lau JD, Kleiber D, Lawless S, Cohen PJ (2021) Gender equality in climate policy and practice hindered by assumptions. Nat Clim Chang 11(3):186–192. https://doi.org/10.1038/s41558-021-00999-7
Le Masson V, Benoudji C, Reyes SC, Bernard G (2019) How violence against women and girls undermines resilience to climate risks in Chad. Disasters 43:245–270
Lopez DE, Bailey A, Farnworth CR, Rietveld A, Gartaula H (2023) Designing for change through "reflecting and doing": the CGIAR community of practice on gender-transformative research methodologies. Front Sustain Food Syst 7:1179503. https://cgspace.cgiar.org/bitstream/handle/10568/132048/fsufs-071179503.pdf?sequence=1
MacGregor S (2010) Gender and climate change': from impacts to discourses. J Indian Ocean Reg 6(2):223–238
MacGregor S (2020) Gender matters in environmental justice. In: Coolsaet B (ed) Environmental justice, 1st edn. Routledge, Abingdon, pp 234–248
Memon FS (2020) Climate change and violence against women: study of a flood-affected population in the rural area of Sindh, Pakistan. Pak J Women's Stud Alam-e-Niswan 7(1):65–85
Mersha AA, van Laerhoven F (2019) Gender and climate policy: a discursive, institutional analysis of Ethiopia's climate resilient strategy. Reg Environ Change 19(2):429–440
Momsen JH (2006) Women, men and fieldwork: Gender relations and power structures. In: Doing development research, pp 44–51
Mullinax M, Hart J, Garcia AV (2018) Using research for gender-transformative change: principles and practice. International Development Research Center (IDRC) and American Jewish World Service (AJWS)
Njuki JM, Wyatt A, Baltenweck I, Yount K, Null C, Ramakrishnan U, Webb Girard A, Sreenath S (2016) An exploratory study of dairying intensification, women's decision making, and time use and implications for child nutrition in Kenya. Eur J Dev Res 28:722–740

Nyasimi M et al (2018) Inclusion of gender in Africa's climate change policies and strategies. In: Climate change and gender in Africa, Springer, Berlin, pp 171–185. https://doi.org/10.1007/978-3-319-69837-3
Ouedraogo M, Partey ST, Zougmore R, Derigubah M, Sanogo D, Boureima M (2018) Mainstreaming gender and social differentiation into CCAFS research activities in West Africa: lessons learned and perspectives. CCAFS Info Note. Wageningen, the Netherlands: CGIAR Research Program on Climate Change, Agriculture and Food Security (CCAFS)
Pearse R (2017) Gender and climate change. Wires Clim Change 8(2):e451
Plan International (2019) Climate change: focus on girls & young women. Plan International Position Paper. https://plan-international.org/publications/climate-change-focus-girls-and-young-women#download-options. Accessed 17 Feb 2025
Rao N (2019) From abandonment to autonomy: Gendered strategies for coping with climate change, Isiolo County, Kenya. Geoforum 102:27–37. https://doi.org/10.1016/j.geoforum.2019.03.017
Ravera F, Iniesta-Arandia I, Martín-López B, Pascual U, Bose P (2016) Gender perspectives in resilience, vulnerability and adaptation to global environmental change. Ambio 45:235–247
Resurrección BP, Bee BA, Dankelman I, Park CM, Haldar M, McMullen CP (2019) Gender-transformative climate change adaptation: advancing social equity. Paper commissioned by the Global Commission on Adaptation (GCA)
Rigaud KK, de Sherbinin A, Jones B, Bergmann J, Clement V, Ober K et al (2018) Groundswell: Preparing for Internal Climate Migration. World Bank, Washington, DC. https://openknowledge.worldbank.org/bitstream/handle/10986/29461/WBG_ClimateChange_Final.pdf
Rodrigues M (2022) How climate change could drive an increase in gender-based violence. Nature
Schipper ELF et al (2020) Turbulent transformation: abrupt societal disruption and climate-resilient development. Clim Dev. https://doi.org/10.1080/17565529.2020.1799738
Seager J et al (2016) Global gender and environment outlook. United Nations Environment Programme (UNEP), Nairobi, Kenya
Shaw A, Kristjanson P (2013) Catalysing adaptive capacity for development and climate change: an exploration of social learning and social differentiation in the CGIAR. CCAFS working paper no. 43. CGIAR research programme on climate change, agriculture and food security (CCAFS), Copenhagen, Denmark
Singh C, Solomon D, Rao N (2021) How does climate change adaptation policy in India consider gender? An analysis of 28 state action plans. Clim Policy 21(7):958–975. https://doi.org/10.1080/14693062.2021.1953434
Staples K, Natcher DC (2015) Gender, decision making, and natural resource co-management in Yukon. Arctic 68(3):356–366. http://www.jstor.org/stable/43872254
Thomas K, Hardy RD, Lazarus H, Mendez M, Orlove B, Rivera-Collazo I et al (2019) Explaining differential vulnerability to climate change: a social science review. Wires Clim Change 10(2):e565
Tuana N (2013) Gendering climate knowledge for justice: catalyzing a new research agenda. In: Research, action and policy: addressing the gendered impacts of climate change, pp 17–31
Uhls YT, Greenfield PM (2011) The rise of fame: a historical content analysis. Cyberpsychology 5(1):1
UN Women (2020) Women's economic empowerment in fisheries in the blue economy of the Indian Ocean Rim: a baseline report. https://www.unwomen.org/en/digital-library/publications/2020/10/womens-economic-empowerment-in-the-indian-ocean-rim
UN Women and African Development Bank (2021) Green jobs for women in Africa. https://africa.unwomen.org/sites/default/files/Field%20Office%20Africa/Attachments/Publications/2021/11/20211206_UN%20Women_Green%20Jobs_report_ENG%20webpages.pdf
UNFCCC (2022) Decision-/CP.27 intermediate review of the implementation of the gender action plan. Advance unedited version. https://unfccc.int/documents/624406. Accessed 17 Feb 2025
UNFCCC (2015) Strengthening gender considerations in adaptation planning and implementation in the least developed countries. LDC expert group. UNFCCC Publishing, Bonn, Germany. http://unfccc.int/files/adaptation/application/pdf/21673_unfccc_leg_gender_low_v5.pdf

UNFPA (2022) Child marriage and environmental crises: an evidence brief. United Nations Population Fund

van Daalen K, Jung L, Dhatt R, Phelan AL (2020) Climate change and gender-based health disparities. Lancet Planet Health 4(2):e44–e45

Werners SE, Wise RM, Butler JR, Totin E, Vincent K (2021) Adaptation pathways: a review of approaches and a learning framework. Environ Sci Policy 116:266–275

Zaman K (2020) Patriarchal gambles: gendered risk regimes and women's empowerment in Bangladesh. Dissertation, The University of Wisconsin-Madison

# Chapter 2
# Exploring Gendered Dimensions of Climate-smart Agriculture for Food Security in the Global South: A Systematic Literature Review

**Neelam Singh, Dibakar Sahoo, and Prasanta Moharaj**

**Abstract** Agricultural communities in the Global South are disproportionately vulnerable to climate change due to their reliance on climate-sensitive agricultural livelihood. These vulnerabilities threaten food security and necessitate transformative approaches to building resilience in agricultural systems. Climate-smart agriculture (CSA) practices have emerged as pivotal interventions, offering pathways to mitigate production risks, enhance food security, and foster sustainable farming practices. However, the adoption of CSA practices is significantly shaped by gendered disparities, with systemic inequalities heightening women's exposure to climate risks. Based on a systematic review of the literature and bibliometric analysis, the findings of this study reveal pronounced gender disparities in the adoption of CSA practices. Male farmers demonstrate a higher propensity for adopting resource-intensive practices such as soil and irrigation management, whereas female farmers predominantly engage with seed selection and fertilizer application, reflecting their constrained access to capital and inputs. Structural barriers, including financial limitations, restricted access to agricultural resources, educational inequalities, and the disproportionate burden of domestic responsibilities, significantly hinder women's capacity to adopt advanced CSA practices. Conversely, enabling factors such as increased age and greater empowerment in agricultural decision-making emerge as critical drivers that enhance women's participation in CSA adoption, offering pathways to reduce gender inequities and foster resilience in agricultural systems. This study emphasizes the critical need for gender-sensitive CSA policies to foster inclusive climate adaptation, directly contributing to SDGs 5 (Gender Equality), 13 (Climate Action), and 2 (Zero Hunger) by offering actionable recommendations that

N. Singh · D. Sahoo (✉)
Department of Economics, Faculty of Social Sciences, Banaras Hindu University, Varanasi, Uttar Pradesh, India
e-mail: dibakar@bhu.ac.in

P. Moharaj
School of Law, Dayananda Sagar University, Bangalore, India
e-mail: moharaj.prasant@gmail.com

 
A. Sarkar and N. Bandyopadhyay (eds.), *Gender-Transformative Approaches for Climate Change Adaptation*, Springer Climate,
https://doi.org/10.1007/978-3-031-90236-9_2

align agricultural innovation with gender equity, climate resilience, and food security goals in the Global South.

**Keywords** Climate-smart agriculture · Gender · Climate change · Food security · Global South · Systematic literature review

## Introduction

Climate change is one of the most pressing global challenges, significantly affecting agricultural productivity and food security, especially in the Global South (IPCC 2014; FAO 2017). Smallholder farmers in these regions, where agriculture is central to livelihoods and economies, disproportionately bear the brunt of rising temperatures, erratic rainfall, and frequent extreme weather events (Mendelsohn and Dinar 2009; Lobell et al. 2011). In response, Climate-Smart Agricultural (CSA) practices have emerged as vital strategies to mitigate these effects while enhancing resilience and food security (Lipper et al. 2014; FAO 2013). These practices encompass interventions aimed at boosting productivity, adapting to climate change, and reducing greenhouse gas emissions (Mbow et al. 2020). The growing impact of climate change on agriculture poses significant challenges to global food security (Dogondaji 2013). Recent studies indicate that these effects may be broader than previously recognized, with serious implications for food production, supply, and accessibility worldwide (Lake et al. 2012; Thornton et al. 2018). Of particular concern is the threat to agricultural productivity in regions dominated by small-scale and subsistence farming (Rao et al. 2022). The increasing frequency and intensity of extreme weather events, such as droughts, floods, and heat waves, can devastate crop yields and livestock production, disrupting food supply chains and driving up food prices (Dogondaji 2013). Additionally, climate change can indirectly impact food security by affecting income, employment, and access to nutritious food (Von Braun 2020).

Mitigating these risks necessitates the widespread adoption of CSA practices (Von Braun 2020; Selvaraju et al. 2011; Dogondaji 2013). These practices, which include the development of heat-tolerant crop varieties, improved irrigation systems, and sustainable land management techniques, are designed to increase productivity, build resilience, and reduce greenhouse gas emissions (Ahmed et al. 2019). However, the implementation and outcomes of CSA practices are not uniform across different segments of the population. Gender significantly influences how individuals experience climate change and access CSA practices (Doss 2014; Goh 2012). Women, who make up a large portion of the agricultural workforce in the Global South, often face barriers such as limited access to resources, decision-making power, and knowledge, which can hinder their ability to adopt and benefit from CSA practices (Quisumbing et al. 2014; Nelson and Stathers 2009). Despite this, the gendered dimensions of CSA adoption and its implications for food security remain underexplored in the existing literature (Jost et al. 2016; Kristjanson et al. 2017).

The aim of this systematic literature review and bibliometric analysis is to fill this knowledge gap by analysing how gender influences the adoption and effectiveness of CSA practices in the Global South. By synthesizing existing literature through a gendered lens, this study seeks to provide insights that can inform more equitable and effective climate adaptation strategies. This approach aligns with Carr and Thompson's (2014) framework, which emphasizes the importance of gender-sensitive strategies in climate change adaptation for sustainable development. The study has four objectives. (1) To assess the role of CSA practices in improving food security within the Global South; (2) To evaluate the socio-economic and gender-related factors influencing the adoption of CSA practices; (3) To review existing CSA practices and identify gender-related barriers that hinder women's participation in these initiatives; and (4) To propose evidence-based policy recommendations that promote gender equity and enhance the effectiveness of CSA strategies in mitigating and adapting to climate change impacts.

## Overview of CSA Practices

Sustainable land management practices, such as conservation agriculture, are central to CSA as they help maintain soil health, improve water retention, and reduce erosion. These practices not only enhance crop yields but also contribute to the long-term sustainability of agricultural systems. Agroforestry, which integrates trees into farming landscapes, provides multiple benefits, including carbon sequestration, improved soil fertility, and increased biodiversity, making it a vital component of CSA (Garrity et al. 2010). Improved water management techniques, such as drip irrigation and rainwater harvesting, are essential in regions facing water scarcity. These methods enhance water use efficiency, reduce dependency on unpredictable rainfall, and support the cultivation of high-value crops even under water-stressed conditions (Pretty 2011). Crop diversification, involving the cultivation of a variety of crops within a farming system, enhances resilience by reducing the risk of total crop failure due to climate-related stresses. This practice also contributes to dietary diversity and nutritional security, particularly in rural communities (Frison et al. 2011).

The development and adoption of stress-tolerant crop varieties—such as drought-resistant maize or flood-tolerant rice—are critical for adapting to the changing climate. These varieties are bred to withstand specific environmental stresses, thereby securing yields and ensuring food supply even under adverse conditions (Tesfaye et al. 2017). Additionally, practices like integrated pest management and organic farming contribute to CSA by promoting ecological balance and reducing reliance on chemical inputs, which can exacerbate environmental degradation and greenhouse gas emissions (Altieri et al. 2005).

## CSA Practices, Food Security, and its Adoption

Climate change significantly threatens agriculture and food security, particularly in the Global South, where small-scale and subsistence farming predominates (Dogondaji 2013). Extreme weather events such as droughts, floods, and heatwaves increasingly disrupt crop yields and livestock production, causing food supply issues and price surges (Rao et al. 2022; Dogondaji 2013). These disruptions also indirectly affect food security by reducing income, employment, and access to nutritious food (Von Braun 2020). To address these challenges, the adoption of CSA practices is crucial (Von Braun 2020; Selvaraju et al. 2011). CSA practices, which include the development of heat-tolerant crop varieties, improved irrigation systems, and sustainable land management techniques, aim to enhance productivity, and resilience, and reduce greenhouse gas emissions (Ahmed et al. 2019). Despite their potential, the effectiveness and adoption of CSA practices are influenced by various socio-economic, cultural, and institutional factors.

Despite the potential of CSA practices to enhance resilience and productivity, their adoption is influenced by a variety of socio-economic, cultural, and institutional factors. Access to resources such as land, credit, and technology is crucial for the successful implementation of CSA practices. However, in many parts of the Global South, these resources are unevenly distributed, often disadvantaging smallholder farmers, particularly women (Quisumbing et al. 2014). Women, who constitute a significant portion of the agricultural workforce, frequently face barriers such as limited access to resources, decision-making power, and knowledge, which can hinder their ability to adopt and benefit from CSA practices (Nelson and Stathers 2009). Cultural norms and social structures further restrict women's participation in agricultural decision-making and limit their access to extension services and training opportunities (Doss 2014; Goh 2012). Additionally, the labour-intensive nature of some CSA practices, combined with existing domestic and caregiving responsibilities, exacerbates these challenges (Quisumbing et al. 2014).

Addressing these gendered disparities and improving access to resources and knowledge are essential for the successful implementation of CSA practices and for achieving equitable improvements in productivity, income, and food security. Integrating climate science with gender-sensitive strategies can help build resilience and ensure the long-term viability of global food systems in the face of climate change (Selvaraju et al. 2011). Gender inequality is a persistent issue in agriculture, limiting women's opportunities. They frequently lack ownership or control over productive assets and resources, including land, information, finance, technology, and inputs, and are further burdened with domestic and caregiving responsibilities obligations (Ranjitha and Mishra 2024).

These gendered disparities are exacerbated by cultural norms and social structures that limit women's participation in agricultural decision-making and restrict their access to extension services and training opportunities (Doss 2014; Goh 2012). As a result, women are often less able to take advantage of the benefits offered by CSA practices, leading to unequal outcomes in terms of productivity, income, and food security. Moreover, the labour-intensive nature of some CSA practices, combined with women's existing responsibilities in household and caregiving tasks, can further impede their adoption.

## Material and Methods

Our research distinguishes itself from previous reviews by placing gender at the core of the analysis, rather than treating it as a peripheral issue. While foundational reviews have acknowledged the importance of gender in agricultural adaptation (Beuchelt and Badstue 2013; Carr and Thompson 2014; FAO 2017; Jost et al. 2016; Kristjanson et al. 2017), they often lack a systematic examination of CSA practices from a gender perspective. This study builds upon these earlier works by providing a detailed and focused review of gender dynamics within CSA adoption and its impact on food security in the Global South.

In addition to focusing on the gendered dimensions of CSA, our study also addresses the broader goal of promoting gender equality within agricultural communities, which is crucial for achieving sustainable climate resilience. By integrating gender as a central element in CSA analysis, this study contributes to a more nuanced understanding of how CSA practices can be optimized to benefit both men and women, ensuring that sustainable agricultural development efforts are equitable and inclusive.

The novelty of this research lies in its comprehensive approach to integrating gender as a central element of CSA analysis. While earlier reviews have made significant contributions, they often fall short of delivering a systematic, gender-focused examination of CSA practices. This study builds upon and extends these foundational works by offering a novel perspective that explicitly addresses how gender disparities shape CSA outcomes and how these practices can be optimized to benefit both men and women (Aguilar 2012). In focusing on the gendered dimensions of CSA, our study also addresses the broader goal of promoting gender equality within agricultural communities, which is crucial for achieving sustainable climate resilience. By integrating gender as a central element in CSA analysis, this study contributes to a more nuanced understanding of how CSA practices can be optimized to benefit both men and women, ensuring that sustainable agricultural development efforts are equitable and inclusive.

This systematic review and bibliometric analysis aim to fill this gap by analysing the role of CSA practices in mitigating and adapting to climate change and ensuring food security through a gendered lens. By examining empirical evidence from the Global South, this review seeks to provide insights into how CSA interventions can be tailored to address gender disparities, thereby promoting more inclusive and sustainable agricultural development in the face of climate change (Twyman et al. 2014; Aguilar 2012).

## *Methodology*

Systematic reviews are increasingly recognized in social sciences for their rigorous, transparent, and methodical approach to summarizing existing literature (Petticrew and Roberts 2008; Siddaway et al. 2019). They enable the identification of research gaps and facilitate the development of a comprehensive research agenda (Müller-Bloch and Kranz 2015). To ensure a robust and reliable systematic review, the study adheres to four critical principles (Victor 2008) like. The first is comprehensiveness. This is achieved by leveraging extensive databases like Scopus and Web of Science, applying detailed inclusion criteria, and utilizing a PRISMA diagram to visually represent the screening process. The second is quality assurance, which is ensured by focusing on peer-reviewed, open-access articles from relevant fields and analysing citation networks to identify impactful studies. The third is transparency, which is maintained through meticulous documentation of the methodology, including clear descriptions of search strategies, inclusion criteria, and data synthesis steps presented in tables and figures. The fourth and last is methodological rigour, which is demonstrated by adhering to established systematic review protocols, conducting bibliometric analyses for deeper insights, and employing thematic maps to identify research trends and gaps.

Complementing this systematic review, the bibliometric analysis further enriches the study by quantitatively examining publication patterns, citation networks, and research trends over time, providing insights into the intellectual structure and evolution of the field (Aria and Cuccurullo 2017; Donthu et al. 2021). By analysing a comprehensive data set of peer-reviewed articles, this approach allows for the identification of key themes, influential studies, and gaps in the existing research, thereby contributing to a deeper understanding of how CSA practices are being explored and implemented across different socio-economic and gender contexts.

We have made use of Scopus and Web of Science databases in our study. We have presented the different steps of the search and screening process in the PRISMA flow diagram in Fig. 2.1. The search string used has been presented in Table 2.1. In the Web of Science database, we use TS (Topic) as a field tag with Boolean operators AND, OR, and NOT in search queries.

Table 2.2 presents the inclusion and exclusion criteria adopted for articles taken in the present study. The initial search string resulted in many articles from different subject domains like physical sciences, biological sciences, law, management, political sciences, etc., which were not relevant to the need of the current study and hence were excluded. Moreover, we took only articles which were available in open access for our study.

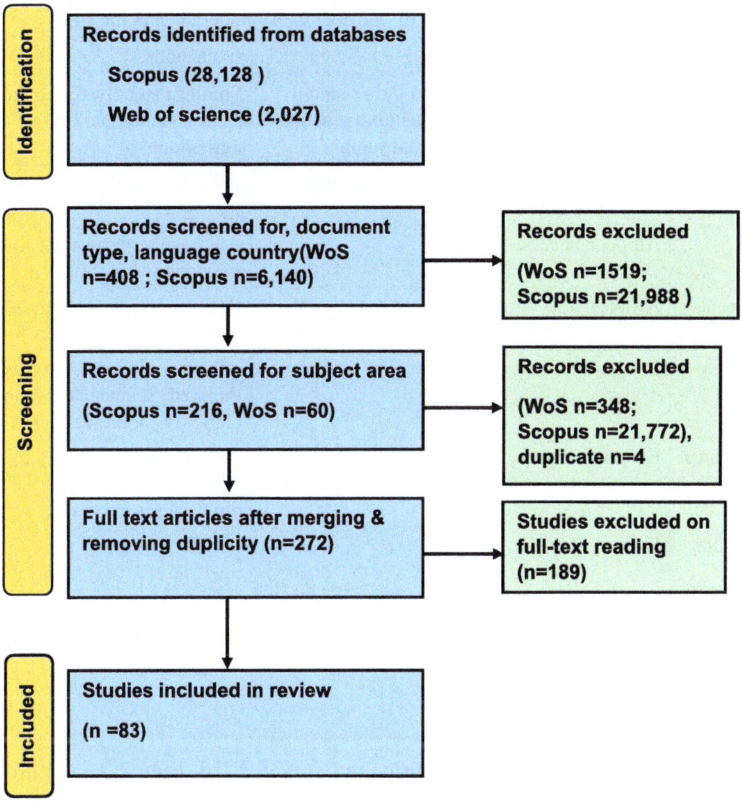


Source: Prepared by the authors

**Fig. 2.1** Preferred reporting items for systematic reviews and meta-analysis (PRISMA) flow diagram. *Source* Prepared by the authors

**Table 2.1** Search strings used in the database search

| Database | Search string |
|---|---|
| Web of Science | TS = (( ( "CSA*" OR "CLIMATE SMART AGRICULTURE*") AND ( "GENDER*" OR "WOMEN*" OR "FEMALE*")) AND ( ( "CSA*" OR "CLIMATE SMART AGRICULTURE*") AND ( "IMPACT*" OR "EFFECT*" OR "FOOD SECURITY*" OR "HUNGER*" OR "FOOD VULNERABILITY*"))) |
| Scopus | (("CSA" OR "CLIMATE-SMART AGRICULTURE") AND ("GENDER*" OR "WOMEN*" OR "FEMALE*")) AND (("CSA" OR "CLIMATE-SMART AGRICULTURE") AND ("IMPACT*" OR "EFFECT*" OR "FOOD SECURITY*" OR "HUNGER*" OR "FOOD VULNERABILITY*")) |

*Source* Compiled by authors

**Table 2.2** Inclusion and exclusion criteria for the documents undertaken in the study

| Parameters | Inclusion | Exclusion |
|---|---|---|
| Article title | Studies that revolved around the gendered aspect of climate-smart agriculture practices and food security in the global south | Studies that delved into broader aspects of climate change on agriculture |
| Region | Countries of the Global South (excluding developed countries) | All other developed countries |
| access | All open access | All the other articles that were not open access |
| Document types | Articles | Reviews, Book chapters, etc. |
| Databases taken | Scopus and Web of Science | All other databases such as Pubmed, Science Direct, and Google Scholar were excluded |
| Subject area | Economics, social sciences | All other fields such as articles from biological sciences and physical sciences were excluded |
| Publication stage | Final | Articles in Press |

*Source* Compiled by authors

## Results and Discussion

The literature on CSA has grown significantly since 2015. Figure 2.2 highlights an upward trend in publications, with notable growth from 2019 to 2020, reflecting heightened awareness of climate change. Although there was a slight decline from 2021 to 2023, the overall trends remain positive. This indicates sustained research interest and underscores the importance of addressing gender-specific barriers and socio-economic factors in CSA adoption to enhance food security in the Global South.

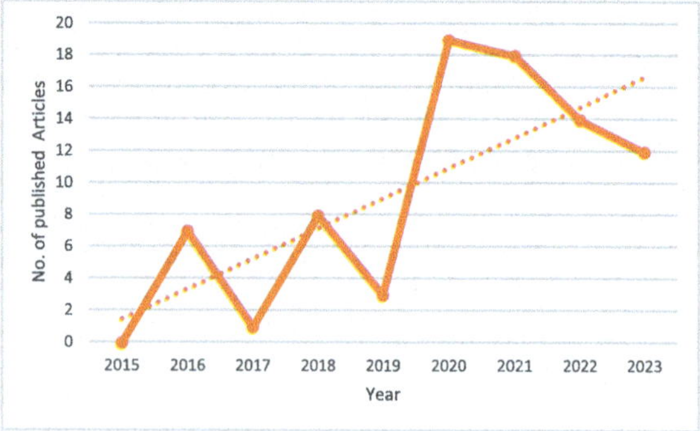


**Fig. 2.2** Trend of published articles included in the sample. *Source* Prepared by the authors

Source: Prepared by the authors

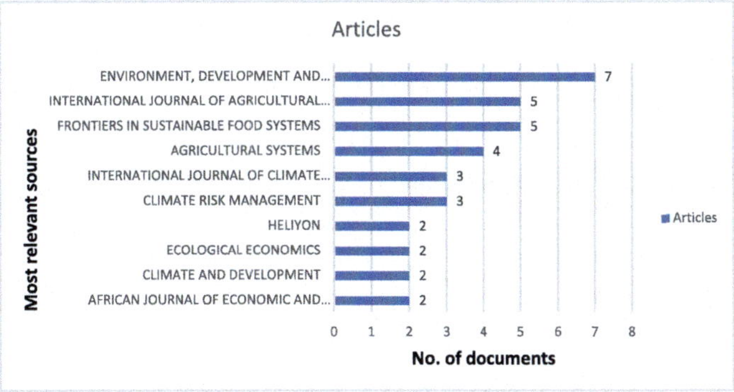


Source: Author's compilation using BIblioshiny

**Fig. 2.3** Most relevant sources based on the number of documents. *Source* Author's compilation using Biblioshiny

Figure 2.3 identifies the most relevant academic sources in the systematic literature review on CSA, with a focus on gendered analysis in the Global South. Environment, Development, and Sustainability is a key source, underscoring its pivotal role in advancing research on sustainability and CSA in developing regions. Frontiers in Sustainable Food Systems and the International Journal of Agricultural Sustainability are also influential, shaping critical discussions on sustainable agriculture. Agricultural Systems contributes essential insights into the technical and systemic aspects of CSA, while Climate Risk Management and the International Journal of Climate Change Strategies highlight the significance of addressing climate risks within CSA frameworks.

The citation analysis (Table 2.3) reveals that recent studies, particularly Makate et al. (2019) in the *Journal of Environmental Management* and Azadi et al. (2021) in the *Journal of Cleaner Production*, have achieved high total citations and citation rates, indicating significant recent impact and relevance in climate-smart agriculture. Papers like Kpadonou et al. (2017) and Kristjanson (2017) also show substantial influence with steady citation counts over time. In contrast, older papers or those with fewer citations, such as Chandra (2017) and Makate et al. (2019) in *Environment and Development*, demonstrate a more modest impact. This highlights the growing prominence of recent research in shaping the discourse on CSA and underscores the evolving nature of research influence in this field.

Figures 2.4 and 2.5 illustrate that gender studies and research on climate change have started evolving during the initials of the second decade of this twenty-first century but the studies in the context of food security in the backdrop of climate change and gender have only come into the documented literature in the recent year, i.e. after the year 2018.

**Table 2.3** Most cited documents

| Paper | Total citations | TC per year | Normalized TC |
|---|---|---|---|
| Makate et al. (2019), J Environ Manage | 149 | 24.83 | 2.90 |
| Kpadonou et al. (2017), Land Use Policy | 111 | 13.88 | 1.59 |
| Kristjanson et al. (2017), Int J Agric Sustain | 107 | 13.38 | 1.53 |
| Mwongera et al. (2017), Agric Syst | 95 | 11.88 | 1.36 |
| Mutenje et al. (2019), Ecol Econ | 91 | 15.17 | 1.77 |
| Adenle (2017), Ecol Econ | 74 | 9.25 | 1.06 |
| Azadi et al. (2021), J Clean Prod | 58 | 14.50 | 4.39 |
| Chandra (2017), J Rural Stud | 58 | 7.25 | 0.83 |
| Makate et al. (2019), Environ Dev | 50 | 8.33 | 0.97 |
| Tran et al. (2020), Int J Clim Change Str | 41 | 6.83 | 0.80 |

*Source* Author's compilation using Biblioshiny

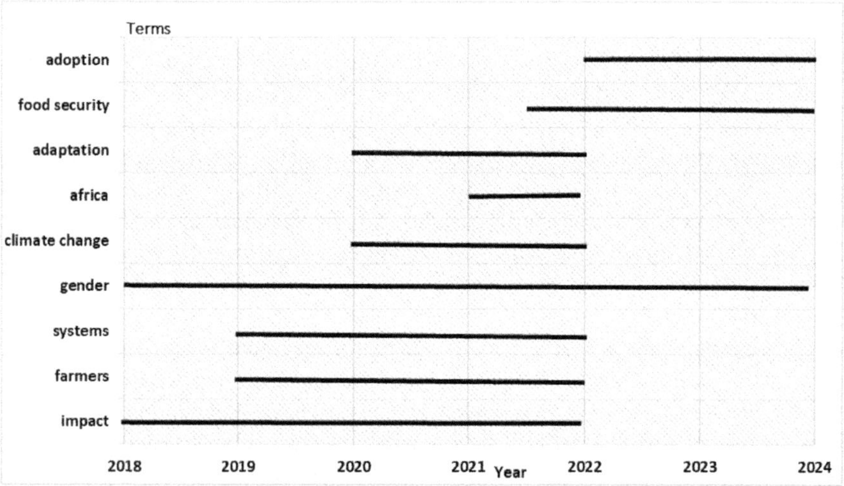


Source: Authors' compilation using Biblioshiny

**Fig. 2.4** Trend of topics covered under the CSA studies. *Source* Authors' compilation using Biblioshiny

## *Region-Wise Description of the Literature on CSA*

To meet the demands of producers and consumers, Thornton et al. (2018) underlined that CSA priority must also take spatial and temporal aspects into account. The 83 documents taken under review in the present study pertain to the Global South countries and for ease of analysis, these studies are divided region-wise. Hence we

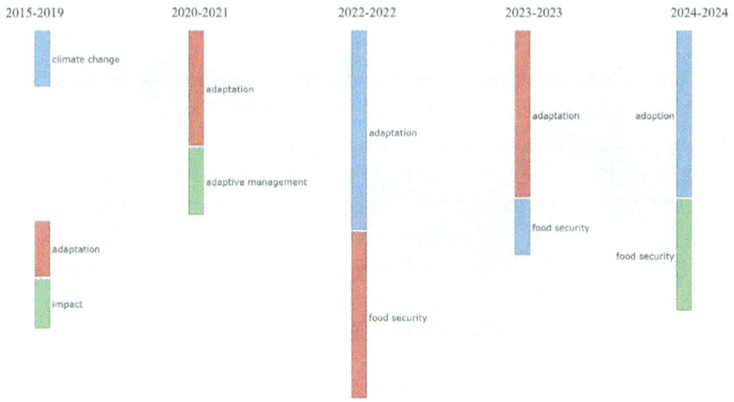


Source: Authors' compilation using Biblioshiny

**Fig. 2.5** Thematic evolution map of the keywords included in the study. *Source* Authors' compilation using Biblioshiny

have categorized the paper into five regions as per the country of study as Sub-Saharan Africa (SSA), South Asia (SA), Southeast Asia (SEA), Latin American and Caribbean countries (LAC), and North African countries and Central Asia.

### Sub-Saharan Africa (SSA)

In Sub-Saharan Africa, the adoption of CSA practices is influenced by various socio-economic and environmental factors. Studies indicate that temperature and rainfall anomalies are adversely affecting agricultural production in the region, leading to increased urbanization rates, particularly in areas where cities act as manufacturing hubs rather than relying on agriculture (Tilahun et al. 2023). In Burkina Faso, gender-specific challenges are evident, with women employing unique strategies to adapt to climate variability (Diendere and Oyodraogo 2023). In Kenya, the adoption of CSA is significantly influenced by sustainable agricultural intensification practices and social networks, which enhance smallholder farmers' welfare and knowledge dissemination (Autio et al. 2021; Kamau et al. 2020). Gender differences in climate change perception and adaptation are prominent among cowpea growers in Mali, reflecting diverse approaches between male and female farmers (Diarra et al. 2021). In Nigeria, gender plays a crucial role in CSA adoption, with men generally more empowered and likely to adopt crop rotation practices, while female plot managers engage more with green manure and agroforestry (Oyawole et al. 2021; Edewor et al. 2023). Additionally, socio-economic factors and social capital are significant determinants of CSA adoption, with some farmers in Southeast Nigeria reluctant to pay for CSA practices due to poverty (Olive and David 2021).

### South Asia (SA)

In South Asia, CSA adoption is shaped by socio-economic disparities and climatic challenges. In India, the National Rural Livelihoods Mission (NRLM) has significantly diversified household livelihoods and increased women's labour force participation, particularly among marginalized communities (Pandey and Gupta 2022). Mobile-based climate services have improved farmers' risk management, though gender disparities in access to modern agricultural inputs persist (Mittal and Hariharan 2018). In Pakistan, market imperfections and gender-based wage disparities limit female labour participation, and warming is inducing out-migration from rural areas by debilitating rural livelihoods (Khan et al. 2021; Makhdum and Kousar 2021).

### Southeast Asia (SEA)

Southeast Asia exhibits diverse responses to CSA adoption influenced by socio-economic and climatic factors. In Vietnam, increasing trends of climate risks have posed challenges to the smallholder farmers' livelihood and food security (Tran et al. 2020). In Vietnam, farmers are facing a multitude of hardships germinating from climate changes like soil degradation, weeds, diseases, insect pests, high production costs, flooding, droughts, changing rainfall pattern, and unreliable rainfall distribution (Mishra and Pede 2017). In these areas, adoption of CSA has reduced the male out-migration (Tran et al. 2020). Cambodia's adaptation constraints in lowland rice-based farming systems necessitate tailored research and interventions (Touch et al. 2023). In the Philippines, extreme climate events in conflict-prone agrarian communities appear to subject women to forced migration, increased discrimination, loss of customary rights to land, resource poverty, and food insecurity (Chandra et al. 2017).

### Latin America and Caribbean (LAC)

In Latin America, CSA adoption is influenced by socio-economic factors and historical migration patterns. In Brazil, the adoption of sustainable agricultural practices is driven by a complex interplay of socio-economic and psychological factors, as highlighted by logistic models (Foguesatto and Machado 2022). In Latin America, research emphasizes the role of women in CSA and how socio-economic factors hinder their access to these technologies. Acosta et al. (2021) show that in rural regions of Latin America, women have limited access to CSA due to structural inequalities, including limited financial autonomy and cultural barriers.

## North African Countries and Central Asia

In the Central Asian region, economic reasons outweigh the social and ethical concerns in the adoption of innovative agricultural practices (Veisi et al.2017). In Turkey, the percentage of illiterate women is 38% in rural areas, even though 90% of the agricultural workforce is made up of women who are at or below the primary school level. In the rural areas of Turkey, women are not viewed as producers. It is generally accepted that women serve as helpers to men in agricultural production (Rad and Kosar 2021). Still, education has been one of the key reasons for adopting CSA practices (Everest 2021).

## Access to CSA Practices, Constraints, and Food Security

Access to CSA practices is critical for enhancing food security, especially in regions vulnerable to climate change. The benefits of CSA innovations, such as improved legumes, drought-tolerant maize, and conservation agriculture, are amplified when these practices are adopted in complementary combinations (Makate et al. 2019). However, various socio-economic and gender-based constraints limit adoption, resulting in uneven food security outcomes. These challenges are particularly pronounced among smallholder farmers and marginalized groups who face resource constraints and limited access to climate information and financial support (Azadi et al. 2021).

Gender disparities significantly influence CSA adoption. In Burkina Faso, 72% of women farmers adopted at least one climate change adaptation strategy, yet their broader engagement was restricted by heavy domestic responsibilities and high input costs (Diendere and Oyodraogo 2023). Similar barriers exist in Malawi and Burundi, where male dominance in land ownership and limited access to agricultural information prevent women from fully adopting CSA practices (Nchanji et al. 2022, 2023). In Ethiopia, gender, land size, and access to climate information were found to be critical factors affecting CSA adoption, with male-dominated communities exhibiting lower adoption rates (Balcha et al. 2023). Across Africa, Kwapong et al. (2024) noted that CSA adoption remains low, partly due to inadequate understanding of decision-making and innovation processes.

Socio-economic challenges further compound these gender disparities. High agricultural input costs, as observed in Burkina Faso, and traditional gender roles, which limit women's access to essential tools and fertilizers in Eastern Zambia, exacerbate adoption barriers (Umar 2021). Marginalized households often lack the financial and social capital needed to invest in adaptation measures, and land inequality enables top landowners to adopt CSA while leaving smaller farmers behind (Coniglio and Pesce 2015; Nawrotzki and Bakhtsiyarava 2017). Conversely, targeted interventions such as credit access, agricultural extension services, and social capital development have proven effective in addressing these barriers. For example, CSA intensification has significantly improved rice yields in China (Vatsa et al. 2023), and social capital, including networks and training, has enhanced CSA adoption in Southern China (Laurent et al. 2023).

The positive impacts of CSA practices on food security are well-documented. In Southwest Ethiopia, comprehensive CSA adoption improved food security through higher per capita food expenditure and reduced food insecurity (Tilahun et al. 2023). In Nigeria, practices like crop diversification and agroforestry contributed to reducing food insecurity, with male-headed households faring better overall (Omotoso and Omotayo 2024). Similarly, in Kenya, Sustainable Agricultural Intensification (SAI) practices boosted crop yields and farmer welfare, although benefits varied by region and crop system (Nyarindo et al. 2024). In Ghana, access to credit significantly increased CSA adoption, improving household income and food security (Agbenyo et al. 2022). Finally, resilience to food insecurity in Niger was found to depend on factors such as household size, coping strategies, farming experience, and livestock volume (Ado et al. 2019).

Addressing the multifaceted challenges of CSA adoption requires comprehensive strategies that integrate gender-sensitive approaches, resource accessibility, financial support, and agricultural training. By overcoming these barriers, particularly in regions with significant socio-economic and gender disparities, CSA practices can contribute to sustainable and equitable food systems, ensuring resilience in the face of climate change. We have presented the conceptual framework in Fig. 2.6.

## *Local Contextual Factors of CSA Adoption*

CSA aims to enhance resilience to climate change, but its effectiveness is highly contingent upon local contexts. Local contextual factors significantly influence the adoption of CSA practices, especially among women. These factors include socio-economic conditions, cultural norms, and infrastructural limitations specific to a region, which collectively shape the accessibility and effectiveness of CSA initiatives.

### Access to Financial Resources and Extension Services

Access to credit plays a pivotal role in facilitating CSA adoption. In Nigeria, farmers' willingness to invest in technologies like drip irrigation and weather-based advisories underscores the importance of financial resources (Anugwa et al. 2022). Similarly, in Ghana, access to credit significantly boosts the adoption of irrigation and crop insurance, demonstrating that financial support can drive CSA implementation effectively (Agbenyo et al. 2022). However, the effectiveness of credit alone is often overshadowed by the necessity of integrating other resources. For instance, in Malawi and Zimbabwe, agricultural extension services prove more effective than credit alone in promoting CSA (Makate et al. 2019). Additionally, mobile-based credit is particularly significant for CSA adoption among smallholder farmers in Kenya, highlighting

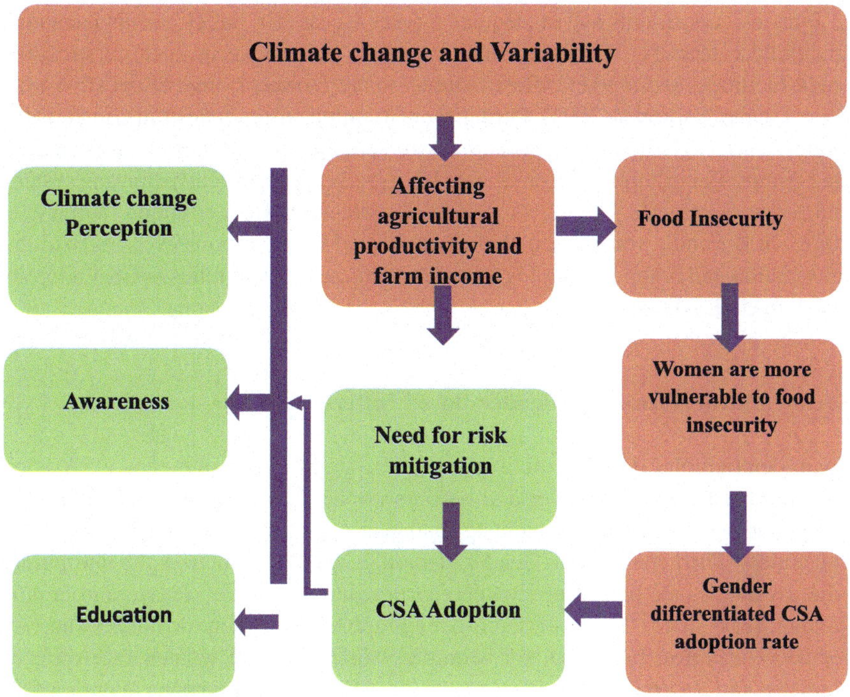


Source: Prepared by authors

**Fig. 2.6** Conceptual framework. *Source* Prepared by authors

the importance of both financial resources and access to information (Andati et al. 2022).

Extension services are critical for promoting CSA adoption, particularly when coupled with financial support. In Malawi and Zimbabwe, the superior effectiveness of extension services compared to credit alone emphasizes the importance of these services in CSA adoption (Makate et al. 2019). This is supported by findings in India and Ghana, where extension services enhance farmers' awareness and engagement with CSA practices (Mittal and Hariharan, 2018; Asante et al. 2024). Nevertheless, the effectiveness of extension services can be constrained by infrastructure and resource limitations. For example, in Zambia, despite the presence of extension services, challenges related to labour demands, and resource access impede the adoption of Conservation Agriculture (Umar 2021). Obossou et al.2023 report that women tend to rely on peers, family, and friends for any information on CSA practices as they have less access to extension services directly. The need for improved extension services is also evident in Uganda, where supporting smallholder farmers through enhanced agricultural extension can significantly improve food production and security (Atube et al. 2022).

Financial constraints and inadequate market access also limit CSA uptake among smallholder farmers. Without financial support mechanisms—such as subsidies, credit facilities, and market infrastructure—many farmers cannot adopt CSA practices at scale (Autio et al. 2021). Furthermore, training programmes and climate information services (CIS) often fail to address the specific needs of women and marginalized groups. For example, in Kenya, men have greater access to early warning systems and advisory services, leaving women disadvantaged in CSA adoption (Ngigi et al. 2022). In Burundi, women face challenges related to unequal access to agricultural information and mobile technology, perpetuating gender inequities in CSA adoption (Tilahun et al. 2023).

### Socio-cultural Norms and Gender-Based Differences

Socio-demographic factors, including gender, significantly influence CSA adoption. Gender-based preferences and barriers impact CSA practices, as observed in Kenya and Nigeria, where different age and gender groups exhibit varying needs and constraints (Phiri et al. 2022). Gender disparities are also evident in the Philippines, where women face heightened vulnerabilities due to climate change and conflict (Chandra et al. 2017). However, socio-demographic factors alone do not fully account for adoption rates. In Ghana, for example, while men and women face different challenges in accessing resources, gender does not solely determine adoption rates (Dwomoh et al. 2023). Similarly, in Ethiopia, socio-demographic factors like education and land size also influence CSA adoption, indicating that a broader set of factors beyond gender is involved (Teklu et al. 2023). Gender dynamics in agriculture, including labour supply decisions and market imperfections in Pakistan, further illustrate the complex interplay of socio-economic and gender factors influencing CSA adoption (Makhdum and Kausar 2021). In Malawi, where 96% of households are male-headed, land ownership remains culturally skewed towards men, restricting women's ability to adopt CSA practices (Nchanji et al. 2022).

Cultural and social norms play a significant role in shaping CSA adoption and migration responses. Traditional gender roles and social norms affect the vulnerability and capacity to adapt among marginalized groups. In the Philippines and India, e.g. caste-based discrimination and gender-specific challenges exacerbate vulnerabilities and hinder CSA adoption (Chandra et al. 2017; Barooah et al. 2023). However, cultural norms are not the sole determinants of CSA adoption. In Zambia, traditional roles impact CA practices, but adoption challenges are also related to practical barriers like labour demands and resource limitations (Umar 2021). The review of gender and conservation agriculture in Sub-Saharan Africa further highlights the significant barriers women face in CSA adoption, including limited access to land, machinery, and credit, while also noting the potential for improved gender relations through increased women's participation (Wekesah et al. 2019).

### Infrastructure and Institutional Support

Infrastructure and institutional support are crucial for CSA adoption. In Ethiopia, the presence of market infrastructure and frequent extension services positively impacts CSA adoption rates (Teklu et al. 2023). In Ghana, larger household sizes and membership in farmer-based organizations enhance the use of Digital Agricultural Services (DAS), facilitating CSA practices adoption (Asante et al. 2024). However, the effectiveness of infrastructure and institutional support can be limited by broader systemic issues. In India, structural barriers such as caste-based discrimination and inadequate collateral for credit hinder CSA adoption, indicating that infrastructure alone may not be sufficient without addressing underlying systemic inequalities (Barooah et al. 2023). Similarly, in Turkey, socio-economic factors and previous experiences with climate-related challenges drive farmers' readiness to adopt CSA practices, though infrastructure and institutional support play a role in facilitating this readiness (Everest 2021).

## Policy Gaps and Recommendations

Despite the growing recognition of CSA as essential for mitigating climate change and enhancing food security, significant gaps remain, particularly in addressing gender disparities. Women, who constitute a substantial part of the agricultural workforce in the Global South, face systemic barriers such as limited access to land, credit, and agricultural inputs. To address these challenges, policy reforms must prioritize inclusivity, resource accessibility, and targeted financial support.

Firstly, the governments should include Gender-Responsive policies to enhance CSA adoption, particularly concerning implementing land reforms to ensure equitable access for women, the introduction of financial mechanisms like subsidies, microfinance, and credit schemes specially targeted towards women farmers, and designing and delivering training programmes that makes special provisions to include women and take specific note of their needs and constraints. Secondly, resource accessibility should be enhanced by strengthening agricultural extension services to reach marginalized groups, particularly women. Information on CSA practices and its schemes should be disseminated through gender-sensitive channels such as mobile platforms and community networks. In the long run, invest in infrastructure, such as roads and market linkages, to support smallholder farmers. Collaborative community engagement can also increase the adoption of CSA. This can be done by promoting community-driven approaches to CSA adoption, enabling collective resource management and knowledge sharing. This will go a long way in empowering local women's groups to act as champions for CSA practices. This needs to be simultaneously done with strengthening institutional support like developing public–private partnerships to scale CSA initiatives integrate CSA strategies into broader climate resilience policies and establish monitoring frameworks to track gender-specific impacts and ensure accountability. Lastly and most importantly, regional and context-specific solutions are needed on the ground for effective CSA interventions. It should address localized constraints to enhance the adaptation capabilities of the local population and promote sustainable agricultural practices using existing local knowledge and community networks.

## Conclusion and Future Research Directions

This systematic literature review and bibliometric analysis underscore the critical role of CSA in addressing the dual challenges of climate change and food insecurity in the Global South. However, it also highlights significant gender disparities in the adoption and implementation of CSA practices, which if left unaddressed, will continue to undermine the effectiveness of CSA interventions. The findings indicate that while CSA holds immense potential for enhancing agricultural resilience, its benefits are not equitably distributed, particularly among women, who face systemic barriers to adoption due to socio-economic, cultural, and institutional factors. One of the most striking conclusions drawn from the analysis is the persistent gender gap in access to resources such as land, credit, and agricultural inputs.

Women, despite their crucial role in agriculture, are frequently marginalized in terms of decision-making power and access to necessary tools and technologies. Moreover, the literature points to the importance of socio-economic support mechanisms, such as financial incentives, subsidies, and improved market access, in facilitating the adoption of CSA practices. Smallholder farmers, who constitute most of the agricultural workforce in the Global South, are particularly vulnerable to the economic risks associated with climate change. Without adequate financial support, these farmers are unlikely to invest in CSA practices despite their long-term benefits. Future research should explore the development of innovative financing mechanisms, such as climate-smart subsidies or microcredit schemes, that can help bridge the gap between smallholder farmers and CSA practices.

Additionally, the role of education and training cannot be overstated. However, socio-cultural norms and logistical barriers often exclude women farmers from accessing essential knowledge and extension services. Sedebo et al. (2021) underscore that farmers with adequate training are more likely to adopt CSA practices, yet women remain underserved. Future efforts must prioritize inclusive training models that address the unique needs of women and marginalized groups, ensuring accessibility and equitable participation in CSA capacity-building initiatives. Finally, the dissemination of climate information services (CIS) is crucial for the widespread adoption of CSA practices. The literature shows that gender disparities in access to CIS hinder the ability of women farmers to make informed decisions about agricultural practices.

Future research should investigate the most effective ways to disseminate climate information to women, considering the socio-cultural and logistical challenges they face. By leveraging digital technologies and ICT tools, CIS can be disseminated more widely and equitably, ensuring that all farmers have access to the information they need to adapt to climate change and enhance food security. In conclusion, while CSA offers a promising pathway for addressing climate change and ensuring food security, its success will depend on the development of gender-responsive policies, inclusive training programmes, and equitable access to resources and information. Future research should continue to explore these critical areas, with a particular focus on addressing the systemic barriers that hinder women's participation in CSA adoption.

Note: Additional supporting documents are available in the below link:

[Click here for additional file].

## References

Acosta M, Riley S, Bonilla-Findji O, Martínez-Barón D, Howland F, Huyer S, Chanana N (2021) Exploring women's differentiated access to climate-smart agricultural interventions in selected climate-smart villages of Latin America. Sustainability, 13(19):10951.https://doi.org/10.3390/su131910951

Adenle AA, Ford JD, Morton J, Twomlow S, Alverson K, Cattaneo A, Ebinger JO (2017) Managing climate change risks in Africa-A global perspective. Ecol Econ 141:190–201. https://doi.org/10.1016/j.ecolecon.2017.06.004

Ado AM, Leshan J, Savadogo P, Bo L, Shah AA (2019) Farmers' awareness and perception of climate change impacts: a case study of Aguie district in Niger. Environ Dev Sustain 21:2963–2977. https://doi.org/10.1007/s10668-018-0173-4

Agbenyo W, Jiang Y, Jia X, Wang J, Ntim-Amo G, Dunya R, Twumasi MA (2022) Does the adoption of climate-smart agricultural practices impact farmers' income? Evidence from Ghana. Int J Environ Res Public Health 19(7):3804.https://doi.org/10.3390/ijerph19073804

Aguilar L (2012) Women and climate change: vulnerabilities and adaptive capacities. In: State of the World 2009. Routledge, pp 59–62

Ahmed I, Ullah A, Rahman MHU, Ahmad B, Wajid SA, Ahmad A, Ahmed S (2019) Climate change impacts and adaptation strategies for agronomic crops. In: Climate change and agriculture, pp 1–14. https://doi.org/10.5772/intechopen.82697

Altieri MA, Ponti L, Nicholls CI (2005) Manipulating vineyard biodiversity for improved insect pest management: case studies from northern California. Int J Biodiv Sci Manag 1(4):191–203. https://doi.org/10.1080/17451590509618092

Andati P, Majiwa E, Ngigi M, Mbeche R, Ateka J (2022) Determinants of adoption of climate-smart agricultural technologies among potato farmers in Kenya: does entrepreneurial orientation play a role? Sustain Technol Entrepren 1(2):100017. https://doi.org/10.1016/j.stae.2022.100017

Anugwa IQ, Onwubuya EA, Chah JM, Abonyi CC, Nduka EK (2022) Farmers' preferences and willingness to pay for climate-smart agricultural technologies on rice production in Nigeria. Clim Policy 22(1):112–131. https://doi.org/10.1080/14693062.2021.1953435

Aria M, Cuccurullo C (2017) Bibliometrix: an R-tool for comprehensive science mapping analysis. J Informet 11(4):959–975. https://doi.org/10.1016/j.joi.2017.08.007

Asante BO, Ma W, Prah S, Temoso O (2024) Promoting the adoption of climate-smart agricultural technologies among maize farmers in Ghana: using digital advisory services. Mitig Adapt Strat Glob Change 29(3):19. https://doi.org/10.1007/s11027-024-10116-6

Atube F, Okello DM, Malinga GM, Nyeko M, Okello-Uma I (2022) Farmers' adaptation to climate change and crop yield: a case of Amuru and Apac districts of Northern Uganda. Int J Agric Sustain 20(5):967–981. https://doi.org/10.1080/14735903.2022.2028400

Autio A, Johansson T, Motaroki L, Minoia P, Pellikka P (2021) Constraints for adopting climate-smart agricultural practices among smallholder farmers in Southeast Kenya. Agric Syst 194:103284. https://doi.org/10.1016/j.agsy.2021.103284

Azadi H, Moghaddam SM, Burkart S, Mahmoudi H, Van Passel S, Kurban A, Lopez-Carr D (2021) Rethinking resilient agriculture: from climate-smart agriculture to vulnerable-smart agriculture. J Clean Prod 319:128602.https://doi.org/10.1016/j.jclepro.2021.128602

Balcha E, Menghistu HT, Zenebe A, Teferi T, Hadush B (2023) Climate-smart agricultural practices: a case of dairy cooperative farmers in Agula and Maychew, Northern Ethiopia. Carbon Manag 14(1): 2271880. https://doi.org/10.1080/17583004.2023.2271880

Barooah P, Alvi M, Ringler C, Pathak V (2023) Gender, agriculture policies, and climate-smart agriculture in India. Agric Syst 212:103751. https://doi.org/10.1016/j.agsy.2023.103751

Beuchelt TD, Badstue L (2013) Gender, nutrition and climate-smart food production: Opportunities and trade-offs. Food Secur 5:709–721. https://doi.org/10.1007/s12571-013-0290-8

Carr ER, Thompson MC (2014) Gender and climate change adaptation in agrarian settings: current thinking, new directions, and research frontiers. Geogr Compass 8(3):182–197. https://doi.org/10.1111/gec3.12121

Chandra A, McNamara KE, Dargusch P, Caspe AM, Dalabajan D (2017) Gendered vulnerabilities of smallholder farmers to climate change in conflict-prone areas: a case study from Mindanao, Philippines. J Rural Stud 50:45–59. https://doi.org/10.1016/j.jrurstud.2016.12.011

Chandra A (2017) Climate-smart agriculture in practice: insights from smallholder farmers, Timor-Leste and the Philippines, Southeast Asia. https://doi.org/10.14264/uql.2017.880

Coniglio ND, Pesce G (2015) Climate variability and international migration: an empirical analysis. Environ Dev Econ 20(4):434–468. https://doi.org/10.1017/s1355770x14000722

Diarra FB, Ouédraogo M, Zougmoré RB, Partey ST, Houessionon P, Mensah A (2021) Are perception and adaptation to climate variability and change of cowpea growers in Mali gender differentiated? Environ Dev Sustain 1–17. https://doi.org/10.1007/s10668-021-01242-1

Diendéré AA, Ouédraogo D (2023) Women farmers and climate change: empirical evidence from Burkina Faso. Agric Resour Econ Rev 52(3):451–475. https://doi.org/10.1017/age.2023.12

Dogondaji MS (2013) Towards mitigating the impacts of climate change on food security: a global perspective||. Acad J Interdisc Stud 2(6):167. https://doi.org/10.5901/ajis.2013.v2n6p167

Donthu N, Kumar S, Mukherjee D, Pandey N, Lim WM (2021) How to conduct a bibliometric analysis: an overview and guidelines. J Bus Res 133:285–296. https://doi.org/10.1016/j.jbusres.2021.04.070

Doss C (2014) If women hold up half the sky, how much of the world's food do they produce? Springer Netherlands, pp 69–88. https://doi.org/10.1007/978-94-017-8616-4_4

Dwomoh D, Agyabeng K, Tuffour HO, Tetteh A, Godi A, Aryeetey R (2023) Modeling inequality in access to agricultural productive resources and socioeconomic determinants of household food security in Ghana: a cross-sectional study. Agric Food Econ 11(1):24. https://doi.org/10.1186/s40100-023-00267-6

Edewor SE, Shittu AM, Olaoye IJ (2023) Gender analysis of land ownership influence on climate-smart agriculture adoption among farming households in Nigeria. Int J Sustain Agric Manag Inform 9(2):159–181. https://doi.org/10.1504/ijsami.2023.129860

Everest B (2021) Farmers' adaptation to climate-smart agriculture (CSA) in NW Turkey. Environ Dev Sustain 23(3):4215–4235. https://doi.org/10.1007/s10668-020-00767-1

FAO (2013) The State of Food Insecurity in the World 2013: the multiple dimensions of food security Food and Agriculture Organization of the United Nations Rome, 2013. https://www.fao.org/4/i3434e/i3434e.pdf

FAO (2017) Regional overview of food security and nutrition: Asia and the Pacific, Bangkok, vol 64. https://openknowledge.fao.org/server/api/core/bitstreams/fff30c4e-4364-4fb0-a518-4d3f11a642fc/content

Foguesatto CR, Machado JAD (2022) Adoption of sustainable agricultural practices in Brazil: understanding the influence of socioeconomic and psychological factors. J Agribus Dev Emerg Econ 12(2):204–222. https://doi.org/10.1108/jadee-11-2020-0256

Frison EA, Cherfas J, Hodgkin T (2011) Agricultural biodiversity is essential for a sustainable improvement in food and nutrition security. Sustainability 3(1):238–253. https://doi.org/10.3390/su3010238

Garrity DP, Akinnifesi FK, Ajayi OC, Weldesemayat SG, Mowo JG, Kalinganire A, Bayala J (2010) Evergreen agriculture: a robust approach to sustainable food security in Africa. Food Secur 2:197–214. https://doi.org/10.1007/s12571-010-0070-7

Goh AH (2012) A literature review of the gender-differentiated impacts of climate change on women's and men's assets and well-being in developing countries. https://doi.org/10.2499/capriwp106

IPCC (2014) Climate change 2014: impacts, adaptation, and vulnerability. Contribution of working group II to the fifth assessment report of the intergovernmental panel on climate change. Cambridge University Press, Cambridge

Jost C, Kyazze F, Naab J, Neelormi S, Kinyangi J, Zougmore R, Kristjanson P (2016) Understanding gender dimensions of agriculture and climate change in smallholder farming communities. Clim Dev 8(2):133–144. https://doi.org/10.1080/17565529.2015.1050978

Kamau MW, Ndung'u CK, Muriu Ng'ang'a FW (2020) Factors influencing farmers' adaptations to climate variability and extreme events in dry Kitui, Kenya. Int J Environ Sci Nat Res 24(1):1–10. https://doi.org/10.19080/ijesnr.2020.23.556126

Khan NA, Gong Z, Shah AA, Abid M, Khanal U (2021) Farm-level autonomous adaptation to climate change and its impact on crop productivity: evidence from Pakistan. Environ Dev Sustainability 1–26. https://doi.org/10.1007/s10668-021-01978-w

Kpadonou RAB, Owiyo T, Barbier B, Denton F, Rutabingwa F, Kiema A (2017) Advancing climate-smart-agriculture in developing drylands: Joint analysis of the adoption of multiple on-farm soil and water conservation technologies in West African Sahel. Land Use Policy 61:196–207. https://doi.org/10.1016/j.landusepol.2016.10.050

Kristjanson P, Bryan E, Bernier Q, Twyman J, Meinzen-Dick R, Kieran C, Doss C (2017) Addressing gender in agricultural research for development in the face of a changing climate: where are we and where should we be going? Int J Agric Sustain 15(5):482–500. https://doi.org/10.1080/14735903.2017.1336411

Kwapong NA, Whitfield S, Ambuko J, Ankrah DA, Swanepoel F (2024) Using participatory videos in understanding farmers' experiences with climate smart agricultural practices: reflections from Ghana. Front Sustain Food Syst 7:1282993. https://doi.org/10.3389/fsufs.2023.1282993

Lake IR, Hooper L, Abdelhamid A, Bentham G, Boxall AB, Draper A, Waldron KW (2012) Climate change and food security: health impacts in developed countries. Environ Health Perspect 120(11):1520–1526. https://doi.org/10.1289/ehp.1104424

Laurent L, Ullmann A, Castel T (2023) Regional climate warming increases occurrence and intensity of winter wheat drought risk. Available at SSRN 4813337. https://doi.org/10.2139/ssrn.4813337

Lipper L, Thornton P, Campbell BM, Baedeker T, Braimoh A, Bwalya M, Torquebiau EF (2014) Climate-smart agriculture for food security. Nat Clim Chang 4(12):1068–1072. https://doi.org/10.1038/nclimate2437

Lobell DB, Schlenker W, Costa-Roberts J (2011) Climate trends and global crop production since 1980. Science 333(6042):616–620. https://doi.org/10.1126/science.1204531

Makate C, Makate M, Mango N, Siziba S (2019) Increasing resilience of smallholder farmers to climate change through multiple adoption of proven climate-smart agriculture innovations. Lessons from Southern Africa. J Environ Manage 231:858–868. https://doi.org/10.1016/j.jenvman.2018.10.069

Makhdum MSA, Kousar R (2021) Gender in agriculture: determinants of female labor supply decisions among rural households in the context of market imperfections in Pakistan. Agric Econ 52(2):235–248. https://doi.org/10.1111/agec.12616

Mbow C, Toensmeier E, Brandt M, Skole D, Dieng M, Garrity D, Poulter B (2020) Agroforestry as a solution for multiple climate change challenges in Africa. In: Climate change and agriculture. Burleigh Dodds Science Publishing, pp 339–374. https://doi.org/10.19103/as.2020.0064.17

Mendelsohn RO, Dinar A (2009) Climate change and agriculture: an economic analysis of global impacts, adaptation and distributional effects. Edward Elgar Publishing. https://doi.org/10.4337/9781849802239

Mishra AK, Pede VO (2017) Perception of climate change and adaptation strategies in Vietnam: are there intra-household gender differences? Int J Clim Change Strat Manag 9(4):501–516. https://doi.org/10.1108/ijccsm-01-2017-0014

Mittal S, Hariharan VK (2018) Mobile-based climate services impact on farmers risk management ability in India. Clim Risk Manag 22:42–51. https://doi.org/10.1016/j.crm.2018.08.003

Müller-Bloch C, Kranz J (2015) A framework for rigorously identifying research gaps in qualitative literature reviews

Mutenje MJ, Farnworth CR, Stirling C, Thierfelder C, Mupangwa W, Nyagumbo I (2019) A cost-benefit analysis of climate-smart agriculture options in Southern Africa: balancing gender and technology. Ecol Econ 163:126–137. https://doi.org/10.1016/j.ecolecon.2019.05.013

Mwongera C, Shikuku KM, Twyman J, Läderach P, Ampaire E, Van Asten P, Winowiecki L A (2017) Climate smart agriculture rapid appraisal (CSA-RA): A tool for prioritizing context-specific climate smart agriculture technologies. Agric Syst 151:192–203. https://doi.org/10.1016/j.agsy.2016.05.009

Nawrotzki RJ, Bakhtsiyarava M (2017) International climate migration: Evidence for the climate inhibitor mechanism and the agricultural pathway. Popul Space Place 23(4):e2033. https://doi.org/10.1002/psp.2033

Nchanji EB, Kabuli H, Nyamolo VO, Cosmas L, Chisale V, Matumba A (2022) Gender differences in climate-smart adaptation practices amongst bean-producing farmers in Malawi: the case of Linthipe Extension Planning Area. Front Sustain Food Syst 6:1001152. https://doi.org/10.3389/fsufs.2022.1001152

Nchanji E, Nduwarugira E, Ndabashinze B, Bararyenya A, Hakizimana MB, Nyamolo V, Lutomia C (2023) Gender norms and differences in access and use of climate-smart agricultural technology in Burundi. Front Sustain Food Syst 7:1040977. https://doi.org/10.3389/fsufs.2023.1040977

Nelson V, Stathers T (2009) Resilience, power, culture, and climate: a case study from semi-arid Tanzania, and new research directions. Gender Dev 17(1):81–94. https://doi.org/10.1080/13552070802696946

Ngigi MW, Muange EN (2022) Access to climate information services and climate-smart agriculture in Kenya: a gender-based analysis. Clim Change 174(3):21. https://doi.org/10.1007/s10584-022-03445-5

Nyarindo WN, Mugera A, Hailu A, Obare GA (2024) Do combined sustainable agricultural intensification practices improve smallholder farmers welfare? Evidence from eastern and western Kenya. Agric Econ 55(2):296–312. https://doi.org/10.1111/agec.12816

Obossou EA, Chah JM, Anugwa IQ, Reyes-Garcia V (2023) Gender dimensions in the adoption of climate-smart agriculture technologies in response to climate change extremes in Benin. Reg Environ Change 23(3):93. https://doi.org/10.1007/s10113-023-02085-4

Olive OO, David UK (2021) Valuation of rice farmers' preferences and willingness to pay for climate-smart agricultural technologies in southeast, Nigeria. Asian J Econ Model 9(1):48–57. https://doi.org/10.18488/journal.8.2021.91.48.57

Omotoso AB, Omotayo AO (2024) Enhancing dietary diversity and food security through the adoption of climate-smart agricultural practices in Nigeria: a micro level evidence. Environ Dev Sustain 1–18. https://doi.org/10.1007/s10668-024-04681-8

Oyawole FP, Shittu A, Kehinde M, Ogunnaike G, Akinjobi LT (2021) Women empowerment and adoption of climate-smart agricultural practices in Nigeria. Afr J Econ Manag Stud 12(1):105–119. https://doi.org/10.1108/ajems-04-2020-0137

Pandey V, Gupta A (2022) Can multi-sectoral development interventions boost livelihoods and women's labor supply? Evidence from NRLM in India. Fem Econ 28(2):217–246. https://doi.org/10.1080/13545701.2022.2037684

Petticrew M, Roberts H (2008) Systematic reviews in the social sciences: a practical guide. Wiley. https://doi.org/10.1002/9780470754887

Phiri AT, Charimbu M, Edewor SE, Gaveta E (2022) Sustainable scaling of climate-smart agricultural technologies and practices in Sub-Saharan Africa: The Case Of Kenya, Malawi, and Nigeria. Sustainability 14(22):14709. https://doi.org/10.3390/su142214709

Pretty J (2011) Interdisciplinary progress in approaches to address social-ecological and ecocultural systems. Environ Conserv 38(2):127–139. https://doi.org/10.1017/s0376892910000937

Quisumbing AR, Meinzen-Dick R, Raney TL, Croppenstedt A, Behrman JA, Peterman A (2014) Closing the knowledge gap on gender in agriculture. Gender Agric Closing Knowl Gap 3–27.https://doi.org/10.1007/978-94-017-8616-4_1

Rad ST, Koşar A (2021) The role of producers' markets in empowering rural women. Tarım Ekonomisi Dergisi 27(1):1–8

Ranjitha P, Mishra A (2024) Learning agenda: climate-smart agriculture and gender. Nairobi, Kenya: CGIAR GENDER Impact Platform. https://cgspace.cgiar.org/server/api/core/bitstreams/232522c3-567a-4dd8-8fdb-91cd0f720f77/content

Rao CR, Raju BMK, Josily S, Rao AVMS, Kumar RN, Rao MS, Singh VK (2022) Impact of climate change on productivity of food crops: a sub-national level assessment for India. Environ Res Commun 4(9):095001. https://doi.org/10.1088/2515-7620/ac8b68

Sedebo DA, Li GC, Abebe KA, Etea BG, Ahiakpa JK, Ouattara NB, Frimpong S (2021) Smallholder farmers' climate change adaptation practices contribute to crop production efficiency in southern Ethiopia. Agron J 113(6):4627–4638. https://doi.org/10.1002/agj2.20900

Selvaraju R, Gommes R, Bernardi M (2011) Climate science in support of sustainable agriculture and food security. Clim Res 47(1–2):95–110. https://doi.org/10.3354/cr00954

Siddaway AP, Wood AM, Hedges LV (2019) How to do a systematic review: a best practice guide for conducting and reporting narrative reviews, meta-analyses, and meta-syntheses. Ann Rev Psychol 70(1):747–770. https://doi.org/10.1146/annurev-psych-010418-102803

Teklu A, Simane B, Bezabih M (2023) Multiple adoption of climate-smart agriculture innovation for agricultural sustainability: empirical evidence from the Upper Blue Nile Highlands of Ethiopia. Clim Risk Manag 39:100477. https://doi.org/10.1016/j.crm.2023.100477

Tesfaye K, Zaidi PH, Gbegbelegbe S, Boeber C, Rahut DB, Getaneh F, Stirling C (2017) Climate change impacts and potential benefits of heat-tolerant maize in South Asia. Theor Appl Climatol 130:959–970. https://doi.org/10.1007/s00704-016-1931-6

Thornton PK, Whitbread A, Baedeker T, Cairns J, Claessens L, Baethgen W, Keating B (2018) A framework for priority-setting in climate smart agriculture research. Agric Syst 167:161–175. https://doi.org/10.1016/j.agsy.2018.09.009

Tilahun G, Bantider A, Yayeh D (2023) Impact of adoption of climate-smart agriculture on food security in the tropical moist montane ecosystem: the case of Geshy watershed, Southwest Ethiopia. Heliyon 9(12). https://doi.org/10.1016/j.heliyon.2023.e22620

Touch V, Cross R, Grünbühel C, Van Ogtrop F, Ampt P, Yorn T, Tan DK (2023) Adaptation constraints and prospects for future research priorities in lowland rice-based farming systems: learning experiences from Northwest Cambodia. Environ Dev Sustain 1–32. https://doi.org/10.1007/s10668-023-03566-6

Tran NLD, Rañola JrRF, Ole Sander B, Reiner W, Nguyen DT, Nong NKN (2020) Determinants of adoption of climate-smart agriculture technologies in rice production in Vietnam. Int J Clim Change Strat Manag 12(2):238–256. https://doi.org/10.1108/ijccsm-01-2019-0003

Twyman J, Green M, Bernier Q, Kristjanson PM, Russo S, Tall A, Badiane Ndour NY (2014) Adaptation actions in Africa: evidence that gender matters. CCAFS working paper

Umar BB (2021) Adapting to climate change through conservation agriculture: a gendered analysis of eastern Zambia. Front Sustain Food Syst 5:748300. https://doi.org/10.3389/fsufs.2021.748300

Vatsa P, Ma W, Zheng H, Li J (2023) Climate-smart agricultural practices for promoting sustainable agrifood production: yield impacts and implications for food security. Food Policy 121:102551. https://doi.org/10.1016/j.foodpol.2023.102551

Veisi H, Carolan MS, Alipour A (2017) Exploring the motivations and problems of farmers for conversion to organic farming in Iran. Int J Agric Sustain15(3):303–320.https://doi.org/10.1080/14735903.2017.1312095

Victor L (2008) Systematic reviewing in the social sciences: outcomes and explanation. Enquire 1(1):32–46

Von Braun J (2020) Climate change risks for agriculture, health, and nutrition. In: Health of people, health of planet and our responsibility: climate change, air pollution and health, pp 135–148. https://doi.org/10.1007/978-3-030-31125-4_11

Wekesah FM, Mutua EN, Izugbara CO (2019) Gender and conservation agriculture in sub-Saharan Africa: a systematic review. Int J Agric Sustain 17(1):78–91. https://doi.org/10.1080/14735903.2019.1567245

# Chapter 3
# The Intersection of Climate Change and Gender-Based Violence in Sub-Saharan Africa: A Synthesis

**Douglas Nyathi, Joram Ndlovu, Stanley Maphosa, Thembelihle Nyathi, and Nkosilamandla Kunene**

**Abstract** One of the most urgent global concerns of the twenty-first century is climate change, which disproportionately affects vulnerable groups in Sub-Saharan Africa, particularly women and girls. This study investigates how environmental stresses due to climate change like droughts, floods, and resource scarcity, worsen existing gender inequities and trigger violence. This chapter particularly explores the relationship between climate change and gender-based violence (GBV) in Sub-Saharan Africa. Climate-related disruptions, including community displacement and livelihood losses, frequently result in heightened economic strain, social unrest, and competition for resources, all of which increase the likelihood of GBV including domestic abuse, sexual assault, and human trafficking. Women, particularly those in rural settings, bear most of these environmental burdens because of their roles as primary caregivers and providers of water and food. As the climate crisis intensifies, women encounter greater challenges in managing household resources, and their freedom of movement becomes increasingly constrained, rendering them more susceptible to violence, both domestically and in the public sphere. Moreover, the collapse of social structures caused by climate-induced shocks such as migration or displacement can cause women to increase exploitation and abuse in refugee camps or informal settlements. This chapter recommends response mechanisms and policy

D. Nyathi (✉)
School of Public Management, Governance and Public Policy, University of Johannesburg, Gauteng, South Africa
e-mail: dnyathi@uj.ac.za

J. Ndlovu
School Social Sciences, Howard College, University of KwaZulu-Natal, Glenwood, South Africa
e-mail: Ndlovuj1@ukzn.ac.za

S. Maphosa
Institute of Pan African Thought, University of Johannesburg, Guateng, South Africa
e-mail: director@gfsa.org.za

T. Nyathi · N. Kunene
Department of Development Studies, Lupane State University, Lupane, Zimbabwe



A. Sarkar and N. Bandyopadhyay (eds.), *Gender-Transformative Approaches for Climate Change Adaptation*, Springer Climate,
https://doi.org/10.1007/978-3-031-90236-9_3

frameworks for addressing both climate change and GBV in Sub-Saharan Africa. It emphasizes the necessity for integrated approaches that acknowledge the dual impact of environmental and gendered vulnerabilities. This underscores the significance of gender-responsive climate adaptation strategies that prioritize women's empowerment and protection, ensuring that efforts to combat climate change also mitigate GBV risk. It also advocates for enhanced cross-sectoral collaboration to ensure that climate justice and gender equality are at the forefront of both policy and practice in the region.

**Keywords** Climate change · Gender-based violence · Vulnerability · Intersectionality

## Introduction

The two most urgent global challenges and sustainable development issues we face today are climate change and violence against women. Violence Against Women and Girls (VAWG) impacts over a third of women during their lives, making it one of the most widespread and persistent human rights violations globally (AfDB 2019; WHO 2021). Research indicates that in 2020 alone, one woman was slain every 11 min, which accounts for 81,000 deaths among women and girls (United Nations 2021; WHO 2021). The sustainability of our world is seriously threatened by climate change, which has far-reaching social, cultural, economic, health, and human rights repercussions that disproportionately affect women and girls, especially in marginalized groups. Additionally, it is a significant aggravating factor for different types of VAWG (Dankelman 2010a; CARE 2017). The 'shadow pandemic,' or spike in VAWG during the COVID-19 pandemic, was a prime example of the upsurge in violence that happens during emergencies and natural catastrophes. As women's environmental human rights defenders work to defend their communities, ecosystems, and territories against the effects of pollution, climate change, habitat loss, and degradation brought on by humans, they are more vulnerable to threats, violence, and even femicide (Dankelman 2010b). Attempts to mitigate, adapt, and increase resilience to climatic catastrophes are seriously jeopardized if VAWG is not eradicated. Simultaneously, overlooking the threats and challenges of gender-based violence (GBV) in climate action could heighten risks to the safety of women and girls and undermine their right to live free from violence, as climate change exacerbates the occurrence of VAWG (WHO 2021; UN Women 2022a).

The linkages between climate change and VAWG have received much less attention, despite a decade-long surge in international attention toward comprehending and solving gender-climate concerns in policies and programs (UNFPA 2024). Developing successful interventions that safeguard the rights and welfare of women and marginalized groups requires an understanding of this relationship. A gender-responsive approach to policymaking and implementation is becoming increasingly necessary as climate-related disasters increase in frequency and severity. By incorporating gender considerations into climate change policies, we can enhance community resilience and work towards eliminating the systemic injustices that fuel GBV. This chapter explores the relationship between climate change and GBV, examines intersectionality, gender, and power dynamics, and then outlines the theoretical frameworks and methodology used. The chapter explores the factors contributing to the increase in GBV and climate change. It aims to connect climate change, forced migration, and GBV. In its concluding section, the chapter addresses critical considerations for creating climate action plans that are sensitive to GBV, along with conclusions and policy suggestions.

## Climate Change and Gender-Based Violence: Conceptual Positioning

The effects of climate change in Sub-Saharan Africa are extensive and varied, with particularly significant consequences for women. The region's dependence on rainfall for agriculture implies that shifting weather patterns, including extended dry spells and erratic precipitation, have resulted in considerable crop losses and food shortages (Niang et al. 2014). For instance, prolonged droughts in Kenya have necessitated women to travel further to water, increasing their exposure to potential violence (Danish Refugee Council (DRC) 2022).

The economic strain caused by climate change further amplifies power disparities within households and communities. Resource scarcity often leads to disputes over control and access, with women frequently bearing the brunt of these conflicts. These women frequently face barriers to accessing legal and social support systems, which worsen their circumstances. Women face a heightened risk of land-grabbing as they take steps to safeguard their land and natural resources (Hart and Lovey 2022).

The environmental pressures compound existing social and economic vulnerabilities, rendering women more susceptible to GBV. Climate change can result in forced displacement, which is linked to rising instances of intimate partner violence (Adeola et al. 2024). Additionally, it can contribute to a rise in sexual violence and has been associated with increased rates of early or forced marriages. Mental health is crucial in understanding the connection between climate change and GBV. As environmental changes threaten livelihoods, men may experience increased stress and frustration, potentially manifesting as domestic abuse (WHO 2021; UNFPA 2024). In Zambia, financial instability due to failed harvests is associated with an increase in Intimate Partner Violence (IPV) (Simona et al. 2015).

The concept of intersectionality highlights how various forms of discrimination such as those related to gender, ethnicity, and socioeconomic status interact to create unique experiences of oppression (Crenshaw 1989). This indicates that women from marginalized communities face a greater vulnerability to the impacts of climate change. For example, resource shortages brought on by climate change disproportionately affect women in pastoral communities in dry and semi-arid Sub-Saharan Africa, increasing the prevalence of GBV (Kagunda et al. 2024).

The psychological impact of environmental degradation and resource scarcity can result in mental health issues such as anxiety, depression, and post-traumatic stress disorder (PTSD) (Njeru et al. 2022). In Nigeria, women who have encountered GBV due to climate-related displacement report increased levels of anxiety and depression, negatively affecting their overall well-being and capacity to manage environmental hardships (Atwoli et al. 2022). Existing research indicates that women who experience GBV in the context of climate-induced displacement report higher levels of mental health issues, affecting their overall well-being and ability to cope with environmental challenges (Goudet et al. 2024; Fisher 2010).

Climate change can lead to increase in sexual exploitation and abuse. It is imperative to recognize the diverse vulnerabilities and contributions of women to fully comprehend the intertwined issues of climate change and GBV. By examining the lived experiences of women in Sub-Saharan Africa, one can gain insight into how environmental changes and socioeconomic pressures intersect to exacerbate gender inequalities and violence. For example, in Malawi, women have reported increased domestic violence linked to the stress of food insecurity and economic hardship caused by climate change (Jafry et al. 2022).

## Material and Methods

### *Theoretical Frameworks and Methodology*

This research is based on the frameworks of intersectionality, gender theory, and power theory. Intersectionality, as introduced by Crenshaw (1989), highlights how overlapping discrimination rooted in gender, race, and socioeconomic status shapes distinct experiences of oppression. This framework is essential for grasping the disproportionate impact of climate change on women, particularly those belonging to marginalized communities. The compounded vulnerabilities faced by these women arise from their intersecting social identities, which increases their exposure to both environmental and social challenges (Kabeer 2016; Collins 2015). For instance, the interaction between gender and socioeconomic status can restrict women's access to resources and decision-making, deepening their vulnerability to the effects of climate change (Hankivsky 2014). Moreover, intersectionality highlights the unequal impact of climate change on women due to their societal roles.

In many Sub-Saharan African communities, women are primarily responsible for sourcing water and food. The increased strain on these duties due to climate-induced resource scarcity underlines the intersectional nature of their vulnerabilities (UNICEF 2019). Research in South Africa and Ethiopia suggests that these compounded responsibilities constrain women's capacity to engage in other productive pursuits, thus perpetuating cycles of poverty and dependence (FAO 2017; Tadesse 2018). This intersectional analysis is essential for grasping the complex impacts of climate change on women, underscoring the importance of gender-sensitive strategies for climate adaptation and mitigation (Arora-Jonsson 2011).

Gender and power theory investigates how power dynamics within households and communities are shaped by gender roles and relations. Climate change intensifies these dynamics by amplifying economic pressures and resource scarcity. As environmental shifts threaten livelihoods, the ensuing stress and frustration can manifest as domestic violence (WHO 2021; Jewkes et al. 2010). This theory underscores how environmental stressors can exacerbate existing socioeconomic vulnerabilities, rendering women more susceptible to GBV. Thus, power imbalances within households and communities are amplified by economic and social disruptions brought about by climate change (Agarwal 1997; Chigwanda 2016).

This study utilized a systematic desk-based approach to examine the impact of climate change on GBV in Sub-Saharan Africa. It emphasizes secondary research, gathering, and analysing existing literature from reputable sources like peer-reviewed journals, reports, and government publications to aggregate pertinent information. The focus was on collecting and evaluating recent academic articles, primarily published between 2016 and 2024, to ensure that the analysis was based on current and relevant data that reflected contemporary trends and issues. The literature review was conducted using a range of academic databases, such as Google Scholar, JSTOR, and Scopus, all of which provide access to scholarly publications. The search utilized targeted keywords such as climate change, GBV, vulnerability, patriarchy, and resilience to find relevant studies for analysis. Additionally, reports from international organizations like the Food and Agriculture Organization (FAO) and the Intergovernmental Panel on Climate Change (IPCC) have been included to offer a broader policy perspective.

To ensure that only the most pertinent studies were reviewed, the selection criteria were implemented. Articles were selected based on their publication date, relevance to the sub-Saharan African context, and focus on climate change and GBV. The data extraction process entailed pinpointing key themes from the literature, including gender-specific vulnerabilities to climate change, the connections between climate change and gender issues, and the relationship between climate change, forced displacement, and GBV. A thematic analysis approach was utilized to categorize and synthesize findings from different studies, enabling a thorough exploration of how climate change affects GBV in sub-Saharan Africa.

## Results and Discussions

### *Addressing Gender-Based Violence Through Climate Action*

The ongoing climate crisis is intensifying and impacting diverse populations and communities through extended periods of drought, flooding, and gradual environmental changes. These effects compromise food security, resource accessibility, and dependability, efforts to reduce poverty, and national development objectives (FAO 2017; Tadesse 2018; UNFPA 2024). Research on climate change and related disasters has shown gender-based differences in survival rates, access to aid and recovery support, and participation in climate-related planning and policymaking (UNFCCC 2020). These findings indicate that the impacts of climate change differ based on an individual's gender. Climate change intensifies existing gender inequalities through various direct and indirect effects and other risk factors for GBV, including poverty, harmful societal norms, interpersonal and community conflicts, and displacement (Castañeda et al. 2020).

Women, girls, members of various Sexual Orientation, Gender Identity and Expression, and Sex Characteristics (SOGIESC), and other marginalized groups can exercise agency to make self-determined life choices, and they are better able to withstand the impacts of climate change when they have bodily autonomy and are liberated from oppression, stigma, violence, and coercion, including all types of GBV. Tackling GBV advances gender equality, empowers women, and supports sustainable development goals, while also playing a crucial role in maintaining healthy and sustainable environments in communities facing prolonged impacts of climate change (Lima and Guedes 2024). Issues including environmental and natural resource management, land and property rights, energy and infrastructure, water security, sanitation, and hygiene are intimately linked to tackling gender inequality and GBV. Gender inequality is worsened by climate change, making women especially susceptible to its effects. Social and cultural norms may limit women's ability to influence household decisions regarding adaptation and mitigation methods for climate change, including relocation (Fruttero et al. 2023).

In the recent years, integrating gender considerations has gained significant importance within the UNFCCC programmes and policies. Furthermore, the incorporation of gender issues into climate policy and action has been officially mandated by recent gender-related resolutions made by the Conference of Parties (COP). Gender equality and gender balance are now persistent agenda topics with specific rules. The lack of comprehensive gender mainstreaming is the reason for the notable discrepancy between national climate policy and action commitments, and the implementation of gender-responsive climate policy.

## *Gendered Vulnerability to Climate Change and Variability*

The impacts of climate change are not experienced equally around the world. The rising frequency and intensity of extreme weather events like hurricanes, floods, droughts, and heavy rainfall lead to notable societal and economic effects. Predictions suggest that existing gender norms and inequalities will result in climate change affecting men and women in different ways (Fruttero et al. 2023). Gender plays a significant role in determining various coping and adaptive strategies (Guszkowska et al. 2015).

Studies conducted in countries such as South Africa, Botswana, Namibia, and Mozambique reveal that women's contributions to climate change adaptation and mitigation remain largely overlooked, insufficiently explored, and poorly documented (Mulinge and Getu 2013). They also imply that major weather events cause damage to dwellings and interfere with basic utilities, such as energy, water, and sanitation. To compensate for losses, households must increase their income-generating activities while also spending more time on household chores (Fruttero et al. 2023).

Women in pastoral communities are particularly vulnerable and disproportionately impacted by climate change, a situation worsened by various social, economic, and geographical factors (UN Women 2022a). Mulinge and Getu (2013) studied how climate change affects pastoral women in Ethiopia and their coping strategies, revealing a notable knowledge gap about the consequences of climate change and adaptive measures within pastoral communities throughout Africa. Water scarcity has transformed the livelihood systems and coping strategies of Maasai women pastoralists in Oldonyo-orok, Kenya (Huho et al. 2011). Research in Tanzania shows that although both men and women pastoralists are affected by climate change and variability, women are particularly more vulnerable due to the gender-specific division of labour in households, unequal resource ownership, and restricted participation in decision-making (Mushi 2013).

Research conducted in the Borena and South Omo zones of southern Ethiopia reveals that the socio-cultural dynamics within pastoral communities have placed women at a disadvantage, stemming from the unequal power dynamics between men and women regarding resource control and access (Amsalu and Wana 2013). Gender inequality among Afar pastoralists in Northeastern Ethiopia highlights how traditional customs (Adda) exacerbate women's vulnerability to the impacts of drought. Additionally, the study found that activities like labour migration, petty trading, and individual wealth accumulation previously uncommon in Afar culture have emerged as responses to climate change (Balehegn and Tafere 2013). This division of labour heightens women's vulnerability when climate shocks disrupt food production, water availability, and access to fuel.

Deep-rooted gender inequalities hinder women's capacity to adapt and heighten their vulnerability to climate shocks (Fruttero et al. 2023). Women are often more vulnerable to the impacts of climate change and have fewer resources to address them, owing to factors like lower wages and workforce participation, their prevalence in agriculture and informal sectors, limited land ownership, unequal access to off-farm employment, restricted credit access, heightened family caregiving responsibilities, and various constraining social norms in many SSA countries (Deere and Doss 2006; Eastin 2018). Additionally, their limited access to assets, economic stability, and legal protections intensifies their susceptibility to climate change (Eastin 2018).

Several factors, including gender, age, caste, socioeconomic status, and disability, influence an individual's vulnerability to climate change (Rao and Morchain 2016). A study conducted in the Chivi District of Southern Zimbabwe indicates that the impacts of climate change are shaped by gender and differ depending on household dynamics and the wider socio-cultural and economic context of agro-pastoralists' lives (Belle et al 2024). People with impairments, those living in poverty, and members of particular castes are particularly susceptible to the effects of climate change (Rao et al. 2017).

Research in Uganda shows that climate change and variability lead to an increased workload for women as men migrate to seek alternative income-generating opportunities (Kabonesa and Kindi 2013). It is widely recognized that due to resource shortages such as food, water, and firewood often brought on by climate change, women and girls are more likely to take on household responsibilities (Doss et al. 2018; Eastin 2018). To access these resources, they must go further, which may put them at greater risk for security issues such as sexual assault and harassment. The increased workload often leads families to withdraw their daughters from school to help at home, which can harm their prospects (Niang et al. 2014).

Climate change can worsen health outcomes, particularly for those already marginalized (Adeola et al. 2024). Increased heat waves and disease outbreaks (like malaria) can disproportionately affect women who may have limited access to health care.

Raidimi (2014) contends that women frequently take on a vital role in managing resources necessary for family survival through activities like subsistence farming, fuel gathering, and water collection, while men usually concentrate on wage labour or cash crop farming. Limitations in mobility, expertise, and financial means may also limit coping strategies in addition to social standards and shame. Women usually face the harshest economic consequences of crop failures due to their limited access to financial resources. Many women and girls are especially vulnerable, as food shortages heighten their risk of illness (Rao et al. 2017). Moreover, the lower socioeconomic status of women and girls complicates their access to and affordability of health care. Tackling gender disparities is essential not only as a necessary action but also as a potential objective for effective climate change strategies and poverty alleviation, given that men and women experience the impacts of climate change differently (UN Women 2020b).

## *The Link Between Climate Change and Gender-Based Violence*

The sharing of duties, responsibilities, and interactions between men and women differs across various socioeconomic groups, age levels, education levels, and marital statuses, emphasizing that gender dynamics encompass both men and women (Rao and Morchain 2016). Consequently, these arrangements affect the perception of dangers and vulnerabilities, which may differ depending on time, location, and social standing (Rao and Morchain 2016). Significantly, climate change could exacerbate existing gender inequalities, making it increasingly challenging for women and girls in affected regions to combat discrimination, marginalization, and dispossession (UN Women 2022b). This disproportionately threatens the livelihoods of women and girls by hindering their capacity to adapt to or recover from climate-related events. For example, women and girls in the Sahel region are more vulnerable to sexual harassment and assault when they must travel farther to seek drinking water because of the region's growing water scarcity due to drought (McOmber 2020). As resources become scarcer, social tensions can rise, leading to increased violence against women and girls. In Kenya, e.g. during times of drought, there have been reports of heightened violence and discrimination against women and girls as communities compete for limited resources (United Nations 2022). As climate change impacts livelihoods, women who lack education or access to financial resources may resort to unsafe coping mechanisms, increasing their vulnerability to GBV (Adeola et al. 2024). For example, in Ethiopia, many women engaged in subsistence farming face educational barriers, limiting their ability to adapt to climate challenges, and thereby increasing their susceptibility to exploitation and violence.

In East Africa, the risk of GBV such as sexual assault, domestic abuse, and early marriages has risen for women and girls due to gradual climate events like widespread drought and persistent famine, which are directly tied to climate change (Castañeda et al. 2020). For instance, it has been claimed that girls in Kenya who are experiencing financial difficulties as a result of drought are either coerced into early marriages by their families or participate in transactional sex (IFRC 2015; Castañeda et al. 2020). Similarly, sexual exploitation and intimate partner abuse have increased in Southern Africa as a result of erratic droughts and unexpected floods (IFRC 2015; UN Women 2020b). A study examining data from 19 Demographic and Health Surveys in Sub-Saharan Africa found that women residing in regions impacted by severe drought were more likely to experience physical and sexual violence from intimate partners compared to those in unaffected areas (Epstein et al. 2020; OCHA 2020).

In some instances, the heightened workload of women and girls can hinder their ability to meet the domestic responsibilities typically assigned to male family members, creating tension at home that may ultimately result in violence. To secure food and water for their families, women and girls may need to travel longer distances, increasing their vulnerability to sexual assault (Niang et al. 2014; Tallman et al. 2022). In resource-limited environments, women and girls face a higher risk of being coerced into sexual exploitation. Likewise, studies show that girls are significantly more likely to marry when their families struggle to meet basic needs (Atiqul Haq et al. 2024).

## *Climate Change, Forced Displacement, and Gendered Violence*

Gender significantly influences how individuals experience climate change, disasters, and displacement (Nyathi et al. 2024). For those in vulnerable or marginalized positions, climate-related events not only increase the risk of displacement but also intensify the dangers faced during such situations. Displacement caused by climate change leads to vulnerabilities that are specific to gender, which can worsen existing inequalities, heighten the incidence of gender-based and sexual violence, and hinder access to vital services and livelihoods (Castañeda et al. 2020). Reducing displacement risks and impacts, promoting quicker recovery, and improving climate change resilience can be significantly aided by recognizing and resolving gender imbalances. Although gender issues in disaster and displacement contexts and the negative impacts of climate change have been extensively studied (Allen et al. 2023), research on gender-related challenges in climate-induced displacement scenarios is still in its infancy. It is evident that existing gender disparities, which primarily affect women, are exacerbated by the interplay of climate change, natural disasters, and displacement. Furthermore, disaster situations can lead to the breakdown of existing protective mechanisms, such as law enforcement, leaving certain individuals, particularly women and girls vulnerable (Le Masson 2022). Women and girls around the world experience increased vulnerability to GBV, especially in emergency shelters, where they frequently face barriers to accessing essential resources, services, and discrimination (Gaillard et al. 2017).

Women face higher mortality rates during extreme weather events than men, and female survivors often have a reduced life expectancy. During crises, occurrences of sexual and GBV, including domestic abuse, often rise as a result of displacement (Omotoso et al. 2023). Additionally, many women experience heightened pressure to provide for their families, contributing to mental health issues and stress-related disorders (Rao and Morchain 2016). A weakened health infrastructure negatively impacts sexual and reproductive health services, potentially resulting in unsafe abortion practices and unintended pregnancies. In addition to directly endangering their lives, disasters expose women to other protection concerns, which are more severe in situations when they are displaced. The disruption of vital services, including sexual and reproductive health care and social protection, especially sudden-onset disasters can heighten the vulnerability of women and girls, restricting their access to essential aid, even as these disasters elevate their risk of GBV and other safety concerns (Dominey-Howes et al. 2014).

In camps, almost 85% of displaced people cook using biomass obtained in an unsustainable manner, such as firewood. In addition to the many health and life-threatening hazards posed by this activity, such as indoor air pollution and conflicts with neighbouring communities make women and children especially susceptible to GBV during wood harvesting activities (Dida et al. 2022).

Women who are displaced by catastrophes often end up living in overcrowded evacuation centres and informal settlements, which increases their vulnerability to sexual and GBV (Nyathi et al. 2024). In certain cases, women who chose to stay where they were to escape living conditions that were socially unacceptable in displacement sites, including living alongside men, died in disasters. Women with disabilities are greatly affected by violence and conflict (Rohwerder 2017). This diminished resilience may result in the adoption of detrimental coping strategies, including child marriage and transactional sex (CARE 2017). The risk of falling victim to trafficking and organized crime may also increase (IFRC 2015; Castaneda et al. 2020).

Lesbian, gay, bisexual, transgender, intersex, queer/questioning, asexual, and many other people may be particularly vulnerable to negative displacement outcomes such as exclusion from humanitarian assistance and ostracism, as they already face various forms of discrimination. This could force them to settle in informal shelters, making them more vulnerable to violence (Gaillard et al. 2017).

## *Climate Change, Men, and Masculinity*

In the context of Africa, grasping the effects of patriarchy, which enforces male dominance and creates imbalanced power dynamics between genders is essential for identifying the underlying causes and developing effective strategies to tackle climate change responses (Nguyen and Rydstrom 2018). In Africa, these analyses have led to deeper understanding and more tailored solutions in critical areas such as combating violence against women, enhancing women's economic empowerment, and promoting sexual and reproductive health and rights. For example, initiatives like microfinance programmes specifically for women have not only provided financial independence but also reduced domestic violence in communities by shifting power dynamics (Maldonado-Casto et al. 2024). Similarly, campaigns that educate women about their reproductive rights have empowered them to make informed choices, ultimately improving health outcomes for entire families (UN Women 2020b).

In Africa, suicide rates among rural men have risen alarmingly, now exceeding those of rural women and both urban men and women. This tragic trend can be attributed to the interplay of climate variability, economic hardship, and declining agricultural productivity (Adeola et al. 2024). For instance, many rural communities rely heavily on farming for their livelihoods, and unpredictable weather patterns such as prolonged droughts or flooding have devastated crops, leading to financial instability. The pressure of not being able to support their families, combined with restricted access to mental health services, has fuelled this escalating crisis. Addressing these interconnected issues is crucial for restoring mental well-being and stability in these communities (Alston 2012). Men's experiences with climate change may not receive sufficient attention, which can lead to an inadequate understanding of how these changes affect their families and communities.

The compounding stress of climate impacts and societal discrimination can significantly affect mental health. In Africa, failing to address men's specific needs in disaster response and humanitarian aid can have dire consequences, exacerbating the collapse of livelihoods and social structures. This oversight can lead to increased substance abuse as men struggle with feelings of inadequacy and loss of purpose (Adeola et al. 2024). Additionally, the resulting stress and frustration can manifest in higher rates of domestic violence and mistreatment of women and children, as men may resort to harmful coping mechanisms. Comprehensive disaster response strategies that include targeted support for men are essential to rebuilding strong, healthy communities post-crisis (Castaneda et al. 2020).

## Governance and Environmental Policy Related to Climate Change Action

A wide variety of players working at various levels, from local to global, are involved in climate-change policies and actions. Climate activists, environmental lawyers, financiers, local and indigenous communities, regulatory bodies, and environmental NGOs are important participants (Lin and Peel 2024). These factors add to the multifaceted and complex nature of climate change governance. The variety of ways that various parties in the Global South influence legal activities reflects the polycentricism of climate change litigation (Lin and Peel 2024). Global actors and policies have a significant impact on how Sub-Saharan Africa responds to climate change.

Many African countries lack coherent climate and gender policies, leading to disjointed approaches that fail to address the unique vulnerabilities of women and marginalized groups (Omotoso et al. 2023). For example, in countries like Uganda, climate adaptation strategies often overlook the specific needs of women farmers, who play a critical role in agriculture but face significant barriers due to gender biases. While global frameworks exist, their implementation in Sub-Saharan Africa faces challenges due to poor governance and inadequate policy coherence (Dan-Woniowei et al. 2022; Ikeme 2003).

Poor governance can result in weak emergency response mechanisms during climate-related disasters. For instance, during the 2019 Cyclone Idai in Mozambique, inadequate shelter provisions and safety measures in displacement camps led to increased reports of GBV, particularly against women and girls, due to a lack of privacy and security. In many African countries, laws addressing GBV are inadequately enforced due to poor governance (Omotoso et al. 2023). For instance, despite legal frameworks against GBV in countries like South Africa, systemic corruption and lack of resources often result in low prosecution rates and inadequate protection for survivors, particularly in rural areas affected by climate-related migration. Poor governance can exacerbate resource scarcity due to climate change, leading to increased competition and conflict. In regions like the Sahel, where desertification is worsening, competition for land and water has led to increased violence, including GBV, particularly against displaced women. The lack of effective governance responses to these tensions can further entrench cycles of violence (Omotoso et al. 2023).

Governments have the power to ratify political agendas and agreements for climate change at all levels, from international to local (Petzold et al. 2023). They also point out that while civil society helps coordinate responses, governments are frequently involved in planning. Governance issues can lead to the misallocation of funds meant for climate resilience and GBV prevention. In many African nations, corruption diverts resources away from effective climate adaptation initiatives (Adeola et al. 2024). For example, in Zimbabwe, funds intended for climate resilience projects have sometimes been siphoned off, leaving communities without necessary support and increasing their vulnerability to climate impacts and violence.

According to Lebenya et al. (2024), local governments are essential for helping states to meet their climate goals. Inadequate governance often marginalizes women's voices in decision-making processes. This exclusion means that climate policies may not reflect the realities of women's experiences, leading to ineffective or harmful outcomes (Adeola et al. 2024). For instance, in Kenya, women's organizations have highlighted that local climate adaptation strategies frequently fail to consider women's roles in managing natural resources, which can limit their economic opportunities and increase their risk of GBV. Non-governmental organizations (NGOs) are also essential in climate change action, contributing to innovation, advocacy, and grassroots initiatives. Poor governance often results in inadequate investment in education and resources for women.

## Key Issues to Consider in Designing GBV-Sensitive Climate Action Plans

GBV tends to be more prevalent during periods of environmental stress such as natural disasters, resource scarcity, or displacement. Climate action plans must include comprehensive knowledge of the interaction between gender inequality, GBV, and the impacts of climate change to successfully address these challenges (Mhaskar 2010; UNFCCC 2020). Gender analysis is a crucial component in creating climate action plans that respond to GBV. This entailed assessing how climate change affects men, women, girls, and boys.

According to the African Development Bank (2019), applying a gender perspective to development issues necessitates analysing the gender division of labour, identifying the needs and priorities of men and women, recognizing existing opportunities and barriers to achieving development goals, and selecting an intervention strategy to address these factors. Gender analysis promotes gender equality and mitigates GBV. Nelson et al. (n.d) posited that gender analysis ensures that women's specific needs and perspectives are considered, leading to more effective climate responses. Women and other marginalized groups must be involved in the policymaking process to guarantee that their needs, preferences, and vulnerabilities are considered when designing and implementing programs (Alston 2013; Bryan et al. 2016). Diverse stakeholder participation in decision-making procedures guarantees a fair representation of climate governance frameworks. This equitable engagement helps to dismantle the stigmas associated with GBV, supports legislation that addresses GBV, and aids in the prevention and addressing of GBV concerns. GBV-sensitive climate action plans require an integrated and holistic strategy that recognizes the complexity of gender inequality and climate change (UNFPA 2020). Gender equality can be promoted, GBV prevention and response mechanisms can be integrated into climate action plans, and women and marginalized genders can be actively included in decision-making processes to create more equitable and resilient communities that are better prepared to face the challenges of a changing climate (UNFPA 2020).

## Conclusion and Recommendations

Climate change disproportionately affects women, especially in patriarchal societies where they have fewer rights and protections. Despite ongoing knowledge and data shortages in many sectors, there is strong qualitative and quantitative evidence to clarify the links between GBV and climate change. Several studies recommend immediate action, investment, and innovation to address this relationship. Multi-level decision-making and policymaking processes across sectors present critical opportunities for effecting change. Women, who are often the primary caregivers and managers of households, have valuable knowledge about sustainable practices. However, patriarchal structures frequently exclude them from decision-making processes.

Understanding the various context-specific facets of GBV-climate linkages is crucial for creating successful policies, plans, and interventions across sectors, as these problems can obstruct or even reverse progress when they interact. Patriarchal societies often perpetuate harmful gender norms that discourage environmental stewardship among men, who may be socialized to prioritize economic development over ecological concerns. In patriarchal societies, men often dominate decision-making regarding resource management, leading to practices that prioritize short-term gains over sustainable use. GBV prevention and response and environmental conservation and sustainable development issues are often seen as intersecting but rarely linked in many international policy frameworks, donor and aid finance mechanism priorities, and sustainable development organizations' strategies. This obscures the potential risk of escalating violence and/or environmental degradation. Setting these links as top priorities provides an opportunity to explore innovative and fresh approaches to achieving international sustainable development and human rights obligations.

Patriarchy contributes to economic disparities, limiting women's access to education, resources, and financial tools necessary for engaging in sustainable practices. The advancement of this viewpoint requires involving men as both persons who are susceptible to climate change-related calamities and change agents who can work with female activists. Patriarchy can create resistance to progressive climate policies that promote gender equality and sustainable practices. The difficulty lies in bringing boys and men into this process as a radical departure from dogmatic ideas of masculinity that are frequently predicated on dominance, control, and conquest. Women are often underrepresented in formal climate negotiations and policymaking platforms, which can lead to policies that do not reflect the needs and perspectives of half the population. Experts have emphasized the need for a deeper comprehension of how destructive masculine ideas, such as the demand for control over others with 'others' being defined as women, other men, children, and the environment, maintain environmental destruction.

## References


Adeola O, Evans O, Ngare I (2024) Gender and climate issues in Africa. In: Gender equality, climate action, and technological innovation for sustainable development in Africa. Sustainable Development Goals Series. Palgrave Macmillan, Cham. https://doi.org/10.1007/978-3-031-40124-4_2

African Development Bank (AfDB) (2019) The African development bank group gender strategy 2021–2025. Investing in Africa's women to accelerate inclusive growth. https://www.afdb.org/sites/default/files/documents/strategy-documents/african_development_bank_group_gender_strategy_-_2021-2025.pdf. Accessed 15 Mar 2023

Agarwal B (1997) Bargaining and gender relations: within and beyond the household. Fem Econ 3(1):1–51. https://doi.org/10.1080/135457097338799

Allen EM, Munala L, Ward-Rannow J (2023) Do Gender-based violence interventions consider the impacts of climate change? A systematic review. Trauma Viol Abuse 25(3):2421–2435. https://doi.org/10.1177/15248380231214793

Alston M (2012) Rural male suicide in Australia. Social Sci Med 74(4):515–522. https://doi.org/10.1016/j.socscimed.2010.04.036

Alston M (2013) Gender mainstreaming and climate change: a review of the literature. Clim Dev 5(1):1–12. https://doi.org/10.1016/j.wsif.2013.01.016

Amsalu A, Wana D (2013) Climate change impacts on pastoral women in Ethiopia: some evidence from the Southern Lowlands. PHE publishers, Addis Ababa

Arora-Jonsson S (2011) Virtue and vulnerability: discourses on women, gender and climate change. Global Environ Change 21(2):744–751. https://doi.org/10.1016/j.gloenvcha.2011.01.005

Atiqul Haq SM, Ahmed MNQ, Lalin SAA, Arno AT, Ahmed KJ (2024) Early marriage of girls in the context of the COVID-19 pandemic: a literature review. Human Soc Sci Commun 11:697. https://doi.org/10.1057/s41599-024-03085-3

Atwoli L, Ndetei DM, Khasakhala LI (2022) Climate change, displacement, and mental health: the Nigerian experience. Int J Environ Health Res 34(5):510–512. https://doi.org/10.1080/09540261.2022.2093626

Balehegn M, Tafere K (2013) Gendered impacts and adaptation mechanisms to climate change among Afar pastoralists in Northeastern Ethiopia. In Mulinge MM, Getu M (Eds) Impacts of climate change and variability on pastoral women in sub-Saharan Africa (Addis Ababa, Ethiopia, Organization of Social Science Research in Eastern and Southern Africa (OSSREA)), pp 83–120. https://doi.org/10.1186/s13570-018-0129-1

Belle J, Mapingure T, Owolabi S (2024). Factors influencing rural women's adoption of climate change adaptation strategies: evidence from the Chivi District of Zimbabwe. Climate 12(11):191. https://doi.org/10.3390/cli12110191
Bryan E, Bernier Q, Espinal M, Ringler C (2016) Integrating gender into climate change adaptation programs: a research and capacity needs assessment for Sub-Saharan Africa. CCAFS working paper no 163. CGIAR research program on climate change, agriculture and food security. CCAFS, Copenhagen, Denmark
CARE International (2017) Hope dries up? Women and girls coping with drought and climate change in Mozambique. http://careclimatechange.org/wp-content/uploads/2016/11/El_Nino_Mozambique_Report_final.pdf. Accessed 15 Mar 2023
Castañeda CI, Sabater L, Owren C, Boyer EA (2020) Gender-based violence and environment linkages: the violence of inequality, IUCN. https://portals.iucn.org/library/node/48969. Accessed 10 Mar 2023
Chigwanda E (2016) Women's experiences of climate change and gender-based violence in Zimbabwe. J Int Women's Stud 17(4):1–13
Collins PH (2015) Intersectionality's definitional dilemmas. Ann Rev Sociol 41:1–20. https://doi.org/10.1146/annurev-soc-073014-112142
Crenshaw K (1989) Demarginalizing the intersection of race and sex: a black feminist critique of antidiscrimination doctrine, feminist theory and antiracist politics. Univ Chicago Legal Forum 1989(1):139–167. https://chicagounbound.uchicago.edu/uclf/vol1989/iss1/8. Accessed 10 Feb 2023
Dan-Woniowei F, Ikporukpo I, Kalagbo S (2022) Climate change governance and ecosystem sustainability in Sub-Saharan Africa. Adv Social Sci Res J 9(8):561–574. https://doi.org/10.14738/assrj.98.12791
Danish Refugee Council (DRC) (2022) Drought-affected populations in Northern Kenya. DRC assessment shows the rapid deterioration of the protection environment for pastoralist communities in Northern Kenya. https://drc.ngo/media/p2ap4yxl/drc-kenya-drought-needs-assessment-oct-22.pdf. Accessed 10 Feb 2023
Dankelman I (2010a) Gender and climate change: an introduction. Earthscan, London
Dankelman I (2010b) Introduction: exploring gender, environment, and climate change. In Dankelman I (ed) Gender and climate change: an introduction. Routledge, London
Deere CD, Doss CR (2006) Gender and the distribution of wealth in developing countries, WIDER working paper series RP2006-115, World Institute for Development Economic Research (UNU-WIDER). Helsinki. https://hdl.handle.net/10419/63540. Accessed 10 Feb 2023
Dida GO, Lutta PO, Abuom PO, Mestrovic T (2022) Anyona DN (2022) Factors predisposing women and children to indoor air pollution in rural villages, Western Kenya. Arch Public Health 80:46. https://doi.org/10.1186/s13690-022-00791-9
Dominey-Howes D, Gorman-Murray A, McKinnon S (2014) Queering disasters: on the need to account for LGBTI experiences in natural disaster contexts. Gender Place Cult 21(7):905–918. https://doi.org/10.1080/0966369X.2013.802673
Doss C, Meinzen-Dick R, Quisumbing A, Theis S (2018) Women in agriculture: four myths. Glob Food Sec 16(2018):69–74. https://doi.org/10.1016/j.gfs.2017.10.001
Eastin J (2018) Climate change and gender equality in developing states. World Dev 107(C):289–305. https://doi.org/10.1016/j.worlddev.2018.02.021
Epstein A, Bendavid E, Nash D, Charlebois ED, Weiser SD (2020). Drought and intimate partner violence towards women in 19 countries in sub-Saharan Africa during 2011–2018: a population-based study. PLoS Med 17(3):e1003064. https://doi.org/10.1371/journal.pmed.1003064
FAO (2017) The impact of disasters and crises on agriculture and food security. Rome: food and agriculture organization of the United Nations. http://www.fao.org/3/I8656EN/i8656en.pdf. Accessed 10 Feb 2023
Fisher S (2010) Violence against women and natural disasters: findings from post-tsunami Sri Lanka. Routledge, London

Fruttero A, Halim D, Broccolini C, Coelho B, Gninafon H, Muller N (2023) Gendered impacts of climate change evidence from weather shocks. Policy research working paper 10442. World Bank Group

Gaillard JC, Sanz K, Balgos BC (2017) People's response to disasters: vulnerability, capacities, and resilience. In: Wisner B, Gaillard JC, Kelman I (eds) Handbook of hazards and disaster risk reduction and management. Routledge, Abingdon, pp 414–423

Goudet JM, Binte Arif F, Owais H, Uddin Ahmed H, Ridde V (2024) Climate change and women's mental health in two vulnerable communities of Bangladesh: an ethnographic study. PLOS Glob Public Health 4(6):e0002080. https://doi.org/10.1371/journal.pgph.0002080

Guszkowska M, Zagórska-Pachucka A, Kuk A, Skwarek K (2015) Gender as a factor in differentiating strategies of coping with stress used by physical education students. Health Psychol Rep 4. https://doi.org/10.5114/hpr.2016.57681

Hankivsky O (2014) Intersectionality 101, Institute for intersectionality research and policy, Simon Fraser University. https://www.sfu.ca/iirp/documents/resources/101_Final.pdf. Accessed 10 Mar 2023

Hart C, Lovey S (2022) Groundbreakers: women overcome bias and lead sustainable use of land and resources. https://www.climatelinks.org/blog/groundbreakers-women-overcome-bias-and-lead-sustainable-use-land-and-resources. Accessed 10 Mar 2023

Huho J, Ngaira J, Ogindo H (2011) Living with drought: the case of the Maasai pastoralists of northern Kenya. Educ Res 2(1):779–789

IFRC (2015) Unseen, unheard: gender-based violence in disasters. https://www.ifrc.org/Global/Documents/Secretariat/201511/1297700_GBV_in_Disasters_EN_LR2.pdf. Accessed 10 Feb 2023

Ikeme J (2003) Equity, environmental justice and sustainability: incomplete approaches in climate change politics. Glob Environ Chang 3(3):195–206. https://doi.org/10.1016/S0959-3780(03)00047-5

Jafry T, Watson SE, Mikulewicz M, Cassidy G, Chirambo GB, Gondwe P, Singini W, Chirwa K, Gondwe D, Matola B, Mkandawire C, Nyirenda J, Nyirenda W, Mkandawire H (2022) The inter-relationship between climate change, mental health and gender-based violence in Malawi. Glasgow Caledonian University

Jewkes R, Levin JB, Penn-Kekana L (2010) Risk factors for domestic violence: findings from a South African cross-sectional study. Soc Sci Med 55(9):1603–1617. https://doi.org/10.1016/S0277-9536(01)00294-5

Kabeer N (2016) Gender equality, economic growth, and women's agency: the endless variety and monotonous similarity of patriarchal constraints. Fem Econ 22(1):295–321. https://doi.org/10.1080/13545701.2015.1090009

Kabonesa C, Kindi FI (2013) Effects of climate change and variability on pastoral communities: coping and adaptive strategies for women and men in Uganda. In: Mulinge MM, Getu M (eds) Impacts of climate change and variability on pastoralist women in sub-Saharan Africa. Fountain Publishers, Kampala, pp 223–264

Kagunda B, Muthoni L, Wanjiku K (2024) Climate change and gender-based violence in pastoralist communities. J Clim Soc 12(3):45–67

Lebenya M, Mubangizi B, Ojogiwa O (2024) Participatory communication for sustainable rural livelihoods: insights from Matatiele Local Municipality. J Admin Sci 4(2). https://doi.org/10.54201/iajas.110

Le Masson V (2022) Disasters, climate change, and violence against women and girls. Oxford Research Encyclopedia of Natural Hazard Science. https://doi.org/10.1093/acrefore/9780199389407.001.0001/acrefore-9780199389407-e-393. Accessed 10 Mar 2023

Lima R, Guedes G (2024) Sustainable development goals and gender equality: a social design approach on gender-based violence. Sustainability 16(2):914. https://doi.org/10.3390/su16020914

Lin J, Peel J (2024) Actors in global south climate change litigation mobilization efforts, pp 147–202. https://doi.org/10.1093/9780191926525.003.0005. Accessed 10 Jan 2025

Maldonado-Castro J, Gallego-Losada R, Montero-Navarro A (2024).Mapping the intellectual structure of microfinance and women's empowerment: a bibliometric analysis. Heliyon 10(20). https://doi.org/10.1016/j.heliyon.2024.e39563
McOmber C (2020) Women and climate change in the Sahel, West African papers, no. 27. OECD Publishing, Paris
Mhaskar A (2010) Gender and climate change: a review of the literature. Clim Dev 2(1):1–12
Mulinge MM, Getu M (2013) Impacts of climate change and variability on pastoralist women in sub-Saharan Africa. African Books Collective
Mushi V (2013) Women pastoralists and climate change impacts in Kilosa District, Tanzania
Nelson V, Meadows K, Cannon T, Morton J, Martin A (n.d) Uncertain predictions, invisible impacts and the need to mainstream gender in climate change adaptations. Gender Dev 10(2). https://doi.org/10.1080/13552070215911
Nguyen H, Rydström H (2018) Climate disaster, gender, and violence: men's infliction of harm upon women in the Philippines and Vietnam. Women's Stud Int Forum 71:56–62. https://doi.org/10.1016/j.wsif.2018.09.001
Niang I, Ruppel OC, Abdrabo MA, Essel A, Lennard C, Padgham J, Urquhart P (2014) Africa. In: Barros VR, Field CB, Dokken DJ, Mastrandrea MD, Mach KJ, Bilir TE, Chatterjee M, Ebi KL, Estrada YO, Genova RC, Girma B, Kissel ES, Levy AN, MacCracken S, Mastrandrea PR, White LL (eds) Climate change 2014: impacts, adaptation, and vulnerability. Part B: regional aspects. Contribution of working group II to the fifth assessment report of the intergovernmental panel on climate change. Cambridge University Press, Cambridge, pp 1199–1265
Njeru J, Kimani M, Wambui N (2022) Mental health impacts of climate change on women in Sub-Saharan Africa. Afr J Environ Psychol 8(2):112–130
Nyathi D, Ndlovu J, Mare A, Dzvimbo M, Ndlovu M (2024) Women's vulnerability and adaptive capacity to climate change in Agrarian settings of Zimbabwe. In: Climate change management, vol Part F2915. Springer Science and Business Media Deutschland GmbH, pp 541–559. https://doi.org/10.1007/978-3-031-58261-5_23
OCHA (2020) Greater Sahel: USG Lowcock remarks on how to avert a looming tragedy. https://www.unocha.org/story/greater-sahel-usg-lowcock-remarks-how-avert-looming-tragedy. Accessed 10 Mar 2023
Omotoso BA, Letsoalo S, Olagunju OK, Tshwene SC, Omotayo OA (2023) Climate change and variability in sub-Saharan Africa: a systematic review of trends and impacts on agriculture, J Clean Prod 137487. https://doi.org/10.1016/j.jclepro.2023.137487.
Petzold J, Hawxwell T, Jantke K et al (2023) A global assessment of actors and their roles in climate change adaptation. Nat Clim Chang 13:1250–1257. https://doi.org/10.1038/s41558-023-01824-z
Raidimi EN (2014) The roles and activities of women in the six selected agricultural projects in Thulamela Local Municipality of Vhembe District Municipality in the Limpopo Province. South Afr J Agric Exten 42(2):10–23
Rao N, Lawson ET, Raditloaneng WN, Solomon D, Angula MN (2017) Gendered vulnerabilities to climate change: insights from the semi-arid regions of Africa and Asia. Clim Dev 11(1):14–26. https://doi.org/10.1080/17565529.2017.1372266
Rao N, Morchain D (2016) Gendered vulnerabilities to climate change: insights from the semiarid regions of Africa and Asia. Clim Dev 11(1). https://doi.org/10.1080/17565529.2017.1372266
Rohwerder B (2017) Women and girls with disabilities in conflict and crises. K4D Helpdesk Report. Institute of Development Studies, Brighton, UK
Simona SJ, Muchindu M, Ntalasha H (2015) Intimate partner violence (IPV) in Zambia: sociodemographic determinants and association with use of maternal health care. USAID. DHS working papers no 121
Tadesse M (2018) Women and climate change in Africa: Impact and adaptation, Addis Ababa: African Union
Tallman PS, Collins S, Salmon-Mulanovich G, Rusyidi B, Kothadia A, Cole S (2022) Water insecurity and gender-based violence: a global review of the evidence. WIREs

United Nations (2021) UNODC research: 2020 saw a woman or girl being killed by someone in their family every 11 minutes. United Nations Office on Drugs and Crime

United Nations (2022) Kenya: UN steps up protection for drought-hit women and girls

United Nations Children's Fund (UNICEF) (2019) Children uprooted in a changing climate: Turning challenges into opportunities with and for young people on the move. UNICEF, New York

United Nations Framework Convention on Climate Change (UNFCCC) (2020) Introduction to gender and climate change. https://unfccc.int/gender. Accessed 10 Mar 2023

United Nations Population Fund (UNFPA) (2024) Climate change impacts and intimate partner violence in sub-Saharan Africa. United Nations Population Fund

United Nations Women (2022a) Tackling Violence against women and girls in the context of climate change. CSW66, IUCN, Generation Equality, UN Women

United Nations Women (2022b) Explainer: how gender inequality and climate change are interconnected. IUCN, Generation Equality, UN Women

World Health Organization (WHO) (2021) Violence against women prevalence estimates, 2018. World Health Organization, Geneva

# Chapter 4
# Climate-Related Disasters and Gender Inequalities: Unveiling Key Elements Through Disaggregated Data Tool

**Gabriela Azevedo Couto, Maurício Trentin, Miguel Angel Trejo-Rangel, Debora Pignatari Drucker, and Carlos A. Nobre**

**Abstract** Globally, climate-related disasters are gender-sensitive, exacerbating the vulnerabilities that women and other marginalized groups already face in a world marked by social, environmental, and climate inequalities. Even so, women play crucial roles as agents and leaders in driving collaborative initiatives for climate adaptation and disaster risk reduction. In Brazil, gender and ethnic inequalities are deeply rooted and unfair, and despite socially constructed disparities, women survive more from landslides, storms, and extreme temperatures than men. In this work, we cover the potential of disaggregated data on disaster victims through a gender lens to unveil important hidden elements for disaster risk reduction. We present a remote platform called Sobrevidas©, a user-friendly responsive online platform that collects qualitative and quantitative data, as well as provides valuable insights on potential human resources and capabilities of women in contexts of threat. We advocate for Sobrevidas© as an insightful tool for decision-makers to support gender grassroots

G. A. Couto (✉)
National Institute for Space Research, São José Dos Campos, Brazil
e-mail: gabriela@sobrevidas.com.br

M. Trentin
2MONOPÓLIO, INCODE Tech, TIMESAFE S.a.a.S, BITBOT, São Paulo, Brazil
e-mail: info@mauriciotrentin.com

M. A. Trejo-Rangel
School of Architecture, Planning and Environmental Policy, University College Dublin, Dublin, Ireland
e-mail: miguel.trejorangel@ucd.ie

D. P. Drucker
Embrapa Digital Agriculture, EMBRAPA, Campinas, Brazil
e-mail: debora.drucker@embrapa.br

C. A. Nobre
Institute of Advanced Studies, University of São Paulo, São Paulo, SP, Brazil



A. Sarkar and N. Bandyopadhyay (eds.), *Gender-Transformative Approaches for Climate Change Adaptation*, Springer Climate,
https://doi.org/10.1007/978-3-031-90236-9_4

initiatives, improve public policies to reduce inequalities, enhance women's safety, and prevent disasters.

**Keywords** Climate-related disasters · Gender inequalities · Intersectionality · Disaggregated data analysis · Sobrevidas© · Women capacities · Leadership

## Introduction

The climate crisis is underway, with $CO_2$ concentration at higher levels in at least two million years and higher temperatures in the most recent multi-century warm period (IPCC 2022). The potential of reaching no reversible changes in the climate and environmental systems is at risk (Lenton et al. 2019; Hoegh-Guldberg et al. 2018), intensifying the people's challenges and vulnerabilities they already experience in their daily lives (Wallemacq and House 2018).

The increment in frequency and intensity of natural hazards, including storms, floods, landslides, droughts, extreme temperatures, and their extremes, worsens the severity of large-scale disasters. According to the United Nations (Centre for Research on the Epidemiology of Disasters of United Nations Office for Disaster Risk Reduction), over the last twenty years, those events killed an average of 60,000 people per year and affected more than four billion people (UNDRR 2020). Although fatalities due to climate-related disasters have decreased over the last 40 years worldwide (WMO 2021), data on deaths affected by disasters show high mortality rates for women, children, and other marginalized.

The international literature on gender and disaster recognizes the established social imbalances and lower socio-economic status of women, and their intersectionalities, as the main ones responsible for exacerbating vulnerabilities and reducing capacities to respond to threats and recover from disasters (Pathak and Emah 2017).

## Gender Approach to Disaster Studies

The impact of disasters does not affect people equally and depends on how vulnerable and exposed a human system is and its interaction with socioeconomic status, gender, ethnicity, and other inequalities (Neumayer and Plumber 2007). Unequal structures determine why women and girls are more vulnerable before, during, and after disasters, and exacerbate and strengthen permanent gender inequalities, propagating this vicious circle. The macro-level of gender vulnerabilities is established by social, economic, and political inequalities, and institutional structures of power and privilege that shape norms and social roles (De Silva and Jayathilaka 2014), which determine how differently men and women suffer from and recover from disasters. These include low decision-making power, social taboos (Reyes and Lu 2017; Aryal 2014), disparities in access to health services and legal systems (Sultana 2010), and higher poverty rates (Cannon 2002). These conditions influence micro-level vulnerabilities such as lower confidence in government institutions (Fatti and Patel 2013; Wachinger et al. 2013), individual skills (Kirsch et al. 2012), and gender-specific traditional clothing (Kantamaneni et al. 2022) and reduce the capacities to respond to threats and recover.

Vulnerabilities that occur before disasters are influenced by societal restrictions (Myeni and Wentink 2020; Lunga and Musarurwa 2016; De Silva and Jayathilaka 2014; Soares and Licea 2013), limited access to information (Aryal 2014), and lower levels of awareness about disasters (Nirupama and Maula 2013). During disasters, these vulnerabilities are associated with restricted mobility (Madurai 2016), and in the post-disaster period, women often assume multiple roles and responsibilities (Chineka et al. 2019), such as becoming heads of households (Ajaero 2017), without the authority to make decisions (Ndlovu and Mjimba 2021). They may also face violence and sexual exploitation (Valerio 2014), leading to excessive post-traumatic stress responses (Akekar and Fordham 2017; Hamama-Raz et al. 2015).

The analyses associate vulnerabilities with both the micro- and macro levels of society distinguished by the phase of disasters (Raza 2017) and also recognize the capacities to respond to threats and recover from disasters. In disaster studies, a gender lens as a category of analysis originates from recognizing that women have been and are more impacted than men in terms of safety, health, housing, women's rights, and life itself in disaster situations. Originally, disaster research focused on men as the central object of study, with research questions and theoretical perspectives oriented toward men's social roles as "heroes" in disaster situations, as "responsible for the family," and as "decision-makers" (Enarson and Meyreles 2024).

Over time, women have been recognized as key agents of cooperative initiatives for sustainable development (UN 2018; UNISDR 2015, UN WOMEN 2014) and as leaders who play vital roles in disaster risk reduction (Aryal 2014) and recovery (Clissold et al. 2020). In the early warning phase, women promote emotional support and collective action to prepare for a disaster (Story et al. 2020), provide and maintain vital conditions such as potable drinking water, agricultural livestock support, health, and relief during disasters, and engage in community and family reconstruction to provide new income and dignity in the post-disaster period (Rakib et al. 2017).

## *Gender and Disasters in Brazil*

Brazil is a large country with a continental dimension of 8.5 million square kilometers. It comprises five political-administrative regions, six large-scale ecosystems, and five climate zones (BRASIL 2021). Each region has multi-climate characteristics, varying in humidity, temperature, and rainfall extremes through the seasons and over the territory (Debortoli et al. 2017). Brazil is home to more than 200 million people (IBGE 2023) and is characterized by significant socioeconomic inequalities that, when associated with climate-extreme events, create conditions for disasters (Perez et al. 2020). In Brazil, disaster records are gradually increasing, totaling

67.230 reported between 1995 and 2024, affecting almost 233 million people, generating around US$ 24 billion in damages and US$ 31 billion in loss (BRASIL 2024). Climate models indicate that in the future scenarios of warming above 2.0 °C, highly exposed regions will become more vulnerable, and disasters will become more severe (Marengo et al. 2021).

Over the last 30 years, women achieved higher educational levels, increased their participation in the labor market, and families headed by women grew (Cavenaghi and Alves 2018) in Brazil. However, gender and ethnic inequalities are still structural in Brazil. According to the IBGE (2021), Brazilian Institute of Geography and Statistics, gender refers to the different socially constructed attributes and opportunities that determine social interactions and relationships between men and women in society. Understanding gender inequalities in Brazil requires acknowledging the existence of gender and ethnic social disparities in combination (Campos et al. 2018) which are further supported by data that shows inequalities (Hirata 2014). These inequalities highlight white men as most privileged, and black women at the base of the social hierarchy. This is apparent with black women with inappropriate employment, lowest wages, and household income (Passos and Wajnman 2021; Leão et al. 2017).

Despite the socially constructed gender and ethnic disparities (Freitas 2010), over the last forty years, women survived more floods, storms, landslides, and extreme temperatures (Couto et al. 2023) than men. Some of the international literature justifies men dying more than women due to a predisposition toward risk-taking (Salvati et al. 2018; Diakakis and Deligiannakis 2017; Zagheni et al. 2015; Coates 1999). However, important questions remain unanswered: In Brazil, are men most of the victims, or do women die less than men? Are women less vulnerable than men to disasters, or are they more capable of dealing with them? The differences in the sentences must be subtle, but the answers can indicate the vulnerabilities and capacities involved in surviving hazards extremes, and disasters in Brazil.

Despite the new insights on data fatalities from the study "Natural hazards fatalities in Brazil, 1979–2019″ (Couto et al. 2023), in Brazil all disaster impacts are measured by numbers. Official disaster data that comes from different sources[1] is represented by several records, categories of hazard, millions of fatalities, millions of people affected, displaced, homeless, monetary damages, and loss. Monitoring and reporting events and their impacts are fundamental. However, reliable and available intersectional data on disaster victims by gender, age, race, educational level, income, disabilities, and other characteristics is a key element to recognizing the most affected by disasters, the impact on marginalized groups, and the capacities involved to respond and survive (Picardi 2021; Saulnier et al. 2019). New information and knowledge to promote and empower initiatives are underway on disaster risk reduction and climate change adaptation (Beltrame and Medina 2022; WMO 2021; UNDRR 2020; Brown et al. 2019).

[1] The International Disaster Database EM-DAT (https://www.emdat.be) and "Atlas Digital de Desastres no Brasil" (https://atlasdigital.mdr.gov.br/paginas/index.xhtml)—the official governmental data source for disasters in Brazil.

In this perspective, we discuss why and how disaggregated data on victims and survivors of disasters, along with social inequalities across different scales, can unveil hidden elements in disaster monitoring to reduce the risks of disasters. We propose a new tool designed to enhance research knowledge on gender and disasters while also improving strategic information for governmental agencies to manage threats and empower grassroots initiatives. We recognize the challenges associated with the adoption of new instruments for established systems but show that a different road exists and can be taken.

## The Rationale of the Study

Science acknowledges the challenges in understanding why and how gender intersects with other variables, such as race, age, geographical location, and other social factors, and calls for new intersectional methods to comprehend these relationships. It also recognizes that research that integrates gender intersectionalities influences and is influenced by science (Tannenbaum et al. 2019). So, collecting and analyzing gender-disaggregated data is crucial for making any concrete decision and making plans.

### *The Need for Time-Sensitive Data*

Disaster conditions differ according to the moment, place, circumstances in which people live, how society is organized, and the typology of hazards. Despite this, different social actors such as non-governmental organizations, grassroots initiatives, academia, and international organizations recognize the interdependency between climate-related vulnerabilities, the root causes of social inequalities, and gender bias, and highlight how crucial it is to investigate the role of gender in the context of disasters.

An intersectional approach integrating gender, generational, ethnic, and socioeconomic characteristics constitutes a valuable model to examine the multidimensionality of identities and can also be taken toward understanding oppressions or reducing inequalities (Bentley et al. 2023). A gender intersection lens provides a valuable model for more equitable and inclusive initiatives on disaster risk reduction (UNDRR 2024), and for better gender-responsive climate action programs and policies (UNFCCC 2024). However, new knowledge production is required with new tools, methodologies, and research based on disaggregated data collection and use, to identify different needs, and recognize key actors in disaster risk reduction (Picard 2021) and climate change adaptation.

Conducting activities on the field immediately after a disaster can be highly sensitive, particularly when conducting conventional face-to-face household surveys. This approach can hinder real-time knowledge exchange (Jones and Ballon 2020) and make it difficult to capture survivors' stories, which may otherwise be lost. The simple presence of an observer and the sex of the researcher can also impact and alter the response of the research participants (Tannenbaum et al. 2019).

According to the World Bank, the availability of mobile-based social networks and affordable handsets has opened new possibilities, facilitating real-time monitoring and opinion polling, even in developing countries (Dabalen et al. 2016). During the 2010 Haiti earthquake, a content analysis using Big Data from platforms such as Twitter (X) allowed a detailed investigation of organizational shifts in communication strategies and how messaging was adapted during disaster response (Gurman and Ellenberger 2015). The analysis drew upon extensive data, providing valuable insights into disaster communication practices.

## *The Power of Qualitative and Quantitative Data Analysis*

Big Data, allied with artificial intelligence algorithms, can be an extremely helpful tool to guide disaster response and reduce risks (Bolland et al. 2023). However, although the volume of data in this field is increasing and academic research is rapidly growing, we face limitations such as the lack of theoretical frameworks in this discipline and the underrepresentation of studies from the Global South (Kondraganti et al. 2024). The most commonly used Big Data sources in disaster research are remote sensing data, social media, crowdsourcing, and mobile phone data sources (Kondraganti et al. 2024). Nevertheless, the prevailing challenges are the availability and ability to comprehensively represent the population studied through the disaster phases. For instance, Yan and Pedraza-Martinez (2019) show that "information on social engagement increases from preparedness to response and decreases from response to recovery." That is, there is a need to advance the use of such evolving technologies to support disaster operations, also focusing on responsive and recovery approaches.

The over-reliance on quantitative data, no matter how granular, causes known problems and already recognized pitfalls, such as the McNamara Fallacy, which involves decisions made solely based on quantitative observations (Cukier and Mayer-Schönberger 2013; Yankelovich 1971). According to the terminology's creator, "The fallacy is: If you're confronted by a complex problem that is full of intangibles, you decide to measure only those aspects of the problem that lend themselves to easy quantification, either because you find other aspects difficult to measure or because you assume that they cannot be very important or do not even exist" (Yankelovich 1971, p. 27). It is crucial to hear the survivors, their stories, challenges, and strategies to inform research and the development of future approaches, rather than relying solely on victim data or structured numerical data (quantitative research), to acknowledge challenges and capacities to deal with disasters.

## *The Importance of Disaggregate Data Collection and Its Challenges*

The scientific community and international agencies recognize gender equality initiatives in reducing the risk and differential impacts of disasters. Recently, international agencies are produced reports and guidelines, promoting programs and action plans to encourage nations to secure ownership and promote the participation of diverse social groups in decision-making processes for disaster risk reduction (UNDRR 2021).

Data disaggregation and its dissemination are considered a priority by the United Nations, as they help recognize not only those fatally affected but, more importantly, those who were able to survive. Measuring mortality and survival requires specific methodologies and tools, and it is essential to have comparable estimates across countries. However, each country has its own strategy for measuring impact, official data sources, and availability, and not all countries disaggregate data or account for both quantitative and qualitative data (Saulnier et al. 2019).

Disaggregated data, categorized separately by gender and its intersectionalities, along with gender statistics, go beyond being just a quantitative tool for post-disasters. They offer recognition of the socio-economic and political factors and provide valuable insights into the participation and leadership roles of women in disaster risk reduction (DRR) and climate change adaptation strategies, at local, national, and potentially international levels (Picard 2021).

A lack of comparable data on how disaster risk and impacts differ by gender exists, which can lead to inadvertently exacerbating gender inequalities (UNDRR 2024). Common challenges include: i) the failure to collect disaggregated data, ii) the omission of disaggregation in the data that is collected, iii) the use of outdated data, and iv) the exclusion of available disaggregated data (UNDRR 2021).

## Key Objective

The main objective of this chapter is to present a new research tool, along with its potential, for collecting remote gender data and its intersectionalities directly from disaster victims on a large scale going beyond traditional fieldwork or a single case study. It introduces a new methodology for remote gender research capable of recognizing local realities, gender-based social inequalities, and the differential impacts of climate change and disasters at the community level. The study focuses on the benefits of using this tool and emphasizes that the use of this tool by women at the grassroots has the potential to identify women's roles as community leaders and as key actors in decision-making processes to prevent future disasters and enhance women's safety.

## Findings and Discussions

### *Sobrevidas©: A New Tool for Gender Transformative Research*

In this chapter, we present a new digital tool called Sobrevidas© created to collect data remotely, tested a decade after a serious disaster event in Brazil—the Serrana Region tragedy, a single catastrophic event on January 11 and 12, 2011 The disaster took place in the highlands Serrana region in the Brazilian State of Rio de Janeiro, when heavy rain triggered flooding and landslides in seven counties. It led to 905 fatalities, left around 300 people missing, affected more than 300,000 and left tens of thousands homeless and displaced (Banco Mundial 2012; Coates and Garmany 2017; Dourado et al. 2012; Marengo and Alves 2012). The disaster equally affected residents of low and high incomes, urban and agricultural zones, public infrastructure, and buildings (CEPED 2020), with significant economic losses nearing $835.2 million USD (Alvalá and Barbieri 2017) within 24 h.

The digital tool resulted from the need to conduct remote methods to collect data during COVID-19 times (between 2021 and 2022), where physical distancing was key and engaging methods needed to adapt, in addition to a low-budget condition and a short period to reinvent participatory methods. It is a remote research tool designed to collect data and stories from disaster survivors remotely, as much information as possible, and enable them to share their survival stories and the loss of loved ones.

Sobrevidas© is designed as a user-friendly, responsive online platform with a simple interface, with a clean layout optimized for mobile devices, and ensures dynamic navigation to facilitate engagement for those unfamiliar with a complex topic. It contains animations controlled by screen scrolling and offers accessible information, including details about the survey, instructions on how to participate, team and contact information, as well as images, videos, and texts (Fig. 4.1).

The questionnaire in this tool consists of multiple-choice, closed-ended questions to identify gender, generational, ethnic, educational, income, disability, and socioeconomic characteristics, as well as the typology of hazards. Open-ended questions are also included in the questionnaire and are associated with the moment, place, circumstances, and emotional experience, allowing free narratives as answers.

It includes interaction options through social media profiles (Twitter, Facebook, Instagram), a dedicated mobile phone contact, and a Gmail account, all supported by an editable cookie policy to ensure legal compliance. The platform features an embedded Google Form for dynamic content updates, allowing for remote editing and real-time adjustments. A QR code links directly to the site, and participation is confirmed through a single click on an electronic informed consent (e-IC) button (Fig. 4.2).

As a 100% remote research tool, Sobrevidas© is continuously available, depending exclusively on digital mobile access. It focuses on disaster survivors who have experienced a disaster and invites them to individually and privately respond to a questionnaire and share their experiences. The period between exposure to a hazardous event and data collection is independent and self-defined by the disaster

**Fig. 4.1** Images produced for Sobrevidas© platform and social networks. *Source* Elaborated by the author (Couto 2023)

survivors. It does not require a special research project budget, fieldwork planning and development, face-to-face household surveys, or governmental readiness.

The tool can be replicated across different nations and regions, considering gender, culture, political context, and local realities. It must be ensured that data is collected using standardized questions respecting cultural norms and organized into historical, recurring series to guarantee comparability.

However, as a remote research tool, it places a few challenges particularly the limitation of securing participants' time. It also has issues related to trust. Researchers may not trust the authenticity of respondents and the respondents may not trust the motive of the data collectors and researchers who will interact through such digital platforms. Such remote digital tools cannot be comparable to the nuances of information collected by face-to-face interviews, maintaining continuous attention and availability of respondents, and balancing emotional engagement. To ensure the

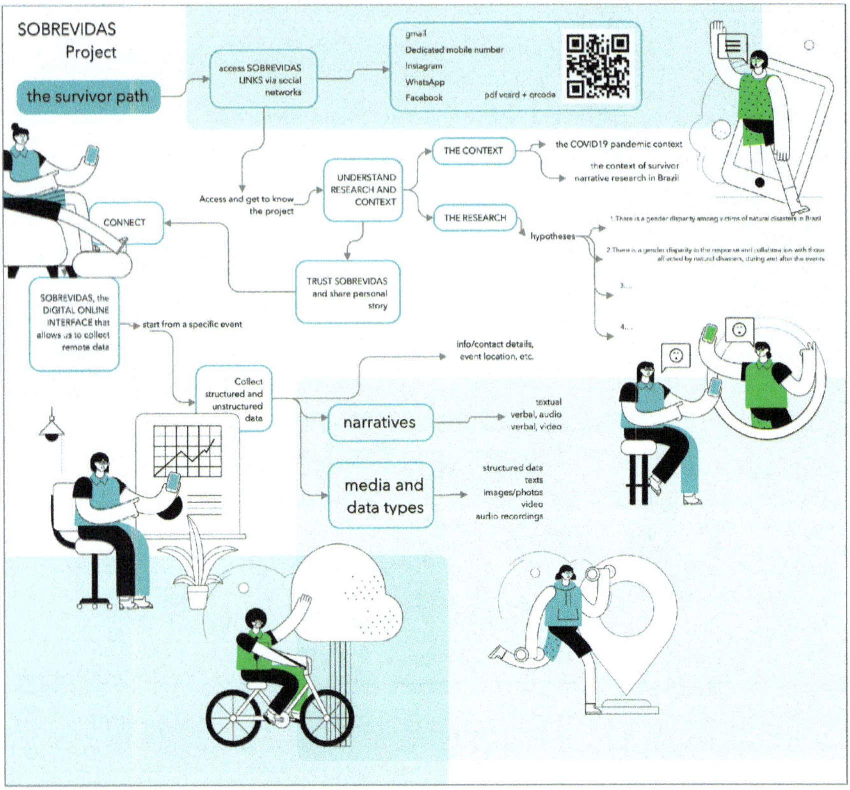


**Fig. 4.2** Sobrevidas© project. 2024 *Source* Elaborated by the author Mauricio Tretin

project gains scale, interest, and participation, effective digital marketing strategies are required to overcome these barriers that mostly depend on individual willingness to engage.

## *Innovative Approaches: Key Features of a Research Tool*

Sobrevidas© is innovative to the extent it aligns with the call from international agencies and research experts on gender and disasters to new research tools, to gender-disaggregated data, aiming for more equitable policies and practices to support gender grassroots initiatives, improve public policies to reduce inequalities, enhance women's safety, and prevent disasters.

Along with the years 2021 and 2020, throughout Couto's (2023) research, the Sobrevidas© platform successfully collected narratives, primarily from women participants, comprising approximately 84% of the total respondents, with an average

age of 48 years old (35 years old in 2011). Demographics include equal representation of Black (50%) and White women (50%). Through the use of the platform, outreach was extended across eight different cities and 21 distinct neighborhoods. The platform enabled participation from a diverse range of individuals across various professions and backgrounds, including bus fare collectors, grocery store owners, cleaners, factory workers, Jeep drivers, members of permanent housing movements, volunteers, nursing technicians, retirees, house cleaners, amateur radio operators, and others. This broad participation provided a wealth of perspectives, offering valuable insights into community strengths and capacities across different regions. (Couto 2023).

So far, these narratives provided valuable insights into critical questions, such as how local communities were warned about the disaster, the responses of individuals in assisting their neighbors, and the specific roles women played in the aftermath. In that regard, the tool provided data to understand how women experienced landslides in Brazil in January 2011 when they were warned just before the tragedy.

*When it was a little after midnight, my cousin, a health agent in the Cuiabá Valley, was responsible for taking care of the siren because she did that in 2008 [previous event]. She turned it on and said, "Maria [Alternative woman's name], tell everyone in the yard to stay alert because a lot of water is coming down a lot! So, place things in a higher place already because a lot of water is coming down." (Anonymous Participant).*

Based on survivors' stories, it was also possible to identify how people supported each other and how they responded by organizing themselves in volunteer groups.

*The first thing that struck me was the number of volunteers. I met some friends from school and asked them, "Do you need volunteers?" "No, we have a lot of people, we accept donations, but we don't need any more people working here at the moment." (Anonymous Participant).*

In this case, women played a pivotal role in these efforts, providing essential support to those in need of assistance.

*A girl was going blind. She needed to go to another country, so we held a raffle, and I sold a thousand reais worth of raffle tickets with the help of this friend from the state government, that's what we get out of this life. If it were a man in this situation, he would have given up on all this, because we are women, we have determination, and we have that thing of putting ourselves in someone else's shoes, we have that. (Anonymous Participant).*

These stories are key to understanding how survivors act during disasters. These narratives are critical sources of relevant, specific and local information, and not just numbers and data. Moreover, they play a crucial role in recognizing women's capacity for collaborative change through local, regional, and national initiatives on women and science. They can amplify the voices of Indigenous communities and civil society advocacy to guarantee rights and life in a changing world.

As a functional and low-cost-effective tool, it enables large-scale research on countries with continental dimensions, with access to remote or underserved areas, and with accessibility to a diverse participant. Continuous access offers opportunities to extensive datasets, whether two-dimensional or three-dimensional dataset. The pursuit of more information and details, the higher the speed of interpretation and understanding (Trentin 2024).

## *The Potential Use of Artificial Intelligence (AI)*

Content analysis is key to analyzing data from survivors, however, it may be time-consuming. The use of artificial intelligence (AI) systems in the analysis (specialized and generic digital agents) can generate speed and accuracy, freeing up team experts to formulate hypotheses and verify diagrammatic relationships in data series, local changes and synchronic and diachronic trends, locally, and in comparison, with international data.

AI analysis systems can investigate large volumes of structured data and identify relevant patterns such as gender, social class, level of education, family size, and other characteristics of disaster survivors, providing mechanisms to identify subpopulations, reduce gender bias and enhance social equality (Tannenbaum et al. 2019).

The adoption of AI and automation in the platform simplifies and streamlines speech analysis, allowing the investigation of transcripts, and the production of a series of detailed results. With artificial intelligence (AI), it is possible to automate part of the discourse analysis by systematically examining written records and transcripts of narratives from individuals who experienced disasters, leading to a series of detailed outcomes. These include the creation of concise summaries of the survivors' narratives highlighting central points and the identification of patterns of practical solutions employed by survivors to cope with extreme situations. It allows the recognition of common difficulties faced by survivors, along with an analysis of the environmental and social factors or conditions that contribute to survival or exacerbate challenges. Furthermore, recognizing the language patterns of disaster survivors can be achieved through the content of their use of language, including emotional expressions, common metaphors, and narrative structures that emerge in the transcripts. It is possible to rapidly systematize a large volume of structured data regarding survivors of extreme natural events and climate refugees to identify relevant individual patterns and behaviors.

It is possible to formulate a hypothesis based on available data and observations, and elaborate questions based on data of particular aspects such as: How were people warned before the disaster event? Did women perceive it differently from men? How did people support each other and organize themselves during the disasters? Why do women play a pivotal role in supporting those in need of assistance after the event?

But most significantly, it can recognize local initiatives, decision-making participation, and leadership and capabilities in local communities to deal with threats, on a larger scale and with a huge amount of knowledge. The development of a user-friendly interface for data collection reduces complexity and improves user experiences enabling individual questions to be viewed more intuitively. Additionally, exploring multiple channels for submitting reports, data, audio, video, and images might enhance the overall flow of information.

### *How Can We Expand the Impact of Sobrevidas© in the Future?*

Sobrevidas©, the interface, should undergo significant improvements and include a new and better system for collecting structured and unstructured data, as well as technological innovations to simplify data mining and speed up analysis. Capturing recurring data has the potential to create a faithful and verifiable record of the evolution of events and the coherent response of survivors, individuals, and managers.

Sobrevidas©, as a gender transformative research tool, can grow as a recurring research project: a multidisciplinary team that recognizes gender and its intersectionalities inequalities, which mines and verifies new data, gathering data already found by other researchers and institutes, trying to map risks, identifying women' responsibilities, survival strategies, and resilience and overcoming disasters. It can work with grassroots initiatives on disaster risk prevention and reduction, and climate justice, to scale up and acknowledge social relationships in the local sphere, and human capacities and skills for safer and resilient spaces.

Potential improvements seek to scale up data collection for greater reach and enhanced participant engagement, aiming to identify community strengths as resources for policy planning. This information can support the creation of public policies developed with and by local actors, recognizing them as proactive agents capable of shaping their future rather than solely as vulnerable groups (Fig. 4.3). This approach emphasizes the need to reassess public policies to ensure they effectively support community-driven disaster knowledge and practices (Trejo-Rangel et al. 2023).

## Conclusions

It is clear and well-recognized that our planet is under pressure. Climate change, biodiversity loss, extreme heat, severe drought and floods, multi-hazards extreme events that are mutually intensifying, threatening planetary boundaries, and the safety, resilience, and survivability of humanity. Due to unequal structures, different social groups are affected differently and exhibit distinct conditions and capacity to respond and recover to threats caused by climate change. The role that different social actors play in adaptation to climate change and disaster risk reduction is an emerging area of study and ambition for international policies and practices. People have started recognizing the importance of the role that women play in climate events,

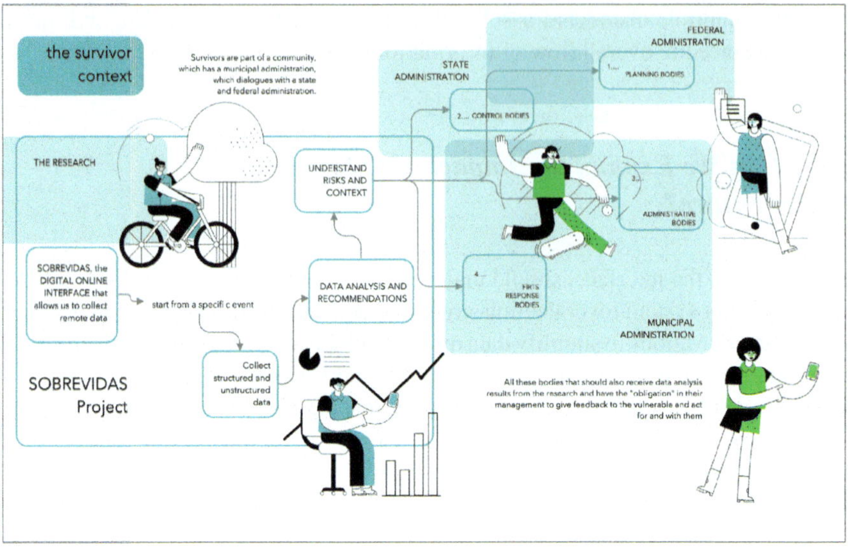


**Fig. 4.3** Sobrevidas© project: Survivor context. 2024 *Source* Elaborated by the author Mauricio Treitin

demonstrating specific knowledge, resilience, capacities, and skills when exposed to situations of threats.

However, important gaps in research knowledge remain. There is a lack of standardized tools and methodologies for understanding mortality, response, and recovery from disasters. There is also limited knowledge on how individuals and different social groups at risk manage to survive disasters when they occur.

Gender-disaggregated data collection presents itself as a power mechanism that generates knowledge at scale, allowing it to identify gender bias on disaster impact and recovery, recognize gender grassroots initiatives, and enable comparisons and statistics that support the recognition of a whole set of social relations that occur at the community level.

Sobrevidas© is the research tool that recognizes the intersectionality of knowledge on gender, ethnicity, age, class, and territory as an instrument to acknowledge the uncertainties and challenges faced by the Global South. It led to the acknowledgement by society of the importance of gender dimensions in disasters and the recognition of innovative ways to approach the affected people during difficult times while being aware of how sensible it has to be.

While the platform is presently on standby, with some new funding, evolutionary improvements and a dedicated team, it would certainly be a great opportunity for recurring research. It can be useful for mining structured data and mainly collecting unstructured data, reports, and videos, and creating a faithful documented narrative evolution of survivors, their insights and local challenges.

In the context of climate change and the increment of disasters policy and decision-makers must understand and recognize women's role not only recognizing the vulnerability increased by a context of social disadvantage, but highlighting the capacities, skills and leadership of women in contexts of threat. Opportunities to challenge historical patriarchal relations can emerge from disaster research, elevating women's recognition in positions of leadership, within their community, nations and globally.

## Competing Interest

The authors have non-financial interests directly or indirectly related to this chapter and no competing interests to disclose.

**Funding** This study was partially supported by CAPES Brazil (Coordenação de Aperfeiçoamento de Pessoal de Nível Superior)—Finance Code 001.

## References

Ajaero CK (2017) A gender perspective on the impact of flood on the food security of households in rural communities of Anambra State, Nigeria. Food Secur 9:685–695. https://doi.org/10.1007/s12571-017-0695-x

Akekar S, Fordham M (2017) Gender, place and mental health recovery in disasters: addressing issues of equality and difference. Int J Disaster Risk Reduct 23:218–230. https://doi.org/10.1016/j.ijdrr.2017.03.014

Alvalá RCS, Barbieri A (2017) Desastres naturais. In: Nobre C, Marengo J (eds) Mudanças Climáticas em Rede: um olhar interdisciplinar. Contribuições do Instituto Nacional de Ciência e Tecnologia para Mudanças Climáticas. Canal 6, Bauru, pp 203–230

Aryal K (2014) Women's empowerment in building disaster resilient communities. Asian J Women's Stud 20:164–174. https://doi.org/10.1080/12259276.2014.11666178

Beltrame D, Medina MDP (2022) Towards gender equality and women's leadership for resilience to disaster Risks in Latin America and the Caribbean. United Nations Entity for Gender Equality and the Empowerment of Women. United Nations Office for Disaster Risk Reduction—Regional Office for the Americas and the Caribbean. https://www.undrr.org/publication/towards-gender-equality-and-womens-leadership-resilience-disaster-risks-latin-america

Bentley C, Muyoya C, Vannini S, Oman S, Jimenez A (2023) Intersectional approaches to data: the importance of an articulation mindset for intersectional data science. Big Data Soc 10:2. https://doi.org/10.1177/20539517231203667

Bolland J, Shanker N, Reinen-Hamill R, Fakhruddin B (2023) WorldFAIR project (D12.2) disaster risk reduction domain-specific FAIR vocabularies (Version 1). Zenodo. https://doi.org/10.5281/zenodo.8110630

BRASIL (2021) Ministério da Ciência, Tecnologia e Inovações. Secretaria de Pesquisa e Formação Científica. Quarta Comunicação Nacional do Brasil à Convenção Quadro das Nações Unidas sobre Mudança do Clima/Secretaria de Pesquisa e Formação Científica. MCTI, Brasília. https://repositorio.mctic.gov.br/handle/mctic/4782

BRASIL (2024) Ministério do Desenvolvimento Regional. Secretaria de Proteção e Defesa Civil. Universidade Federal de Santa Catarina. Centro de Estudos e Pesquisas em Engenharia e Defesa

Civil. Atlas digital de desastres no Brasil. MDR, Brasília. https://atlasdigital.mdr.gov.br/paginas/index.xhtml
Brown S, Budimir M, Upadhyay-Crawford S, Clements R, Sneddon A (2019) Gender and age inequality of disaster risk: r research paper. UNICEF and UN Women. https://www.unicef.org/documents/gender-and-age-inequality-disaster-risk
Campos L, França D, Junior P (2018) Relatório das desigualdades de raça, gênero e classe. GEMAA 2:1-18. https://gemaa.iesp.uerj.br/relatorios/relatorio-das-desigualdades-gemaa-no-1
Cannon T (2002) Gender and climate hazards in Bangladesh. Gender Dev 10:45–50. https://doi.org/10.1080/13552070215906
Cavenaghi S, Alves JED (2018) Mulheres chefes de família no Brasil: avanços e desafios. ENS-CPES, Rio de Janeiro, p 120
CEPED—Centro de Estudos e Pesquisas em Engenharia e Defesa Civil (2020) Relatório de danos materiais e prejuízos decorrentes de desastres naturais no Brasil: 1995–2019, 2nd edn. Florianópolis. https://www.gov.br/mdr/pt-br/centrais-de-conteudo/publicacoes/protecao-e-defesa-civil-sedec/danos_e_prejuizos_versao_em_revisao.pdf
Chineka J, Musyoki A, Kori E, Chikoore H (2019) Gender mainstreaming: a lasting solution to disaster risk reduction. Jàmbá J Disaster Risk Stud 11:1–6. https://doi.org/10.4102/jamba.v11i3.723
Clissold R, Westoby R, Mcnamara KE (2020) Women as recovery enablers in the face of disasters in Vanuatu. Geoforum 113:101–110. https://doi.org/10.1016/j.geoforum.2020.05.003
Coates L (1999) Flood fatalities in Australia, 1788–1996. Aust Geogr 30:391–408. https://doi.org/10.1080/00049189993657
Coates R, Garmany J (2017) The ecology of citizenship: understanding vulnerability in urban Brazil. Int Dev Plan Rev 39:37–56. https://doi.org/10.3828/idpr.2017.3
Couto GA, Sanchez A, Alvalá RCS, Nobre CA (2023) Natural hazards fatalities in Brazil, 1979–2019. Nat Hazards 118:1487–1514. https://doi.org/10.1007/s11069-023-06059-x
Couto GA (2023) Climate related disasters in Brazil through a gender lens: Sobrevidas@ and female voices. 119 p. Tese (Doutorado em Ciência do Sistema Terrestre)—Instituto Nacional de Pesquisas Espaciais (INPE), São José dos Campos, 2023. http://urlib.net/ibi/8JMKD3MGP3W34T/48MCGEP
Cukier K, Mayer-Schönberger V (2013) The Dictatorship of data. MIT Technology Review. https://www.technologyreview.com/2013/05/31/178263/the-dictatorship-of-data/
Dabalen A, Etang A, Hoogeveen J, Mushi E, Schipper Y, von Engelhardt J (2016) Mobile phone panel surveys in developing countries: a practical guide for microdata collection. Directions in Development. World Bank, Washington, DC. https://doi.org/10.1596/978-1-4648-0904-0 (License: Creative Commons Attribution CC BY 3.0 IGO)
De Silva K, Jayathilaka R (2014) Gender in the context of disaster risk reduction; a case study of a flood risk reduction project in the Gampaha District in Sri Lanka. Proc Econ Finance 18:873–881. https://doi.org/10.1016/S2212-5671(14)01013-2
Debortoli NS, Camarinha PIM, Marengo JA, Rodrigues RR (2017) An index of Brazil's vulnerability to expected increases in natural flash flooding and landslide disasters in the context of climate change. Nat Hazards 86:557–582. https://doi.org/10.1007/s11069-016-2705-2
Diakakis M, Deligiannakis G (2017) Flood fatalities in Greece: 1970–2010. J Flood Risk Manag 10:115–123. https://doi.org/10.1111/jfr3.12166
Dourado F, Coutinho Arraes T, Fernandes Silva M (2012) O megadesastre da Região Serrana do Rio de Janeiro: as causas do evento, os mecanismos dos movimentos de massa e a distribuição espacial dos investimentos de reconstrução no pós-desastre. Anuário Do Instituto De Geociências 35:43–54
Enarson E, Meyreles L (2024) International perspectives on gender and disaster: differences and possibilities. Int J Sociol Soc Policy 24(10/11):49–93. https://doi.org/10.1108/01443330410791064
Fatti CE, Patel Z (2013) Perceptions and responses to urban flood risk: implications for climate governance in the South. Appl Geogr 36:13–22. https://doi.org/10.1016/j.apgeog.2012.06.011

Freitas RdCM (2010) A construção de uma agenda para as questões de gênero, desastres socioambientais e desenvolvimento. Rev Estud Fem 18(3):889. https://doi.org/10.1590/S0104-026X2010000300014

Gurman TA, Ellenberger N (2015) Reaching the global community during disasters: findings from a content analysis of the organizational use of Twitter after the 2010 Haiti earthquake. J Health Commun 20(6):687–696. https://doi.org/10.1080/10810730.2015.1018566

Hamama-Raz Y, Palgi Y, Shrira A, Goodwin R, Kaniasty K, Ben-Ezra M (2015) Gender differences in psychological reactions to Hurricane Sandy among New York metropolitan area residents. Psychiatr Q 86(2):285–296. https://doi.org/10.1007/s11126-014-9333-3

Hirata H (2014) Gênero, classe e raça: interseccionalidade e consubstancialidade das relações sociais. Tempo Social 26:61–73

Hoegh-Guldberg O, Jacob D, Taylor M, Bindi M, Brown S, Camilloni I, Diedhiou A, Djalante R, Ebi KL, Engelbrecht F, Guiot J, Hijioka Y, Mehrotra S, Payne A, Seneviratne SI, Thomas A, Warren R, Zhou G (2018) Impacts of 1.5°C global warming on natural and human systems. In: Masson-Delmotte V, Zhai P, Pörtner H-O, Roberts D, Skea J, Shukla PR, Pirani A, Moufouma-Okia W, Péan C, Pidcock R, Connors S, Matthews JBR, Chen Y, Zhou X, Gomis MI, Lonnoy E, Maycock T, Tignor M, Waterfield T (eds) Global warming of 1.5°C. An IPCC special report. Cambridge University Press, Cambridge, UK, NY, USA, pp 175–312. https://doi.org/10.1017/9781009157940.005

IBGE—Instituto Brasileiro de Geografia e Estatística (2021) Estatísticas de gênero indicadores sociais das mulheres no Brasil. 2a ed. IBGE, Rio de Janeiro. https://www.ibge.gov.br/estatisticas/multidominio/genero/20163-estatisticas-de-genero-indicadores-sociais-das-mulheres-no-brasil.html?=&t=publicacoes

IBGE—Instituto Brasileiro de Geografia e Estatística (2023) Censo Demográfico 2022: população e domicílios: primeiros resultados. IBGE, Rio de Janeiro. https://biblioteca.ibge.gov.br/index.php/biblioteca-catalogo?view=detalhes&id=2102011

IPCC—International Panel on Climate Change (2022) Climate change 2022: impacts, adaptation, and vulnerability. In: Pörtner H-O, Roberts DC, Tignor M, Poloczanska ES, Mintenbeck K, Alegría A, Craig M, Langsdorf S, Löschke S, Möller V, Okem A, Rama B (eds) Contribution of working group II to the sixth assessment report of the Intergovernmental Panel on Climate Change. Cambridge University Press (in press). https://www.ipcc.ch/report/sixth-assessment-report-working-group-ii/

Jones L, Ballon P (2020) Tracking changes in resilience and recovery after natural hazards: Insights from a high-frequency mobile-phone panel survey. Glob Environ Change 62:102053. https://doi.org/10.1016/j.gloenvcha.2020.102053

Kantamaneni K, Panneer S, Sudha Rani NNV, Palaniswamy U, Bhat LD, Jimenez-Bescos C, Rice L (2022) Impact of coastal disasters on women in urban slums: a new index. Sustainability 14(6):3472. https://doi.org/10.3390/su14063472

Kirsch TD, Wadhwani C, Sauer L, Doocy S, Catlett C (2012) Impact of the 2010 Pakistan floods on rural and urban populations at six months. PLoS Curr Disasters

Kondraganti A, Narayanamurthy G, Sharifi H (2024) A systematic literature review on the use of big data analytics in humanitarian and disaster operations. Ann Oper Res 335:1015–1052. https://doi.org/10.1007/s10479-022-04904-z

Leão N, Candido MR, Campos LA, Feres Jr J (2017). Relatório das desigualdades de raça, gênero e classe. Gemaa, vol 1, pp 1–21. https://gemaa.iesp.uerj.br/relatorios/relatorio-das-desigualdades-gemaa-no-1/

Lenton TM, Rockström J, Gaffney O, Rahmstorf S, Richardson K, Steffen W, Schellnhuber HJ (2019) Climate tipping points—too risky to bet against. Nature 575:592–595. https://doi.org/10.1038/s41586-019-1654-3

Lunga W, Musarurwa C (2016) Indigenous food security revival strategies at the village level: the gender factor implications. Jàmbá J Disast Risk Stud 8:1–7. https://doi.org/10.4102/jamba.v8i2.175

Maduri (2016) The impact of flooding in Bihar, India on women: a qualitative study. Asian Women 32(1):31–52. https://doi.org/10.14431/aw.2016.03.32.1.31

Marengo JA, Alves LM (2012) The 2011 intense rainfall and floods in Rio de Janeiro. Bull Am Meteorol Soc 93:S176. https://doi.org/10.1175/2012BAMSStateoftheClimate.1

Marengo JA, Camarinha PI, Alves LM, Diniz F, Betts RA (2021) Extreme rainfall and hydro-geo-meteorological disaster risk in 1.5, 2.0, and 4.0 °C global warming scenarios: an analysis for Brazil. Front Clim 3:610433. https://doi.org/10.3389/fclim.2021.610433

Banco Mundial (2012) Avaliação de Perdas e Danos: Inundações e Deslizamentos na Região Serrana do Rio de Janeiro - Janeiro de 2011. Relatório elaborado pelo Banco Mundial com apoio do Governo do Estado do Rio de Janeiro. https://documents.worldbank.org/en/publication/documents-reports/documentdetail/260891468222895493/brazil-avaliacao-de-perdas-e-danos-inundacoes-e-deslizamentos-na-regiao-serrana-do-rio-de-janeiro-janeiro-de-2011

Myeni S, Wentink GJ (2020) A gendered approach to drought-coping mechanisms: a case of the Lubombo region, Eswatini. TD J Transdisc Res Southern Afr 17:1–12. https://doi.org/10.4102/td.v17i1.744

Ndlovu T, Mjimba V (2021) Drought risk-reduction and gender dynamics in communal cattle farming in southern Zimbabwe. Int J Disaster Risk Reduct 58:102203. https://doi.org/10.1016/j.ijdrr.2021.102203

Neumayer E, Plümper T (2007) The gendered nature of natural disasters: the impact of catastrophic events on the gender gap in life expectancy, 1981–2002. Ann Assoc Am Geogr 97:551–566. https://doi.org/10.1111/j.1467-8306.2007.00563.x

Nirupama N, Maula A (2013) Engaging public for building resilient communities to reduce disaster impact. Nat Hazards 66:51–59. https://doi.org/10.1007/s11069-011-0045-9

Passos L, Wajnman S (2021) Diferencial salarial entre as mulheres: a dupla penalidade das negras. In: Fusco W, Myrrha LJD, Jesus JC (Org) Migração, trabalho e gênero: textos selecionados. ABEP, Belo Horizonte, pp 192–206. https://www.researchgate.net/profile/Wilson-Fusco/publication/351109230_Migracao_trabalho_e_genero_textos_selecionados/links/6087f05b2fb9097c0c0fdc3a/Migracao-trabalho-e-genero-textos-selecionados.pdf

Pathak S, Emah IE (2017) Gendered approach towards disaster recovery: experiences from 2011 floods in Pathumthani province, Thailand. Int J Disaster Risk Reduc 24:129–134. https://doi.org/10.1016/j.ijdrr.2017.06.007

Perez LP, Rodrigues-Filho S, Marengo JA, Santos DV, Mikosze L (2020) Climate change and disasters: analysis of the Brazilian regional inequality. Sustain Debate 11(3):260–296. https://doi.org/10.18472/SustDeb.v11n3.2020.33813

Picard M (2021) Beyond vulnerability to gender equality and women's empowerment and leadership in disaster risk reduction: critical actions for the United Nations System. UN Women, United Nations Entity for Gender Equality and the Empowerment of Women. https://www.unwomen.org/en/digital-library/publications/2021/11/research-paper-beyond-vulnerability-to-gender-equality

Rakib MA, Islam S, Nikolaos I, Bodrud-Doza M, Bhuiyan MA (2017) Flood vulnerability, local perception and gender role judgment using multivariate analysis: a problem- based "participatory action to future skill management" to cope with flood impacts. Weather Clim Extrem 18:29–43. https://doi.org/10.1016/j.wace.2017.10.002

Raza H (2017) Using a mixed method approach to discuss the intersectionalities of class, education, and gender in natural disasters for rural vulnerable communities in Pakistan. J Rural Commun Dev 12(1):128–148 https://journals.brandonu.ca/jrcd/article/view/1338

Reyes DD, Lu JL (2017) Gender dimensions and women's vulnerability in disaster situations: a case study of flood prone areas impacting women in Malabon City, Metro Manila. J Int Women's Stud 18(4):69–88. https://doi.org/10.1016/j.ijdrr.2015.11.001

Salvati P, Petrucci O, Rossi M, Bianchi C, Pasqua AA, Guzzetti F (2018) Gender, age and circumstances analysis of flood and landslide fatalities in Italy. Sci Total Environ 610–611:867–879. https://doi.org/10.1016/j.scitotenv.2017.08.064

Saulnier DD, Green HK, Ismail R, Chhorvann C, Bin Mohamed N, Waite TD, Murray V (2019) Disaster risk reduction: why do we need accurate disaster mortality data to strengthen policy and practice? Disaster Prev Manag 28:846–861. https://doi.org/10.1108/DPM-09-2019-0296

Soares D, -Licea Murillo D (2013) Gestión de riesgo de desastres, género y cambio climático. Percepciones sociales en Yucatán, México. Cuadernos de Desarrollo Rural 10(72):181–199. https://revistas.javeriana.edu.co/index.php/desarrolloRural/article/view/7029

Story WT, Engidawork B, Tadesse M et al (2020) Social capital and disaster preparedness in Oromia, Ethiopia: an evaluation of the "women empowered" approach. Soc Sci Med 222:131–138. https://doi.org/10.1016/j.socscimed.2018.08.027

Sultana F (2010) Living in hazardous waterscapes: gendered vulnerabilities and experiences of floods and disasters. Environ Hazards 9:43–53. https://doi.org/10.3763/ehaz.2010.SI02

Tannenbaum C, Ellis RP, Eyssel F et al (2019) Sex and gender analysis improves science and engineering. Nature 575:137–146. https://doi.org/10.1038/s41586-019-1657-6

Trejo-Rangel MA, Marchezini V, Rodriguez DA et al. (2023) Incorporating social innovations in the elaboration of disaster risk mitigation policies. Int J Disaster Risk Reduct 84:103450. https://doi.org/10.1016/j.ijdrr.2022.103450

Trentin M (2024) Colaborando com o inimigo: simbioses possíveis em tempos de redes adversárias. Notas sobre entes, funções e agentes digitais. In: Encontro DAT—design, art and technology, 25 jun. 2024, São Paulo. Anais [...]. São Paulo

UN WOMEN (2014) The world survey on the role of women in development: gender equality and sustainable development. https://sdgs.un.org/publications/world-survey-role-women-development-33037

UN—United Nations (2018) General recommendation on the gender-related dimensions of disaster risk reduction in the context of climate change. Convention on the Elimination of All Forms of Discrimination against Women, Committee on the Elimination of Discrimination against Women, 13 March. https://digitallibrary.un.org/record/1626306?ln=en&v=pdf

UNDRR—United Nations Office for Disaster Risk Reduction (2021) Sendai framework monitor (SFM): sex, age, and disability disaggregated data (SADDD). https://www.undrr.org/media/76930/download?startDownload=20250126

UNDRR—United Nations Office for Disaster Risk Reduction (2024) Gender action plan to support implementation of the sendai framework for disaster risk reduction 2015–2030. https://www.preventionweb.net/sendai-framework/sendai-framework-for-disaster-risk-reduction/SendaiGAP

UNDRR—United Nations Office for Disaster Risk Reduction (2020) The human cost of disasters: an overview of the last 20 years (2000–2019). United Nations Office for Disaster Risk Reduction. https://www.undrr.org/publication/human-cost-disasters-overview-last-20-years-2000-2019

UNFCCC (2024) Progress, challenges, gaps and priorities in implementing the gender action plan, and future work to be undertaken on gender and climate change. Synthesis report. FCCC/SBI/2024/11, UNFCCC Secretariat. https://unfccc.int/documents/638589

UNISDR—United Nations International Strategy for Disaster Reduction (2015) Mobilizing women's leadership in disaster risk reduction. High-level multi-stakeholder partnership dialogue. https://www.wcdrr.org/uploads/Mobilizing-Women's-Leadership-in-Disaster-Risk-Reduction.pdf

Valerio KA (2014) Storm of violence, surge of struggle: women in the aftermath of Typhoon Haiyan (Yolanda). Asian J Women's Stud 20(1):148–163. https://doi.org/10.1080/12259276.2014.11666177

Wachinger G, Renn O, Begg C, Kuhlicke C (2013) The risk perception paradox—implications for governance and communication of natural hazards. Risk Anal 33(6):1049–1065. https://doi.org/10.1111/risa.12015

Wallemacq P, House R (2018) Economic losses, poverty & disasters: 1998–2017. United Nations Office for Disaster Risk Reduction. https://www.undrr.org/publication/economic-losses-poverty-disasters-1998-2017

WMO—World Meteorological Organization (2021) WMO Atlas of mortality and economic losses from weather, climate and water extremes (1970–2019). https://wmo.int/publication-series/atlas-of-mortality-and-economic-losses-from-weather-climate-and-water-related-hazards-1970-2021

Yan L, Pedraza-Martinez AJ (2019) Social media for disaster management: operational value of the social conversation. Prod Oper Manag 28(10):2514–2532. https://doi.org/10.1111/poms.13064

Yankelovich D (1971) Interpreting the new life styles. sales management: the marketing magazine 107:26–27. https://archive.org/details/sim_sales-management_1971-11-15_107_11/page/26/mode/2up?view=theater

Zagheni E, Muttarak R, Striessnig E (2015) Differential mortality patterns from hydro-meteorological disasters: Evidence from cause-of-death data by age and sex. Vienna Yearbook Popul Res 13:47–70. https://doi.org/10.1553/populationyearbook2015s047

# Part II
# Vulnerability to Climate Change and Disproportionate Burden on Women

# Chapter 5
# Climate Extremes and Disproportionate Burden on Women in the Global South: A Way Forward

**Mumtaz Alam and Kunal Avishek Gounder**

**Abstract** This study analyzes the unequal effects of climate change on women in the Global South, focusing on the intersections of gender inequality, socioeconomic limitations, and climate vulnerability. The study utilizes a qualitative integrative literature review methodology, examining sources from 2009 to 2024, which encompass national policy documents, climate action plans, environmental assessment reports, and scholarly publications from leading academic databases and international organizations. The chapter examines how existing gender disparities intensify women's susceptibility to climate extremes, demonstrating that women in the Global South encounter considerably greater obstacles due to restricted access to resources, decision-making roles, and economic opportunities. Research indicates that climate change exacerbates the pre-existing challenges faced by women, especially regarding food and water security, agricultural productivity, and community resilience. The findings indicate that women exhibit greater vulnerability to climate-induced disasters as a result of socioeconomic limitations, gender discrimination, and unequal involvement in climate adaptation initiatives. The study advocates for a multi-tiered strategy to tackle these challenges, encompassing gender-responsive national policies, engagement of women at the community level in climate action, and fostering international collaboration. It also calls for a comprehensive strategy that emphasizes women's leadership, provides targeted support, and dismantles cultural barriers to address these disproportionate impacts and help in the achievement of sustainable development goals.

**Keywords** Global South · Women · SDG · Extreme climate · Gender inequality · Climate change

---

M. Alam (✉)
Department of Social Sciences, College of Humanities and Education, Fiji National University, Suva, Fiji
e-mail: mumtaz.alam@fnu.ac.fj

K. A. Gounder
College of Humanities and Education, Fiji National University, Suva, Fiji



A. Sarkar and N. Bandyopadhyay (eds.), *Gender-Transformative Approaches for Climate Change Adaptation*, Springer Climate,
https://doi.org/10.1007/978-3-031-90236-9_5

## Introduction

Climate change is a global issue with its impact felt drastically across various regions around the globe, especially in the Global South, as it has pre-existing challenges. Much research has proven that climate change will become more adverse even if there is a full stop to carbon emissions today. This is because of the amount of carbon emissions already stored in the atmosphere (Pindyck 2021). Therefore, increasing the carbon sink is also necessary to reduce the catastrophic impacts of climate change. As climate change is expected to increase the intensity and frequency of natural disasters, vulnerability, and exposure are also expected to increase (Field et al. 2012). The impacts of climate change, such as frequent extreme events like cyclones and heat waves and slow onset events like rising sea levels are predicted to increase, making people vulnerable, especially the population of developing countries. Thus people of the global south will bear the brunt of most of the negative impacts despite contributing the least toward Greenhouse Gas (GHG) emissions. Althor et al. (2016) opined that the inequality between the country contributing GHGs versus the country affected by climate change will significantly worsen by 2030, as carbon-emitting countries are not active in reducing GHG emissions.

The main aim of this study is to thoroughly examine the disproportionate effects of climate change on women in the Global South, particularly how existing gender disparities intersect with and intensify climate vulnerability. The research seeks to rigorously analyze the complex problems encountered by women encompassing financial limitations, restricted access to resources, and less involvement in decision-making processes while coping with the negative effects of climate change. It investigates the disproportionate impact of climate extremes on their lives. The study aims to synthesize evidence from various sources to highlight the specific vulnerabilities of women in developing countries, assess the efficacy of existing climate adaptation strategies, and recommend innovative policy interventions and community-level initiatives that can strengthen women's resilience, advance gender equality, and foster sustainable development in the context of climate change.

## Overview of Climate Change Impacts on the Global South Population

Climate change has significantly affected vulnerable communities in the Global South, particularly women and children, who endure unparalleled abuse, deprivation of privacy and protection, and psychological anguish. The population of the Global South are particularly vulnerable to the impacts of climate change due to socioeconomic constraints, gender inequality, poor governance and ineffective climate mitigation and adaptation measures (Ngcamu 2023). The most vulnerable populations in the Global South include women, children, and people with disabilities. These populations have seen unparalleled abuse, deprivation of privacy and protection, and psychological suffering during and after climate-induced calamities. Effectively designed policies addressing climate change, together with the incorporation and participation of women in disaster risk reduction (DRR) activities, may significantly drive transformation (Ngcamu 2023).

The abilities of vulnerable populations may inform and enhance disaster risk reduction measures, preparation efforts, and decision-making processes. Nevertheless, empirical research about the impacts of climate change on these vulnerable groups is limited. The elder generation often disregards early warning systems, and their historical inadequacies, lack of comprehension, and illiteracy exacerbate their opposition to these systems. Empirical research by Malak et al. (2020) revealed that elderly women in Bangladesh's coastal regions are more susceptible than their male counterparts owing to socioeconomic factors and gender discrimination within a patriarchal culture.

The intersectional perspective has been advocated to understand marginalized populations comprehensively. Sams (2019) determined that women are more susceptible than males owing to socioeconomic factors and gender discrimination within a patriarchal culture. Memon (2020a, b) analyzed Gender-Based Violence (GBV) against women in flood settlement camps, advocating for the formulation of gender-sensitive policies and climate change strategies to mitigate their risk and vulnerability. It evaluated vulnerable populations' adaptive and mitigative capacities in response to climate change, scrutinized the efficacy of international mechanisms and policies designed to engage these groups in proactive climate adaptation initiatives, and analyzed vulnerable populations' constraints in participating in community-level climate and environmental change efforts. Incorporating and involving women in climate change projects may effectively tackle these difficulties and enhance resilience against climate change.

The complexity of cultural, political, and economic opportunities has hindered the advancement of women in the Global South over time. Women have been given limited access to fundamental opportunities like education, healthcare, and equal rights. Despite many awareness and efforts to bring gender equality, there are still wide gender gaps in the Global South, mainly due to socio-cultural norms. Paludi and Krysa (2023) argue that gender inequality has existed in the workplace in the Global South since World War II. Though firms hire many women in the workforce, the decision-making positions are limited for women (Paludi and Krysa 2023). Poverty is a substantial barrier for women since those in developing nations are more susceptible to its impacts owing to limited economic prospects and reduced wages. They also endure a disparate proportion of caregiving responsibilities, which restricts their capacity to engage in paid employment. Restricted access to educational opportunities is a significant problem since women in underdeveloped nations are less likely to attain literacy and pursue further education. The deficiency in education restricts women's economic prospects and hinders their access to vital services. Moreover, women in the Global South have substantial obstacles to political involvement since they are markedly underrepresented in national parliaments across almost all nations in this region, therefore limiting their capacity to influence legislation and policies that impact their lives (Tokas 2024).

## Materials and Methods

This chapter undertook a narrative literature review methodology to explore the unequal effects of climate extremes on women in the Global South. The research utilized a qualitative design and an integrative review approach to synthesize evidence from the existing literature from diverse sources. The main data sources included national policy documents, climate action plans, environmental assessment reports, and publications from international organizations such as UN Women, WHO, UNICEF, and the World Bank. The secondary sources included regional assessments of climate change vulnerability, case studies, reports from non-governmental organizations, and scholarly articles that have undergone peer review. The literature search was executed across prominent academic databases like Web of Science, Scopus, Google Scholar, and international organization databases like UN Data, World Bank, and Open Data. The study incorporated literature published from 2009 to 2024, focusing particularly on analyses of gender-specific climate impacts and policy interventions relevant to countries or regions in the Global South. The examination was carried out following a three-tier framework that assessed contextual factors (pre-existing challenges), impact evaluation (direct and indirect consequences), and response analysis (policy frameworks and intervention strategies). A multitude of sources were meticulously cross-referenced to ensure reliability and validity, emphasizing peer-reviewed and globally acknowledged references while maintaining a balanced geographical representation across diverse regions of the Global South. The authors, however, recognize few limitations. The swift progression of climate change effects demands ongoing revisions of research outcomes, the possibility of publication bias favoring English-language materials, restricted access to localized data in specific situations, and the inconsistent quality and availability of gender-disaggregated data across different areas. The methodology provided an extensive framework for analyzing the complex interconnections between gender-specific vulnerabilities and the effects of climate change in the Global South, notwithstanding these limitations.

## Results and Discussion

### *Impact of Climate Extremes on Women in the Global South*

The impact of climate extreme on the women in the Global South is far more than the men due to existing structural inequalities. These inequalities exist in the form of gendered roles and responsibilities undertaken by women, which makes them more vulnerable. Generally, women are responsible for gathering water and food for the family, and due to climate extremes, food and water security are at risk, increasing women's efforts in securing food and water for the family. Patel et al. (2020) study demonstrates a distinct correlation between climate change and the vulnerability of women in South Asia. Climate change has considerable socioeconomic effects on women, disproportionately impacting them across several areas, including agriculture, livelihoods, food security, physical and mental health, and water and sanitation in South Asia (Patel et al. 2020).

## Cultural and Social Impacts

According to UN Women (2022), climate change acts as a "threat multiplier," exacerbating social, political, and economic pressures in conflict-affected regions, especially for women and girls. It heightens susceptibility to gender-based violence, including sexual assault, human trafficking, and child marriage. Disasters increase the danger of survival and harm for women owing to entrenched gender disparities (Erman et al. 2021). These discrepancies restrict access to resources and training, jeopardizing their livelihoods and well-being (Boateng 2021). Climate change and catastrophes jeopardize women's health by restricting access to services and healthcare, hence elevating risks associated with maternal and child health. Severe heat increases the risk of stillbirths, while climate change exacerbates the proliferation of vector-borne diseases such as malaria, dengue fever, and Zika virus, which are associated with worse maternal and newborn outcomes (UN Women 2022; Alam and Ali 2022).

Desai and Zhang (2021) explain that in very isolated regions, the patriarchal structure of societies exacerbated gender discrimination and violence against women after natural catastrophes that devastated agricultural harvests and reduced family income. They were often recognized as the last individuals to dine in the home, permitting the males and youngsters to eat beforehand. De Wit (2021) further adds that gender is often associated with climate change since women rely directly on natural resources for their sustenance, including the collecting of water and firewood. This increases risks and responsibilities from climate change since they serve as the main childcare providers. Women possess less potential to adapt to environmental changes and natural hazards due to restricted access to resources, justice, mobility, and influence in decision-making and policy formulation. Furthermore, women possess less economic, political, and legal influence, rendering them more vulnerable to the detrimental impacts of climate change (Goodrich et al. 2017).

Poor climate action has led to many women's movements in the Global South as they aimed to protect their environment. The Chipko Movement in 1974 of Reni Village in India is an example of how indigenous women grouped and guarded trees to prevent deforestation. The movement was recognized globally, leading to a discussion on the importance of nature conservation and women's environmental action during the 1985 Nairobi Conference on Women from the Third World Countries (World Economic Forum 2023). Another example is Women in Zapotalito, a town in southern Mexico, who are collaboratively undertaking the restoration of a natural canal in the Coaxaca mangrove region while confronting environmental adversities like the presence of dead fish, an ammonia odor, and deteriorating air and water quality attributed to a dam project and pollution from a lime oil refinery. Notwithstanding these problems, they persist in their everyday activities, which include cooking, preparing tortillas, cleaning residences, nurturing children, and fishing (World Economic Forum 2023).

## Natural Resource Scarcity

Residents of the Global South rely heavily on natural resources for their livelihoods and survival. Therefore, they become more vulnerable as natural resources are at risk of destruction and depletion due to climate extremes. Mirza et al. (2003) found that environmental changes caused by climate change in South Asia are aggravating natural resource depletion. Almulhim et al. (2024) in a recent study found that though water problems have escalated worldwide with the intensification of the negative impact of climate change, most cases of water scarcity and drought were located in the Global South. Water scarcity has invariably placed a huge burden on women's responsibilities, making them walk longer distances in search of water. They also face hardships in food security as agricultural production declines due to scarcity of water. In another similar study, UNEP also reported that water scarcity is a critical issue for a few countries, especially the Global South (United Nations Environment Programme 2024). According to a UN report, there are 15,906 water desalination plants worldwide, producing 95 million cubic meters of desalinated water for human consumption, from which 48% is produced in West Asia and North Africa (United Nations Environment Programme 2024). The issue associated with water desalination is that it is costly to run and cannot be afforded by the rural community, making it a challenge for rural women.

In many parts of the Global South, deforestation has resulted in a shortage of fuelwood, the most prevalent solid fuel for the poor. In impoverished communities throughout the Global South, women and girls are responsible for gathering traditional fuels, a physically exhausting endeavor that may consume from 2 to over 20 h weekly (UN WomenWatch 2009). Climate change and deforestation have resulted in its shortage forcing women and girls to walk greater distances for its collection. Increasing collection time for water and fuelwood gives women less time to fulfil family obligations, generate income, participate in governance or other public endeavors, pursue literacy or gain more skills, or even relax. Girls are sometimes retained at home to assist with fuel collection, prolonging the disempowerment cycle. Furthermore, as the degradation of the environment compels people to seek supplies at greater distances, women and girls become more susceptible to accidents from transporting heavy loads over long distances and also encounter heightened risks of domestic violence and abuse. A study by Scheid et al. (2019) found that women in Sub-Saharan Africa are facing the problem of a shortage of firewood as traditionally, they are responsible for collecting the firewood. As women try to find a mitigation solution to this issue, it is beyond their reach to address the root of the issue (Scheid et al. 2019).

Approximately 80% of the global population resides in Sub-Saharan Africa, South Asia, and Southeast Asia, regions where agricultural families are markedly impoverished and susceptible to marginalization (Suri 2022). Climate change induces crop failures and hunger in these areas, with millions more at risk of descending into poverty because of loss of production and income, lack of options for livelihood diversification and impoverishment. Climate change adversely affects agricultural productivity in areas with little water resources due to diminished water availability, heightened severe weather occurrences, thermal stress, and increased pests and diseases (Muzammal et al. 2024; Cai et al. 2015). Increasing temperatures, water scarcity, droughts, floods, and elevated atmospheric $CO_2$ levels impact crop production globally. Suri (2022) highlights that climate variability accounts for 80% of unpredictable crop yields in Africa's Sahel. Elevating sea levels provide a distinct challenge to food security in nations such as Bangladesh and Vietnam, where even little flooding may yield substantial consequences (Suri 2022).

## Health Implications

Climate change results in heightened morbidity and death from heat waves, floods, storms, wildfires, and droughts (Majedul Islam 2022). Environmental dangers increase the likelihood of developing severe diseases. Climate affects women's health via water shortages and pollution, while the emergence and spread of infectious illnesses are associated with climate and weather patterns. This increases the prevalence of illnesses such as cholera, malaria, and dengue fever owing to the prolonged risk of exposure due to changing seasonality and expanded geographical range of disease vectors (Bell et al. 1995; WHO 2023). Weather significantly affects the timing and intensity of epidemics, especially mosquito-borne illnesses, which are highly responsive to climatic fluctuations. Increasing temperatures may expedite the lifespan of the malaria parasite and disseminate it into new regions (Filho et al. 2019; Anikeeva et al. 2024). Floods escalating due to climate change may heighten the incidence of water-related illnesses, especially waterborne and vector-borne diseases, impacting millions of impoverished individuals yearly. An escalation in illness may exacerbate women's caregiving responsibilities for sick family and community members (Cissé 2019).

WHO (2023) elaborates that climate change is a threat to human health, making women more vulnerable to extreme climate events, resulting in poor maternal health, hypertension, and heat exhaustion (Afifi et al. 2024; Alam et al. 2022). As women's responsibilities do not halt during pregnancy, the struggle of gathering food and water puts them at risk of pregnancy complications. The leading cause of hypertension in pregnant women is the consumption of high levels of salt in drinking water sources in rural coastal Bangladesh, which poses a public health concern and challenge for the government, donor communities, and non-governmental organizations. Pregnant women's sodium intake is above WHO/FAO-recommended levels, and the problem may be exacerbated by future sea-level rise and environmental change (Khan et al. 2011). Hypertension in pregnancy is associated with increased rates of adverse maternal and fetal outcomes, including impaired liver function, low platelet count, intrauterine growth retardation, preterm birth, and maternal and perinatal deaths. Women with superimposed (pre)eclampsia develop significantly more adverse outcomes (Cífková 2023) and this raises concern in areas that are prone to salinity due to sea-level rise.

The World Health Organization (WHO) and the Intergovernmental Panel on Climate Change (IPCC) Fourth Assessment Report have determined that the adverse effects of climate change on agricultural output, food security, and undernutrition in developing nations represent the most significant detrimental impact on health (Tirado et al. 2013). Climate change exacerbates the danger of hunger and undernutrition, undermining human rights to health and sufficient nourishment. Climate change influences nutritional security via several channels, including food security, livelihoods, family food accessibility, maternity and childcare, health, water and sanitation, and socioeconomic determinants (Alam and Gounder 2025). The poorest and most susceptible populations, particularly women, face the most risk owing to their significant exposure to natural hazards, reliance on climate-sensitive resources, and restricted adaptive ability (Desai and Zhang 2021). Smallholder and subsistence farmers, pastoralists, and artisanal fishers experience intricate, localized effects of climate change. Failure to meet the nutrition needs of women, especially pregnant women, puts them at risk. Women also face psychological issues as they overthink and are concerned about the challenges faced due to climate change. Failure to gather food and water will risk their family being hungry and may also lead to domestic violence (Fan and Zlatnik 2023).

## *Women's Rights*

Women's rights have been impacted in the Global South in the form of limited access to resources to deal with climate change, limited climate risk approaches and political involvement. Women globally are more susceptible to poverty owing to unequal access to economic resources, financial opportunities, and decision-making power. They also perceive the negative impacts variably due to their dependence on climate-sensitive agriculture-related industries. In Africa, severe climatic phenomena such as heat waves, floods, and droughts have specific gendered impacts (Addaney and Moyo 2018).

Kumar et al. (2022) also add that women are the first to experience poverty due to limited adaptation opportunities to climate change and the lack of women's recognition in national planning. For example, India's Nationally Determined Contributions (NDCs) fail to account for the gender-specific vulnerabilities exacerbated by climate change. Despite asserting a commitment to allocating more resources for advancing gender parity and empowerment, this commitment remains vague and lacks specificity in its broader application. India's sixth Nationally Determined Contribution (NDC) for mitigating sectoral vulnerabilities emphasizes investments in development programs most susceptible to climate change, namely in agriculture, water and natural resources, health, and innovation and technology. Nevertheless, it fails to delineate sector-specific risks and does not specifically mention strategies to address the specific needs of women. Instead, it employs a "one-size-fits-all" methodology. Indigenous knowledge of marginalized groups, particularly women, is not prevalent despite being included in the Paris Agreement and India's NDC policy framework (Kumarnt et al. 2022). UN Women (2022) adds that climate change impacts agricultural production, putting pressure on women to work harder and for longer hours to increase production. This also results in girls leaving school and joining their mothers to manage the increased burden of climate change.

## *Need for Gender-Responsive Climate Action*

### Importance of Women's Involvement in the Decision-Making Process

Women's involvement in the decision-making process for climate action is a need, as women are more dependent on the environment and are disproportionately affected by climate change. So far, limited policies and frameworks have been adopted in Global South to empower women's involvement in climate action. After natural disasters like floods and cyclones, women are typically the primary responders and leaders in post-disaster recovery efforts. Nevertheless, they are often marginalized when formulating policies, strategies, and programs to mitigate catastrophic risk and enhance resilience. The active participation of women in decision-making is essential for recovery and rebuilding initiatives (Erman et al. 2021). Women and girls, including students, mothers, indigenous women, and celebrities, have been at the forefront of global and national climate campaigns, underscoring the urgency of the climate problem. They are crucial in instigating change as customers, employees, political figures, and corporate leaders. Women are more inclined to recycle, reduce trash, purchase organic and eco-labelled items, and save water and energy within the family (United Nations n.d.).

Studies have demonstrated a definitive correlation between women's leadership and initiatives to address climate change. Countries with more significant percentages of women in parliament are more inclined to adopt international environmental accords and implement harsher climate policies (United Nations n.d). Gender-diverse enterprises often exhibit superior environmental reporting and climate governance compared to their counterparts, and they allocate more expenditures toward renewable energy production and enhancements in energy efficiency (Abd Majid and Jaaffar 2023). Women are progressively contributing to the renewable energy sector, comprising 40% of the workforce in solar energy, a much more significant proportion than in the oil and gas industry (United Nations n.d.). Investing in women as a formidable catalyst for change is imperative, guiding the transition to a more sustainable future.

Shahbaz et al. (2022) elaborate that women constitute more than fifty percent of the worldwide agricultural workforce, and their participation in decision-making may enhance the implementation of Climate-Smart Agriculture (CSA) at the farm level in poor nations, particularly in the Global South. Nevertheless, female farmers possess insufficient authority in critical choices about agricultural operations such as land transactions, equipment acquisitions, and loan use. They possess enough authority regarding crops, inputs, financing access, and equipment use. Female agriculturists are inventive and eager to embrace novel agricultural techniques. Implementing CSA provides social, economic, and environmental advantages for farmers and society. Empowering women in decision-making may improve the adoption of CSA and promote economic growth in rural regions. Emphasizing women's empowerment in national policies and acknowledging their contributions to national planning objectives may enhance the adoption of CSA.

Singh et al. (2022) research emphasizes the growing participation of women in environmental management, enhancing their adaptive capacity and capability to offset the effects of climate change. Women exhibit high sociability, maintain robust networks, and diversify revenue streams to enhance community resilience. The patriarchal structure of society, characterized by conventional gender role divisions, impedes women's contributions. Women may profoundly impact community-oriented climate change adaptation initiatives with support and acknowledgement. Enhancing women's adaptive potential necessitates collaboration with men to transform cultural norms and beliefs related to gender roles. Traditional authorities and religious community leaders significantly impact decision-making processes.

## *Strategies for Addressing the Disproportionate Burden on Women*

Addressing the disproportionate burden on the women in the Global South is fundamental as the impact of extreme weather has been predicted to worsen over time. To begin with, a comprehensive approach to gender-responsive climate finance is necessary as a social protection system to assist women's vulnerability to extreme climate impacts. UNICEF (2024a, b) highlights that social assistance supports families in meeting their fundamental needs and enhances coping strategies, fostering climate adaptive practices and mitigating social challenges such as child marriage. Integrating social protection measures, including labor market regulation, agricultural skills development, and climate insurance, may alleviate the inequalities that women and girls face throughout the green transition. Nevertheless, gender-responsive and age-sensitive adaptation measures are necessary. UNICEF, in collaboration with the German Federal Ministry for Economic Cooperation and Development, has initiated a five-year evidence-gathering project titled ASPECT—Adaptive Social Protection: Evidence for Child Outcomes in Fragile Settings (United Nations International Children's Emergency Fund 2024a, b). The project aims to demonstrate the impact of adaptive social protection on strengthening family resilience to climate-related events, thus enhancing prospects for children and their families.

Furthermore, women in the Global South are primarily involved in agriculture as a means to gather food for their families. Therefore, an inclusive and gender-responsive agrifood system is essential. In South Asia, 71% of women are employed in agrifood systems, while 47% of males are engaged in the same sector. Should half of small-scale producers gain from women's empowerment, the earnings of 58 million individuals and the resilience of 235 million individuals may be improved (United Nations International Children's Emergency Fund 2024a, b). Gender-responsive strategies in agrifood systems effectively enhance resilience. Community-oriented strategies promote capacity building customized to women's requirements and improve access to information, resources, funding, and collective empowerment. Digitalization may mitigate mobility limitations and enhance financial independence. The ASPA + initiative of the International Fund for Agricultural Development implements sustainable technology for gender-sensitive, climate-resilient agricultural practices, guaranteeing that women and girls represent at least 50% of beneficiaries (United Nations International Children's Emergency Fund 2024).

Women are skilled with knowledge and experience that can make a difference in mitigating the challenges of extreme weather events if they are well recognized and supported by others, including men, government, and other organizations. Women are pictured as the victims of climate change but not as agents who have experienced the real impacts and understand the situation. For a long time, women have been involved in water harvesting, food preservation practices, early warning and management of natural resources. Women are the factor that needs to be involved in planning and decision-making to mitigate the impacts of climate change. In this way, an effective and efficient strategy can be discovered that can address the challenges faced by women. For example, in Africa, elderly women embody reservoirs of wisdom via their acquired knowledge and experience about early warning systems and catastrophe mitigation. The knowledge and experience transmitted across generations will effectively enhance local adaptation ability and preserve communal livelihoods (Osman-Elasha 2007).

## *Achieving Sustainable Development Goals*

Addressing the disproportionate burden on women in the Global South is crucial to achieving the SDGs, especially when gender inequality exists. Targeted strategies and interventions can assist Global South in reducing the burden faced by women and progress toward achieving multiple SDGs simultaneously. To begin with a comprehensive approach, gender-responsive climate finance is necessary as a social protection system to assist women's vulnerability to extreme climate impacts. UNICEF (2024a, b) highlights that social assistance supports families in meeting their fundamental needs and enhancing coping strategies, fostering climate adaptive practices and mitigating social challenges such as child marriage.

Addressing the challenges faced by women will directly aim to achieve SDG 5 (Gender Equality). Supporting women in building climate resilience will empower them to act more in mitigating climate challenges. Women's leadership is also necessary in climate finance and strategic decision-making to remove the traditional gender barriers. As women understand the challenges better and experience the most, it is crucial for women's leadership. For example, when women access climate-smart farming practices, they can boost leadership within their community, defying the prevailing gender stereotypes. This will empower climate action and impact the social structure to recognize gender equality.

Providing support and resources to women in the Global South can simultaneously achieve SDG 1 (No Poverty) and SDG 2 (Zero Hunger). Women in the Global South depend on agriculture and natural resources for food and income. Climate adaptation strategies to assist women, such as access to resources, finance and training, will help women build resilience, mitigate the extreme climate burden, and increase the production level. Increasing food production will reduce food insecurity, helping achieve SDG 2, and increased production of agriculture will assist women in commercializing their produce, allowing the achievement of SDG 1. When women are trained in the knowledge of crop nutrients, they can grow crops to ensure family nutrition and assist in achieving SDG 3 (Good Health and Well-being).

## Recommendations

The issue of climate extremes and their disproportionate burden on women in the Global South requires various strategies, such as national policies, community involvement and international cooperation. These strategies must align with each other to achieve the intended outcome of relieving the disproportionate burden on women and achieving SDGs in the Global South.

### *Policy Intervention*

At the national level, policies prioritizing the development of gender-responsive climate action plans that recognize women and support their vulnerabilities must be implemented. According to the World Bank, gender must be integrated into developing climate response programs to achieve synergistic goals related to gender equality, poverty alleviation, and climate resilience (Deininger et al. 2023). These goals co-relate to the Nationally Determined Contributions and pledges to the Paris Agreement. The government must involve women during the policy consultation to ensure priorities as part of community-led climate action. Involving women should be done by promoting women's leadership and decision-making in climate projects. This can be done from planning to implementation levels at national and community levels, empowering women's contribution and effort. For instance, Costa Rica is proactively tackling the issues of declining forest areas and land use constraints by emphasizing the involvement of women and Indigenous Peoples in sustainable land management efforts. The nation has instituted initiatives that link women's income generation to conservation efforts, underscoring their vital role in safeguarding forest ecosystems. Costa Rica, as a leader among tropical countries in the REDD + project, has instituted prizes for female conservationists, recognizing their significant achievements in the management of natural resources (Deininger et al. 2023).

### *Community-Level Involvement*

Community-level involvement is vital to successfully implement policies and strategies. Local government, community settlements, and community groups must be involved so that women's opinions from the ground level are heard and they are involved in the implementation. Community-based women's groups can be initiated so that all women can carry out local climate planning and mitigation efforts. Since women understand the community and its challenges better, allowing them to be part of their community-based climate planning is fundamental. However, the government and other stakeholders need to support these groups, as women have limited resources to initiate actions. For instance, The World Bank's Enhancing Community Resilience and Local Governance Project in South Sudan seeks to amplify women's voice and agency by augmenting their representation in decision-making bodies, altering views about women's roles, and improving their job prospects (Deininger et al. 2023).

Building community capacity is essential for effective impact. Therefore, training women as leaders in climate resilience and creating a peer-learning network will allow women across communities to exchange knowledge and ideas for achieving climate resilience. Integrating technical training initiatives for women will help them better understand the extreme climate situation and train them in environmental management, enhancing individual and community resilience. Other projects, such as natural resource management and renewable energy, must be established with the active participation of women.

### *International Cooperation and Support*

International cooperation and support are essential for effectively executing gender and climate projects due to limited resources, budgets and experts in the Global South. International support must prioritize the establishment of projects and support to women groups who are involved in climate projects. This will ease women's disproportionate burden due to climate change and support them in climate-resilient action. Technology transfer and information dissemination are critical elements of foreign assistance, including access to climate-smart technology tailored to women's needs and South-South collaboration on gender-responsive solutions. Global networks of female climate leaders may promote information sharing and enhance capacity development. International partnership development enhances these initiatives by facilitating links among women's groups and establishing multi-stakeholder forums for gender and climate action. Implementation support must include technical help for formulating gender-responsive policies and enhancing national institutions' capacities, as well as providing direction on successful gender mainstreaming in climate initiatives. Resource mobilization needs synchronized donor assistance, inventive finance strategies, and consolidated financial frameworks. International advocacy and awareness bolster these initiatives via worldwide campaigns and women's involvement in international climate conferences. The World Bank's activities in Zambia seek to bolster climate resilience and assist disadvantaged populations, especially female-headed families, via diverse adaption techniques and enhanced governance. The Farmer Income Support Program has markedly enhanced agricultural output, as shown by a 20% in the value of agriculture added per farmer (Deininger et al. 2023).

## Conclusion

This study elucidates the significant and complex issues faced by women in the Global South about climate change. The research reveals a significant connection between prevailing gender inequalities and the intensifying effects of climatic extremes, illustrating how pre-existing socioeconomic gaps are substantially exacerbated by environmental changes. Women in these places experience compounded vulnerability transcending natural concerns, including structural obstacles in education, economic involvement, political representation, and social agency.

The findings highlight that climate change is not a neutral occurrence but rather a problem that disproportionately impacts women, especially in poor nations. Women are both the most susceptible to climate-induced disruptions and the most essential agents of community resilience. Their conventional responsibilities in resource management—mainly collecting water, obtaining food, and ensuring household stability become progressively risky as climatic extremes escalate, resulting in augmented workloads, elevated health risks, and intensified economic marginalization.

The research highlights the critical need for a holistic, multi-faceted strategy for climate adaptation that goes beyond conventional environmental solutions. This strategy must fundamentally include gender perspectives into policy frameworks, ensuring women are beneficiaries of climate action and active architects of resilience measures. The proposed solutions encompass national policy reforms, community empowerment, and strong international engagement, highlighting the need to eliminate cultural and structural obstacles that have traditionally marginalized women in essential decision-making roles.

By endorsing gender-responsive climate policies, fostering women's leadership, and establishing supporting ecosystems for their involvement, the Global South can convert a possible crisis into an opportunity for significant societal transformation. The way forward necessitates not only the mitigation of climate hazards but also the concurrent resolution of the fundamental gender inequities that exacerbate women's susceptibility. This comprehensive strategy is essential for attaining sustainable development objectives and signifies a fundamental reconfiguration of climate resilience that prioritizes gender equity. The research advocates for a paradigm shift—acknowledging women not as passive victims of climate change but as formidable, imaginative leaders who can spearhead sustainable solutions when afforded the resources, recognition, and chances. The future of climate adaptation in the Global South relies on our collective capacity to eliminate systematic obstacles and empower women as equal collaborators in addressing one of the most pressing challenges of our era.

## References

Abd Majid N, Jaaffar AH (2023) The effect of women's leadership on carbon disclosure by the top 100 global energy leaders. Sustainability 15(11):8491. https://doi.org/10.3390/su15118491

Addaney M, Moyo CG (2018) Women's rights, gender and climate change law in Africa: advancing an equity agenda. J Law Soc Dev 5(1):1–18. https://doi.org/10.25159/2520-9515/7066

Afifi M, Malkawi M, El-Adawy M. (2024) Women's health and climate change: a threat or an opportunity. East Mediterr Health J 30(5):330–332. https://doi.org/10.26719/2024.30.5.330

Alam M, Ali MF (2022) Health and wellbeing and quality of life in the changing urban environment: challenges and government response in Fiji Island. J Positive School Psychol 6(9):2853–2863. https://journalppw.com/index.php/jpsp/article/view/12783/8284

Alam M, Gounder KA (2025) Interwoven realities: gender, environment, and human rights. In: Chakraborty S, Satapathy S (ed) Gender, environment, and human rights: an intersectional exploration. IGI Global, pp 1–16. https://doi.org/10.4018/979-8-3693-6069-9.ch001

Alam M, Ali MF, Kundra S, Nabobo-Baba U, Alam MA (2022) Climate change and health impacts in the South Pacific: a systematic review. In: Chatterjee U, Akanwa AO, Kumar S, Singh SK, Dutta Roy A (eds) Ecological footprints of climate change. Springer Climate. Springer, Cham. https://doi.org/10.1007/978-3-031-15501-7_29

Almulhim AI, Alverio GN, Sharifi A, Shaw R, Huq S, Mahmud MJ et al (2024) Climate-induced migration in the Global South: an in-depth analysis. NPJ Clim Action 3(1):47. https://doi.org/10.1038/s44168-024-00133-1

Althor G, Watson JE, Fuller RA (2016) Global mismatch between greenhouse gas emissions and the burden of climate change. Sci Rep 6(1):20281. https://doi.org/10.1038/srep20281

Anikeeva O, Hansen A, Varghese B, Borg M, Zhang Y, Xiang J, Bi P (2024) The impact of increasing temperatures due to climate change on infectious diseases. BMJ 387. https://doi.org/10.1136/bmj-2024-079343

Bell JE, Brown CL, Conlon K, Herring S, Kunkel KE, Lawrimore J, Luber G, Schreck C, Smith A, Uejio C (2018) Changes in extreme events and the potential impacts on human health. J Air Waste Manag Assoc 68(4):265–287. https://doi.org/10.1080/10962247.2017.1401017

Boateng GO (2021) Resource insecurity and gendered inequalities in health: a challenge to sustainable livelihood. In: Geography, health and sustainability, pp 127–144. Routledge. https://doi.org/10.4324/9780367743918-10

Cai X, Zhang X, Noël PH, Shafiee-Jood M (2015) Impacts of climate change on agricultural water management: a review. Wiley Interdiscp Rev Water 2(5):439–455. https://doi.org/10.1002/wat2.1089

Cífková R (2023) Hypertension in pregnancy: a diagnostic and therapeutic overview. High Blood Press Cardiovasc Prev 30(4):289–303. https://doi.org/10.1007/s40292-023-00582-5

Cissé G (2019) Food-borne and water-borne diseases under climate change in low- and middle-income countries: further efforts needed for reducing environmental health exposure risks. Acta Trop 194:181–188. https://doi.org/10.1016/j.actatropica.2019.03.012

De Wit S (2021) Gender and climate change as new development tropes of vulnerability for the Global South: essentializing gender discourses in Maasailand, Tanzania. Tapuya Latin Am Sci Technol Soc 4(1):1984638

Deininger F, Woodhouse A, Kuriakose AT, Gren A, Liaqat S (2023) World Bank group gender thematic policy notes series: issues and practice note. World Bank Group. https://documents1.worldbank.org/curated/en/099718102062367591/pdf/IDU08c737dd00f858041 2b0aed90fce874ab09b0.pdf. Accessed 25 Oct 2024

Desai Z, Zhang Y (2021) Climate change and women's health: a scoping review. Geohealth 5(9):e2021GH000386. https://doi.org/10.1029/2021GH000386

Erman A, De Vries Robbe SA, Thies SF, Kabir K, Maruo M. (2021) Gender dimensions of disaster risk and resilience: existing evidence. https://wrd.unwomen.org/sites/default/files/2021-11/Gender-Dimensions-of-Disaster-Risk-and-Resilience-Existing-Evidence.pdf Accessed 10 Nov 2024

Fan W, Zlatnik MG (2023) Climate change and pregnancy: risks, mitigation, adaptation, and resilience. Obstetr Gynaecol Surv 78(4):223–236. https://doi.org/10.1097/OGX.0000000000001116

Field CB, Barros V, Stocker TF, Dahe Q (eds) (2012) Managing the risks of extreme events and disasters to advance climate change adaptation: special report of the intergovernmental panel on climate change. Cambridge University Press

Filho WL, Scheday S, Boenecke J, Gogoi A, Maharaj KS (2019) Climate change, health and mosquito-borne diseases: trends and implications to the Pacific Region. Int J Environ Res Public Health 16(24):5114. https://doi.org/10.3390/ijerph16245114

Goodrich CG, Manjari Mehta MM, Suman Bisht SB (2017) Status of gender, vulnerabilities and adaptation to climate change in the Hindu Kush Himalaya: impacts and implications for livelihoods, and sustainable mountain development. https://lib.icimod.org/record/32618/files/icimodWP3-017.pdf

Khan AE, Ireson A, Kovats S, Mojumder SK, Khusru A, Rahman A, Vineis P (2011) Drinking water salinity and maternal health in coastal Bangladesh: implications of climate change. Environ Health Perspect 119(9):1328–1332. https://doi.org/10.1289/ehp.1002804

Kumar A, Sinha VN, Basu K (2022) Gender equality in climate action: women at the core. In: PwC. https://www.pwc.in/assets/pdfs/consulting/esg/gender-equality-in-climate-action.pdf Accessed 10 Nov 2024 (Retrieved October 21, 2024)

Majedul Islam MM (2022) Threats to humanity from climate change. In: Climate change: the social and scientific construct. Springer International Publishing, Cham, pp 21–36. https://doi.org/10.1007/978-3-030-86290-9_2

Malak MA, Sajib AM, Quader MA, Anjum H (2020) "We are feeling older than our age": vulnerability and adaptive strategies of ageing people to cyclones in coastal Bangladesh. Int J Disast Risk Reduct 48:101595

Memon FS (2020a) Climate change and violence against women: study of a flood-affected population in the rural area of Sindh, Pakistan

Memon FS (2020b) Climate change and violence against women: study of a flood-affected population in the rural area of Sindh, Pakistan. Pak J Women's Stud Alam-E-Niswan 27(1):65–85

Mirza MMQ, Warrick RA, Ericksen NJ (2003) The implications of climate change on floods of the Ganges, Brahmaputra and Meghna rivers in Bangladesh. Clim Change 57(3):287–318

Muzammal H, Zaman M, Safdar M, Adnan Shahid MK, Khil A et al (2024) Climate change impacts on water resources and implications for agricultural management. In: Transforming agricultural management for a sustainable future: climate change and machine learning perspectives. Springer Nature Switzerland, Cham, pp 21–45. https://doi.org/10.1007/978-3-031-63430-7_2

Ngcamu BS (2023) Climate change effects on vulnerable populations in the Global South: a systematic review. Nat Hazards 118(2):977–991

Osman-Elasha B (2007) Women. In: The shadow of climate change. United Nations. https://www.un.org/en/chronicle/article/womenin-shadow-climate-change#:~:text=Women%20should%20be%20part%20of,technology%20transfer%20and%20capacity%20building. Accessed 23 Oct 2024

Paludi MI, Krysa I (2023) Exploring development, gender inequality, and the global south: challenges and opportunities. In: Businesses' contributions to sustainable development goal 5: gender equality across B corps in Latin America and the Caribbean, pp 9–24

Patel SK, Agrawal G, Mathew B, Patel S, Mohanty B, Singh A (2020) Climate change and women in South Asia: a review and future policy implications. World J Sci Technol Sustain Dev 17(2):145–166

Pindyck RS (2021) What we know and don't know about climate change and implications for policy. Environ Energy Policy Econ 2(1):4–43

Sams IS (2019) Impacts of climate change induced migration on gender: a qualitative study from the Southwest Coastal Region of Bangladesh. Int J Soc Sci Stud 7:57

Scheid A, Hafner JM, Hoffmann H, Kächele H, Uckert G, Sieber S, Rybak C (2019) Adapting to fuelwood scarcity: the farmers' perspective. Front Sustain Food Syst 3:28

Shahbaz P, Ul Haq S, Abbas A, Batool Z, Alotaibi BA, Nayak RK (2022) Adoption of climate smart agricultural practices through women involvement in decision making process: exploring the role of empowerment and innovativeness. Agriculture 12(8):1161

Singh P, Tabe T, Martin T (2022) The role of women in community resilience to climate change: a case study of an Indigenous Fijian community. In: Women's studies international forum, vol 90. Pergamon, p 102550

Suri S (2022) Climate change and food security in the Global South. Observer Research Foundation. https://www.orfonline.org/expert-speak/climate-change-and-food-security-in-the-global-south. Accessed 19 Oct 2024

Tirado MC, Crahay P, Mahy L, Zanev C, Neira M, Msangi S et al (2013) Climate change and nutrition: creating a climate for nutrition security. Food Nutr Bull 34(4):533–547

Tokas M (2024) The women of the Global South in the 21st century. Global South studies series. https://globalsouthseries.in/2024/01/26/the-women-of-the-global-south-in-the-21st-century/#:~:text=Additionally%2C%20women%20in%20the%20global,is%20limited%20access%20to%20education. Accessed 17 Oct 2024

UN Women (2022) Explainer: how gender inequality and climate change are interconnected | UN Women—Headquarters. https://www.unwomen.org/en/news-stories/explainer/2022/02/explainer-how-gender-inequality-and-climate-change-are-interconnected. Accessed 18 Oct 2024

UN WomenWatch (2009) Women, gender equality and climate change. In: UN WomenWatch. https://www.un.org/womenwatch/feature/climate_change/downloads/Women_and_Climate_Change_Factsheet.pdf. Accessed 19 Oct 2024

United Nations Environment Programme (2024) As shortages mount, countries hunt for novel sources of water. UNEP. https://www.unep.org/news-and-stories/story/shortages-mount-countries-hunt-novel-sources-water#:~:text=%22Water%20scarcity%20has%20become%20a,particularly%20in%20the%20Global%20South.%22&text=Several%20nations%2C%20such%20as%20Bahamas,drinking%20water%20comes%20from%20it. Accessed 19 Oct 2024

United Nations International Children's Emergency Fund (2024a) Why women are key to climate action | United Nations. United Nations. https://www.un.org/en/climatechange/science/climate-issues/women#:~:text=Having%20women%20actively%20involved%20in,security%20for%20families%20and%20communities. Accessed 21 Oct 2024

United Nations International Children's Emergency Fund (2024b) Tackling gender inequality in a climate-changed world. UNICEF. https://www.unicef.org/blog/tackling-gender-inequality-climate-changed-world#:~:text=Long%2Dterm%20investments%20in%20social,kind%20assistance%20alongside%20livelihood%20diversification. Accessed 25 Oct 2024

United Nations (n.d.) Why women are key to climate action | United Nations. https://www.un.org/en/climatechange/science/climate-issues/women. Accessed 21 Oct 2024

WHO (2023) Climate change and health. https://www.who.int/news-room/fact-sheets/detail/climate-change-and-health. Accessed 12 Oct 2023

World Economic Forum (2023) How women in the Global South are leading the fight against climate change. https://www.weforum.org/agenda/2023/10/women-global-south-climate-change/. Accessed 18 Oct 2024

# Chapter 6
# Understanding 'Gendered' Impacts of Climate Change and the Role of Women in Adaptation: Evidence from the Indian Sundarbans

**Pratyay Raha, Amrita Sen, and Debarchana Biswas**

**Abstract** The Intergovernmental Panel on Climate Change (IPCC) forecasts that, in the coming years, climate change will substantially exacerbate the livelihood challenges already confronting coastal communities. Upscaling resilience and adaptation strategies is critical for overstressed ecosystems like the erosion-prone coasts of the Indian Sundarbans Delta (ISD), which endure frequent land transformations and biodiversity fragmentation. This chapter examines the unequal impacts of climate shocks on gender groups, focusing particularly on the adaptation strategies employed by women in two Sundarbans villages in India. Women face disproportionate challenges in the aftermath of disasters, including income loss, increased domestic burdens, and the arduous task of securing freshwater under harsh conditions. Using qualitative methods, including semi-structured interviews, focused group discussions, and participatory observations, the study documents women's lived experiences and adaptation mechanisms while analyzing gaps in policy and institutional responses. While their livelihood practices are severely disrupted, women employ innovative coping mechanisms such as collective resource-sharing, participation in self-help groups, and diversifying income streams to sustain their households. However, these efforts are often unsupported by state disaster management policies, which lack gender-sensitive principles. The research highlights the absence of provisions addressing women's specific needs, particularly in health care, where the unpreparedness of systems aggravates post-disaster vulnerabilities. Additionally, the chapter underscores the urgent need for policies that recognize and support women's roles in fostering resilience within their communities. By documenting

P. Raha
School of Art, College of Design and Social Context, RMIT University, Melbourne, Australia
e-mail: pratyayraha2010@gmail.com

A. Sen (✉)
Department of Humanities and Social Sciences, Indian Institute of Technology, Kharagpur, India
e-mail: amrita.sen@hss.iitkgp.ac.in

D. Biswas
Department of Humanities and Social Sciences, Indian Institute of Technology, Guwahati, India



A. Sarkar and N. Bandyopadhyay (eds.), *Gender-Transformative Approaches for Climate Change Adaptation*, Springer Climate,
https://doi.org/10.1007/978-3-031-90236-9_6

these challenges and strategies, the study aims to amplify women's voices in adaptation discourse and bridge policy gaps to foster equitable resilience-building in coastal areas.

**Keywords** Gender · Climate disasters · Sundarbans · Coastal communities · Adaptation

## Introduction

In the long history of class struggle, the subaltern marginal subject has always been perceived and engaged with as 'the other'. Hence, the top-down approach of governance has historically neglected the aspirations of the marginal population. Living on the smallest branch and end point of the social tree, these populations live in constant fear of being cut off from the tree. They believe that they are in no socio-economic position to demand policy change and action in favor of basic facilities, development, and modernization of lifestyle. In her seminal essay, 'Can the subaltern speak?', Gayatri Chakravarty Spivak discussed the epistemic violence that entails in the historiography of colonial states (Spivak 1988). She goes on to say that within the marginal subaltern population, women have been further marginalized concerning the spectrum of power dynamics in the ever-evolving socio-political nexus. She writes (Nelson and Grossberg 1988, p. 288) 'within the effaced itinerary of the subaltern subject, the track of sexual difference is doubly affected. The question is not of female participation in the insurgency, or the ground rules of the sexual division of labor, for both of which there is "evidence." It is, rather, that, both as an object of colonialist historiography and as the subject of insurgency, the ideological construction of gender keeps the male dominant. If, in the context of colonial production, the subaltern has no history and cannot speak, the subaltern as female is even more deeply in shadow'.

This marginalization and the resulting positionality have profound and enduring effects on the psyche and cultural conditioning of a population. The title of this chapter hints at the limitations of a reductionist perspective. It means that studying a specific population, place, country or event inherently carries a lot of limitations. It does not cater to the intertwined relationships or the systematic historical epistemic and experiential violence that has been ingrained in our genes through generations. It is time we holistically approach a particular research or ethnographic fieldwork to understand our social dichotomies and interrelated environmental catastrophes. It is time to find out what we are missing out on, the context and content of the gap. Despite numerous research projects and academic publications, why is there negligible change on the ground? Why cannot we shift from writing research papers to ensuring that the research leads to direct and meaningful action? When we categorize or explain a disaster event, we often ignore or neglect the social aspects of the process of disaster. That is the limitation of our research which makes it reductionist.

The Intergovernmental Panel on Climate Change (IPCC) warns that climate change will intensify current livelihood challenges, with women in coastal communities being particularly vulnerable. Women often face unique challenges, such as increased domestic responsibilities and limited access to resources, which heighten their vulnerability. (IPCC 2019). Our study, situated in the Indian Sundarbans Delta (ISD), aims to formulate and implement an action plan that prioritizes understanding the impacts on social resilience and adaptive capacities, with a focus on the specific challenges faced by women. Through a place-based risk management framework, we aim to address the layered impacts of climate change on women, prioritizing their voices and needs in adaptation efforts. This approach aligns with the United Nations' calls to integrate gender perspectives in climate action, aiming to limit losses and damages from multiple disasters. The insights from this research will support global initiatives to enhance adaptive capacities, especially in areas where climate change disproportionately impacts women. By highlighting gender-specific effects, this chapter emphasizes the need to build resilience not only to reduce disaster impacts but also to address the underlying vulnerabilities faced by women in precarious environments globally.

## Background and Context

The Sundarbans, the world's largest mangrove ecosystem, is a vital climate hotspot situated at the confluence of the Padma, Brahmaputra, and Meghna rivers, stretching across southern Bangladesh and the Indian state of West Bengal. (Rogers and Goodbred 2014). Recognized as a UNESCO World Heritage Site, Tiger Reserve, Biosphere Reserve, and Ramsar Wetland of International Importance, the Sundarbans face serious threats from climate change, disproportionately affecting women. (Jena 2018).

According to the IPCC (IPCC 2019), climate change will profoundly affect coastal regions, leading to irreversible degradation of marine ecosystems. These impacts are projected to surpass previous forecasts, leading to extensive degradation of ecosystem resilience, structure, and functionality. Women, who frequently depend on these ecosystems for their livelihoods, are particularly at risk as shifts in seasonal patterns and natural adaptive capacities take place. Projected climate change, alongside non-climatic factors, poses a grave threat to vast expanses of the world's forests, coral reefs, and low-lying coastal wetlands, with direct consequences for women's roles in agriculture and fisheries (IPCC 2019). The swift rise in sea levels jeopardizes coastal infrastructure and settlements, placing an extraordinary burden on women who oversee households and manage vital community resources (IPCC 2022).

Globally, around a billion people—particularly women in small island nations and low-lying regions—face peril from climate hazards specific to coastal areas, driven by shifts in population dynamics. The IPCC (C. C. S. R. IPCC 2023) warns of escalating threats to food and water security stemming from extreme temperatures, erratic rainfall, and drought, predicting a decline in coastal fishing resources. These transformations will intensify risks to livelihoods, settlements, health, and well-being while undermining food and water security and cultural values. Women will shoulder a disproportionate share of these mounting challenges.

Women in the Sundarbans have historically navigated environmental uncertainties, yet climate change and other factors are now heightening their vulnerability. (Ghosh et al. 2018). These changes impact their roles in agriculture, fisheries, land management, and household maintenance. Media, governments, and policymakers often focus on the economic costs and infrastructure damages of climate change, neglecting the daily vulnerabilities and marginalization faced by women in disaster-prone areas. Likewise, conversations surrounding adaptation and resilience often highlight the roles of governments, NGOs, and international agencies, frequently overshadowing the agency and participation of women in these adaptation processes. (Ghosh and Bose 2024).

## Rationale and Problems

This chapter will qualitatively address the 'gendered' impacts of climate disasters in the Indian Sundarbans. We seek to uncover and discuss reasons for the disproportionate impacts through literature review and fieldwork, from conversations with people who have lived experiences of these impacts. One factor that emerges from conversations and literature is the impact of migration. Forced migration driven by environmental stressors must be clearly distinguished from voluntary migration. The prevalence of vague and conflicting definitions results in insufficient or nonexistent regulations for support and compensation (Das and Hazra 2020). The planet, a shared concern encompassing diverse ecosystems, peoples, and cultures, demands a unified approach to its management, protection, and sustainability. However, climate negotiations often resemble a farce within an unequal global economy, where men and powerful nations dictate the terms of their participation and responsibility in addressing escalating environmental crises. In contrast, women and less affluent countries are relegated to the sidelines, possessing little to no power to influence or reshape the direction of these critical discussions (Denton 2002).

It is crucial to recognize that these challenges are unfolding on a global scale. Local research and analysis prompt us to envision and map a broader spectrum of social and environmental justice issues emerging worldwide. While the biological distinctions between males and females are undeniably clear and universally recognized, the varying impacts of climate change on men and women have yet to be fully acknowledged and incorporated into the design and implementation of response strategies. The divergent roles of men and women within communities suggest that climate change will affect them in fundamentally different ways, shaped by their unique responsibilities and varying degrees of access to resources. (Wamukonya and Rukato 2001).

In the article 'Gender, Risk, and Disaster', (Fothergill 1996) points out that research consistently shows that gender differences shape the entire disaster experience. These differences often stem from factors like childcare duties, economic challenges, social networks, traditional gender roles, and discrimination. These recurring patterns highlight key areas of concern and suggest directions for future research and policy recommendations. O'Brien in his article *Rethinking Social Contracts: Building Resilience in a Changing Climate* contended that '*the environmental problems cannot be addressed in isolation of social context*' (O'Brien et al. 2009). The system might move from one basin into another when exposed to external shock or stress or even due to changes in internal structure. The ability to transition from an existing system to a new one when ecological or social conditions render the former untenable is what we refer to as resilience through transformability. This embodied and lived resilience has the potential to reshape narratives and address the historical roots of violence we discussed earlier. By embracing this transformative capacity, communities can not only adapt to change but also actively reframe their histories and futures. There is an increasing recognition that the biophysical effects of climate change are intricately connected to socio-economic, institutional, and demographic factors, interacting in complex and often nonlinear ways that can exceed the adaptive capacity of households and communities. This dynamic is particularly pronounced in contexts where gender disparities and issues of social and economic justice remain unaddressed. Climate change is now understood as a powerful amplifier of existing socio-economic inequalities, exacerbating vulnerabilities and deepening the divides within society (Awiti 2022).

Women in Ghoramara (An island in the Sundarbans delta, 150 km south of Kolkata and named the 'sinking island' by the media, has lost nearly half its area to soil erosion in the past two decades and could disappear if a solution is not found) where the households with seasonal male migrants face disproportionate burdens, as remaining members shoulder the weight of leadership, childcare, and caregiving for the elderly, disabled, and ill. They are responsible for growing crops for household sustenance and managing domestic chores, including tending to the family's betel vine sheds. Their socio-economic status within rural society significantly constrains their mobility, access to financial resources, healthcare, and opportunities for meaningful participation in decision-making processes (Das and Hazra 2020). Though this chapter does not focus on the ecological and social dynamics of Ghoramara island, we find a common thread of transformation and gendered impacts in most of the areas in the Indian Sundarbans Delta. There are some differences related to the facilities and disaster management infrastructure between islands.

## Methodology

The study focuses on selected villages within the Indian Sundarbans, specifically in the blocks of Gosaba, Kakdwip, Sagar, Patharpratima, and Basanti. These five blocks have all suffered significant devastation due to the ferocious impacts of cyclones Amphan and Yaas during 2020 and 2021.

Historically, West Bengal had largely escaped the ravages of cyclones, but Amphan shattered this precedent. Sweeping through South Bengal with terrifying winds of 112–150 mph, it wreaked havoc, especially in the Sundarbans. The storm's ferocity breached embankments, unleashing powerful surges that devastated the landscape and obliterated forest cover. In the aftermath, the scene was one of destruction: uprooted trees, tangled power lines, fallen utility poles, and widespread disruptions to electricity and water supply. Amphan's impact was catastrophic, exacerbated by the fact that most embankments were constructed from earth with few reinforced concrete structures. Only a handful of areas, like Laskarpur in the Basanti block, still boasted any forest cover. The consequences of both Amphan and Yaas were particularly dire, driven by the lack of forest protection and the vulnerable positioning of these villages along the coast.

Consequently, the study focuses on six villages—Satjelia, Kalinagar, Beguakhali, Durgagobindapur, Birajnagar, and Laskarpur—spanning the five blocks previously identified. Fieldwork was conducted continuously from January to November 2023, aimed at uncovering emerging risks, the complexities of current disaster management practices, and mapping community adaptive strategies. This research employs an open-ended qualitative data collection approach, emphasizing the documentation of social interactions through the lenses of the individuals and stakeholders involved. The researchers immersed themselves in the daily lives of these communities, striving to gain a profound understanding of their lived experiences and challenges.

Open-ended semi-structured interviews were conducted in the native Bengali language across each of the five villages, engaging both male and female participants. Most residents of the southern islands possess limited or no landholdings and depend heavily on the forests for their livelihoods, primarily fishing for species such as fish, prawns, and crabs. The inhabitants belong to backward caste categories and have deep roots in the region. To identify additional respondents, including panchayat officials and local power brokers, a snowball sampling method was employed. Community guides—such as NGO workers and local leaders identified during fieldwork—played an essential role in facilitating participant recruitment. Focus Group Discussions (FGDs) were organized to explore critical topics, including community perceptions of climate change impacts, the involvement of government and local leaders in disaster management, and future aspirations for the Sundarbans. These discussions were carefully recorded and documented with informed consent from all participants.

## Gendered Vulnerabilities in Climate Disasters: Insights from the Indian Sundarbans

During our fieldwork in Beguakhali (Block Sagar, District South Twenty-Four Parganas), we observed that women disproportionately bear the impact of natural disasters, emerging as the most affected demographic. With their husbands or male family members often employed as migrant laborers, these women are left to confront the consequences of cyclones independently.

One participant recounted the aftermath of Cyclone Yass, highlighting how increased salinity rendered their agricultural land unusable. This environmental challenge led to heightened tensions and conflicts within her joint family, primarily due to inadequate financial resources. As a result, her family was compelled to separate, finding refuge in a temporary shelter constructed from fragile materials. The economic burden intensified as they resorted to borrowing from money lenders, creating a debt cycle that proved insurmountable. Her husband's temporary employment in southern India further isolated her, leaving her and their children to navigate these adversities alone. The rise in salinity has not only led to a decline in agricultural productivity but has also intensified the struggles of daily life, placing immense stress on family dynamics and highlighting the larger socio-economic challenges confronting women in disaster-stricken areas (Fig. 6.1).

Another narrative sheds light on the escalating struggle for survival among women in the region, who face increasingly limited livelihood options due to environmental changes. With their expertise confined to farming and lacking any other skills or training, they find themselves with no viable alternatives for income generation. As their agricultural land gradually succumbs to salinity, many husbands are forced to migrate elsewhere in search of better job opportunities, leaving women to manage on their own. One woman shared her experience of remaining in the village with her two children and mother-in-law, relying solely on her farming abilities. She emphasized the pressing need for government intervention, suggesting that training programs,

**Fig. 6.1** Voices from the margins, Beguakhali. *Source* Field 2023

such as those for embroidery work, could provide a crucial opportunity for her family. Such initiatives would enable them to become less reliant on her husband's income alone, offering a pathway to greater financial stability.

This narrative highlights the broader socio-economic challenges faced by women in disaster-affected areas and underscores the urgent need for skill development and support programs. By equipping women with diverse skills, these programs could enhance their economic resilience and empower them to navigate the difficulties posed by environmental and economic changes. This leads us to think more about Government policies and disaster management frameworks, as rightly mentioned by (Denton 2002), Gender disparities remain pervasive, particularly in access to land, control over resources, and opportunities for securing paid labor. Women face significant challenges in diversifying their incomes and managing the time they devote to agricultural and forestry activities. Governments must implement gender-sensitive policies that address land tenure issues.

To foster equitable access to land and resources, as well as to enhance productivity and promote environmental and soil conservation, legal and structural obstacles must be dismantled. The issue of landlessness is compelling women to seek alternative income-generating activities to supplement their earnings. Furthermore, the commercialization of agriculture necessitates that women explore additional income sources. For those lacking the educational background to navigate bureaucratic channels for credit and employment opportunities, community resources often become their only means of support.

Increasing evidence indicates that women face distinct challenges during and after climate disasters. Despite global initiatives to address their needs, these efforts have not consistently led to substantial improvements. In some instances, they have even resulted in the reduction of programs specifically aimed at supporting women. Additionally, many villagers express frustration with political parties exploiting those in poverty, forcing them to endure hardships indefinitely. Testimonies from residents in Beguakhali Village reveal deep-seated discontent toward local Panchayat members, particularly regarding the distribution of relief resources during calamities. One resident noted:

> We did not receive any assistance or relief after the cyclone. In the aftermath of Amphan, we were only given a single sheet of paper and some dry food. Members of the local panchayat visited and took pictures of our homes, but as of 2023, we have not received any financial aid for cyclone relief. However, after Yass, we did receive INR 5000 (60USD). It seems that after Amphan, which was a major setback for us, we were excluded from any cyclone relief. Many others received aid ranging from INR 5000 (60 USD) to INR 20000 (230 USD). The reason for this disparity is likely due to a lack of political involvement. Despite being severely affected by Amphan, we were deprived of any cyclone relief. If it weren't for the corrupt Gram Panchayat members in this panchayat, the situation could have been managed better. Although the government has provided all the necessary aid, the local members are only interested in their gain. Additionally, the population in our Gram Panchayat has been steadily increasing, which may explain why resource allocation is not being properly maintained. Proper allocation of resources after a cyclone is essential to ensure that no one is left without relief, as everyone in this area is in need. The distribution of aid should be fair and equal. There is no other alternative strategy at present.

Pregnant women in the Sundarbans region face significant challenges in finding suitable locations for childbirth, particularly in villages devastated by flooding. Many women have been forced to give birth in temporary shelters. The study highlights the inadequacies of the current healthcare system in meeting immediate health needs following climatic shocks. Other women recount harrowing experiences of miscarriage, attributing their losses to prolonged exposure to saline water, as one woman shared. She described suffering from persistent itching and a uterine infection, consequences she believes are linked to her environment. During her pregnancy, with no alternative livelihood options available, she resorted to collecting fingerlings while dressed in sarees. Tragically, during her fourth pregnancy, her amniotic fluid began to leak while she was out gleaning for fish. After being admitted to a local hospital for four days, she returned home and spent over a month bedridden, grappling with the aftermath of her ordeal. A similar study also talks about the impact of prolonged exposure to saline water during fishing activities on the reproductive health of women. (Shamsuddoha et al. 2024). This leads to an unhealthy pregnancy and in turn, causes severe damage to the health of the baby.

A group of researchers carried out a research project through the photovoice method and documented the voices of young mothers in the Indian Sundarbans. (Ghosh et al. 2016) says, "While engaging with the community level decision-makers, mothers countered traditional norms of a patriarchal structure and put forth their health demands. For most of the mothers, this project was the first time they used a digital camera, took part in a group meeting with fellow villagers, and traveled beyond their island. It was also the first time these mothers had engaged with the decision-makers, which was considered a man's job, to discuss community issues. When families objected to their daughter-in-law or wife participating in the photovoice project, fellow women came forward to resolve the issues. This solidarity reflects women's acceptance of photovoice as a tool to present their concerns." It is a powerful gender transformative research approach and brings to the forefront the leadership and community skills of women groups. It additionally brings liberation and autonomy in thinking and democratization of practices.

Similarly, residents of Durgagobindapur expressed concerns about the lack of adequate healthcare infrastructure during disasters. They said,

> Amidst both calamities, a dearth of adequate healthcare infrastructure plagued our surroundings. The establishment of medical camps was confined to distant areas, primarily accessible to the affluent class. Furthermore, the daunting challenge of combating COVID-19 coincided with the ravages of the Amphan cyclone. The government did not adequately plan for this situation, leading to a chaotic rush to shelters where large crowds gathered. The distribution of a mere single mask by the authorities was woefully insufficient. Many households included expectant mothers, yet they were deprived of the necessary support. In our locality, Asha workers (Asha 2011) visited after the cyclone and provided some limited assistance, but the medical camp offered no other substantial aid. Consequently, we often find ourselves relying on quack doctors

The Jharkhali Gram Panchayat stands out as a key entity in tackling the distinct and significant challenges presented by the region's environmental conditions. Situated in an area susceptible to cyclones, rising sea levels, and salinity intrusion, this local governance body assumes a central role in fostering resilience and sustainable development within the community. The Jharkhali Gram Panchayat plays a crucial role in confronting the environmental challenges that characterize its region, and one of its primary strategies involves the proactive organization of training sessions. These sessions play a crucial role in empowering the community with the knowledge and skills necessary to tackle the particular environmental challenges confronting the region. A notable aspect of the Gram Panchayat's approach is its continuous engagement of women in mangrove plantations.

This strategy acknowledges the vital contributions women can make to environmental conservation and the resilience of their communities. This provides a direction toward effective climate adaptation strategies led by women. Mangrove ecosystems are indispensable for safeguarding coastlines and acting as a natural shield against storm surges and erosion. By involving women in mangrove plantations, the Gram Panchayat not only contributes to the restoration of the local environment but also empowers women as active participants in sustainable development. One such woman, who is a mother of three from a small village on the fringes of the forest, now leads a group of 20 women who have planted thousands of mangrove saplings along the riverbanks.

> When I plant these mangroves, I feel like I am planting hope,

Her group not only earns wages for their labor but also gains a deep sense of pride in their contribution to protecting the Sundarbans. Training sessions serve as the foundation for these initiatives, providing women with the necessary knowledge about the ecological importance of mangroves, the techniques of plantation, and the long-term benefits for the community. This educational component ensures that participants understand the significance of their efforts and fosters a sense of ownership and responsibility for the environment. The continuous engagement of women in mangrove plantations brings several benefits. First and foremost, it contributes to the conservation and restoration of mangrove ecosystems, which are critical for the overall health of coastal regions. Mangroves act as nurseries for fish and other marine life, support biodiversity, and provide livelihood opportunities for local communities engaged in fishing. Moreover, involving women in such activities promotes gender inclusivity and empowerment. By breaking traditional gender roles and involving women in environmental initiatives, the Gram Panchayat enhances the resilience of the community from a social perspective. Women become key agents of change, not only in environmental conservation but also in fostering community cohesion and well-being. Additionally, the inclusion of women in mangrove plantations aligns with broader sustainable development goals (Fig. 6.2).

Beyond environmental benefits, this initiative contributes to the economic empowerment of women. It opens avenues for skill development, income generation, and community resilience. Women involved in mangrove plantations become stakeholders in the sustainable development of their community, creating a positive ripple effect on the overall social and economic fabric. These sessions serve as crucial platforms for imparting knowledge and skills that are essential for confronting the distinct environmental threats faced by the community. Disaster preparedness takes

**Fig. 6.2** Resilient roots: Mangrove plantation done by women villagers. *Source* Field 2023

precedence in these training programs, equipping residents with the tools and strategies needed to respond effectively to the impact of cyclones. This includes evacuation plans, emergency shelter management, and first-aid training, creating a community that is better prepared to navigate the challenges of climate disasters.

In addition to disaster preparedness, the Gram Panchayat places a strong emphasis on promoting sustainable agriculture practices. Given the susceptibility of the region to salinity intrusion, the traditional methods of agriculture may become unsustainable. Training sessions cover innovative and adaptive farming techniques that can mitigate the impact of salinity on arable land. This encompasses the implementation of salt-tolerant crop varieties, innovative water management techniques, and effective soil conservation practices.

By disseminating knowledge on sustainable agriculture, the Gram Panchayat contributes to the long-term food security and economic well-being of the community. Furthermore, recognizing the need for diversification of livelihoods, the local body actively supports training initiatives on alternative livelihood options. This entails pinpointing and fostering income-generating activities that are more resilient to the detrimental impacts of environmental conditions. Training sessions can focus on skill development in fields such as handicrafts, aquaculture, and eco-tourism, providing community members with alternative pathways to economic sustainability. By fostering diversification, the Jharkhali Gram Panchayat contributes to reducing the community's dependence on traditional livelihoods that are increasingly threatened by environmental factors.

The engagement of the Gram Panchayat in these training sessions is a testament to its commitment to the holistic development of the community. It recognizes that addressing environmental challenges requires a multifaceted approach that encompasses not only immediate responses to disasters but also long-term strategies for sustainable living. By empowering residents with knowledge and skills, the Gram

Panchayat positions the community as active participants in their resilience and development.

In summary, the Jharkhali Gram Panchayat plays a pivotal role in tackling the environmental hurdles encountered by the community. Utilizing focused training sessions equips residents with the expertise needed to combat specific challenges such as cyclones, increasing sea levels, and salinity intrusion. The prioritization of disaster readiness, sustainable farming practices, and alternative livelihood choices showcases a holistic and progressive strategy for community development. Through active participation in these endeavors, the Gram Panchayat not only addresses current issues but also establishes the groundwork for a resilient and sustainable future for the region's inhabitants.

During our fieldwork in Birajnagar village (Block Gosaba, District South Twenty-Four Parganas), we observed the shifting roles of local women who have become primary providers by taking on fishing responsibilities traditionally held by men. This change occurred after a devastating cyclone forced many men to leave the region in search of employment in other cities or countries. Once the waters stabilized, women began fishing to sustain their families. An elderly woman shared a vivid account of encountering a tiger during an eight-day fishing excursion, underscoring the inherent dangers of their work.

Fishing in this region is an arduous task, requiring women to spend extended periods in the forest with minimal resources. The physical demands and risks involved have led some families to encourage their sons to pursue opportunities in other parts of the country. Additionally, we spoke with local women who perform folk songs and dances. They described the dual burden they face performing for tourists in the evenings and farming during the day. With many men absent or not contributing due to alcohol abuse, these women shoulder the full responsibility of supporting their households. Despite the uncertainty and meagre earnings from tourism, they persevere, driven by necessity and the hope for more visitors to the Sundarbans.

## Discussion

From literature review and fieldwork, the gendered impacts of climate disasters can be categorized into multiple interconnected networks of social, ecological, and political processes. The cyclone's aftermath, particularly the increased salinity, makes the land unusable, leading to financial difficulties. This economic strain is compounded for women, especially if they are primary caregivers. The tension and conflicts within the joint family suggest that women may face increased domestic stress. The need to separate from the family could also lead to isolation and a loss of support networks. The inadequate shelter made of flimsy paper sheets highlights the precarious living conditions women and children may endure, affecting their safety and well-being. Borrowing money from lenders and the inability to repay adds financial pressure, which can be particularly burdensome for women managing household needs. With the husband working away temporarily, the woman is left to manage these challenges alone, increasing her burden and sense of isolation. The most neglected and ignored impacts are the health issues that women are facing constantly as an aftermath of the climate disasters. The inflammation of reproductive organs, infections of various kinds, poor reproductive health, impact on pregnancy, adverse effects on the health of children, lack of medical centers or hospitals to have an immediate response to these complications, and many other allied factors.

Multiple healthcare programs like the National Anti-Malaria Program, Revised National Tuberculosis Program, and Integrated Child Development Services program have brought an overall decrease in infant and maternal mortality rates. However, the overall efficiency is still very limited due to a variety of factors with the 1st being its difficult geographical position making the healthcare centers in cities inaccessible to the people, lack of trained staff, and better facilities in these healthcare centers. This is extremely important both from the point of view of disasters and everyday medical issues because of increased salinity. In case of disasters, the citizens are the first responders and it is very difficult to reach out to far-away medical centers amidst landfalls, floods, and cyclones. A lot of women living here face multiple gynecological issues due to increased salinity which are not addressed due to the absence of any specialized medical services (World Bank 2014).

In the 'The State Disaster Management Plan', there is a significant absence of gender-sensitive principles in disaster management—In the framework, sensitivity toward aspects of gender is absent. While their livelihood practices of collecting shrimps and crabs or working as agricultural laborers are affected in the post-disaster phase, they bear the additional burden of procuring fresh water for domestic use and taking care of the children in harsh conditions. On the other hand, the burden of transgenders to make ends meet after a disaster event becomes even more difficult in already difficult social circumstances where their access to work is limited due to prevailing social stigma. Hence, gender-sensitive policies in disaster management are crucial to reduce the risk of women and transgenders to disaster events. Women-Centric Disaster Reduction—Although the plan elaborates on the vulnerability of women to disaster events, there is no clear plan for disaster risk reduction to reduce the risks and vulnerability of women.

Stories of how the inhabitants of the Sundarbans are adapting and working toward sustainability on a community level are inspiring. A non-profit organization is collaborating with rural women to restore mangrove cover, with the dual goal of mitigating cyclone impacts and providing these women with a sustainable source of income as they plant and nurture mangrove saplings (Gupta 2021). Local women develop small forests by planting trees, which now shield river embankments. Women have planted mangroves like Avicenna (grey mangrove or white mangrove), which can withstand saline water because the salinity in their area is high (Aastha Ahuja 2022). These initiatives by women's groups are important for the future of the landscape. Preservation and restoration of mangrove cover are critical to all other allied issues that emerge from climate disasters. They said—

> We noticed that wherever we had planted mangroves, those embankments were okay. It came as a lesson on the importance of mangrove plantations. Personally, the devastation was an eye-opener for me too. We as an organisation started wondering what we were doing here. Unless we can make islands resilient whatever livelihood activities we do or houses we build, they will be destroyed if mangroves are not there. (Aastha Ahuja 2022)

This shows the importance of preserving the mangrove covers, restoring areas that are destroyed by severe cyclones, and nurturing the mangrove forests in the Sundarbans. Along with this kind of restoration, it is important to address key gender issues surrounding social justice, political economy, policy-oriented action, health crises, and employment. Apart from a few areas like Jharkhali, women in Sundarbans are absent from the climate disaster decision-making process. Denton (2002) Observing similar trends in Mozambique, "The climate debate has not sought to address the existing marginalization of women, nor their need to be integrated into environmental policies. Nor have the immediacy of global warming, the magnitude of such a phenomenon, or even extreme events such as the floods in Mozambique, succeeded in impressing on decision-makers the importance of placing women at the heart of sustainable development."

Moving forward it is critically important for the inclusion of women and other gender groups in the decision-making process relating to climate change and climate disasters. It is evident from the data coming out of the Jharkhali Gram Panchayat initiatives that placing women at the center of climate risk reduction has resulted in positive outcomes. The state must take decisive action to establish essential infrastructure in the Sundarbans, including access to fresh water, proper sanitation, reliable roads, medical facilities, and educational resources. Facilitating the active involvement of women in the decision-making processes related to the reconstruction and rehabilitation of climate change victims is crucial for addressing gender-specific challenges (Karmakar 2022). Adaptation projects should serve not only as tools for uncovering the various layers of vulnerability across different regions and populations but also as mechanisms to ensure that appropriate incentives are in place. This will empower both women and men to leverage their unique strengths and work together toward building a sustainable future. (Denton 2002).

To grasp the intricate dynamics at play, it is essential to integrate insights from academic institutions, state agencies, civil society, and local practitioners within affected communities. This collaborative approach will foster interaction and the exchange of expertise, ultimately enhancing the planning and execution of grassroots initiatives by communities directly facing the challenges of climate change. (Das and Hazra 2020).

## Conclusions

Here we can again go back to the beginning of this chapter and think about the philosophical background of this exposition. The nonlinearity of climate disaster issues faced by women and other gender groups in the Sundarbans and other parts of the world need careful mapping and detailed network-based assemblage theory (Deleuze and Guattari 2001) approach for better understanding. This level of detailed understanding and mapping will lead to direct action-based research and execution of plans within specific areas. The network-based approach will also guide us to have a common yet differentiated approach for communities based on social position, gender, education, and employment opportunities.

Through our extensive fieldwork, we have observed a significant and steadily growing movement in the Sundarbans, where women are taking the lead in developing innovative climate change adaptation strategies to address ecological and socio-economic challenges. This emergence highlights the resilience and resourcefulness of women within these vulnerable landscapes, who are often at the forefront of dealing with the impacts of climate change. Their strategies, grounded in lived experience and local knowledge, offer a valuable foundation for fostering long-term solutions. However, to ensure these efforts are effective and impactful, it is essential to channel and refine them into structured, sustainable processes that can support community-wide resilience.

To achieve this, integrating diverse technological and methodological approaches is crucial. Tools such as the photovoice method, audio documentation, and other participatory techniques can provide a platform for amplifying marginalized voices, particularly those of women. In parallel, initiatives like mangrove restoration, health-care improvements, and advanced agricultural technologies should be designed with a focus on promoting gender equality. By creating inclusive frameworks, these tools and methods can ensure women are actively involved in decision-making processes, fostering collaboration and equity. Moreover, embedding these approaches within action-based research can bridge the gap between knowledge and practice, driving tangible, community-led change in local adaptation plans.

**Acknowledgements** All the authors gratefully acknowledge funding received from the Indian Council of Social Science Research (ICSSR) via grant no. 02/107/GN/2021-22/ICSSR/MJ/RP. Amrita Sen acknowledges a Junior Core Fellowship (academic year 2024/25) awarded by the Institute for Advanced Study, Central European University, Budapest.

## References

Aastha A (2022) The resilient women of sundarbans are fighting back cyclones by planting mangroves. NDTV. Retrieved from https://swachhindia.ndtv.com/the-resilient-women-of-sundarbans-are-fighting-backcyclones-by-planting-mangroves-69535/

Asha (2011) Association for social and humanitarian action. https://www.asha4change.org/. https://www.asha4change.org/. Accessed 25 Jan 2015

Awiti AO (2022) Climate change and gender in Africa: a review of impact and gender-responsive solutions. Front Climate. 2:1–14. https://www.frontiersin.org/journals/climate/articles/10.3389/fclim.2022.895950/full

Das S, Hazra S (2020) Trapped or Resettled: coastal communities in the Sundarbans Delta, India. Forced Migrat Rev 16:15–17. https://ora.ox.ac.uk/objects/uuid:dc115d95-e756-40b1-99c4-6339295fb9b3/files/sxp68kh35f

Deleuze G, Guattari F (2001) Thousand plateaus. Bloomsbury, Basingstoke

Denton F (2002) Climate change vulnerability, impacts, and adaptation: why does gender matter?. Gender and Developm 17:10–20. https://www.tandfonline.com/doi/epdf/10.1080/13552070215903?needAccess=true

Fothergil A (1996) Gender, risk, and disaster. Int J Mass Emergencies and Disasters 33–56. https://journals.sagepub.com/doi/10.1177/028072709601400103

Ghosh U, Bose S (2024) Bringing together voices to address climate change uncertainty in the Indian Sundarbans. In: IDS policy briefings. The Institute of Development Studies and Partner Organisations. https://opendocs.ids.ac.uk/articles/report/Bringing_Together_Voices_to_Address_Climate_Change_Uncertainty_in_the_Indian_Sundarbans/26440162. Accessed 25 Jan 2025

Ghosh U, Bose S, Bramhachari R, Mandal S (2016) Expressing collective voices on children's health: Photovoice exploration with mothers of young children from the Indian Sundarbans. BMC Health Serv Res 128:119–130. https://bmchealthservres.biomedcentral.com/articles/10.1186/s12913-016-1866-8

Ghosh U, Bose S, Brahmachari R (2018) Living on the edge: climate change and uncertainty in the Indian Sundarbans. ResearchGate 1–60. https://www.researchgate.net/publication/323573020_Sundarbans_Living_on_the_Edge_Climate_Change_and_Uncertainty_in_the_Indian_Sundarbans

Gupta S (2021) Twin benefits of mangrove plantation in Sundarbans: building resilience to climate change and supporting rural livelihood. Retrieved from https://en.gaonconnection.com/reportage/sundarbans-west-bengal-mangrovesenvironment-trees-plantation-women-empowerment-livelihood-climate-changecyclones-amphan-yaas-42523/

IPCC (2019) IPCC special report on the ocean and cryosphere in a changing climate. Cambridge University Press. https://www.ipcc.ch/site/assets/uploads/sites/3/2019/12/SROCC_FullReport_FINAL.pdf.

IPCC (2022) Sea level rise and implications for low-lying islands, coasts and communities. In: IPCC special report on the ocean and cryosphere in a changing climate. https://www.ipcc.ch/site/assets/uploads/sites/3/2022/03/06_SROCC_Ch04_FINAL.pdf. Accessed 25 Jan 2025

IPCC, CCSR (2023) Climate change 2023: synthesis report. https://www.ipcc.ch/report/ar6/syr/downloads/report/IPCC_AR6_SYR_FullVolume.pdf. Accessed 25 Jan 2025

Jena M (2018) Climate change is making sea water in the Sunderbans Saltier, putting women at risk. Scroll. https://scroll.in/article/900558/climate-change-is-making-sea-water-in-the-sunderbans-saltier-putting-women-at-risk (in press)

Karmakar S (2022) Climate change and everyday-life: negotiation of women in the Sundarbans, India. Man, Environment and Society 3(1):49–64. https://doi.org/10.47509/MES.2022.v03i01.04

Nelson C, Grossberg L (1988) Marxism and the interpretation of culture. Macmillan Education, Basingstoke, p 288

O'Brien K, Hayward B, Berkes F (2009) Rethinking social contracts: building resilience in a changing climate. Ecology and Society. Available via ResearchGate. https://www.researchgate.net/publication/42766221_Rethinking_Social_Contracts_Building_Resilience_in_a_Changing_Climate. Accessed 25 Jan 2025

Rogers KG, Goodbred SL (2014) The Sundarbans and Bengal Delta: the World's largest tidal mangrove and delta system. In: Kale S Vishwas (ed) landscapes and landforms of India. World Geomorphological Landscapes. Springer, Dordrecht, pp 181–187

Shamsuddoha M et al. (2024) Impacts of climate change-induced natural hazards on women and their human rights implications: a study in the Southwest Coast of Bangladesh. J Migrat Health. https://www.sciencedirect.com/science/article/pii/S2666623524000114

Spivak GC (1988) Can the Subaltern Speak?. Die Philosophin

Wamukonya N, Rukato H (2001) Climate change implications for Southern Africa: a gendered perspective. Minerals and Energy Policy Centre. https://www.energia.org/assets/2015/06/16-Climate-change-implications-for-Southern-Africa.pdf

World Bank (2014) Building resilience for sustainable development of the Sundarbans: World bank strategy report. https://documents1.worldbank.org/curated/en/879351468259748207/pdf/880610REVISED00ns000Strategy0Report.pdf. Accessed 25 Jan 2025

# Chapter 7
# The Gender Divide: Factors Amplifying Women's Vulnerability During Extreme Events in Bangladesh

**Iffat Ara Jasmin, Atia Sanjida Sushoma, Imran Hosen, and Ananda Kumar Biswas**

**Abstract** Climate change fuels most of Bangladesh's extreme events, but its effects on women are more severe than those on men. This paper reviews the gender dimensions of disaster impacts, focusing on enhanced challenges women face in light of socio-economic and cultural constraints linked to structural inequalities in a patriarchal societal framework. To analyze the socio-economic hardship and cultural barriers, the paper uses feminist political ecology (FPE) and constructivism theory to understand how social constructs and power dynamics intensify the vulnerabilities of women. This study adopted the qualitative phenomenological research approach and data was gathered from women's in-depth interviews and focus group discussions in the northern region of Bangladesh, particularly from five districts (Tangail, Gaibandha, Netrokona, Sirajganj, and Kurigram). The study found that the gender division of labour, the intersection of public–private dichotomy, and societal conflicts due to climate-induced migration exaggerate the socio-economic vulnerabilities of women during extreme events. Regarding cultural barriers, gender stereotypes and social dogmas heighten their struggles leading to malnutrition and health risks curtailing decision-making powers, thus denting disaster preparedness. This paper, therefore, invokes gender-sensitive adaptation practices that acknowledge climate change hits men and women differently due to entrenched gender roles and societal structure. Placing women's experiences at the centre, this paper underlines the urgent

---

I. A. Jasmin · A. S. Sushoma · I. Hosen
Department of International Relations, University of Dhaka, Dhaka, Bangladesh
e-mail: iffat@du.ac.bd

A. S. Sushoma
e-mail: sushoma@du.ac.bd

I. Hosen
e-mail: mdimran-2020512459@ir.du.ac.bd

A. K. Biswas (✉)
Development Studies Discipline, Khulna University, Khulna, Bangladesh
e-mail: Ananda.ku.mds18@gmail.com



A. Sarkar and N. Bandyopadhyay (eds.), *Gender-Transformative Approaches for Climate Change Adaptation*, Springer Climate,
https://doi.org/10.1007/978-3-031-90236-9_7

need for gender-transformative approaches by challenging and changing the underlying social norms and power dynamics that perpetuate structural inequalities. We can foster resilient mechanism of women through a gender-transformative approach that paves the way for risk reduction ensuring gender equality and community development.

**Keywords** Vulnerability · Extreme events · Climate change · Gender roles · Constructivism · Gender · Bangladesh

## Introduction

Climate change, which is a "non-traditional" security concern, is one of the most important issues for national security and socio-economic growth in the twenty-first century. In 1991, the Intergovernmental Panel on Climate Change (IPCC) released a particularly alarming report following scientific evidence to show that rising emissions were causing global warming and the subsequent effects of climate change. Due to its effects on the world's oceans and atmosphere, climate change is a major cause of many extreme events. Natural processes in the atmosphere, hydrosphere, biosphere, or geosphere are the cause of these occurrences (Glade and Alexander 2013). Such disasters pose a significant threat to people's lives and livelihoods exacerbated by anthropogenic causes. Damage, injury, loss, destruction, and disruption are the outcomes of disasters such as floods, river erosion, landslides, cyclones, tropical storms, droughts, wildfires, and others (Fatema et al. 2023). The interplay between social and biophysical determinants of health and welfare, social roles and responsibilities, economic status, education level, societal structure, coping mechanisms, disaster preparedness, and many other factors leads to gender vulnerability, concerning disasters intensified by climate change. The International Institute for Environment and Development (IIED) report highlights that, due to its threat multiplier effect, climate change would mostly affect vulnerable regions and vulnerable populations (Bari et al. 2024).

Bangladesh is known for its rich culture, history, and growing economy. However, climate change is making things very hard for the country. Bangladesh's location in a disaster-prone area significantly increases the likelihood of floods, droughts, river erosions, and storms. Moreover, low sea levels, density of population, poverty, and big floodplains exacerbate the issue. The country faced severe catastrophes such as floods in 1988 and 1998, cyclone in 1991, Aila in 2005, Sidr in 2007, Cyclone Mora in 2017, Cyclone Fani in 2019, Amphan in 2020, Cyclone Remal in 2024, which damaged people's livelihood drastically. Studies have revealed that the floods and droughts in Bangladesh between 1998 and 2014 resulted in an estimated loss of 3 billion USD (Ozaki 2016). Climate change may force 41.8 million people to relocate from shoreline areas due to the increasing salinity in ground and surface water, which is a result of rising sea levels (Abdullah et al. 2022). The country has witnessed 185 extreme weather events between the years 2000 and 2019, making it the seventh most vulnerable country in terms of climate (Haque and Yousuf 2024). The future is even more grim as it is forecasted that tropical cyclones, storm surges, droughts, and flooding will all become more often and severe. The IPCC's fifth assessment report reveals that Bangladesh is extremely vulnerable to climate change, especially sea level rise, which could affect up to 18 million people by 2050 (Chowdhury et al. 2024).

These climate change-related events disproportionately affect women, who comprise nearly half of the country's population. Most of the research statistics showed the number of deaths of women due to extreme events was higher than men, for example, in 1991, during the cyclone and floods in Bangladesh, of the 140,000 people who died, 90% were women (Ikeda 1995). The World Economic Forum currently ranks Bangladesh 99th out of 146th in the 2024 global gender gap index (Samin and Nahr 2024). Women suffer greatly due to their multiple roles, including homemaker, food provider, health protector, economic agent, and caretaker. The patriarchal societal structure consistently disregards women's opportunities and rights, controlling them in all spheres of life both public and private (Chipuriro 2023). Disasters exacerbate the marginalization of women's rights in a society where hunger, poverty, and exclusion already exist. Due to socio-economic and cultural constraints, women often become the primary victims of these catastrophes enduring a life of misery and inhumanity while receiving no special attention or provisions to guarantee their rights and security.

A review of some existing literature reveals that women in Bangladesh face heightened vulnerabilities during extreme events due to a combination of pre-existing gender relations, socio-economic hardships, and cultural constraints. They argued that extended gender roles that are socially constructed contribute to women's poverty and suffering which are exacerbated during disasters by resource constraints, intensified competition, and additional burdens ultimately leading to increased discrimination against women (Ashraf and Azad 2015; Arora-Jonsson 2011). But still, we have limited knowledge of power dynamics, their daily lives, and the complex social and economic problems they face. Moreover, most of the literature pays little attention to gender stereotypes and social dogmas. So, a critical view is needed. This paper analyzes the factors behind the vulnerabilities of women during disasters in terms of socio-economic and cultural aspects. The study also addresses structural inequalities through the lenses of Feminist Political Ecology (FPE) and Constructivism theories, both offering distinct yet interrelated perspectives.

## Materials and Methods

This study has adopted the qualitative phenomenological approach to understand the vulnerabilities of women in disasters. It seeks to understand the gender perspectives on the vulnerabilities of women during climate-related disasters and the effects of climate change. To get an in-depth view of this area, the qualitative research method is the more appropriate approach. Additionally, qualitative research offers thorough explanations of topics of interest (Forbes 2015). Through participant descriptions, phenomenological research examines the human condition (Nieswiadomy and Bailey 2018).

## *Theoretical Framework*

Our qualitative research is based on two theoretical perspectives--FPE and constructivism theories. The empirical analysis from the qualitative survey is examined to understand the validity of these theories in practical life. It offers a nuanced understanding of how gender, social norms, power structures, and environmental factors interact to shape women's vulnerability and resilience in the face of extreme events in Bangladesh. These theories along with field surveys provide a contextual interpretation that goes beyond simply describing a situation revealing the root causes of vulnerability and leading towards more effective disaster preparedness and coping mechanisms for women.

### **Constructivism**

Constructivism is a theory that centres on norms, identities, and social constructs that shape human behaviour and societal results. Unlike those theories that focus on solely material factors, constructivism usually emphasizes the role of socially constructed norms, identities, and power relations in shaping vulnerability. Constructivism posits that active processes create and produce vulnerability, rather than it being a pre-existing condition. This framework is particularly useful in understanding the multi-faceted nature of women's vulnerability in Bangladesh during disasters. Applying constructivism, we can explain women's vulnerability in multiple ways:

Firstly, constructivism highlights how norms assign gender roles to women and men. In Bangladesh, the restriction of women to domestic work limits their access to resources, decision-making, and information. This restricts their ability to prepare, recover, and respond to disasters. For example, women might lack the authority to decide evacuation plans and warning systems, exposing them to harm during disasters. Moreover, women are expected to prioritize children and the elderly over them during crises that put them at subsequently greater risk. These roles are socially constructed and reinforced through cultural practices, religious beliefs, and societal expectations.

Secondly, this showcases how societal rules allow unequal access to power and resources. Women in Bangladesh face huge disparities in land ownership, economic opportunities, and political representation (Nazneen 2024). They participate in low-paying or unpaid domestic work, leaving them with limited resources to prepare for or recover from disasters (Sultana 2022). Societal norms that prioritize male employment and property ownership further fuel this disparity. Such marginalization excludes women from the important processes of decision-making at the community and national levels, which in turn increases their vulnerability.

Thirdly, constructivism focuses on the connectivity of the social categories and how they influence vulnerability. Bangladeshi women face multiple forms of exclusion due to caste, ethnicity, class, and location, which exacerbates their vulnerability during disasters (Jobaer et al. 2024). Poor, minority, and working-class women will have different challenges than women from affluent backgrounds, even though both are indeed women (Roy 2020). Constructivism helps us understand how such interconnected identities mould individual experiences of vulnerability.

Fourthly, it further focuses on how societal norms and power impact the process of recovery. Usually, we undervalue women's contributions to reconstruction after disasters, as they often shoulder the burdens of domestic labour and childcare while participating in rebuilding. This further heightens women's vulnerability during future disasters, as the uneven distribution of labour impedes their ability to recover. Furthermore, social beliefs that disregard women's capabilities can lead to a lack of consideration for their health needs during disaster preparedness. Constructivist theory helps explore how social beliefs shape the policies and practices affecting women's physical security.

Finally, constructivist analysis reveals that the vulnerability of women during calamities is not a natural phenomenon but rather a consequence of socially constructed gender roles, unequal power relations, and practices of exclusion. Understanding these processes is critical for the development of effective and equitable preparedness during disasters and response strategies that truly ensure women's participation, emancipation, and promotion of their resilience.

## Feminist Political Ecology (FPE)

FPE explores the interplay between gender, environment, and social and political factors. It combines feminism and political–ecological ideas. The primary concept that revolves around political ecology, examines how social, political, and economic factors influence environmental issues, with women playing a significant role (Rocheleau et al. 1996). It shows that environmental change, such as climate change, is not gender neutral, but it will affect different groups differently because of the roles, responsibilities, and power dynamics that men and women work within. FPE demonstrates the relationship between environmental and gender justice through the lens of ecofeminism. Bina Agarwal states that men and women experience environmental impacts differently due to societal roles, unequal access to resources, and decision-making power disparities. According to her, women and children are particularly vulnerable to the effects of climate change due to three main factors: pre-existing gender divisions of labour, gender-based resource access inequalities, and limited access to productive resources like property rights and employment (Agarwal 1992). These factors hinder their ability to recover from environmental changes.

This theory includes multi-scale and intersectional analysis. It recognizes diversity not only between men and women but also among women themselves. This approach helps capture the complexity of gender, ecological changes, and cultural dimensions. The FPE approach calls for inclusive policies and programmes that consider women's voices and address their unique challenges.

## *Study Area*

The study area was the northern part of Bangladesh. We selected five different districts for this study. The districts were Tangail, Gaibandha, Netrokona, Sirajganj, and Kurigram. The rural areas were selected for this study. The research site area was particularly susceptible to floods, river erosion, droughts, and heat waves, with women particularly vulnerable. The study considers the risk level indices of the districts to disasters as calculated by Lewis and Hasan (2022). Five districts of Bangladesh were selected for this study purposively where risk levels were high. For example, the risk level of Tangail was calculated at 5.4, Gaibandha at 5.7, Netrokona at 5.6, Sirajganj at 5.8, and Kurigram at 5.8 level (Fig. 7.1).

## *Sampling and Data Collection*

The study used both primary and secondary data to derive its findings. Secondary data was collected from published books, articles, journals, reports, conference papers, and other online sources. We used the study of the Risk Index for Bangladesh, which was jointly conducted by Govt. Bangladesh and the United Nations (Lewis and Hasan 2022).

For the collection of the primary data, this study employed purposive sampling to select respondents for the qualitative survey. To ensure study quality, we collected data using in-depth interviews, focus group discussions (FGDs), and participant observation, triangulating data sources. We selected semi-structured in-depth interviews (IDI) to collect data for this study. All interviews lasted 30–40 min. Based on the estimation from our pilot in each district, around eight to twelve participants participated in each FGD session. All FGDs lasted an hour and followed ethical guidelines. We also conducted participant observations over four months from April to July 2024. To keep homogeneity among the respondents, the optimal size was maintained. We collected participants' observations and quotes during FGDs.

Researchers conducted interviews and FGDs with local ladies in Bengali and transcribed them into English. IIDs were also conducted in Bangla following all the guidelines of research ethics. Researchers transcribe Bangla into English before analyzing the results. All data was transferred anonymously and was marked using different codes. We used NVivo-14, for analyzing the qualitative data.

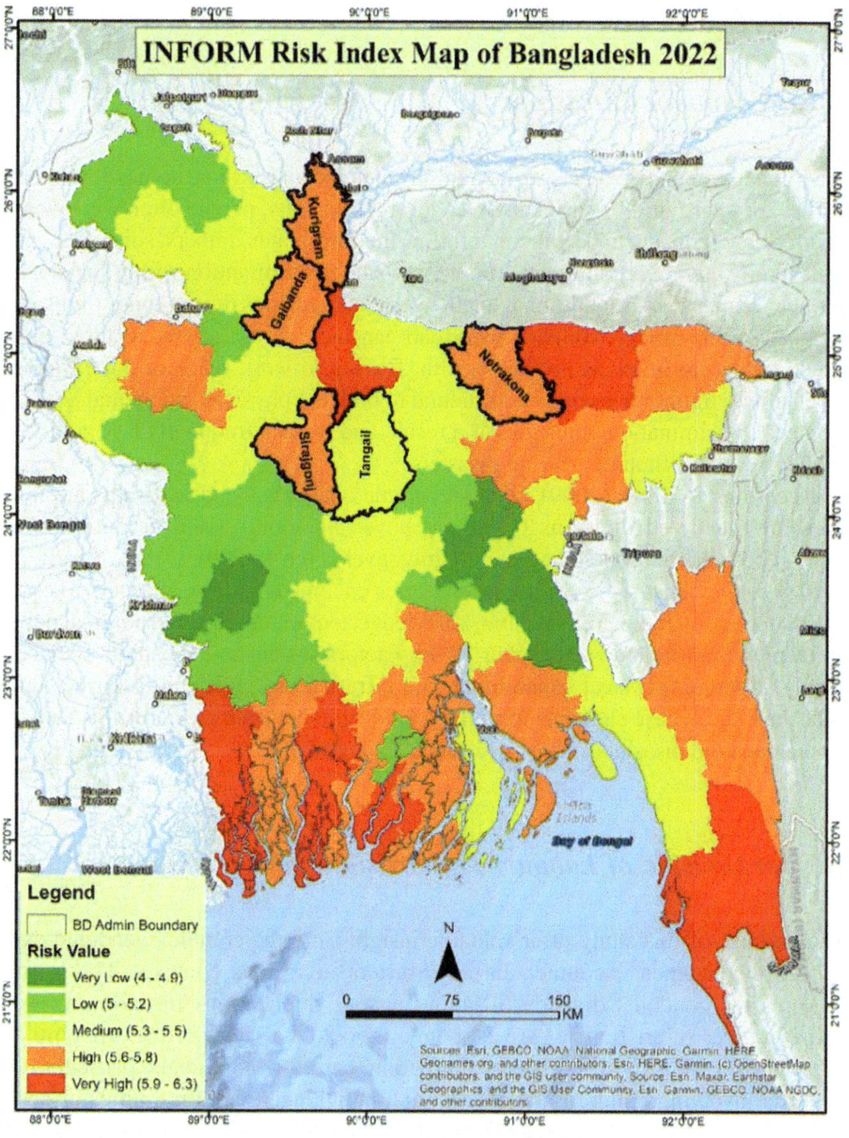


**Fig. 7.1** Sub-national risk index of Bangladesh, 2022. 2022 *Source* Bangladesh Inform Sub-National Risk Index 2022 (Lewis and Hasan)

## Results and Discussions

### Socio-Economic Hardships

Traditionally, society expects women to stay at home and assume domestic responsibilities for their family's survival. In a patriarchal society, people often view women as inferior to men and expect them to follow their husbands, fathers, sons, or as other male family members command. This exacerbates discrimination within society, and women frequently face additional challenges and hardships during times of disaster. When disasters strike, women's workload can increase up to two to three times, which puts a lot of stress on their health. Having to work twice or three times as much after a traumatic event is detrimental to both the physical and mental health of women. Discrimination intensifies as a result, and women frequently face additional challenges and difficulties during disasters.

According to Terry (2009), socio-economic challenges and hardships are influenced by gendered divisions of labour, physical mobility, and decision-making roles at both the household and community levels (Terry 2009). Cutter, Boruff, and Shirley suggest that socio-economic struggles encompass a lack of access to essential resources, including information, knowledge, technology, limited political power, and representation (Dow and Cutter 1998). Factors such as social capital, social and family structure, physical capabilities, and infrastructure availability further influence the issue. These elements form the relationship between the hardships faced by women and the disasters caused by climate change.

### Gender Division of Labour: The Invisible Care Economy

The findings of this study offer valuable insights into the complex gendered experiences that women encounter during disasters in various contexts. This section explores the gendered division of labour and its implications for women's well-being in times of crisis. The care economy refers to the sector of the economy that encompasses all activities related to providing care for individuals, particularly in contexts involving children, the elderly, and individuals with disabilities or health needs. This includes both paid and unpaid labour, encompassing roles such as childcare workers, nurses, home healthcare aides, and domestic workers, as well as unpaid care provided by family members and community members. This care economy is heavily influenced by the gender division of labour.

It was asserted that the assignment of work to men or women shapes societal perceptions of gendered acts (van der Horst 2014). This division and the associated gender norms affect individuals' access to resources, power, decision-making processes, and social services (Siraw et al. 2023). The gender division of labour often contributes to sustained gender inequality, as a study by Matud and the World Health

Organization illuminates (Matud 2017). Women's undervalued and often unrecognized domestic labour is a significant aspect of the invisible care economy. Anthropologist Ester Boserup observed that women carry out around 80 per cent of agricultural tasks, a significant portion of which remains undervalued (Jain and Zarghamee 2023). In feminist economics, the institutional rules, norms, and practices that govern task allocation between genders contribute to this division of labour. The separation of unpaid and paid work between men and women in both private and public spheres most prominently reflects this division (Pérez Fragoso 2022).

Women's roles are typically associated with domestic and family responsibilities, while men engage in public duties as citizens, soldiers, and workers. The aforementioned approach enhances the public roles of men while weakening the socially constructed domestic roles of women. Such facts exacerbate the situation for women, particularly during disasters. The study's results also highlight that women's workload increases during calamitous times, which disables them from participating in paid work. This is a prime example of a pattern that has its roots in patriarchal society. Resources are scarce during calamities, making it difficult to complete all household chores. This puts extra pressure on women, who are usually the main providers. Women usually do all the work around the house, including getting water and fuel, taking care of family members, and handling daily domestic chores. Our findings also show that many mothers-in-law don't help, leaving all the housework to the daughters-in-law. This indicates that even within families, age and hierarchical family relations determine women's burden of domestic chores in addition to the traditional male dominance.

Data from FGDs reveal that many women are restricted from working outside their homes during disasters. Women engaged in various occupations ranging from garment factory workers to labourers before such events were compelled to cease work during the disasters due to prohibitions imposed by family members, especially husbands. During these periods, men often found themselves jobless and spent their time searching for work, thereby increasing the domestic responsibilities assigned to women such as collecting water, repairing homes, caring for children, and preparing meals. Very few women mentioned being permitted to work outside, albeit under the condition of wearing a veil at all times, without which her husband would not allow her to leave. A lot of women have shared similar opinions of their husbands and mothers-in-law that stated that women are only suited for domestic work and criticisms flowed for the dreams of doing "men's work." Such dynamics reinforce the traditional gender role of society, which restricts women to domestic labour over other paid jobs. Some of the women stated as follows:

> I work as a maid on normal days. But during floods, as there was huge work, my husband did not allow me to go outside. He also told me that from now on my work was only to do household chores and I should not cross my boundaries as I am a woman. As I lost work and was burdened with extra household chores at that time, I even did not eat one time a day properly, and could not buy anything according to my choice.

> Due to floods, there was a shortage of food, work, and water and prices of foods were becoming high. So, I want to work outside to share household burdens. But my husband and mother-in-law never allowed me to work outside and also threatened that if I continued to

> ask then my husband would give me divorce. They also said that being a woman, my only work is to raise a child and to do domestic work, nothing else.

Researchers observed that women have to bear a "triple role" during disasters, encompassing reproductive, productive, and community-managing responsibilities (Khan et al. 2023). In the FGDs, women participants revealed that their tasks included child-rearing, household cleaning, washing utensils and clothes, cooking, and occasionally engaging in agricultural activities like harvesting paddy, jute, and various vegetables to prevent crop destruction from floodwaters. The findings further revealed that women in fostering cohesion and social food security often extend their reproductive duties even at the cost of their time and energy. These increased demands often go unrecognized, reinforcing gender disparities and limiting women's well-being.

In terms of productive roles, women often face barriers to outside work, leading them to engage in low-wage activities such as stitching Kantha (traditional quilts), weaving mats, creating handicrafts, making stoves, and other tasks that impose additional burdens on them. Consequently, we can label this productive role as a "double-edged sword" that provides necessary economic assistance while concurrently exacerbating cycles of financial insecurity and vulnerability of women in disaster-prone regions. People often undervalue and view women's community management as voluntary. The notable absence of men in such roles widens the gender division of labour while placing the responsibility of survival solely on women who lack resources as well. Participants stated for instance:

> Before floods, I worked as a maid in a house in our village and helped them by cleaning their homes, and utensils, by sashing spice, and many more. But after the floods, due to household work burdens, I had to quit this job. But our conditions were getting worse day by day. So, I started to stitch kantha (traditional quilts) of other people. Each kantha took me around 7–10 days to complete and they gave me 300-400 taka per kantha (approximately \$3–4 USD) per quilt. Along with this, I had to cook food, take care of my children so that they could not fall into the water, had to collect fuel, and had to take care of elderly people. After the floods, I had to clean my house and also had to repair my home with additional soil. And, I alone had to do all this work and manage the whole family.

> Due to river erosion, we lost our home and wealth and had to rebuild our house. A lot of things were needed in the new home such as utensils, furniture, mats to sleep on, and many more. So, I started to work and my work was to cut soil from the land. The owner paid me 100–120 taka (equivalent to \$1.08–\$1.25) per day. After returning home, again I had to cook food, wash clothes, clean the home, take care of children and adults, collect water, and many more. I had no time for myself. All my time and energy were spent on my family's care.

Therefore, the invisible care economy, which upholds societal norms and perpetuates women's reliance on men, marginalizes women's contributions. As such, the gendered division of labour not only increases the hardships that women face during disasters but also underscores the pressing need for recognition and restructuring of gender roles within both domestic and public spheres to mitigate these vulnerabilities.

## *The Intersection of Public–Private Dichotomy with Disaster Vulnerability*

The dichotomy of public and private means the conceptual and pragmatic division between the public sector that includes governance, formal institutions, and paid labour while the home, family life, and unpaid labour, particularly caregiving and domestic work, comprise the private sphere. This disparity has a profound impact on gender roles, economic structures, and social dynamics. Forced to fend off paid labour, women automatically shoulder more responsibility including preparing food, collecting water, and caring for family members. This burden affects women, especially young married women, leading to increased physical and mental exhaustion.

Responses from women indicate that their autonomy and financial independence are significantly curtailed when external work is prohibited. Many acknowledged that working allowed them to make purchasing decisions and meet their family's needs, contributing to their sense of self-worth and independence. Conversely, when excluded from the workforce, they often experience social exclusion and marginalization from the decision-making process which diminishes their self-respect and self-reliance exacerbating vulnerabilities during disasters. Most of the women reported that their health, happiness, needs, and well-being were ignored by family members. None of the family members asked the women if they were starving. Women who are financially stable and occupy advanced socio-economic positions often experience fewer hardships and more support. For instance, a woman recounted how her cousin, who owned a house, held a respectable family position, received assistance from her mother-in-law during a crisis, and had a voice in family matters, thereby mitigating the issues. In most cases, the absence of female representation in decision-making exacerbates vulnerabilities.

Additionally, the exclusion of men from unpaid domestic work highlights significant structural barriers to women's resilience. This lack of shared responsibility reflects cultural and societal norms that dictate gender roles rather than individual choices. As disasters intensify household workloads, but societal expectations discourage men from participating in domestic tasks, often perceived as feminine. Respondents indicated that even those men who wish to help are frequently dissuaded by family and community pressures, reinforcing the notion that participating in household labour undermines their masculinity. Consequently, these entrenched social expectations further enhance women's vulnerability during disasters, limiting their capacity to cope and recover effectively.

## *Climate-Induced Migration and Vulnerabilities*

Migration due to disasters, such as floods, river erosion, drought, and cyclones, can create vulnerabilities for women by intensifying conflicts. When people migrate from disaster-stricken areas in search of natural resources like food, water, and fuel, they often move to destinations that are already resource-scarce. This migration can strain the limited resources of the host communities, leading to competition and potential conflicts between the migrants and the residents.

In FGDs, women reported frequent experiences of hostility and discrimination when trying to access resources in their new communities. They often have to wait for the residents to address their needs first and face disputes over water access, like waiting hours to fill their jugs. While local women might secure water for multiple jugs, the newcomers are often restricted to filling only one. These women endure rough treatment and humiliation, forced to silently accept the situation as they are outsiders in the destination place. Even those who remain in their original communities also experience similar conflicts over scarce resources.

Conflicts also arise among people who live in the same place but migrated from different communities or places. Women often face power dynamics where influential local figures control access to necessities leading to disputes and further marginalization. As women are the prime agents to manage household activities, so, these types of conflicts caused by disasters exacerbate the vulnerability of women curtailing social harmony and cohesion.

Some of the participants stated as follows:

> Due to river erosion, we lost our home and took shelter in a neighbouring village at a school. But the local people did not cooperate with us. The local women quarrelled with us regarding filling the jug from the river and they never permitted us to fill the jug first. After the local women finished filling their water pots, then we filled our pot. Sometimes we had to wait 2 or 3 hours to fill the pot because after they filled the pot and bath, then we were permitted to fill our jug. Most of the time the local women behaved very roughly with us, taunted and humiliated us but as it is not our village, so we had to listen to their taunts silently.

> Due to river erosion, we lost our home and migrated to my cousin's house and later migrated to the neighbouring village at a school. As we lost our wealth and resources, so we were very helpless at that time. My husband also was looking for a job. I too started to make a stove with soil. I made this stove to sell and per piece, I collected 40-50 taka. But the other women, who came from other villages, were not happy about my work. Sometimes, she stole my stove and when I claimed that it was my stove then she told me that it was her stove and she made it. Then she quarrelled with me. Thus, conflict occurred often between me and her. Fear of stealing my stove, I could not sleep at night peacefully and most of the time I woke up at night.

> Though there were floods, we did not migrate but faced problems and conflicts in our village. There was a shortage of food, fresh water, fuel, and many more in our village. I along with some women went to a distant place to collect fresh water. However, some influential women from our village did not permit us to fill our water pot first. They filled their pots first which would be more than three to four in number. They also took their children to fill many water pots. But when we took our children too to fill the pot, they refused and did not permit us to fill the jug. They quarrelled with us and taunted us. So, because of their fear, we could not fill more than 1 jug. When our reserved water ended then we had to come again to fill our water pot. We would need to make repeated rounds to fetch water each day.

## Cultural Barriers or Challenges

### *Socializing and Social Expectations*

In patriarchal societies like Bangladesh, women are indoctrinated from an early age to be courteous, kind, and nurturing as well as to be pleasant to everyone. On the other hand, society expects boys to be powerful, protect women, command, and dominate others. Culturally, women are also taught to be modest and ashamed of their bodies from a young age, but men are not subjected to the same expectations. As a result, society treats men differently than women. Adolescent girls who are not married are supposed to preserve their "purity" and "virginity" until marriage. Contact is often restricted to family and close friends. Therefore, during disasters, women (especially the poor) face severe hardships due to cultural constraints (Alam et al. 2023). These barriers consequently limit women's emancipation and exacerbate their hardships during disasters such as floods, cyclones, and river erosion. Society also confines women to their homes, causing them even more difficulties during times of crisis compared to men. Moreover, the challenges women face during these disasters are further intensified by socially constructed gender roles and norms, superstitions, and stereotypes.

### *Gender Stereotypes and Social Norms*

Gender stereotypes provide simplistic perspectives on the expected gender roles, characteristics, and behaviours of individuals. In a traditional sense, societal norms state that women should marry, have children, and prioritize their family over themselves. People firmly expect women to embody roles such as compassion, nurturing, and beauty. Conversely, men, perceived as providers, exhibit assertiveness, competitiveness, emotional restraint, and even anticipate initiating sexual relations (Ellemers 2018). Social dogma refers to a set of beliefs uncritically accepted by a group, often perpetuating these stereotypes (Brauer et al. 2001). Such stereotypes and dogmas are inculcated from childhood, shaping women's identities and roles which negatively affects women during disasters.

Studies from fieldwork, especially FGDs, reveal that women learn early on not to eat before their husbands, to consume less than boys, and to save fresh and healthy foods for their sons and husbands. Participants highlighted that during disasters, necessities like food and water become scarce. In these situations, women prioritize feeding the males first, leaving themselves to eat the leftovers. Moreover, women frequently survive on limited resources like water, parched rice, and, at times, gruel rice. In times of food scarcity, sons and husbands typically receive the best available food, while daughters and female family members may only receive one meal a day, compared to the two or three meals that males may have. The gender stereotype of men as future breadwinners is deeply rooted in this striking division, implying that they should receive better nourishment. This disparity often leads to malnutrition among women as food distribution becomes unequal. This makes women even more vulnerable during disasters. For instance, participants stated that-

> My mother always told me that women should eat less than boys. So, during floods, there were shortages of food. So, my mother cooked hotchpotch for my father and brother and gave me parched rice and dry bread. Father sometimes caught fish from flood water but my mother never gave me that. Only father and brother ate those limited fishes. Thus because of the shortage of food, often I had to drink water, eat bread, gruel, and parched rice.

> Due to floods, there were shortages of fresh food and clean water. So, most of the time what we get, had to give my husband and other family members. So, sometimes I had to starve and nobody cared for me whether I ate or not.

Many women reported experiencing difficulties during pregnancy, particularly during floods and river erosion, when food and clean water are scarce. Cultural beliefs restrict pregnant women from consuming certain foods, such as potatoes and cucurbits, due to superstitions that these foods could negatively affect the child's physical attributes. For example, if she eats potatoes then her child's eye will be round and if she eats cucurbit then her child's body will have more hair. Therefore, pregnant women often refrain from consuming these foods. During disasters, as foods like potatoes and cucurbits are stored in entresol areas (high places) of the home, these restrictions contribute to the already limited food supplies, exacerbating malnutrition risks for both the mother and the baby. Such kinds of superstitions or social dogmas increase the vulnerability of women during disasters.

The FGDs reveal that cultural beliefs impose dietary restrictions on women also during menstruation. Specifically, cultural beliefs discourage women from consuming parsley ("shak"), as they believe it darkens menstrual blood, causing stains and supposedly indicating poor health. Similarly, women are restricted from consuming eggs due to their strong odour in menstrual blood. This results in women refraining from consuming such foods during their periods. In times of disaster, when food is scarce, these dietary restrictions pose additional challenges for women causing malnutrition and heightening their risk of health issues.

Menstruation is often viewed through the lens of societal stigmas, often considered "impure" or "bad," which leads to extreme hardships for women during disasters. Some of the participants highlighted that due to disasters, they were displaced from their homes and took shelter in a school. There, they faced difficulties while washing their rag (a piece of cloth) during the daytime as there were many people in the school taking shelter and therefore compelled to manage menstrual hygiene in secrecy. To remain discrete and maintain privacy, women had to wash their menstrual rags at night. Often, they pass sleepless nights filled with anxiety due to the stigma of someone observing them. They further added that limited facilities and fear of stigma often led them to use menstrual clothes for extended periods without changing them which resulted in infections like urinary tract infections or genital tract infections. If left untreated, these conditions can lead to future complications, including infertility. To manage this secretly women face severe mental and physical strain, along with the challenges of displacement.

From the discussions, it is also found that women are often at risk of unwanted attention or harassment in crowded shelters, particularly when they are isolated due to cultural practices during menstruation. This increases their vulnerability to exploitation and psychological distress. For instance, a participant stated that-

> Due to floods, I along with my family took shelter in a school. Without the home, I faced problems during menstruation. As there were many people in the school, I could not wash my sanitary rag (reusable sanitary cloth) during the day. So, I had to wake up in the night and wash and dry the sanitary rag at night. During those days, my children, husband, and mother-in-law slept on another mat and gave me a separate mat to sleep on. I was not allowed to sleep with them at night during menstruation. So, sometimes other men look at me at night and have bad intentions towards my body by seeing me sleep alone.

Cultural beliefs sometimes confine pregnant and postpartum women to specific areas, exacerbating the challenges during disaster conditions. For example, a few women from FGDs shared their experiences that during floods they gave birth to their children and were confined in a room for 40 days due to the natural process of postpartum bleeding (which is viewed negatively as "bad blood"). That period was marked by restrictions on movement and limited interaction with family members, leading them to isolation. Due to floods, water entered their room spreading a bad smell everywhere along with dust, mosquitoes, and other insects. These confined spaces became breeding grounds for insects and other health hazards that increased the risk of disease for the mother and child.

## Conclusions

It has brought into light the fact that Bangladeshi women face gendered vulnerabilities during disasters, putting boundaries on the interplay of socio-economic, cultural, and structural factors within a greater patriarchal framework. The interlinkages of socio-economic adversities such as gendered divisions of labour, cultural constraints, and resource imbalances exacerbate the increased vulnerabilities of women during disasters. Women carry a large and invisible unpaid care economy that restricts their economic independence, reinforces stereotypical roles, and devalues their contributions. Cultural stereotypes, social dogmas, and the public–private dichotomy further exacerbate these issues, subordinating women to a limited role in decision-making and disaster preparedness. These factors thus intersect to leave women suffering disproportionately from the consequences of increased workload, health risks, and social isolation resulting from disasters. Cultural barriers, dietary restrictions, and stigmas related to menstruation and pregnancy add to the list of problems, often resulting in malnutrition and inadequate health care.

Climate-induced disaster migrations introduce additional vulnerabilities and social conflicts related to resource scarcity and social discrimination, further marginalizing a woman who has already been displaced. Guided by a theoretical approach combining constructivist and feminist political ecology, this study underlines that vulnerability should not become an inherent condition but be regarded as one brought about by socially constructed conceptions and power relations. To reduce the vulnerabilities, gender-sensitive adaptation practices are necessary, and the study highlights the need for a gender-transformative approach to make women more resilient during extreme events.

A gender-transformative approach to climate change aims to establish enduring, positive development outcomes that are based on equality among women, men, girls, boys, and gender-diverse individuals. In addition, it endeavours to involve communities in climate action to challenge gender norms, promote social and political influence positions for women, and resolve power imbalances between genders. As a result, gender-transformative climate change programmes can offer a chance to confront the underlying causes of inequality and vulnerability by taking action that challenges the fundamental societal structures in the context of climate change adaptation plans.

The following strategies could be prioritized to achieve gender-transformative outcomes: (1) gender-transformative institutional strengthening (equal engagement of men and women to achieve gender equality); (2) evidence-based policy research and action; (3) robust regional partnership, coordination, collaboration, and networking for women's empowerment and leadership potential. The analytical emphasis of transformation research is shifted from "accommodating change" to the contestation of the underlying social, political, cultural, and economic structures that generate marginalization and inequality. Examples include the gendered responsibilities in farming, where women are typically in charge of planting subsistence crops and caring for small cattle. Women play a crucial part in this industry since they are also significant food and consumable buyers and sellers. While considering the triple burdens of women, transformative approaches that concentrate on cash commodities and agricultural extension can broaden their options, livelihoods, and waged income.

To work effectively, adaptation must address gender regimes that regulate access to, use of, and control over resources, particularly those that define land distribution, labour division, and strategic decision-making power. It is crucial to address issues such as mobility, men's and women's attitudes towards equitable roles in family life, women's participation in public life, and men's participation in domestic tasks, as well as changes in women's productivity, incomes, and access to markets and services. It is also imperative to recognize and address intersectionality to achieve transformative results. This study thus underscores the deeper and micro-level understanding of women's vulnerability during disasters and urges for a gender-transformative approach that not only addresses structural inequalities but also redefines power relations and gender roles. This study has limitations for not conducting comprehensive assessments like the quantitative scenario due to limitations of time and funds. However, it contributes to analyzing the varied reasons for the vulnerability of women tying it to structural inequalities and societal norms. Through its contextualized yet comprehensive narratives, it has created a foundation to explore the unique vulnerabilities that women face while struggling to adapt to various slow onset as well as extreme events associated with the negative impacts of climate change. The study recommends that attention should be paid not only to research but also to reflect in policy frameworks for the holistic development of vulnerable women.

## References

Abdullah AYMd, Bhuian MdH, Kiselev G, Dewan A, Hassan QK, Rafiuddin M (2022) Extreme temperature and rainfall events in Bangladesh: a comparison between coastal and inland areas. Int J Climatol 42(6):3253–3273.https://doi.org/10.1002/joc.6911

Agarwal B (1992) The gender and environment debate: lessons from India. Fem Stud 18(1):119–158. https://doi.org/10.2307/3178217

Alam P, Bolio A, Thakkar N, Marston C (2023) "I have taken your izzath (honour), so you can't betray me": young people's lived experiences of navigating sexual norms and violence in Bangladesh. Cogent Public Health 10:2253575. https://doi.org/10.1080/27707571.2023.2253575

Arora-Jonsson S (2011) Virtue and vulnerability: discourses on women, gender and climate change. Glob Environ Change 21(2):744–751. https://doi.org/10.1016/j.gloenvcha.2011.01.005

Ashraf MA, Azad MdAK (2015) Gender issues in disaster: understanding the relationships of vulnerability, preparedness and capacity. Environ Ecol Res 3(5):136–142. https://doi.org/10.13189/eer.2015.030504

Bari ABMM, Intesar A, Mamun AA, Debnath B, Islam ARMdT, Alam GMM et al (2024) Gender-based vulnerability and adaptive capacity in the disaster-prone coastal areas from an intersectionality perspective. Clim Risk Manag 43:100581. https://doi.org/10.1016/j.crm.2023.100581

Brauer M, Judd CM, Jacquelin V (2001) The communication of social stereotypes: the effects of group discussion and information distribution on stereotypic appraisals. J Pers Soc Psychol 81(3):463–475. https://doi.org/10.1037/0022-3514.81.3.463

Chipuriro R (2023) Vulnerabilities, power, and gendered violence in food systems. Agenda 37(1):62–71. https://doi.org/10.1080/10130950.2023.2179931

Chowdhury KJ, Ali MR, Chowdhury MA, Islam SLU (2024) Climate change induced risks assessment of a coastal area: a socioeconomic and livelihood vulnerability index-based study in coastal Bangladesh. Nat Hazards Res.https://doi.org/10.1016/j.nhres.2024.06.005

Dow K, Cutter SL (1998) Crying wolf: repeat responses to hurricane evacuation orders. Coast Manag 26(4):237–252. https://doi.org/10.1080/08920759809362356

Ellemers N (2018) Gender stereotypes. Annu Rev Psychol 69:275–298. https://doi.org/10.1146/annurev-psych-122216-011719

Fatema SR, East L, Islam S, Usher K (2023) Gender-based vulnerabilities for women during natural disasters in Bangladesh. Front Commun 8:1180406. https://doi.org/10.3389/fcomm.2023.1180406

Forbes D (2015) C. Mutch: Doing educational research: a practitioner's guide to getting started, 2nd ed. NZ J Educ Stud 50(2):307–309. https://doi.org/10.1007/s40841-015-0032-x

Glade T, Alexander DE (2013) Classification of natural disasters. In: Bobrowsky PT (ed) Encyclopedia of natural hazards. Springer, Dordrecht, pp 78–82. https://doi.org/10.1007/978-1-4020-4399-4_61

Haque STM, Yousuf MPH (2024) Bangladesh. In: Fiorino DJ, Eisenstadt TA, Ahluwalia MK (eds) Elgar encyclopedia of climate policy. Edward Elgar Publishing, Cheltenham. https://doi.org/10.4337/9781802209204.ch78

van der Horst M (2014) Gender role attitudes. In: Michalos AC (ed) Encyclopedia of quality of life and well-being research. Springer, Dordrecht, pp 2451–2453. https://doi.org/10.1007/978-94-007-0753-5_1135

Ikeda K (1995) Gender differences in human loss and vulnerability in natural disasters: a case study from Bangladesh. Indian J Gend Stud 2:171–193. https://doi.org/10.1177/097152159500200202

Jain S, Zarghamee H (2023) Invisible women: a psycho-economic exploration of domestic and reproductive labour. Int J Appl Psychoanal Stud 20(4):654–665. https://doi.org/10.1002/aps.1805

Jobaer MdA, Hossain MdZ, Ha-Mim NM, Haque SF, Rahaman KR (2024) Examining the link between marginality and differential climate resilience among disaster-affected communities in southwestern Bangladesh. Climate and Developm 0:1–16.https://doi.org/10.1080/17565529.2024.2329465

Khan MM, Azad MAK, Kamal A-HM, Zahid D, Malak MA (2023) Why is women's leadership important for enhancing disaster resilience to natural perturbations?: Important insights from coastal communities of Bangladesh. Coastal disaster risk management in Bangladesh, 1st edn. Routledge, London, pp 153–170

Lewis G, Hasan K (2022) Bangladesh INFORM Sub-national risk index 2022. United Nations, Bangladesh. Available via UN. https://bangladesh.un.org/en/210194-bangladesh-inform-sub-national-risk-index-2022

Matud MP (2017) Gender and health. In: Alvinius A (ed) Gender differences in different contexts. IntechOpen, London. https://doi.org/10.5772/65410

Nasreen M (2022) Gender and disaster in Bangladesh. Oxford Research Encyclopedia of Natural Hazard Science.https://doi.org/10.1093/acrefore/9780199389407.013.380

Nazneen S (2024) Women's struggles for empowerment in Bangladesh. Current History 123:135–140. https://doi.org/10.1525/curh.2024.123.852.135

Nieswiadomy RM, Bailey C (2018) Foundations of nursing research, 7th edn. Pearson, New York

Ozaki M (2016) Disaster risk financing in Bangladesh

Pérez Fragoso L (2022) A feminist-economics analysis of Latin American budgets shows an urgent need for redistribution of income and care work. Gend Dev 30(1–2):311–320. https://doi.org/10.1080/13552074.2022.2072003

Rocheleau D, Thomas-Slayter B, Wangari E (eds) (1996) Feminist political ecology: global issues and local experience. Routledge, New York.https://doi.org/10.4324/9780203352205

Roy S (2020) Transforming gendered lives and livelihoods in post-disaster settings in the Bangladesh Sundarbans Forest. In: Leal Filho W, Nagy GJ, Borga M, Chávez Muñoz PD, Magnuszewski A (eds) Climate change, hazards and adaptation options: handling the impacts of a changing climate. Springer International Publishing, Cham, pp 463–479

Samin M, Nahr S (2024) Bangladesh slips 40 notches in global gender equality ranking. The Business Standard. https://www.tbsnews.net/bangladesh/bangladesh-slips-40-notches-global-gender-equality-ranking-875646

Siraw G, Zewude B, Habtegiorgis T (2023) Attitude towards gender division of labour and convictions regarding the nature of spousal relationships among never married youth in Ethiopia. Cogent Soc Sci 9(1):2196104. https://doi.org/10.1080/23311886.2023.2196104

Sultana N (2022) Understanding the economic dimensions of women's vulnerability during cyclones: the Bangladesh perspective. Int J Disaster Risk Reduct 70:102730. https://doi.org/10.1016/j.ijdrr.2021.102730

Terry G (2009) No climate justice without gender justice: an overview of the issues. Gend Dev 17(1):5–18. https://doi.org/10.1080/13552070802696839

# Chapter 8
# Climate Change and Food Security in Sub-Saharan Africa: An Analysis from Gendered Lens

**Chandani Tiwari, Sakshi Bahuguna, and Aparna P. Krishnan**

**Abstract** Sub-Saharan Africa is grappling with significant challenges related to climate change and food security which are magnified by the gender disparities. The convergence of climate change and food security should be addressed through a gendered lens. The contribution of women to agriculture and food production is significant and hence improving their productivity will help in achieving better levels of food security. To examine and establish the correlation between food security and gender in Sub-Saharan Africa, this chapter analyses regions based on the availability of data from 21 countries. By combining both quantitative data and qualitative insights, this chapter analyses the challenges as well as opportunities faced by women in the agricultural system. The generalized least squares approach is used for running a panel data regression and to empirically establish a relationship between agricultural productivity and gender metrics. The results indicate a significant relationship between productivity and gender in Sub-Saharan Africa. This chapter reaffirms that women make essential contributions to agriculture and its productivity, and hence gender-sensitive initiatives will bring forth desired results on food security in the context of climate change.

**Keywords** Food security · Gender · Culture · Empowerment · Inequality · Climate change · Policy intervention

C. Tiwari · A. P. Krishnan (✉)
Symbiosis School of Economics, Symbiosis International University, Pune, India
e-mail: aparna.krishnan.2022@sse.ac.in

C. Tiwari
e-mail: chandani.tiwari@sse.ac.in

S. Bahuguna
Abu Dhabi University, Abu Dhabi, United Arab Emirates
e-mail: sakshi.b-adjunct@adu.ac.ae



A. Sarkar and N. Bandyopadhyay (eds.), *Gender-Transformative Approaches for Climate Change Adaptation*, Springer Climate,
https://doi.org/10.1007/978-3-031-90236-9_8

## Introduction

Africa is a continent known for its rich resource pool, vast agricultural potential, mineral wealth, and biodiversity. Despite all these, Africa, particularly Sub-Saharan Africa (SSA), struggles with acute poverty and food insecurity owing to the long-standing climate change challenges (Kemoe et al. 2022). SSA includes countries located in the south of the Sahara Desert, known for continuously grappling with climate change and food security issues. The region faces rising temperatures, unpredictable rainfall, and extreme weather events that eventually put food and agricultural systems at risk (Adeola et al. 2024). Climate change is one of the biggest threats to countries in Africa, especially SSA, which faces climate-related shocks like floods, droughts, and storms more than any other region in the world (Ogwang et al. 2018). This will in turn endanger the health and productivity of the labor force and community in general.

Africa has some of the world's most vulnerable populations, who even though contribute very little to global carbon emissions, shoulder the impacts of climate change (Ogwang et al. 2018). These communities who are already grappling with food insecurity and poverty are among the hardest hit by environmental challenges. This shows that those least responsible for climate crises are suffering the most severe aftereffects. The long-standing climate shocks hinder the climate adaptation efforts made by the African community, and it is believed that since these communities are mainly reliant on rain-fed agriculture and already facing malnutrition and poverty issues, they are likely to face more climate change impacts when compared to other wealthier nations around the world (Adeola et al. 2024). These changes reduce agricultural productivity and crop yield in the region which heightens the risk of hunger and food insecurity.

According to the Global Hunger Index (GHI) 2023, SSA and South Asia have the highest levels of hunger in the world with both regions having a GHI score of 27.0 indicating serious hunger. After many years of steady progress till 2015, progress against hunger in SSA has largely stalled (GHI 2023). The climate shocks, geopolitical tensions, wars, the COVID-19 pandemic, and other economic disruptions have contributed to lowering the food security levels (FAO 2023). These crises put a strain on food supply chains and increased food inflation, making it more challenging to achieve the Sustainable Development Goal of eradicating hunger. The 2022 Global Hunger Index shows that eight out of nine countries having alarming levels of hunger are in Africa, particularly concentrated in SSA (GHI 2022). Many countries in SSA have low to medium Human Development Index (HDI) ranking, indicating the existing developmental issues and challenges (HDI 2022). The proportion of undernourished people in Africa between 2010–2012 and 2020–2022 has increased from 14.9% to 19.3% (FAO et al. 2023). This shows the escalating food insecurity which has impacted the livelihoods and overall well-being of people across the region. This impact is more severe and disproportionate for women, owing to deepened climate vulnerability and unequal access to land, finance, and lack of decision-making power.

The optimal policy mix to mitigate food insecurity and build climate change resilience will differ among countries within SSA.

The overlap of climate change and food insecurity can also be addressed with a gendered lens. Men and women experience varied levels of climate vulnerability owing to the nature of their responsibilities and rigid social norms. During climate disasters, households headed by women experience higher income losses and increased workloads compared to those headed by men (Chanana-Nag and Aggarwal 2020). In most low- and middle-income countries, food production and household management are handled mainly by women. But instead of being empowered, certain social norms create gender-based discrimination that results in relatively less resource access for women. Their limited access to land, climate finance, and lack of decisive power further prevent them from coping with climate change. This gender gap can deteriorate human development levels and lead to low agricultural productivity and food security (Brody et al. 2008). Gender blindness is common because climate change policies, funding mechanisms, and programs are often mistakenly seen as gender-neutral even though in reality they are based on male-biased perspectives (Otzelberger 2011). Thus, gendered opportunities and risks are overlooked while formulating climate adaptation policies.

To effectively address these challenges, a gendered approach to climate adaptation can be adopted for formulating superior climate policies and building a strong and resilient community in regions of SSA. The gendered vulnerabilities can be transformed to empower women and for better climate adaptation. Better access to resources and infrastructure like land, housing, transportation, and credit could improve their standing in society as well as their adaptation against climate change. This could boost agricultural productivity and food security in SSA and could also improve their GHI and HDI rankings globally.

This chapter proposes a context-based approach toward climate change policies and agricultural practices from a gendered perspective. The comprehensive relationship between food security levels and gender metrics is studied with the help of indicators like agricultural productivity, gender development index, and gender inequality index. These gender metrics cover certain livelihood indicators like education, health, and income and hence are considered wholesome for evaluation. Improving HDI factors, especially women's health and education can lead to better food security levels. Thus, integration of food security initiatives with broader development strategies can be achieved. This paper addresses a significant research gap in the existing literature by focusing on the overlapping of food insecurity and climate change adaptation in SSA through a gendered lens. This transformative approach highlights how empowered women can build up food security and enhance resilience to climate change. Therefore, positive discrimination, while framing policies and considering the gendered aspect, could be a strategy that could be initiated.

## Theoretical Framework

Especially in the context of climate change, the theoretical perspective on the correlation between women and the environment majorly ignores the interplay between gender roles, environmental vulnerabilities, and adaptive capacities. Women play a critical role in the agricultural economy of developing countries, contributing significantly to food production and household food security. However, climate change acts as a threat multiplier, aggravating existing challenges and disproportionately affecting their livelihoods. Hence, addressing climate change as well as food security should inherently involve considering gender issues, which have traditionally been sidelined or treated as a control variable in environmental psychology. This understanding of gender pushes us to think about when, why, and how gender matters while referring to climate change and food security. Thus, this paper tries to locate the aspect of gender in the current climate change and food security intersection in the larger people-environment interaction. Its absence leads to many unexplored questions on nature-health interactions. Hence, given the lack of a gender lens in the field, the attempt has been to bring it back to present a better understanding and to propose the appropriate point of intervention that could be beneficial for the whole community or society.

Agriculture is the largest employer of women in Oceania, South Asia, and sub-Saharan Africa. It employs almost 60% of the women workforce. Despite this data, their participation in agriculture is characterized by a worldwide gender gap concerning vulnerabilities, access and control to resources, decision-making power, and productivity. For addressing gender inequalities in agriculture concerning climate change and food production, only erasing the inequities concerning resources would not solve the large purpose of bringing them back into the picture. Policies and technologies that support resilience and adaptation to climate change can, in the long run, promote women's empowerment in a true sense (Huyer 2016).

Nevertheless, the agency of women, as a result, becomes a vital area of research that highlights their critical role in agriculture or specifically their role in food production, management of resources, and adaptive or sustainable agricultural practices. They represent a substantial portion of the global agricultural workforce, simultaneously facing systemic barriers and trying to find out the meaning of empowerment. The contradiction arises when they play such a significant role, but when it comes to climate change and food production, they disappear completely. Hence, the question of the agency of women in agriculture, the challenges faced by them, and policy implications becomes very crucial. The lack of recognition for women in these areas also compromises the idea of their empowerment.

The understanding of the above-mentioned context through a gendered lens itself highlights the unique vulnerabilities and roles played by women in agriculture. Food security, climate change, and gender are a complex intersection that needs policy intervention. The challenges caused by climate change and the problems of food security further complicate things for women, who often serve as key caregivers and play essential roles in food production in many societies. The situation further gets worse in developing countries as they are often responsible for overall household nutrition and food production (Parvin and Ahsan 2013). Few studies also highlight the importance of giving more access to women as it enhances their agency and leads to better agricultural outcomes. It has been highlighted that women's involvement in cooperatives where their access to resources, markets, and knowledge results in increased productivity and income generation (Chebet 2023).

The understanding of socio-economic characteristics is also crucial for developing effective adaptive strategies; if not taken into consideration, they make women more vulnerable (Chanana-Nag and Aggarwal 2020). It can also be argued that an application of universal initiative will also compromise the impact on societies as the situation varies from place to place. For example, in a few studies, it has been argued that culture also plays a role leading to the differential impact on societies. There is an interplay of gender, access, ownership, decision-making, and culture on selected productive resources (Ajadi et al. 2015). Focus on gender-responsive approaches would enhance women's capacity to address climate challenges. While doing so, sustainable practices could be promoted (Kristjanson et al. 2017).

Thus, for strategies to be effective, the first and foremost need is to have a clear picture of the agency of women in the field of agriculture and the role played by them. The role of socio-economic context, culture, knowledge, etc., exercise different impacts on women as they vary from place to place. Thus, knowing and interpreting the correct target category that needs intervention would be helpful in the long run. The correct initiative and strategies would enable them to make decisions per their needs and requirements. Consequently, better opportunities for them with better decision-making power might lead to their empowerment in the true sense. The role of women in agriculture cannot be ignored, who even in some places, become the first category affected by climate change.

## Materials and Methodology

According to the Food and Agriculture Organization (FAO), food security refers to the situation "when all people, at all times, have physical, social and economic access to sufficient, safe and nutritious food that meets their dietary needs and food preferences for an active and healthy life" (FAO 2010). In the context of rising climate change concerns, food insecurity is one of the most critical challenges faced across the world, particularly in developing countries. SSA also grapples with acute food insecurity and health issues over the decade, and it has greatly affected the livelihood of the community (FAO 2023). Since most of the population relies on rain-fed agriculture, climate concerns further lead to declining productivity and food availability. Agricultural productivity thus plays a crucial role in the food security goals in SSA as it directly influences the overall quantity and quality of food produced in the region (Kidane et al. 2006).

Rosegrant et al. (1995) describe a two-tier system of food security, where wealthier economies usually have access to affordable food supplies while poorer economies face food shortages. It was also suggested that slow-growing and underdeveloped nations must aim at improving their agricultural productivity as it paves the way to achieve food security goals. Food security depends on a lot of socio-economic and demographic factors as well (Rosegrant et al. 1995). Muzumdar (2012) also considers food security to be a multi-dimensional issue and considers agricultural productivity to be the first step toward resolving food security issues. SSA has low productivity compared to other regions and to tackle this issue, a thorough understanding of its causes and obstacles is necessary. Limited access to credit, inadequate access to water, land rights, underdeveloped infrastructure, and lack of investment ventures are a few obstacles found in SSA (Muzumdar 2012).

## *Measuring Agricultural Productivity*

Agricultural productivity often measures the quantity of output generated from a specific quantity of input or inputs. There are many tools to measure agricultural productivity like the conventional ones that compare the volume of output to the inputs used. Zepeda (2001) discusses the partial factor productivity measures which sometimes do not indicate the changes in production and its causes. Changes in productivity can be traced back to many variables, depending on the type of factor input (land or labor) used, which is rather dynamic and complicated. To avoid the potential problems of these partial productivity measures, the superior choice will be Total Factor Productivity (TFP). TFP is a multifactor productivity measure that takes into account all measurable inputs like land, labor, and other purchased units (Alston et al. 2009). It compares total output relative to total inputs or composite set of inputs used. Thus, for this chapter, TFP will be used as a variable to measure agricultural productivity and indirectly also connect to food security. The data for Agricultural Total Factor Productivity (ATFP) is taken from the USDA Economic Research Service's data on International Agricultural Productivity, wherein the TFP for different countries and regions is provided (Fig. 8.1).

## *Food Security and Productivity Under a Gendered Lens*

Women represent a crucial resource in the agriculture sector. Their roles vary depending on the development of many socio-economic factors. In SSA, they are primarily responsible for household management and food preparation (Kolovich et al. 2024). However, they have less access to resources like land and credit compared to men and are usually kept away from decision-making (Rodgers and Akram 2019). Their involvement is necessary not only for agricultural production but also for enhancing food security since they are directly contributing to the nutritional outcomes of their household. In the current scenario of climate shocks and disasters,

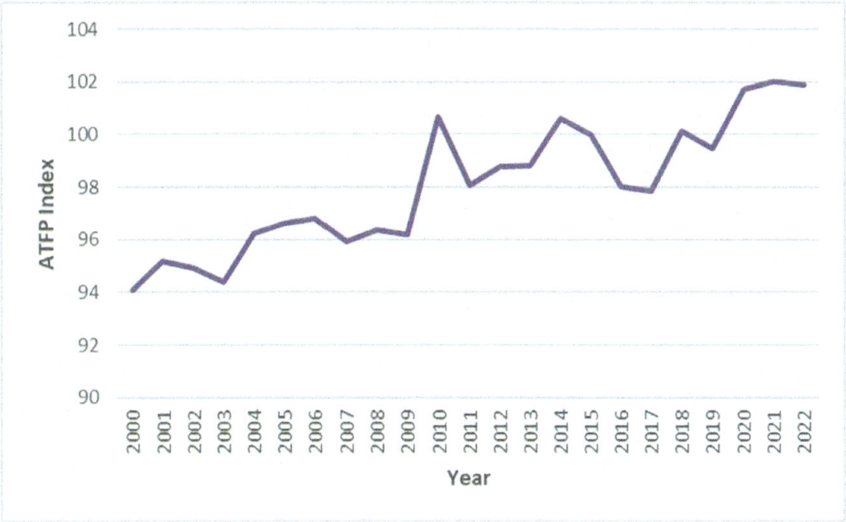


**Fig. 8.1** Agricultural total factor productivity index of SSA (as a whole), 2000–2022. *Source* Author's compilation

women tend to be more vulnerable to the aftereffects. Since women are more vulnerable to climate shocks, climate change adaptation policies and agricultural policies should be driven with a gendered approach in mind. Implementing gender-sensitive agricultural policies not only closes the gender gap in productivity but also leads to better food security outcomes. In SSA, it is believed that women possess Indigenous knowledge that can be used for resource management and agricultural practices (Pidatala and Khan 2003). Thus, their involvement in the decision-making process will help in developing a superior climate-resilient community. It is already discussed that improving productivity is the initial step toward food security and we have seen the importance of gender-sensitive policies. Through this chapter, we empirically try to show the relationship between gender and agricultural productivity. The results will highlight the need for a transformed approach toward food security and climate change.

## *Measuring Gender Disparities*

For measuring gender disparities across SSA, we use the Gender Inequality Index (GII) and Gender Development Index (GDI) developed by the United Nations Development Programme (UNDP). These variables measure gender disparities across various dimensions of human development. GII focuses on gender-based disadvantages in three dimensions namely empowerment, reproductive health, and labor market participation. A value closer to zero shows low inequality between men and women and a value closer to one shows extreme inequality in all dimensions. On the other hand, GDI measures gender inequalities in three fundamental aspects of human development. It takes into account inequality in education, health, and economic resources through measures like life expectancy at birth, expected years of schooling, and estimated income, respectively. Together, these two indices will shed light on the quantitative side of gender sensitivity. While GII focuses on inequalities hindering female progress, GDI focuses on human development outcomes for both genders. These indices reveal disparities across different dimensions of human development that are not inclusive of other gender metrics.

## *Methodology and Empirical Framework*

To understand the food security issues and gender aspects in SSA, we have considered 21 countries from the region based on the data available. The chapter follows a mixed approach involving both quantitative and qualitative methodology. By combining both quantitative data and qualitative insights, this chapter analyses the challenges faced and opportunities for women in the agricultural system. The results will likely provide suggestions for policies that support a gender-inclusive strategy for climate change adaptation. As for the quantitative approach, the focus will be on finding empirical evidence regarding the need for a gendered approach toward food insecurity concerns in SSA. For the same purpose, a panel dataset covering 21 SSA countries from the year 2000 to 2020 is used. Stata is used to evaluate the collected data using a variety of statistical methods.

The computation of descriptive statistics yields a summary of the distribution, central tendency, and variability of the variables (Table 8.1). The peculiarities of the data are analyzed to shed light on broad patterns and trends. The generalized least squares (GLS) approach is used for running a panel data regression and to empirically establish a relationship between agricultural productivity and gender metrics. GLS estimators are used when the Ordinary Least Squares (OLS) estimator is not BLUE (best linear unbiased estimator). Regression is run while taking ATFP as regress and GII and GDI as regressors.

## *Cross-Sectional Dependence Test*

It is essential to examine the cross-sectional dependence of the panel data series, as this would reveal whether there are any spillover effects of any policy among the countries included in the panel. Detecting cross-sectional dependence would help to understand how the countries are interconnected and whether a policy impact in one country has broader effects. Table 8.2 shows the result of Pearson's (2004) test for cross-sectional dependence. The null hypothesis is that there is no cross-sectional dependence. Since the probability value of the test statistic is less than 0.05, the

**Table 8.1** Variables and data sources

| Variables and sources | Definition |
|---|---|
| Gender Development Index (GDI) from UNDP Human Development (Year 2000 to 2022) Reports: Gender Indices (Composite Indices) | • Evaluates gender disparities in three fundamental aspects of human development:<br>• Health (represented by life expectancy at birth for both males and females)<br>• Education (measured through expected years of schooling for boys and girls, and mean years of schooling for men and women aged 25 and above)<br>• Economic resources (assessed based on estimated earned income of males and females) |
| The Gender Inequality Index (GII) from the UNDP Human Development (Year 2000 to 2022) Reports: Gender Indices (Composite Indices) | • GII captures gender-based disadvantages across three key dimensions:<br>• Reproductive health (assessed through maternal mortality ratio and adolescent birth rate)<br>• Empowerment (measured by the proportion of males and females with at least secondary education, and the shares of parliamentary seats held by men and women<br>• Labor market (evaluated using female-male labor force participation rates)<br>• GII highlights the loss in potential human development caused by inequalities in achievements between genders |
| Agricultural Total Factor Productivity (ATFP) from Economic Accounts for Agriculture from Eurostat (Year 2000 to 2022) | • Total factor productivity evaluates the ratio of total outputs to the total inputs employed in the production process. By representing both output and inputs using volume indices, TFP effectively measures productivity growth. TFP reflects the combined impact of various factors such as advancement in technology, efficiency improvements, economies of scale, managerial expertise, and shifts in production organization |

*Source* Compilation by authors

null hypothesis is rejected, and it suggests that there is cross-sectional dependence in panel data. Hence, GLS approach will be used for estimation in this chapter to account for the dependence across the cross-sectional units. To account for cross-sectional dependence, we will specify the *panels (correlated)* option and additionally specify *corr (psar1)* while assuming that the residuals in each panel follow a first-order autoregressive process.

**Table 8.2** Cross-sectional dependency test

| Variable | CD-test | p-value | Average joint T | Mean ρ | Mean abs(ρ) |
|---|---|---|---|---|---|
| ATFP | 2.058 | 0.04 | 23 | 0.03 | 0.4 |
| GII | 50.072 | 0 | 23 | 0.72 | 0.74 |
| GDI | 50.803 | 0 | 23 | 0.73 | 0.73 |

*Source* Calculated by authors

## *GLS Estimation*

GLS aims to provide estimates that are efficient (minimum variance) and unbiased (accurate) even in the presence of heteroscedasticity and serial correlation, ensuring the estimators are optimal (Table 8.3). It adjusts the estimation process by incorporating information about the covariance structure of the errors, allowing for more accurate parameter estimates in the presence of non-constant error variances and correlated errors. Unlike OLS, which assumes homoscedasticity and no serial correlation, GLS relaxes these assumptions by accounting for the covariance structure of the errors, making it a more robust estimator in the presence of complex error patterns. In this chapter, generalized least squares regression is utilized for estimation due to its ability to account for heteroskedasticity and serial correlation in the data, providing more efficient and unbiased coefficient estimates. GLS is particularly suitable for panel data analysis, allowing for the examination of relationships between variables across multiple countries and over time.

$$\mathrm{ATFP} = \mathrm{X}\beta + \eta$$

**Table 8.3** GLS estimation results

| v1 | v2 | v3 | Notes titles |
|---|---|---|---|
| | −1 | −2 | |
| Variables | ATFP | ATFP | Standard errors in parentheses |
| | | | *** $p < 0.01$, ** $p < 0.05$, * $p < 0.10$ |
| GDI | 54.8168*** | 54.8168*** | |
| | (3.1989) | (3.1989) | |
| GII | −19.8284*** | −19.8284*** | |
| | (3.3205) | (3.3205) | |
| Constant | 61.4718*** | 61.4718*** | |
| | (4.1911) | (4.1911) | |
| Observations | 483 | 483 | |
| Number of Country | 21 | 21 | |

*Source* Author's compilation

where "ATFP" is the Agricultural Total Factor Productivity Index. "X" is the matrix of independent variables, which includes GII and GDI as in the Gender Inequality Index and Gender Development Index, respectively. "β" is the vector of regression coefficients to be estimated. "η" is the error term, which is assumed to be normally distributed with a mean of 0 and a variance–covariance matrix V.

## Results and Discussion

Table 8.3 shows the results of the GLS regression. After performing the generalized analysis, GDI and GII are found to influence ATFP. GDI has more influence with a positive beta coefficient and a p-value below 0.05 which shows that it has a significant influence over ATFP. Since the coefficient value is positive, it denotes a positive relationship between GDI and ATFP. As the GDI of a country improves, agricultural productivity will increase. On the other hand, GII has a negative coefficient with p-value below 0.05, which shows its significant influence over ATFP. The negative coefficient indicates that as the GII of a country increases, the agricultural productivity declines.

GDI measures gender inequalities in the dimensions of health, education, and command over economic resources (Fig. 8.2 shows GDI values of the SSA region). A higher index value indicates a low level of gender disparity and better overall human development outcomes for both genders. When gender inequalities are reduced, both women and men can contribute effectively toward production and food security. Greater equality in command over economic resources allows women to contribute more effectively toward productivity and become more resilient. Improved GDI denotes better educational level and health of women when compared to men across countries in SSA. This indicates that empowering women will directly improve the workforce of a country and also bring hikes in productivity levels. Countries that invest in women's education and health will see significant improvement in productivity and food security levels.

The population with some kind of secondary education (male and female levels can be seen separately in Fig. 8.3). It shows the gender gap in education levels of men and women across SSA. In countries where traditional gender roles and social norms are deep-rooted, rapid changes toward gender equality could initially disrupt existing mechanisms. This could temporarily lead to a fall in productivity, but it does not negate the long-term benefits of gender equality. So empowering women to act as agents of change is the superior way toward better future outcomes.

GII, as mentioned earlier, measures gender-based disadvantages in dimensions, namely, empowerment, reproductive health, and labor market participation (Fig. 8.4). A higher index value shows the presence of high levels of gender inequality which shows disparities in male and female achievements. Higher levels of GII could mean lower participation of women in the workforce in the labor market. Figure 8.5 shows the labor force participation rate (men and women). It can be seen that the participation rate of women is low compared to men throughout the years under consideration.

**Fig. 8.2** Aggregate gender development index of SSA, 2000–2022. *Source* Author's compilation

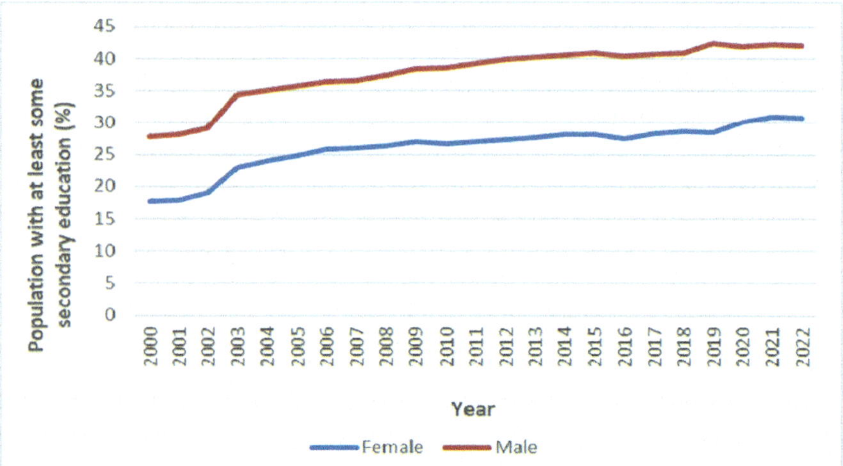


**Fig. 8.3** Population above 25 years of age with some kind of secondary education in SSA, 2000–2022. *Source* Author's compilation

But that does not mean women are not contributing at all. It could mean that women are involved in activities that are not considered "economically active employment" (Raney et al. 2011). Women spend long hours on unpaid care, child rearing, food preparation, and other domestic activities. These activities are not accounted but they are essential for the well-being of the households. When women are excluded from decision-making and leading roles, productivity levels and efficiency will go down. Countries with lower levels of GII will have better productivity levels and thereby better food security outcomes. Thus, promoting gender equality will bring resilience to the population in SSA.

Women in SSA are primary contributors to the agriculture sector and they are responsible for nearly 80 percent of the food production (Schalatek 2022). According to Madsen (2020), many African countries, including those in SSA, have achieved gender parity in primary education and some countries have seen an increasing proportion of women in certain parliaments. These are also indicators measured under GDI and GII. The progress toward gender equality is inconsistent in most of

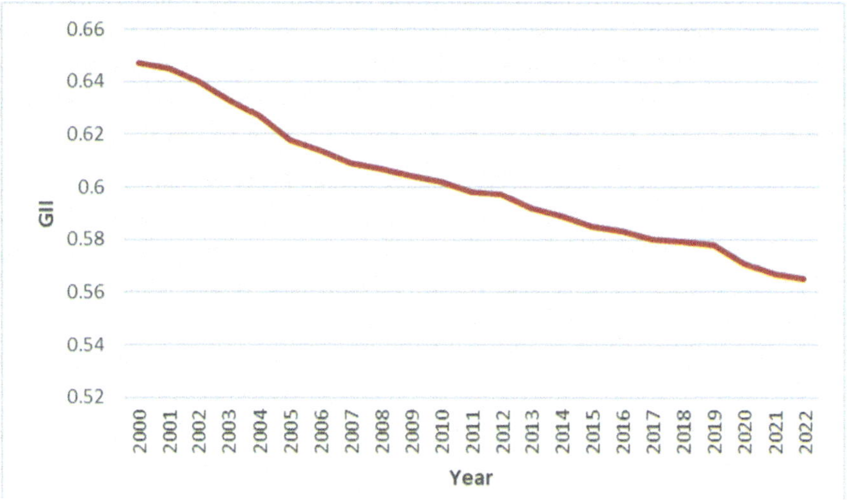


**Fig. 8.4** Gender inequality index of SSA, 2000–2022. *Source* Author's compilation

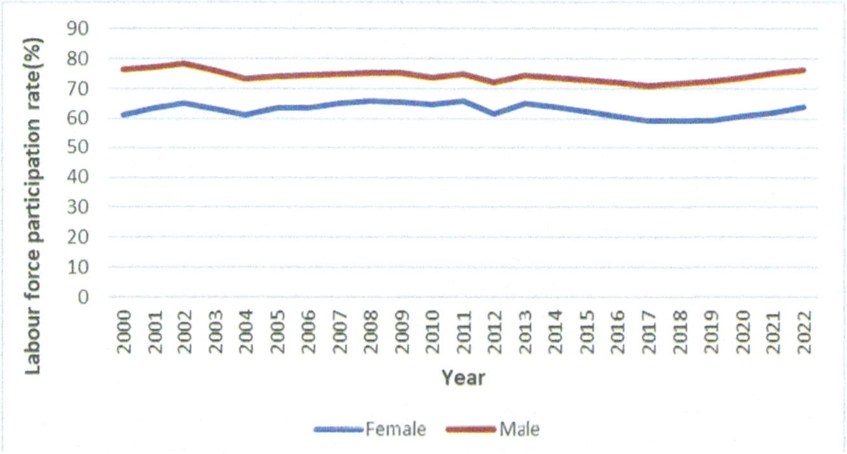


**Fig. 8.5** Labor force participation rate in SSA, 2000–2022. *Source* Author's compilation

the countries and much of the female workforce is unpaid or underpaid, even though women work twice as many hours as men (AfDB 2016). There exists a huge gender gap in the pay scale as well.

## *Key Achievements in SSA Countries*

Gender-responsive climate change actions can be seen in different parts of Africa. Many SSA countries have taken initiatives to integrate gender policies into climate action. Few initiatives have made considerable progress, and certain areas still lack advancement. These actions consist not only of the gender-sensitive programs or policies themselves but also of the political will of governments and changes in the institutional framework and infrastructure that backs up these policies. The United Nations Development Programme (UNDP) has reported on the climate change adaptation practices in SSA countries and on the adaptation gaps as well, which gives an idea as to where the countries stand regarding climate adaptation. The key achievements made by countries, as well as challenges hindering the progress in gender equity, also need to be discussed. We will be looking at certain initiatives of SSA countries that show the progress in the region (Table 8.4).

## *Challenges Faced by Women in Agriculture*

Women in SSA play a significant role in agriculture and their role in climate change adaptation and food security cannot be overlooked. The national governments and international platforms have already taken initiatives to ensure that climate change and food security action plans are gender-transformative (AfDB 2016). Many of these initiatives have seen progress to an extent, yet women face numerous challenges that hinder their productivity and overall contributions to food security.

*Limited Access to Resources*

The existing literature and gender reports have indicated the gender gap in SSA regarding access to essential resources like land, finance, and other agricultural inputs (Rodgers and Akram 2019). The inequitable access to resources hinders the productivity of women farmers and forbids them from exploring their true potential. The stepping stone toward women's empowerment would be giving them access to crucial resources that can improve their resilience and productivity.

*Land Ownership*

In SSA, women typically own smaller and less productive plots of land compared to men, and this limits their agricultural output (Slavchevska et al. 2021). The social norms and in some cases, even the legal frameworks restrict women from owning land and having equal land rights (FAO 2018). This makes them more vulnerable

**Table 8.4** Gender-responsive climate change actions in a few countries of SSA

| Countries | Initiatives taken |
|---|---|
| Cameroon | Created the Ministry of Women's Empowerment and the Family |
| | A National Gender & Climate Change Focal Point was appointed, along with several other Gender Focal Points across various sectoral ministries |
| | The National Gender Policy stipulates a minimum of 30 per cent representation of women on electoral lists for parliament |
| | Established the Centre for the Promotion of Women, the Inter-Ministerial Committee for the Environment, Project Steering and Monitoring committees, Regional Crisis Management committees, and numerous regional councils |
| | Efforts have been undertaken to integrate 'gender' in the state budgeting processes |
| | *Gender and Cameroon's REDD + road map*: Efforts have been undertaken to integrate 'gender' in the state budgeting processes |
| Gambia | Efforts have been undertaken to integrate 'gender' in the state budgeting processes |
| Malawi | Initiated the Women in Climate Action Network |
| | *Gender Climate Change and Agricultural Support Programme (GCCASP):* An initiative launched to address the rising extreme weather conditions in the country, aimed at enhancing climate resilience among rural female farmers. It emphasizes bridging the gender gap in policies by training women in climate-smart agriculture and creating platforms for their empowerment |
| Kenya | Proposed a Gender Responsive Climate Change Act |
| | *Kenya Gender Analysis:* This initiative examined gender disparities in resource access, control, and utilization within the country. It also evaluated the extent to which gender issues are incorporated into Kenya's legal framework, climate policies, and institutions to promote gender-responsive climate action |
| Rwanda | It includes National Gender Policy and National Gender machinery within its governance framework |
| | The Organic Law on State Finances and Property has enabled the institutionalization of the concept of Gender Responsive Budgeting |
| | *Women in Rwandan Energy (WIRE):* The initiative focuses on analyzing gender disparities in the country's renewable energy sector |
| Uganda | Uganda's Gender Monitoring Office is tasked with tracking, analyzing, and updating gender data to enhance accountability for gender equality and women's empowerment |
| | The National Public Finance Act requires gender and equity compliance for funding proposed programs and ensures that budgets are gender-responsive |
| | *Improvement of women's income and resilience to climate change:* This initiative has facilitated the growth of women's farming groups from small-scale to large-scale agriculture. Previously cultivating vegetables in scattered plots, the groups now farm in larger areas with support in sustainable farming practices and access to solar-powered irrigation |
| Zambia | A climate change action plan was created to ensure that all policies adopted are gender transformative |

(continued)

**Table 8.4** (continued)

| Countries | Initiatives taken |
|---|---|
| | *The Cashew Infrastructure Development Project (CIDP):* This initiative aims to reduce poverty by boosting household incomes through enhanced cashew production and value addition |
| Zimbabwe | Developed resilient water resources management with gender-integrated adaptation policies in their design |
| | The updated National Gender Policy and Implementation Plan incorporates gender and climate change, emphasizing the integration of gender in environmental and climate-related policies |
| | There is a gender-inclusive institutional framework in place, including the Zimbabwe Gender Commission, the Ministry of Women Affairs, and a Climate Change Management Department with a dedicated gender focal point |

*Source* Authors' compilation

to losing access to land in case of conflicts, divorce, or the death of a spouse. This inequality in land ownership rights is a significant barrier to increasing productivity in SSA.

*Limited Access to Credit and Financial Services*

Women often face challenges in securing credit and other financial services and many times it is due to the lack of collateral and discriminatory lending practices in place. The climate finance support provided in many countries still does not specifically target women and other vulnerable communities (Achampong 2023). This restricts their ability to make use of superior agricultural technologies or expand their farming operations to a larger scale.

*Time Constraints and Domestic Responsibilities*

Women are usually burdened with household responsibilities and unpaid care alongside their agricultural work. This excess burden limits the time and effort they can dedicate to farming, accessing markets, and training. This unpaid labor for household maintenance can significantly affect their productivity in agriculture.

*Inadequate Support Services*

Many of the agricultural extension services in place do not adequately address the needs of women farmers. These services are male-oriented and fail to provide women with proper training and support for improvement (UNDP 2023). Women's low access to climate-smart agricultural practices worsens their vulnerability, making it harder for them to adapt in times of peril.

Addressing these challenges is crucial for enhancing women's contributions to agriculture and achieving food security in SSA. Empowering women not only benefits them individually but also strengthens their community as a whole by contributing to national food security.

## Conclusion

As Amartya Sen also puts it the capability approach is something that should be the focus of all policy initiatives that would eventually make a difference for women (Vijay and Yadav 2022). The above-mentioned discussion also argues in the same light that SSA has witnessed extreme climate shocks and geopolitical tensions that have aggravated the existing food security concerns. The governments have been introducing several policies and programs targeting this situation, but the limited progress calls for superior climate-related policies. As men and women in SSA experience varied levels of climate vulnerability owing to the nature of their responsibilities and social norms, these challenges can be addressed through a gender lens.

A gendered approach to climate adaptation and food security can be adopted. It is widely believed that enhancing agricultural productivity is the first step toward food security. Productivity in SSA is influenced by gender disparities. This chapter has shown that agricultural productivity measured through ATFP is influenced by GDI and GII in SSA countries. Lowering GII and improving GDI levels will greatly influence the productivity in the region. It could also ensure that women are participating in economically active employment. The countries in SSA individually and collectively have put forward action plans to improve agricultural productivity considering gender concerns. The outcome of a few policies is satisfying and has improved the livelihood of communities. However, proper targeting of resources and policies for women is required to ensure the closure of gender gaps in productivity. Ensuring better participation and providing targeted training to women could boost agricultural productivity and food security in SSA and improve their GHI and HDI ranking globally. If women are excluded from decision-making and lack secure ownership and control over resources, any policy actions are unlikely to have their desired outcomes.

## Policy Recommendations

The recommendations are as follows:

(i) *Promote gender-sensitive agricultural policies:* Integrate gender considerations into agricultural policies and practices, ensuring that both women's and men's roles are properly recognized and supported throughout the processes.

(ii) *Implement climate-smart agricultural practices:* Women can be provided with training in climate-smart agriculture techniques that enhance productivity while minimizing impacts on environmental systems. This could include promoting drought-resistant crops and introducing efficient water management systems in the region. It will be beneficial to the community to ensure that women have access to smart technologies that reduce labor intensity such as energy-efficient tools which will indirectly help them balance their work and household responsibilities in a better way.

(iii) *Enhancing access to land and resources:* The existing situation calls for policies that ensure women have equal access to agricultural inputs and modern technologies. This can be achieved through targeted subsidies for women and programs that specifically address women farmers. It is also desirable to ensure that women have ownership rights and access to land by amending the existing legal frameworks in SSA. This will empower them to give their all to agriculture without the fear of losing their land in the long term.

(iv) *Investment in capacity building:* The introduction of targeted training for women farmers in SSA will improve their skills in modern agricultural practices, and business management and bring forth better levels of financial literacy and economic independence as well. It is crucial to ensure that their voices are heard in decision-making processes and foster leadership opportunities for them within organizations at local and national levels. This can be achieved by improving their representation in the existing systems through reservation or other policies.

(v) *Facilitate access to financial services:* Creating financial products that are tailored to the specific needs of women farmers can improve their standing and ensure greater participation in the labor force. The introduction of microfinance, flexible repayment terms, and collateral-free (unsecured) loans could boost credit availability in SSA and aid financial institutions in adopting gender-transformative lending practices. Promoting the formation of Self-Help Groups (SHG) among women farmers will facilitate savings, and credit availability and create a platform for collective bargaining.

(vi) *Comprehensive gender analysis:* The quality of gender analyses conducted has always been a concern in many countries. Before designing any kind of adaptation policies or programs, a comprehensive gender analysis should be conducted to explore the vulnerabilities and gender gaps that exist. It should focus on ownership, access, and control of resources and unequal power relations in decision-making. The analysis should aim at designing interventions that will close the existing gender gaps in the long run.

(vii) *Enhance data collection on gender disparities:* The establishment of systems or institutions for collecting gender-disaggregated data on agricultural productivity will help in tracking the progress toward the closure of existing gender gaps. The existing datasets are challenging to transcribe and insufficient for fully understanding the nature and extent of women's participation.

The countries in SSA are making significant commitments to policy development and strategic plans aimed at gender-transformative climate adaptation and agricultural practices. However, in many cases, progress may be delayed due to limited resources. To ensure successful policy implementation, specific funds should be allocated to the gender-focal units. It is widely believed that women in SSA possess a great wealth of indigenous knowledge that could improve existing agricultural practices. So, there should be platforms where they can share their ideas and put forward innovative solutions. By harmonizing the efforts of governments and agencies, the gendered action plans can be properly targeted and receive expected outcomes.

## References

Achampong L (2023) Gender-responsive climate finance: the key to just climate action and tackling inequalities. Expert Paper for the 68th session of the Commission on the Status of Women Expert Group Meeting. EuroDad and UN Women

Adeola O, Evans O, Ngare I (2024) Gender equality, climate action, and technological innovation for sustainable development in Africa. Sustainable development goals series. Springer International Publishing.https://doi.org/10.1007/978-3-031-40124-4

AfDB (2016) African development report 2015 growth, poverty and inequality nexus: overcoming barriers to sustainable development. African Development Bank

Ajadi AA, Oladele OI, Ikegami K, Tsuruta T (2015) Rural women's farmers access to productive resources: the moderating effect of culture among Nupe and Yoruba in Nigeria. Agricul Food Sci 4:1–9. https://doi.org/10.1186/S40066-015-0048-Y

Alston JM, Beddow JM, Pardey PG (2009) Agricultural research, productivity, and food prices in the long run. Science 325:1209–1210. https://doi.org/10.1126/SCIENCE.1170451

Brody A, Demetriades J, Esplen E (2008) Gender and climate change: mapping the linkage—a scoping study on knowledge and gaps. UK Department for International Development, Institute of Development Studies (IDS)

Chanana-Nag N, Aggarwal PK (2020) Woman in agriculture, and climate risks: hotspots for development. Clim Change 158:13–27. https://doi.org/10.1007/S10584-018-2233-Z

Chebet N (2023) The role of women on agricultural sector growth. Int J Agricul 8:41–50. https://doi.org/10.47604/ija.1980

FAO (2010) The state of food insecurity in the World. Food and Agriculture Organization of the United Nations

FAO (2018) The gender gap in land rights. Policy brief. Food and Agriculture Organization of the United Nations

FAO, IFAD, UNICEF, WFP, WHO (2023) The state of food security and nutrition in the World 2023: urbanization, agrifood systems transformation and healthy diets across the rural-urban continuum. Rome, FAO.https://doi.org/10.4060/cc3017en

GHI (2022) 2022-Global Hunger Index-Food Systems Transformation and Local Governance.

GHI (2023) 2023 Global hunger index: the power of youth in shaping food systems

Huyer S (2016) Closing the gender gap in agriculture. Gend Technol Dev 20:105–116. https://doi.org/10.1177/0971852416643872

Kemoe L, Lanci L, Spray J, et al (2022) Climate change and chronic food insecurity in Sub-Saharan Africa. Departmental Paper 2022:1–48https://doi.org/10.5089/9798400218507.087

Kidane W, Dardel P, Maetz M (2006) Food security and agricultural development in sub-Saharan Africa: building a case for more public support. Policy Assistance Series 2. Exp Agric 43:526–527. https://doi.org/10.1017/s0014479707005509

Kolovich LL, Newiak M, Alarakhia M, Ahmed ZS, Tanima T (2024) Gender inequality and care work: valuing and investing in care. In: Gender equality and economic development in Sub-Saharan Africa. International Monetary Fund. https://doi.org/10.5089/9798400246968.071

Kristjanson PM, Bryan E, Bernier Q et al (2017) Addressing gender in agricultural research for development in the face of a changing climate: where are we and where should we be going? Int J Agric Sustain 15:482–500. https://doi.org/10.1080/14735903.2017.1336411

Madsen DH (ed) (2020) Gendered institutions and women's political representation in Africa. Africa Rev 17:95–99. https://doi.org/10.1163/09744061-bja10171

Mozumdar L (2012) Agricultural productivity and food security in the developing world. Bangladesh J Agricul Econ 35:53–69

Ogwang BA, Kabengwela MH, Dione C, Kamga A (2018) The state of climate of Africa in 2017. African Centre for Meteorological Applications for Development

Otzelberger A (2011) Gender-responsive strategies on climate change: recent progress and ways forward for donors. Institute of Development Studies (IDS), BRIDGE Development-gender

Parvin GA, Ahsan SMR (2013) Impacts of climate change on food security of rural poor women in Bangladesh. Managem Environ Quality: an Int J 24:802–814. https://doi.org/10.1108/MEQ-04-2013-0033

Pidatala K, Khan AR (2003) Women and indigenous knowledge: a south-south perspective (English). Indigenous Knowledge (IK) Notes; No. 63 Washington, D.C. World Bank Group. http://documents.worldbank.org/curated/en/417911468302957741/Women-and-indigenous-knowledge-a-South-South-perspective

Raney T, Anríquez G, Croppenstedt A, et al. (2011) The role of women in agriculture. ESA Working Papers 289018, Food and Agriculture Organization of the United Nations, Agricultural Development Economics Division (ESA)

Rodgers Y, Akram-Lodhi H (2019) The gender gap in agricultural productivity in sub-Saharan Africa: causes, costs and solutions. Policy Brief 11.https://doi.org/10.7282/00000259

Rosegrant MW, Agcaoili-Sombilla MC, Perez ND (1995) Global food projections to 2020: implications for investment. Food, agriculture, and the environment discussion paper, 2020 discussion paper 5. Washington, DC, International Food Policy Research Institute

Schalatek L, Nakhooda S (2022) Gender and climate finance. Climate finance fundamentals, 10. Heinrich Böll Stiftung, Washington, DC and ODI

Slavchevska V, Doss CR, de la O Campos AP, Brunelli C (2021) Beyond ownership: women's and men's land rights in Sub-Saharan Africa. Oxford Developm Stud 49:2–22.https://doi.org/10.1080/13600818.2020.1818714

UNDP (2023) Gender-responsive climate change actions in Africa. Climate Promise Technical Paper. United Nations Development Programme

Vijay S, Yadav K (2022) Understanding gender justice through the lenses of Amartya Sen's capability approach. Specialusis Ugdymas 1:7491–7496

Zepeda L (2001) Agricultural investment, production capacity, and productivity. FAO economic and social development paper, pp 3–20

# Chapter 9
# Climate Change Migration and Gender Vulnerability: Challenges and Solutions

**Junaid ul Shafi**

**Abstract** This chapter explores the gendered impacts of climate change on migrants and refugees, highlighting the vulnerabilities faced by women and sexual minorities. It emphasizes that climate change is a significant driver of migration, worsening existing gender inequalities that limit access to resources, health care, and decision-making. The chapter advocates gender-responsive legal frameworks and institutional support systems tailored to the unique needs of climate-affected populations. Utilizing a mixed-methods approach, the study analyses existing literature, migration statistics, and policy documents to identify current challenges and propose equitable interventions. The findings call for integrating gender sensitivity into legal and institutional frameworks to enhance protection and support for those displaced by the negative impacts of climate change.

**Keywords** Climate change · Displacement · Migration · Gender vulnerability · Refugees · Gender-based violence · Legal reform

## Introduction

Climate change represents one of the most pressing global challenges of our time, with far-reaching effects on ecosystems, economies, and human societies. Among its myriad impacts, one of the most significant and urgent is climate-induced migration and displacement. As environmental conditions deteriorate exemplified by rising sea levels, increased frequency of extreme weather events, and prolonged droughts, communities across the planet are compelled to relocate. This phenomenon is not only a matter of geography but also bears profound social, political, economic, and gender dynamics. This phenomenon of climate-induced displacement and forced migration presents a complex yet unique web of challenges, particularly when viewed through the lens of gender dynamics.

J. Shafi (✉)
School of Law, BML Munjal University, Gurgaon, India
e-mail: junaidulshafi@gmail.com

 
A. Sarkar and N. Bandyopadhyay (eds.), *Gender-Transformative Approaches for Climate Change Adaptation*, Springer Climate,
https://doi.org/10.1007/978-3-031-90236-9_9

The intersection of climate change migration and gender vulnerability is particularly critical. While climate-induced displacement affects all individuals, the impacts are often disproportionately felt by women and gender minorities (Deshar and Koirala 2020). Gender-specific roles, responsibilities, and vulnerabilities mean that the challenges faced by these groups can differ markedly from those experienced by men. Additionally, gender disparities in access to resources, decision-making authority, and social protection further exacerbate their vulnerability during and after migration (Silchenko and Murray 2023). Gender is often conceptualized through a male–female binary, framed as opposing forces rather than acknowledging the complex interdependencies between different gender identities. This limited perspective can lead to the development of maladaptive policies and interventions that fail to address the nuanced needs of diverse gender experiences, resulting in culturally inappropriate and ineffective solutions (Daoud 2021).

## Material and Methods

This chapter explores the complex relationship between climate change, migration, and gender vulnerability. It examines the intricate relationship between climate change, migration, and gender, with a particular focus on the vulnerabilities faced by women and sexual minorities among climate change migrants and refugees. Through a combination of narrative literature review and empirical case studies, the chapter uncovers the nuances of gendered experiences in climate-induced migration and proposes strategies to address these disparities. The chapter adds knowledge to the important issue of climate-induced displacement and contributes to the growing body of research on climate justice and human rights informing policymakers, legal practitioners, and humanitarian organizations. It advocates for more inclusive and equitable gender-responsive approaches to address the complex challenges faced by climate refugees.

This research employs a mixed-methods approach, combining a comprehensive narrative literature review with both qualitative and quantitative data analysis. It draws upon existing research on climate change, migration, and gender disparities. It analyses migration statistics and climate impact reports, policy documents, statutes, international legal conventions, and institutional practices. The chapter critically examines the scholarly debates surrounding displacement due to climate change, with a focus on the global south. By analysing a broad range of scholarly articles, government reports, and international publications, the study investigates how gaps in data have shaped policy decisions and the discourse surrounding migration. The research specifically focuses on a review of the literature on climate-induced displacement in the Global South.

The chapter is divided into three broad sections. In the first section, the gendered impacts of climate change on migrants and refugees are highlighted with the specific vulnerabilities experienced by women and sexual minorities. The second section analyses the existing legal frameworks and institutional support systems, evaluating their efficacy in addressing the needs of climate change migrants from a gender perspective. The third and final section proposes recommendations for integrating gender sensitivity into legal and institutional frameworks, aiming to promote a more equitable and comprehensive response to climate-induced displacement.

## Climate Change Migration and Displacement

Human migration is inherently complex and varied, encompassing a wide range of situations. However, climate-related disasters are leading to more internal displacement than politically induced factors (Borras-Pentinat 2024). Climate change is emerging as a key driver of global displacement and migration (Dankelman 2010) with its impacts being unevenly distributed across regions and populations. Rising sea levels, extreme weather events, prolonged droughts, and resource scarcity are forcing millions of people to migrate, particularly from vulnerable, low-income countries (Ionesco et al. 2016). This phenomenon is often referred to as "climate-induced displacement" or "environmental migration" (McLeman 2013. When affected populations are unable to adapt to the changing climate conditions and extremes in their areas of origin, they are compelled to relocate from their original locations or countries (Sedova and Kalkuhl 2020). The International Organization for Migration (IOM) defines climate migrants as individuals or groups who, due to sudden or gradual environmental changes that negatively impact their living conditions, are compelled or choose to leave their homes, either temporarily or permanently, and relocate within their country or internationally (IOM 2021).

The interplay between rising sea levels, extreme weather events, disappearing territories (McAdam 2010), and resource scarcity is creating unprecedented migration pressures, particularly for those in the most vulnerable and economically disadvantaged regions (Castles 2017). By 2050, climate change is expected to result in the migration of around 25 million to 1 billion people, both within and across national borders, with an estimated 80% of the displaced population being women and girls (Habtezion 2016a, b). Economic insecurity and poverty have significantly constrained individuals' ability to recover from climate change-related disasters. In 2021, disasters displaced 23.7 million people across 141 countries, with over 94% of these displacements attributed to climate-related disasters. By year-end, nearly 6 million individuals remained displaced in 84 countries and territories due to such disasters. In 2022, climate-related and other anthropogenic disasters forced over 100 million people to flee their homes, often seeking refuge in already vulnerable communities. In 2023, there were 46.9 million new internal displacements, with 56% resulting from disasters (IDMC 2024). By the end of 2023, at least 7.7 million people in 82 countries and territories were reported internally displaced due to disasters. This marks an 11% decrease in the total number of internally displaced persons (IDPs) due to disasters by 2023 as compared to 2022. This situation exacerbates existing risks, including water and food insecurity, disease outbreaks, resource competition, marginalization, and conflict between host and displaced populations (Schuster et al.

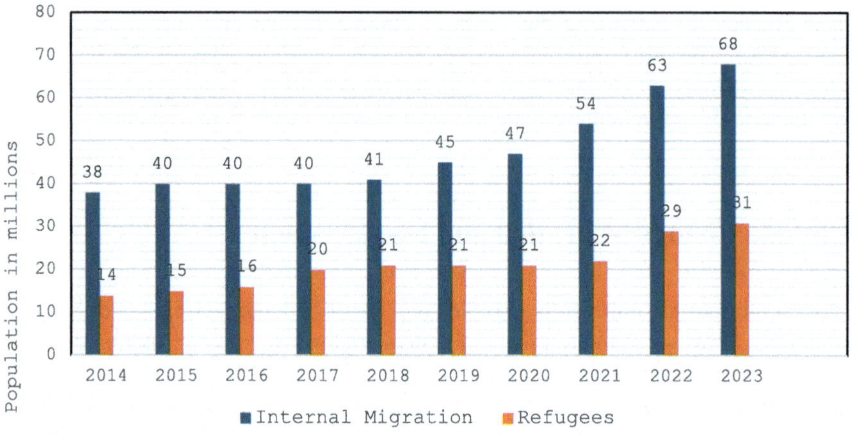


**Fig. 9.1** Global trends of forced displacement 2014–2023. 2023 *Source* UNHCR, Global Trends forced Displacement,Copenhagen

2024). A significant majority of climate-induced migrants and displaced people move to urban centres (Mosel and Jackson 2022) (Fig. 9.1).

As negative impacts of climate change increasingly affect livelihoods and resources, it is anticipated to drive migration. However, the relationship between climate change and migration is complex. Few studies on migration explore the pathways through which climate change may drive people to migrate. Migration may occur as a result of climate impacts on livestock production and destruction of agriculture (Gray and Mueller 2012), as well as from extreme weather events, including floods, cyclones, and prolonged or intensified droughts (Davis et al. 2018; Black et al. 2011). The relationship between climate change and migration varies according to the specific climate scenarios such as droughts and rising temperatures (Thiede and Gray 2017). These events can significantly undermine living standards at both individual and community levels, leading to displacement especially when regions become uninhabitable due to the severe destruction by extreme climate phenomena (Gemenne et al. 2022). Contemporary research has identified climate change as a critical driver of political instability and armed conflict, which can catalyse international migration, as exemplified by the situation in Syria (Abel et al. 2019; Reuveny 2007; Aksu and Sirkeci 2023).

A limited number of studies have explored the relationship between extreme environmental events and migration levels. For example, a spatial regression analysis by Saldaña-Zorrilla et al. (2009) focusing on Mexico reports that emigration rates between 1990 and 2000 were higher in Mexican municipalities experiencing frequent disasters, such as droughts, floods, and hurricanes, compared to neighbouring areas. Carvajal and Pereira (2010) provide a more robust argument, showing that households affected by heavy rains from Hurricane Mitch in Nicaragua were significantly more likely to migrate than those unaffected by such rainfall. Additionally, various case studies (Tinoco 2023; Asad 2015) on post-disaster migration underscore the complexity of migration patterns rather than indicating straightforward correlations (Piguet 2010).

## *Gendered Causes of Climate Migration*

It is widely acknowledged that climate change has distinct gendered impacts (Dankelman 2010). Migration, whether voluntary or forced, is rarely a gender-neutral process. Gender is a crucial factor in migration, influencing the drivers, impacts, and outcomes of the process (Christou and Kofman 2022). A gender-sensitive examination of climate change migration, as noted by feminist scholars such as Chant (2013) and Gaard (2015), illuminates the complex ways gender shapes migration patterns and responses to environmental degradation. Although environmental factors affect entire populations, their impacts are disproportionately experienced by women and marginalized gender groups due to existing socio-economic inequalities, cultural norms, roles, expectations, and structural power imbalances (Nagel 2015; De Jong 2000). Women, men, and non-binary individuals often encounter distinct challenges and opportunities as they navigate displacement, resettlement, and integration, influenced by socio-cultural, legal, and economic structures (Brettell 2017).

Historically, migration studies focused primarily on male migrants, particularly labour migration, sidelining the experiences of women, and other gender minorities (DeLaet 1999). However, contemporary scholarship has highlighted the need for a more nuanced understanding of how gender shapes migration flows and patterns. Women, in particular, have become increasingly visible as autonomous migrants, whether as labourers, refugees, or asylum seekers (Ghorashi 2010). Their migration trajectories are often motivated by both economic necessity and gender-based violence, which complicates the binary categorization of migrants as either economic or forced migrants. Gender relations, rather than biology, shape migration patterns, including who migrates, in what proportions, and with what goals and motives, as well as the paths and organization of migration streams (Donato and Gabaccia 2015). Migration impacts family life, parenting, childcare, power relationships between spouses, and even migrant mother–child relationships (Nawyn 2019).

One major gendered driver of climate-induced migration is the heightened socio-economic vulnerability of women, particularly in developing countries (Williams 2015). Women in agrarian communities are more reliant on natural resources, which are increasingly threatened by climate change (Gaard 2015). Climate changes such as droughts, desertification, and unpredictable rainfall patterns disrupt food security and income sources, especially for women responsible for agricultural work (Alfthan et al. 2011). Furthermore, scholars such as Denton (2002) argue that women's limited access to critical resources, such as land and credit, diminishes their resilience to climate shocks and weakens their capacity to recover. In comparison, men often have better access to migration channels, which positions them to migrate in search of employment, while women are left behind to manage precarious household economies. Approximately 2/3rd of the 1.3 billion people living in poverty are women, many of whom rely on increasingly threatened natural resources for their survival as climate impacts intensify (Habtezion 2016b: Jones and Presler-Marshall 2012).

Gendered social norms restrict women's mobility, increasing their vulnerability to climate change, particularly during sudden-onset events (Castillo Betancourt and Zickgraf 2024). In many communities, women are bound by caregiving responsibilities for children, the elderly, and other dependents, which limits their ability to migrate, even in the face of environmental crises (Chant 2013). This "immobility paradox" traps women in environmentally degraded areas, heightening their exposure to resource scarcity and climate-related risks (Findlay 2011; Fischer and Malmberg 2001). For example, rural women who are mostly reliant on agriculture are severely impacted by climate change (Brown et al. 2012). These socio-economic vulnerabilities not only hinder women's ability to migrate but also increase their dependence on unstable local resources, compelling them to remain in environmentally degraded areas despite the escalating risks posed by climate change.

## *Climate Change and Vulnerability: The Disproportionate Impact on Women*

Climate refugees who are women, and children, and belong to sexual minorities face heightened vulnerabilities (Masika 2002). Evidence shows that women and children are 14 times more likely to lose their lives in climate-related disasters than men (Aguilar 2009; UNHCHR 2022) and 80% more likely to be displaced (Mastrorillo et al. 2016). For instance, in the 1991 cyclone and flood in Bangladesh, women comprised 90% of the victims due to several factors: lack of warning information directed at women, their confinement to homes, lack of swimming skills, and caregiving responsibilities which hindered their escape (Cannon 2002). According to the United Nations Development Program (UNDP) and the Global Gender and Climate Alliance, women and children represent nearly 80% of all individuals displaced by climate-related events (UNDP and GGCA 2016). Women often experience exacerbated vulnerabilities in the context of climate-induced migration due to socio-economic and cultural factors that limit their access to resources and decision-making processes (Enarson 2012).

Cultural, social, and religious norms related to gender roles can restrict women's ability to make timely decisions during disasters (Rashid and Michaud 2000; Logie and Daniel 2016). Additionally, factors such as clothing norms and childcare responsibilities may impede their mobility in emergencies. Economic and social disparities, along with limited access to productive resources, exacerbate climate vulnerabilities and further entrench discriminatory practices, thereby affecting women's capacity to migrate.

Gender-specific cultural norms can severely restrict women's freedom of movement and limit their ability to seek employment or integration into new environments. Becker (2019) highlights that such norms can impede women's access to essential services and community activities, resulting in increased isolation (Alston 2015). Social isolation is further exacerbated by the lack of support networks in new locations, which impairs women's ability to secure employment, access services, and effectively rebuild their lives. The decline of social capital following climate-related displacement complicates individuals' ability to manage migration challenges and adapt to new environments (Chowdhury et al. 2020). Research indicates that the relocation of migrants undermines social structures, such as extended household systems, and diminishes adherence to traditional laws, particularly for women (McMichael et al. 2023). Establishing formal spaces, such as community centres, can significantly improve migrants' well-being by providing opportunities for communal engagement and addressing shared concerns (Yang et al. 2024). Women also bear a heavier burden in caregiving responsibilities, which can be aggravated by displacement.

The loss of livelihoods and community support networks intensifies these burdens, making it even more challenging for women to rebuild their lives (Alston 2013). Additionally, women face obstacles in securing employment and educational opportunities, often due to discrimination, unrecognized qualifications, or cultural barriers (Djoudi et al. 2016). Women from poor rural families who migrate to cities often end up in low-paying jobs, such as domestic labour or sweatshop employment, due to a lack of education and skills (Kakissis 2010). Similar patterns are observed in the Philippines, where women from fishing communities, affected by climate change, migrate to work as domestic workers for affluent families (Rúa 2010). Similarly, the prolonged drought in the Horn of Africa has exacerbated migration pressures, with notable impacts on women and sexual minorities ((Gómez-Álvaro and Caro-Carretero 2024). This perpetuates cycles of poverty and dependency, hindering their ability to adapt and thrive in new environments.

Women often face barriers in accessing direct relief aid due to not being recognized as the "head of the household" (Spring 2008), leaving them more vulnerable. After Cyclone Aila in 2009, which struck the climate-vulnerable Indian Sundarbans, many men migrated due to loss of livelihood, leaving women behind. These women, facing gender discrimination and rigid social roles, had limited alternative income sources. As a result, many of them migrated to the red-light district of Kolkata, contributing to a 20–25% increase in the population of migratory sex workers in that area (EPW Engage 2020). Similarly, in the drought-prone regions of Gujarat Western India, the rural population is profoundly impacted by intense and recurrent droughts, resulting in substantial financial losses, job displacement, widespread migration, a rise in prostitution, and an alarming increase in suicide rates (Bandyopadhyay et al. 2020). In the aftermath of annual flooding in Assam, Northeast India, many women and girls were compelled into child slavery or forced marriages as a means of economic survival (Bharadwaj et al. 2021).

Climate change migrants and refugees often struggle to access essential resources and services in host communities (McGregor 1994). These challenges are exacerbated for women and children due to discriminatory practices, language barriers, and inadequate gender-sensitive support systems (Chindarkar 2012). Access to health care, particularly sexual and reproductive health services, can be particularly problematic, resulting in adverse health outcomes and heightened vulnerability (Ngcamu 2023). The sexual and reproductive health needs of women are frequently neglected in displacement situations (Onyango and Heidari 2017). Disruptions to supply chains and limited access to contraception contribute to higher rates of unplanned pregnancies under precarious conditions (Carballo et al. 2005). Even when healthcare services are accessible, displaced women and girls often face significant barriers that increase their vulnerability to unwanted pregnancies, unsafe abortions, and reproductive health-related complications (Stark et al.2015). Additionally, restrictive gender norms surrounding sexual activity can further stigmatize and hinder the use of sexual and reproductive health services (Schuster et al. 2024).

Women are more vulnerable to physical and mental health issues due to displacement and inadequate healthcare access (Lori and Boyle 2015; Thomas and Thomas 2004). Women also experience the psychosocial impacts of natural disasters more acutely than men, with the disintegration of social ties and family separation worsening their distress (Enarson 2016). A study in The *Lancet* (2021) indicates that the migration process often worsens pre-existing health conditions and introduces new health risks, including complications related to maternal and reproductive health.

The forced displacement of humans, especially women and children, amplifies the risk of sexual violence and exploitation. This issue of heightened exposure to gender-based violence (GBV) and sexual and gender-based violence (SGBV) has been a recurring theme in the literature (Piguet et al. 2011). The International Union for the Conservation of Nature has established that heightened environmental stress and resource scarcity are correlated with an increase in GBV (Harvey 2020). This violence encompasses various forms, including domestic abuse, sexual assault, rape, forced prostitution, forced marriages, and an elevated incidence of human trafficking, particularly in regions experiencing environmental distress.

Stone et al. (2022) highlight that GBV tends to increase following extreme weather events. Typhoon Haiyan, which devastated the Philippines in 2013, serves as a compelling case study for examining the gendered impacts of climate-induced migration (Valerio 2014). This disaster displaced millions of individuals and disproportionately affected women, who encountered significant challenges including heightened risks of GBV and severe economic hardship (Tummala-Narra 2021). The Action Aid and IDS reports highlight that inadequate safe shelters during evacuation or forced migration pose significant risks. In temporary shelters, refugee camps, or urban informal settlements where many climate migrants take shelter experience overcrowding and a lack of privacy. This can create environments conducive to sexual harassment and assault (Mitchell et al. 2007). Enarson and Morrow (1998) found that risks of GBV, including sexual assault and trafficking, are heightened due to dangerous living conditions. Women and children often endure abuse, including SGBV, lack of privacy, and mental trauma, with perpetrators frequently remaining unpunished.

Neira et al. (2023) note that displacement can lead to scarce resources and uncertain legal status, which may compel some individuals to engage in sex work for survival (Jafree and Mukhopadhyay 2020). This situation also results in increased rates of kidnapping, rape, sexual exploitation, sexually transmitted infections (STIs), and unintended pregnancies. For example, in Vanuatu, Cyclone Vania and Cyclone Atu in 2011 triggered a 300% increase in GBV against women.

A joint study by the United Nations International Centre for Integrated Mountain Development and the Centre for International Climate and Environmental Research found that the trafficking of girls and women, along with other forms of SGBV, has significantly increased in Southern Asia after extreme weather events and disasters (Desai and Mandal 2021). Ramachandran (2005) in a study reported that undocumented women migrants from Bangladesh were particularly at risk of trafficking and exploitation due to their illegal status.

## *Climate Change-Induced Displacement, Gender Intersectionality, and Their Vulnerabilities*

The term "women"encompasses a diverse group, with everyone's social, economic, and environmental circumstances shaped by numerous factors: ethnicity, income, education, class, sexuality, ability, age, nationality, and motherhood. The intersectionality of gender, race, and class further complicates climate-induced migration for women. Drawing on the work of Crenshaw (2018) and Kaijser and Kronsell (2014), it is evident that women from marginalized racial or ethnic groups face compounded disadvantages in both their home communities and host regions. These intersecting identities limit their access to resources, employment, and legal protection, reinforcing systemic inequalities.

Climate change-induced displacement significantly impacts gender minority individuals, exposing them to heightened risks and discrimination. As extreme weather events, rising sea levels, and droughts increasingly compel people to leave their homes, individuals with diverse sexual orientations, gender identities, or expressions (SOGIE) encounter particularly severe challenges. These individuals often face heightened risks of persecution, including physical and sexual violence, abuse, and discrimination as documented in many countries (Alessi et al. 2017). Displacement not only heightens existing inequalities but also introduces new forms of marginalization and exploitation for these individuals. Sexual minorities face compounded vulnerabilities due to intersecting forms of oppression related to both their sexual orientation and their status as displaced persons. Research also shows that authorities disproportionately target, arrest, detain, and displace LGBTQ + individuals during global and national crises (Thoreson 2020).

Sexual minorities often experience dual layers of marginalization: one from the effects of climate change and another from societal and legal discrimination based on their sexuality. For LGBTQ individuals displaced by climate change, the transit phase of migration is particularly complex and perilous (Yarwood et al. 2022). The journey is often nonlinear, with transit points sometimes becoming permanent destinations (Ansems et al. 2016). Migrants frequently face irregular border crossings, and lack of access to official migration services, and may only claim asylum upon reaching a final destination. This period is fraught with dangers such as smugglers, traffickers, armed groups, arbitrary detention, and homelessness. For many individuals, especially those from marginalized groups, the journey from their place of origin to their destination often involves various forms of abuse, exploitation, sexual violence, and, in some cases, even kidnapping and torture (Campana 2018). Migrants with diverse SOGIE often face heightened risks of abuse, discrimination, and limited access to services during transit compared to the general migrant population (Hunter 2010). These challenges arise from both cisgender, heteronormative forced migrants, and authorities or locals, especially if their identity is unintentionally exposed. Women and LGBTQ + individuals face elevated risks of exploitation and violence within refugee camps (Mastrorillo et al. 2016).

Legal and institutional barriers further compound these vulnerabilities. Many countries lack legal protections for sexual minorities, and existing frameworks often fail to address their specific challenges. This lack of recognition can limit access to essential health services and social support, increasing susceptibility to exploitation and health crises. A study of LGBTQ + refugees in Congo and Uganda highlighted significant invisibility and dehumanization, with participants reporting marginalization and their identities perceived as "haram" (Ndiritu 2021). In many regions, LGBTQ individuals face both societal stigma and legal barriers, which are exacerbated in unfamiliar environments where their identities may be further stigmatized or criminalized. In several countries, sexual minorities are subjected to severe punishments under laws that include death sentences (Cheney 2012). LGBTQ + or SOGIE communities grapple with violence, harassment, and significant barriers to accessing essential healthcare. During displacement, LGBTQ + individuals often endure violence and harassment, with barriers to accessing essential healthcare and institutional protection. Upon arrival at their destination, they continue to face pervasive discrimination. For example, a survey of LGBTQ + refugees in Kenya found alarming mistreatment rates with 93% of respondents reporting verbal insults, 83% experiencing physical assault, and 88% with denial of police assistance (Ndiritu 2021).

Furthermore, the challenges faced by transgender and gender-diverse individuals are compounded by the unique vulnerabilities they encounter during displacement, necessitating a closer examination of the specific forms of abuse and discrimination they endure. Transgender and gender-diverse individuals are particularly vulnerable to severe abuse, including beatings, rape, and even murder. Navigating the asylum process adds to these challenges. LGBTQ + individuals may struggle to prove their sexual orientation or gender identity, which can expose them to additional fears of persecution or being socially ostracized (Alessi et al. 2020; Gray et al. 2015). LGBQ women face significantly higher rates of physical and sexual assault compared to heterosexual women (Balsam et al. 2005; Morocco et al. 2007), and those who are more visibly "out" are at greater risk of violence (Huebner et al. 2004).

## Towards Inclusive Legal Frameworks: Protecting Climate-Displaced Individuals and Marginalized Groups

To effectively address the multifaceted needs of climate-displaced populations, including women and sexual minorities, legal frameworks must integrate gender sensitivity and tackle structural inequalities. Current legal frameworks and statistical data often fail to integrate a gender perspective, perpetuating the invisibility and lack of protection for gender minorities in climate-induced migration (Borras-Pentinat 2024). Legal regimes have historically adapted to shifting international norms, but they have made little progress in recognizing climate refugees as a legal category (Williams 2008). Existing policies and legal frameworks fail to differentiate and protect climate migrants or address the specific needs of women and LGBTQ migrants, perpetuating inequalities and vulnerabilities. This legal gap also obscures the responsibilities of industrialized nations, particularly in the Global North, while often criminalizing victims from the Global South who flee climate-induced disasters in search of safety and dignity (Fernández and Peek 2024; Shafi 2024; Anand 2017). As climate change accelerates and impacts socio-economic conditions, there is an urgent need for legal protections tailored to climate refugees. These protections must address ecological and gender inequalities to prevent perpetuating existing disparities and violence. They should consider how climate change affects different regions and incorporate gender as a fundamental organizing principle, influencing mobility patterns.

The 1979 Convention on the Elimination of All Forms of Discrimination against Women underscores the critical need for gender equality and protection of women's rights in the face of climate change. While the 1995 Beijing Platform does not explicitly address legal protections, it emphasizes the inclusion of women in decision-making and environmental stewardship. The Sustainable Development Goals (SDGs) prioritize gender equality, recognizing women's key role in development. Similarly, UN Security Council Resolution 1325 highlights the importance of safeguarding women during climate crises. The International Bar Association affirms that gender equality is central to climate justice, as climate change heightens migration, food insecurity, and women's vulnerability to violence (Wilson and Tisdell 2001). Integrating gender sensitivity into legal and institutional frameworks is essential for effectively addressing the complex challenges faced by displaced populations affected by climate change. Gender-responsive approaches acknowledge that women, sexual minorities, and other marginalized groups encounter distinct vulnerabilities and challenges that require tailored interventions and policies.

Research demonstrates that gender disparities significantly impact the effectiveness of disaster response and migration management. For instance, Alston (2013) highlights that without gender-sensitive policies, women and marginalized groups are frequently excluded from crucial support services and protections. Implementing gender-responsive frameworks is vital for ensuring equitable access to resources and protecting the rights of all displaced individuals, thereby improving the overall effectiveness of disaster response and migration policies (Enarson 2012). Moreover, adopting gender-sensitive approaches helps address structural inequalities that intensify the impacts of climate change. By embedding gender considerations into policy and legal frameworks, governments and institutions can effectively respond to the specific needs of women and sexual minorities. This approach not only protects them from violence and exploitation but also promotes their full participation in decision-making processes (Mastrorillo et al. 2016).

The 2016 UN General Assembly New York Declaration for Refugees and Migrants acknowledges the unique needs of women in transit and establishes specific commitments to integrating a gender-responsive perspective into appropriate responses. The Declaration emphasizes the promotion of gender equality and the empowerment of all women and girls, ensuring that their human rights are fully respected and protected. It specifically commits to addressing SGBV and ensuring access to sexual and reproductive health care services. Furthermore, the Declaration addresses the multiple and intersecting forms of discrimination that affect refugee and migrant women and girls. Additionally, the Declaration highlights the significant contributions and leadership of refugee and migrant women, underscoring the importance of their full and meaningful participation in the development of local solutions. This commitment reflects a broader recognition of the essential role that women play in addressing and resolving the challenges associated with displacement and migration. The UN Framework Convention on Climate Change (UNFCCC) established the Task Force on Displacement within the Warsaw International Mechanism to recommend strategies for addressing displacement linked to climate change. Social protection, a critical aspect in this context, gained prominence during the COVID-19 pandemic (Drolet et al. 2020) and could play a pivotal role in implementing UNFCCC decisions on Loss and Damage funding as highlighted at COP27 (UNFCCC 2022). The commitment to gender-sensitive approaches in climate migration policies signifies a transformative shift towards recognizing the interconnectedness of social justice and environmental sustainability. Thus, incorporating these perspectives into legal frameworks is essential for addressing structural inequalities and ensuring that all voices are effectively represented in the development of climate policies.

## *The Invisibility of Climate Migrants: Legal Challenges and Implications*

Though climate change is acknowledged as a significant driver of migration and displacement, individuals displaced by climate-related factors are not recognized as refugees under current international law (McAdam 2012). The 1951 Refugee Convention and its 1967 Protocol define refugees as persons who have fled their country due to a well-founded fear of persecution based on race, religion, nationality, membership in a particular social group, or political opinion. This definition excludes those displaced by environmental degradation or climate-related events (Williams 2008). There is no consensus on whether the definition of "refugee" should be expanded to include individuals displaced by climate change. Some scholars argue that broadening the definition may not effectively address the needs of those displaced by climate-induced events.

Moniruzzaman et al. (2018) argue that climate migrants may face vulnerabilities exceeding those of individuals recognized under the UN Refugee Convention. Despite their heightened susceptibility to adverse conditions, climate migrants lack formal recognition and protection. This gap underscores the inadequacy of existing international legal frameworks, such as the 1951 Refugee Convention, to address the realities of climate-induced displacement. Black (2001) supports this perspective by differentiating between "environmental refugees" and traditional categories such as "internally displaced persons" (IDPs) or those displaced by development projects. Black's analysis highlights the limitations of the current refugee definition and calls for a broader understanding of displacement that includes environmental factors. The absence of legal recognition for climate-displaced individuals has significant implications (Atapattu 2020). Without formal refugee status, climate migrants are often excluded from international protection mechanisms and humanitarian aid designed for refugees (Behrman and Kent 2018). This lack of recognition can exacerbate their vulnerability, leaving them without access to essential services and legal resources.

## A New Legal Paradigm: Protecting Climate-Induced Displaced Persons Through Reform

To address these emerging challenges, there is an increasing call within the academic and legal communities for the establishment of novel legal frameworks that encompass climate-induced displacement. Proposals suggest either expanding the scope of the Refugee Convention or developing supplementary international agreements specifically tailored to climate migrants. Such measures are advocated as necessary to offer a more robust and nuanced approach to safeguarding the rights and addressing the needs of those affected by climate change. An effective response to the issues faced by individuals displaced by climate change necessitates a legal framework that explicitly incorporates these individuals and acknowledges their distinct vulnerabilities within the international legal system.

A critical aspect of this legal framework involves integrating gender-sensitive provisions to address the unique vulnerabilities faced by women and sexual minorities in the context of climate-induced displacement. This framework should involve several key reforms. First, expanding the 1951 Refugee Convention or creating a complementary protocol to encompass climate-induced displacement, thus broadening the scope of protection to include those displaced by environmental degradation and extreme weather events. Second, it is essential to integrate gender-sensitive provisions into both international and national laws to address the specific needs of women and sexual minorities, including their heightened risk of GBV and the need for tailored support services such as safe shelters, legal aid, and health care

(Mitchell et al. 2007; Enarson and Morrow 1998). Third, the policy development process should actively involve women and sexual minorities to ensure their perspectives are represented, thereby enhancing the effectiveness of legal frameworks and support services. Fourth the policy must include provisos for social protection of those displaced, particularly children, women, and sexual minorities. While research often focuses on migrants' access to social protection (Vera Espinoza et al. 2021) and migration as a coping strategy (Hagen-Zanker and Himmelstine 2013; Schüring et al. 2017; Long and Sabates-Wheeler 2017) less attention has been given to social protection as an alternative to migration. Fourth, the plight of LGBTQ + individuals during and after displacement underscores the urgent need for tailored support and protective measures. Their compounded vulnerabilities, exacerbated by violence, harassment, and systemic discrimination, highlight the critical gaps in humanitarian aid and legal frameworks. Addressing these issues requires a concerted effort from governments, NGOs, and international bodies to ensure that LGBTQ + displaced persons receive the safety, respect, and resources they deserve.

Finally, fostering international cooperation to establish and implement guidelines and agreements specifically for climate-induced displacement is crucial, including providing financial and technical support to countries hosting displaced individuals. Thus, an inclusive legal framework is needed that broadens refugee definitions, incorporates gender-sensitive measures, engages affected communities, and promotes international collaboration that is essential to adequately protect and support climate change-induced displaced populations. This legal recognition will help in improving their resilience and well-being and enhancing the overall efficacy and fairness of international protection mechanisms. In other words, the development of an inclusive legal framework is imperative for addressing the needs of those displaced by climate change. Expanding the Refugee Convention, integrating gender-sensitive provisions, involving affected communities in policy development, and enhancing international cooperation are crucial steps towards providing comprehensive protection. By adopting these measures, the international community can better safeguard the rights and well-being of climate-displaced populations, ensuring a more effective and equitable response to the evolving challenges posed by climate change.

## Conclusion

The gender impacts of climate change on migrants and refugees are multifaceted and deeply rooted in existing social, economic, gender, and cultural inequalities. Recognizing and addressing these specific vulnerabilities is crucial and urgent for developing effective, inclusive, and equitable responses to climate-induced displacement. This urgency is amplified in contexts of severe climate impacts, which exacerbate violations of women's rights. Pre-existing gender discrimination places women in disparate positions when confronting forced climate-induced migration, further compounded by intersecting vulnerabilities. Therefore, addressing the migratory ramifications of climate change is essential.

It is imperative to recognize that women and sexual minorities are not merely fleeing the adverse effects of climate change but also escaping from conditions that are influenced or exacerbated by persecution and socio-economic exclusion. The safeguarding of individual civil rights in the face of systemic discrimination, multidimensional inequality, and GBV especially against women and girls is always paramount but it becomes even more critical during instances of forced displacement.

Assessing individual vulnerabilities and climate risks is crucial to ensure the protection of fundamental rights, including freedom from cruel or degrading treatment and the right to life, particularly for women and LGBTQ individuals. As climate change increasingly drives migration, particularly affecting populations in the Global North, adapting the Geneva Convention and applying the principle of non-refoulement to include climate migrants becomes essential. Innovative legal and policy responses are necessary to address the unique humanitarian challenges posed by climate change. This includes integrating gender-sensitive measures and recognizing intersectional vulnerabilities within existing refugee and migration frameworks. States must embrace these international developments, incorporating gender-responsive strategies to safeguard the rights of migrant women and girls. Legal reforms should focus on recognizing climate change as a migration driver and implementing protective measures that address gender-based vulnerabilities through both mitigation and adaptation efforts.

As the global community continues to grapple with the increasing reality of climate-induced displacement, gender considerations must remain at the forefront of policy discussions and practical interventions. Gender-sensitive legal rights must ensure protection from violence, exploitation, and discrimination, while institutional support should provide targeted assistance and foster inclusive decision-making processes. Only through a concerted effort to integrate gender sensitivity at all levels can we hope to create a more just and sustainable future for all those affected by climate change and forced migration.

## References

Abel GJ, Brottrager M, Crespo Cuaresma J, Muttarak R (2019) Climate, conflict and forced migration. Glob Environ Chang 54:239–249. https://doi.org/10.1016/j.gloenvcha.2018.12.003

Aguilar L (2009) Women and climate change: women as agents of change. Training Manual on Gender and Climate Change. International Union for Conservation of Nature (IUCN)

Aksu Kargın İ, Sirkeci I (2023) Understanding Syrian refugees in Turkey from an environment of insecurity and the conflict model of migration perspective. Third World Q 44:856–871. https://doi.org/10.1080/01436597.2022.2161358

Alessi EJ, Kahn S, Van Der Horn R (2017) A qualitative exploration of the Premigration victimization experiences of sexual and gender minority refugees and Asylees in the United States and Canada. The Journal of Sex Research 54:936–948. https://doi.org/10.1080/00224499.2016.1229738

Alessi EJ, Kahn S, Greenfield B, Woolner L, Manning D (2020) A qualitative exploration of the integration experiences of LGBTQ refugees who fled from the Middle East, North Africa, and

Central and South Asia to Austria and the Netherlands. Sexuality Res Soc Policy 17:13–26. https://doi.org/10.1007/s13178-018-0364-7
Alfthan B, Skaalvik J, Johnsen K, Sevaldsen P (2011) Women at the frontline of climate change: gender risks and hopes
Alston M (2013) Women and adaptation. Wires Clim Change 4:351–358. https://doi.org/10.1002/wcc.232
Alston M (2015) Women and climate change in Bangladesh. Routledge
Anand R (2017) International environmental justice: a North-South dimension. Taylor and Francis
Ansems de Vries L, Carrera S, Guild E (2016) Documenting the migration crisis in the Mediterranean: spaces of transit, migration management and migrant agency. CEPS Pap Lib Secur Eur 94
Asad AL (2015) Contexts of reception, post-disaster migration, and socioeconomic mobility. Popul Environ 36:279–310. https://doi.org/10.1007/s11111-014-0221-4
Atapattu S (2020) Climate change and displacement: protecting 'climate refugees' within a framework of justice and human rights. J Hum Rights Environ 11:86–113. https://doi.org/10.4337/jhre.2020.01.04
Balsam KF, Rothblum ED, Beauchaine TP (2005) Victimization over the life span: a comparison of lesbian, gay, bisexual, and heterosexual siblings. J Consult Clin Psychol 73:477–487. https://doi.org/10.1037/0022-006X.73.3.477
Bandyopadhyay N, Bhuiyan C, Saha AK (2020) Drought mitigation: critical analysis and proposal for a new drought policy with special reference to Gujarat (India). Progress in Disaster Sci 5:100049. https://doi.org/10.1016/j.pdisas.2019.100049
Becker A (2019) On the economic origins of restrictions on women's sexuality. SSRN Electron J. https://doi.org/10.2139/ssrn.3432818
Behrman S (2018) Climate refugees. Routledge Routledge https://doi.org/10.4324/9781315109619 (Accessed 26 Oct 2024)
Bharadwaj R, Bishop D, Hazra S, Pufaa E, Kofi Annan J, Hawkey S, Clarkson House T (2021) Climate-induced migration and modern slavery
Black R, Bennett SRG, Thomas SM, Beddington JR (2011) Migration as adaptation. Nature 478:447–449. https://doi.org/10.1038/478477a
Black R (2001) Environmental refugees: myth or reality??. http://eprints.soas.ac.uk/18197/ (Accessed 26 Oct 2024).
Borràs-Pentinat S (2024) Climate migration: a gendered perspective. Environ Policy and Law 53:385–399. https://doi.org/10.3233/EPL-239008
Brettell CB (2017) Gender and migration. Wiley and Sons
Brown D, Chanakira RR, Chatiza K, Dhliwayo M, Dodman D, Masiiwa M, Muchadenyika D, Mugabe P, Zvigadza S (2012) Climate change impacts, vulnerability and adaptation in Zimbabwe
Campana P (2018) Out of Africa: the organization of migrant smuggling across the Mediterranean. Eur J Criminol 15:481–502. https://doi.org/10.1177/1477370817749179
Cannon T (2002) Gender and climate hazards in Bangladesh. Gend Dev 10:45–50. https://doi.org/10.1080/13552070215906
Carballo M, Hernandez M, Schneider K, Welle E (2005) Impact of the Tsunami on reproductive health. J R Soc Med 98:400–403. https://doi.org/10.1177/014107680509800904
Carvajal L, Pereira I (2010) Evidence on the link between migration, climate shocks, and adaptive capacity. Risk, shocks, and human development. Palgrave Macmillan UK, London, pp 257–283
Castillo Betancourt T, Zickgraf C (2024) It's not just about women: broadening perspectives in gendered environmental mobilities research. Clim Dev 1–22.https://doi.org/10.1080/17565529.2024.2393305
Castles S (2017) International migration at a crossroads. In: Migration, citizenship and identity. Edward Elgar Publishing
Chant S (2013) Cities through a "gender lens": a golden "urban age" for women in the global South? Environ Urban 25:9–29. https://doi.org/10.1177/0956247813477809

Cheney K (2012) Locating Neocolonialism, "Tradition", and human rights in Uganda's "Gay Death Penalty." Afr Stud Rev 55:77–95. https://doi.org/10.1353/arw.2012.0031
Chindarkar N (2012) Gender and climate change-induced migration: proposing a framework for analysis. Environ Res Lett 7:025601. https://doi.org/10.1088/1748-9326/7/2/025601
Chowdhury MA, Hasan MK, Hasan MR, Younos TB (2020) Climate change impacts and adaptations on health of Internally Displaced People (IDP): an exploratory study on coastal areas of Bangladesh. Heliyon 6:e05018. https://doi.org/10.1016/j.heliyon.2020.e05018
Christou A, Kofman E (2022) Gender and migration: an introduction. Cham, Springer International Publishing, pp 1–12. https://doi.org/10.1007/978-3-030-91971-9_1 (Accessed 26 Oct 2024)
Crenshaw K (2018) Demarginalizing the Intersection Of Race And Sex: A Black Feminist Critique Of Antidiscrimination Doctrine, Feminist Theory, And Antiracist Politics [1989]. In: Feminist legal theory. Routledge, pp 57–80
Dankelman I (2010) Gender and climate change: an introduction. Routledge. https://doi.org/10.4324/9781849775274
Daoud M (2021) Is vulnerability to climate change gendered? And how? Insight from Egypt. Reg Environ Change 21:52. https://doi.org/10.1007/s10113-021-01785-z
Davis KF, Bhattachan A, D'Odorico P, Suweis S (2018) A universal model for predicting human migration under climate change: examining future sea level rise in Bangladesh. Environ Res Lett 13:064030. https://doi.org/10.1088/1748-9326/aac4d4
DeLaet DL (1999) Introduction: the invisibility of women in scholarship on international migration. Gender and immigration. Palgrave Macmillan UK, London, pp 1–17
Denton F (2002) Climate change vulnerability, impacts, and adaptation: why does gender matter? Gend Dev 10:10–20. https://doi.org/10.1080/13552070215903
Desai BH, Mandal M (2021) Role of climate change in exacerbating sexual and gender-based violence against women: a new challenge for international law. Environ Policy and Law 51:137–157. https://doi.org/10.3233/EPL-210055
Deshar R, Koirala M (2020) Climate change and gender policy. In: Global climate change and environmental policy. Springer, Singapore, pp 411–422
Djoudi H, Locatelli B, Vaast C, Asher K, Brockhaus M, Basnett Sijapati B (2016) Beyond dichotomies: gender and intersecting inequalities in climate change studies. Ambio 45:248–262. https://doi.org/10.1007/s13280-016-0825-2
Donato KM, Gabaccia D (2015) Gender and international migration. Russell Sage Foundation
Drolet M-J, Désormeaux-Moreau M, Soubeyran M, Thiébaut S (2020) Intergenerational occupational justice: ethically reflecting on climate crisis. J Occup Sci 27:417–431. https://doi.org/10.1080/14427591.2020.1776148
Enarson E, Morrow BH (1998) Why gender? Why women? An introduction to women and disaster. The gendered terrain of disaster: through women's eyes, pp 1–8
Enarson E (2012) Women confronting natural disaster. Lynne Rienner Publishers
Enarson E (2016) Men, masculinities and disaster. In: Men, masculinities and disaster. Routledge, Abingdon, Oxon; New York, N, Routledge is an imprint of the Taylor & Francis Group, an Informa Business | Series: Routledge studies in hazards, disaster risk and climate change, pp 219–233
Engage EPW (2020) COVID-19, domestic abuse and violence: Where do Indian women stand? EPW Engage
Fernández CG, Peek D (2024) Climate justice for the displaced? The path to legal protection of the climate migrant. In: Çiner A et al (eds) Recent research on environmental earth sciences, geomorphology, soil science and paleoenvironments. MedGU 2022. Advances in science, technology & innovation. Springer, Cham. https://doi.org/10.1007/978-3-031-48754-5_6
Findlay AM (2011) Migrant destinations in an era of environmental change. Glob Environ Chang 21:S50–S58. https://doi.org/10.1016/j.gloenvcha.2011.09.004
Fischer PA, Malmberg G (2001) Settled people don't move: on life course and (Im-)mobility in Sweden. Int J Popul Geogr 7:357–371. https://doi.org/10.1002/ijpg.230

Gaard G (2015) Ecofeminism and climate change. Womens Stud Int Forum 49:20–33. https://doi.org/10.1016/J.WSIF.2015.02.004
Gemenne F, Zickgraf C, Hut E, Castillo Betancourt T (2022) Forced displacement related to the impacts of climate change and disasters. orbi.uliege.beF Gemenne, C Zickgraf, E Hut, T Castillo Betancourt9780198786467, 2022•orbi.uliege.be
Ghorashi H (2010) From absolute invisibility to extreme visibility: emancipation trajectory of migrant women in the Netherlands. Fem Rev 94:75–92. https://doi.org/10.1057/fr.2009.38
Gómez-Álvaro G, Caro-Carretero R (2024) Climate change and migration dynamics in the Horn of Africa: a comprehensive review and future research directions. epsir.net. https://doi.org/10.31637/epsir-2024-412
Gray C, Mueller V (2012) Drought and population mobility in rural Ethiopia. World Dev 40:134–145. https://doi.org/10.1016/j.worlddev.2011.05.023
Gray NN, Mendelsohn DM, Omoto AM (2015) Community connectedness, challenges, and resilience among gay Latino immigrants. Am J Community Psychol 55:202–214. https://doi.org/10.1007/s10464-014-9697-4
Habtezion S (2016a) Gender and climate change: overview of linkages between gender and climate change. United Nations Development Programme, New York
Habtezion S (2016b) Overview of linkages between gender and climate change
Hagen-Zanker J, Himmelstine CL (2013) What do we know about the impact of social protection programmes on the decision to migrate? Migr Dev 2:117–131. https://doi.org/10.1080/21632324.2013.778131
Harvey F (2020) Climate breakdown "is increasing violence against women". The Guardian. https://www.theguardian.com/environment/2020/jan/29/climate-breakdown-is-increasing-violence-against-women. Accessed 17 Jan 2025
Huebner DM, Rebchook GM, Kegeles SM (2004) Experiences of harassment, discrimination, and physical violence among young gay and bisexual men. Am J Public Health 94:1200–1203. https://doi.org/10.2105/AJPH.94.7.1200
Hunter M (2010) Love in the time of AIDS: inequality, gender, and rights in South Africa. Indiana University Press
Internal Displacement Monitoring Centre (IDMC) (2024) Global Report on Internal Displacement. Available at: https://www.internal-displacement.org/publications/2024-global-report-on-internal-displacement-grid/ (Accessed 18 Sept 2024)
IOM (2021) World migration report 2022. International Organization for Migration
Ionesco D, Mokhnacheva D, Gemenne F (2016) The atlas of environmental migration. Routledge
Jafree SR, Mukhopadhyay B (2020) Refugee, displaced, and climate-affected women of South Asia and their health challenges. The sociology of South Asian women's health. Springer International Publishing, Cham, pp 101–127
Jones N, Presler-Marshall E (2012) Governance and poverty eradication: applying a gender and social institutions perspective. Public Adminis Develop 32:371–384. https://doi.org/10.1002/pad.1618
De Jong GF (2000) Expectations, gender, and norms in migration decision-making. Popul Stud (NY) 54:307–319.https://doi.org/10.1080/713779089
Kaijser A, Kronsell A (2014) Climate change through the lens of intersectionality. Env Polit 23:417–433. https://doi.org/10.1080/09644016.2013.835203
kakissis J (2010) Environmental refugees unable to return home—The New York Times. https://www.nytimes.com/2010/01/04/world/asia/04migrants.html. Accessed 17 Jan 2025
Logie CH, Daniel C (2016) 'My body is mine': qualitatively exploring agency among internally displaced women participants in a small-group intervention in Leogane, Haiti. Glob Public Health 11:122–134. https://doi.org/10.1080/17441692.2015.1027249
Long K, Sabates-Wheeler R (2017) Migration, forced displacement and social protection. GSDRC Rapid Literature Review Birmingham, University of Birmingham http://www.gsdrcorg/wp-content/uploads/2017/06/migration-forced-displacement-and-social-protection-2017-06-20-blpdf

Lori JR, Boyle JS (2015) Forced migration: health and human rights issues among refugee populations. Nurs Outlook 63:68–76. https://doi.org/10.1016/j.outlook.2014.10.008
Masika R (2002) Gender, development, and climate change. Oxfam
Mastrorillo M, Licker R, Bohra-Mishra P, Fagiolo G, D. Estes L, Oppenheimer M (2016) The influence of climate variability on internal migration flows in South Africa. Global Environ Change 39:155–169.https://doi.org/10.1016/j.gloenvcha.2016.04.014
McAdam J (2012) Climate change, forced migration, and international law. Oxford University Press
Mcadam J (2010) 'Disappearing States', statelessness and the boundaries of international law. In: Climate change and displacement : multidisciplinary perspectives. Hart Publishing, pp 105–130
McGregor J (1994) Climate change and involuntary migration: implications for food security. Food Policy 19:120–132. https://doi.org/10.1016/0306-9192(94)90065-5
McLeman RA (2013) Climate and human migration. Cambridge University Press
McMichael C, Schwerdtle PN, Ayeb-Karlsson S (2023) Waiting for the wave, but missing the tide: case studies of climate-related (im)mobility and health. J Migr Health 7:100147. https://doi.org/10.1016/j.jmh.2022.100147
Mitchell T, Tanner T, Lussier K (2007) 'We know what we need': South Asian women speak out on climate change adaptation
Moniruzzaman Md, Cottrell A, King D, Islam MdN (2018) Environmental migrants in Bangladesh: a case study on climatic change hazards in the Southwestern coastal area. pp 99–117
Moracco KE, Runyan CW, Bowling JM, Earp JAL (2007) Women's experiences with violence: a national study. Women's Health Issues 17:3–12. https://doi.org/10.1016/j.whi.2006.03.007
Mosel I, Jackson A (2022) Sanctuary in the City?: urban displacement and vulnerability in Peshawar, Pakistan
Nagel J (2015) Gender and climate change. Routledge
Nawyn JS (2019) Routledge international handbook of migration studies. 2nd edn. New York, Routledge I Series, Routledge International Handbooks
Ndiritu J (2021) The challenges facing LGBTQI+ refugees in Kakuma Refugee Camp, Kenya. Organization for Refuge, Asylum and Migration (ORAM) and Rainbow Railroad https: //www rainbowrailroad org/wp-content/uploads/2021/10/Rainbow-Railroad-and-ORAM-Report-on-Kakuma-2021 pdf
Neira M, Erguler K, Ahmady-Birgani H, AL-Hmoud ND, Fears R, Gogos C, Hobbhahn N, Koliou M, Kostrikis LG, Lelieveld J, Majeed A, Paz S, Rudich Y, Saad-Hussein A, Shaheen M, Tobias A, Christophides G (2023) Climate change and human health in the Eastern Mediterranean and Middle East: literature review, research priorities and policy suggestions. Environ Res 216:114537. https://doi.org/10.1016/j.envres.2022.114537
Ngcamu BS (2023) Climate change effects on vulnerable populations in the Global South: a systematic review. Nat Hazards 118:977–991. https://doi.org/10.1007/s11069-023-06070-2
Onyango MA, Heidari S (2017) Care with dignity in humanitarian crises: ensuring sexual and reproductive health and rights of displaced populations. Reprod Health Matters 25:1–6. https://doi.org/10.1080/09688080.2017.1411093
Piguet E (2010) Linking climate change, environmental degradation, and migration: a methodological overview. Wires Clim Change 1:517–524. https://doi.org/10.1002/wcc.54
Piguet E, Pecoud A, de Guchteneire P (2011) Migration and climate change: an overview. Refug Surv Q 30:1–23. https://doi.org/10.1093/rsq/hdr006
Ramachandran S (2005) Indifference, impotence, and intolerance: transnational Bangladeshis in India. Global Migration Perspectives, Global Commission on International Migration
Rashid SF, Michaud S (2000) Female adolescents and their sexuality: notions of honour, shame, purity and pollution during the floods. Disasters 24:54–70. https://doi.org/10.1111/1467-7717.00131
Reuveny R (2007) Climate change-induced migration and violent conflict. Polit Geogr 26:656–673. https://doi.org/10.1016/j.polgeo.2007.05.001
Rúa TA (2010) Migration, remittances and development in times of crisis. Pontificia Universidad Católica del Perú, Departamento de Ciencias Sociales

Saldaña-Zorrilla SO, Sandberg K (2009) Spatial econometric model of natural disaster impacts on human migration in vulnerable regions of Mexico. Disasters 33:591–607. https://doi.org/10.1111/j.1467-7717.2008.01089.x

Schüring E, Pearson C, Castro A, Mathebula B (2017) Social protection as an alternative to migration?: An assessment of the role of social protection in reducing push factors for migration in different country

Sedova B, Kalkuhl M (2020) Who are the climate migrants and where do they go? Evidence from rural India. World Dev 129:104848. https://doi.org/10.1016/j.worlddev.2019.104848

Schuster RC, Wachter K, Hussain F, Gartin ML (2024) Gendered effects of climate change and health inequities among forcibly displaced populations: displaced Rohingya women foster resilience through technology. The J Climate Change and Health 18:100303. https://doi.org/10.1016/j.joclim.2024.100303

Shafi JU (2024) Beyond borders: exploring international law and climate-induced migration in small island developing states. Rev Juríd Portucalense 572–598. https://doi.org/10.34625/issn.2183-2705(36)2024.ic-25

Silchenko D, Murray U (2023) Migration and climate change—the role of social protection. Clim Risk Manag 39:100472. https://doi.org/10.1016/j.crm.2022.100472

Spring ÚO (2008) Gender and disasters: human, gender and environmental security

Stark L, DeCormier PW, Horn R, Canavera M (2015) 'He always thinks he is nothing': the psychosocial impact of discrimination on adolescent refugees in urban Uganda. Soc Sci Med 146:173–181. https://doi.org/10.1016/j.socscimed.2015.10.045

Stone K, Blinn N, Spencer R (2022) Mental health impacts of climate change on women: a scoping review. Curr Environ Health Rep 9:228–243. https://doi.org/10.1007/s40572-022-00346-8

Thiede BC, Gray CL (2017) Heterogeneous climate effects on human migration in Indonesia. Popul Environ 39:147–172. https://doi.org/10.1007/s11111-016-0265-8

Thomas SL, Thomas SDM (2004) Displacement and health. Br Med Bull 69:115–127. https://doi.org/10.1093/bmb/ldh009

Thoreson R (2020) Covid-19 backlash targets LGBT people in South Korea. Human Rights Watch 13

Tinoco N (2023) Post-disaster (im)mobility aspiration and capability formation: case study of Southern California wildfire. Popul Environ 45:4. https://doi.org/10.1007/s11111-023-00416-5

Tummala-Narra P (2021) Trauma and racial minority immigrants: turmoil, uncertainty, and resistance. American Psychological Association, Washington

United Nations Development Programme (UNDP) and Global Gender and Climate Alliance (2016) Overview of linkages between gender and climate change, UNDP, New York, pp 1–8

UNFCCC (2022) United Nations Climate Change Annual Report 2022. Available via https://unfccc.int/sites/default/files/resource/UNClimateChange_AnnualReport_2022.pdf (Accessed 14 Dec 2024)

UNHCHR (2022) Gender, displacement and climate change. UNHCHR, Geneva. Available via https://www.unhcr.org/5f21565b4.pdf Accessed 14 Dec 2024

UNHCR (2023) Global trends: forced displacement in 2023. UNHCR. Available via https://www.unhcr.org/sites/default/files/2024-06/global-trends-report-2023.pdf Accessed 14 Dec 2024

Valerio KA (2014) Storm of violence, surge of struggle: women in the aftermath of typhoon Haiyan (Yolanda). Asian J Women Stud 20:148–163. https://doi.org/10.1080/12259276.2014.11666177

Vera Espinoza M, Prieto Rosas V, Zapata GP, Gandini L, Fernández de la Reguera A, Herrera G, López Villamil S, Zamora Gómez CM, Blouin C, Montiel C, Cabezas Gálvez G, Palla I (2021) Towards a typology of social protection for migrants and refugees in Latin America during the COVID-19 pandemic. Comp Migr Stud 9:52. https://doi.org/10.1186/s40878-021-00265-x

Williams M (2015) Gender and climate change financing. Routledge

Williams A (2008) Turning the tide: recognizing climate change refugees in international law. Law Policy 30:502–529.https://doi.org/10.1111/j.1467-9930.2008.00290.x

Wilson C, Tisdell C (2001) Why farmers continue to use pesticides despite environmental, health and sustainability costs. Ecol Econ 39:449–462. https://doi.org/10.1016/S0921-8009(01)00238-5

Yang C, Xu H (2024) Direct and spillover effects: how do community-based organizations impact the social integration of passive migrants? Sustainability 16(11):4530. https://doi.org/10.3390/su16114530

Yarwood V, Checchi F, Lau K, Zimmerman C (2022) LGBTQI + migrants: a systematic review and conceptual framework of health, safety and wellbeing during migration. Int J Environ Res Public Health 19:869. https://doi.org/10.3390/ijerph19020869

# Chapter 10
# Assessment of Women Farmers' Drudgery in Rice Farming with Climate Change: A Case Study from South India

**Dhanya Punnoli, Jayarajan Kunnampalli, and Sreeraj Punnoli**

**Abstract** Although women make up most of the agricultural workforce globally, their labour is largely underestimated. The objectives of this chapter are to obtain insight into the women farmers' drudgery during farm operations and its health consequences, identify potential risks under climate change perspective, and improve understanding to enhance their coping capacities. This work is a part of a participatory research conducted in the Bharathapuzha River basin, in Palakkad, Kerala, a state in India during 2023. Kerala has been experiencing strong climate extremes, especially for the past decades, in the form of floods, droughts, etc. Women farmers perform the most arduous and back-breaking work in the rice farms starting from nursery preparations to sowing, transplanting, weeding, harvesting, threshing, winnowing, and storing grains along with duck or fish and livestock rearing. It found that challenges of physical load and drudgery during three crop seasons such as *kharif/virippu, rabi/mundakan, or summer/puncha crop season* vary based on the varying physical exposures to heavy rainfall, droughts, heat waves, heavy winds, lightning etc. The survey revealed that most of the women farmers had more than twenty years of experience and were exposed to waterborne diseases, skin allergies, heat exhaustion, etc. Mechanization has largely helped to minimize the physical load and exposure to harsh climates, especially during harvesting which was majorly done by women. Weather early warnings and Agromet Advisories have also helped them to face extremities. It is vital to enhance the quality of life and the well-being of the women farmers through planned adaptive interventions under the scenario of rising climate change challenges.

**Keywords** Women farmers' drudgery · Climate change · Weather early warnings · Rice farming · Bharathapuzha basin · Paddy farmers · Adaptation interventions

---

D. Punnoli (✉) · J. Kunnampalli
Department of Geography, Government College Chittur, Palakkad, Kerala, India
e-mail: dhanya.eptri@gmail.com

S. Punnoli
Senior Software Developer, Ubisoft, Halifax, Canada



A. Sarkar and N. Bandyopadhyay (eds.), *Gender-Transformative Approaches for Climate Change Adaptation*, Springer Climate,
https://doi.org/10.1007/978-3-031-90236-9_10

## Introduction

In India, rural populations are constantly under pressure because of the climate catastrophe and socioeconomic challenges affecting their livelihoods. The burden on these communities increases as environmental conditions deteriorate and disasters become more frequent and severe (FAO 2023). The FAO estimates that women make up 43% of the agricultural workforce globally, but still, gender inequality plays a significant role in determining women's adaptive capacity to climate change (FAO 2024). Disasters, as well as subtle variations in the weather, can cause significant losses to agriculture ecosystems, both financially and non-financially. Extreme weather events are occurring all across the planet, wreaking havoc on human well-being and landforms (IPCC 2022). Climate change, particularly in its more obvious extremes, may have an impact on cropping systems in several ways, including crop productivity, yield, rates of photosynthesis and transpiration, moisture availability, and crop quantity and quality (IPCC 2019).

Climate vulnerability assessment has helped to materialize the need to quantify multidimensional issues using indicators related to exposure, sensitivity, and adaptive capacity (Mani Murali et al 2013). At the community level, these vulnerability assessments are rarely carried out. In a nation like India that depends heavily on the monsoon, sustainable agriculture and sustainable communities are a major concern even in the Indian Meteorological Department's (IMD) long-range predictions and risk communication for the monsoon. Government responses to crop losses need to go beyond providing fair compensation and relief. Rather, climate-resilient seeds must be distributed, crop calendars must be updated collaboratively by agricultural and meteorological agencies and give farmers weekly weather information for the planting, sowing, and harvesting times of crops grown in their districts need to be updated, and farmers' capacity to adapt must be strengthened (Prabhu and Vishwas 2024).

It is crucial to pinpoint locations and communities with rising susceptibility to climate change to allocate resources and formulate approaches to tackle the combination of issues facing the most susceptible individuals (Chanana et al. 2020; Chandra et al. 2017). In addition to implementing proactive measures to lessen the impact of climate shocks on agriculture and food security, the Food and Agriculture Organization of the United Nations (FAO) contributes its expertise in agricultural resilience and food security (WMO 2022). This includes generating risk knowledge and customizing early warning systems to the unique needs of farming communities. It is essential to alter farming practices as climate change becomes a harsh reality in many parts of the world for farmers to survive. There are challenges for female farmers in adjusting to climate change. Their essential position as farmers and frequently the family's primary supply of nutrition does not necessarily align with their rights and access to land, loans, inputs (such as better seed and fertilizers), agricultural training, and other information. To achieve sustainable development, these shifts from traditional farm systems to more climate-smart agriculture (CSA) must happen as quickly as possible. Early warning plays a crucial role in minimizing the adverse impacts of climate-induced disasters.

In Western countries, farmscapes are transforming rapidly due to financial and technological improvements. However, in Asian countries, especially in India, rural communities particularly women farmers have particular difficulties in maintaining their traditional rice-growing practices in the face of climate change. The rising number of frequent disasters and extreme events have impacted the physical and mental health along with the socioeconomic well-being of citizens, especially those who are vulnerable. There are multiple dimensions to well-being. As science is always evolving the definition of Health put forth by WHO, 'Health is a state of complete physical, mental and social well-being and not merely the absence of disease or infirmity' that encompasses not only physical health but also mental and social health (WHO 2023).

A vital part of rural development and the fight against poverty is the work of small and marginal farmers (Lipton 2006). Kerala is depending on migrant labourers from West Bengal, Tamil Nadu, and other places to carry out farming operations due to a growing labour shortage. Numerous obstacles confront the state's current farmers, including very variable weather, dispersed plots, limited access to resources and technology, banking services, and market infrastructure (Van Touch 2024). Punnoli et al. (2022) have conveyed implementing prioritized actions based on Agromet advisories can help farmers save crops and lessen the burden of women farmers. ICT tools are powerful to enhance the benefits of scientific findings to reach farmers and key stakeholders more efficiently. It is the responsibility of researchers and ground-level officers to provide real-time promotion of risk management practices through agro-advisories for farmers with limited resources.

Women farmers, even though the biggest workforce on the planet, play multiple roles in their daily lives to meet household, community, and societal demands. Unfortunately, their needs remain unaddressed, and much of their caregiving and domestic work remain unaccounted for. It applies to more than 2.7 billion women in the world. They are caregivers but never get enough time or energy for self-care. Climate change has escalated all these pressing issues. Women farmers manage several chores in the paddy fields every season, even in the study area. Reductions in the number of farmers are not taken seriously, despite statistics showing a sharp decline in the acreage planted for rice crops. Women farmers are mostly ignored, their difficulties are rarely recorded, and they are rarely informed of early warnings, which prevents them from taking proactive adaptation steps. In the rice belts, the geographical variations in particular adaptation or coping mechanism requirements are never assessed. This study is an attempt to evaluate the same as a preliminary step to enhance their resilience.

## Materials and Methods

### *Study Area*

The river Bharathapuzha is the lifeline of the people living in Palakkad, and the rice fields of this granary are nourished by its waters (Fig. 10.1). This is one of the state's longest rivers which significantly influence the surface and groundwater in this area. The region's ecology, socio-cultural conditions, agriculture practices, and water supply all depend on this river. Farming practices, conservation efforts, monsoon, and rainfall patterns all have an impact on water levels.

Farmers rely on the rains from the Southwest monsoon season to start kharif ('viruppu') farming. The basin receives sufficient rainfall primarily due to the presence of Western Ghats; only the eastern and northeastern portions of the basin are covered by clouds. The monsoon rains have an impact on the inflow and outflow of irrigation water from the dam, which in turn impacts rice yield. The basin contains several dams, the most notable of which is Malampuzha. This basin is separated into several smaller basins, namely Thootha Puzha, Gayathri Puzha, Kalpathy Puzha, and Chittur Puzha (Fig. 10.1). This rice granary of Kerala has about 83,000 ha under paddy cultivation, accounting for 40 per cent of the state's total paddy area. The Malampuzha Dam is only and mostly utilized for irrigation. Rabi and Mundakan crops are primarily supported by irrigation.

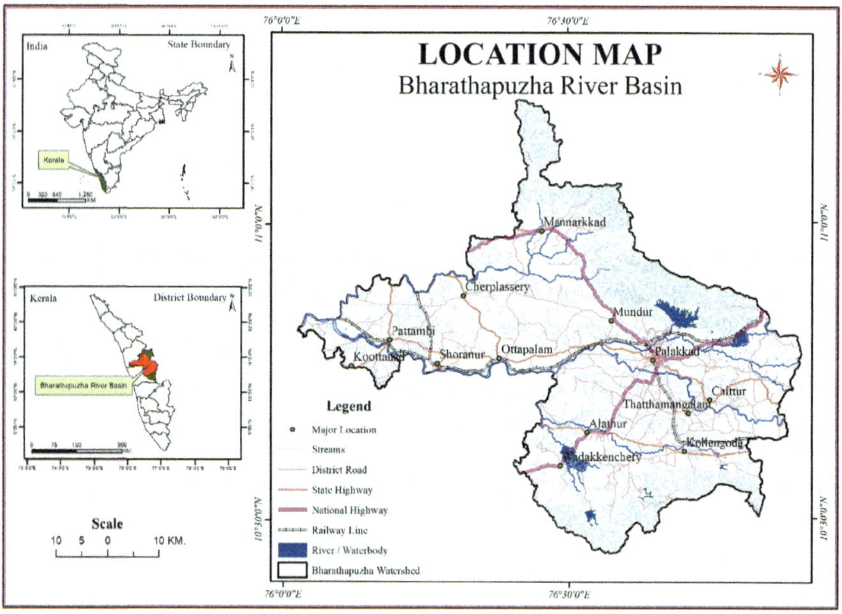


**Fig. 10.1** Location of study area

## *Data and Methodology*

This study was a component of a participatory research program that was carried out in Palakkad in Kerala, a state in south India located along the Bharathapuzha River. Particularly during the last ten years, Kerala has had severe weather extremes in the form of storms, floods, and droughts. It is one of the main rice bowls in the state. The farm season is highly dependent upon the southwest and northeast monsoon season. In the study area, the first rice cropping season, *kharif* locally known as *virippu*, starts with sowing in May/June and is harvested in autumn, and the second crop, *rabi/mundakan* crops are sown in August or September and harvested in spring. The third one is puncha or summer crops. Block-wise crop cultivation area was collected from The Department of Economics and Statistics in Palakkad. The study uses statistical data for the years 2011 to 2016.

It was an interactive community-based descriptive questionnaire survey conducted during July–August, in 2024. Questions related to the farmers' knowledge, attitude, and practice were prepared in advance in the vernacular language 'Malayalam'. Farmers' problems with increasing physical load or drudgery during three crop growing seasons varied due to varying physical exposures to heavy rainfall, droughts, heat waves, heavy winds, lightning, etc. It also helped to provide insights on access to technology adoption, extension services, and off-farm income. The study noted if in these areas the farming communities could access the available agro-advisories services.

The research employed primary data collected through a pretested, well-structured interview schedule. The participatory survey was conducted in July 2024 using a questionnaire prepared in the vernacular language Malayalam. The survey was to identify the key constraints faced by paddy farmers especially female farmers. The constraints faced by them were analysed using the Problem Confrontation Index. One hundred and thirty-six rice farmers were involved in the experimental survey that included 99 female farmers, and 37 male farmers selected randomly from blocks of Bharathapuzha Basin in Palakkad district of Kerala. There are automatic rain gauges fixed at three different locations in the study area as part of the project implementation. The observed real-time temperature and rainfall data of the study area are also incorporated in the article.

Measuring the problem confrontation score of the surveyed farmers was carried out using a 4-point rating scale for measuring the Problem Confrontation Index score (PCI). The farmers were asked to give their responses to 5 selected questions related to impacts.

The weights assigned for each response were: 3 for high confrontation, 2 for medium confrontation, 1 for low confrontation, and 0 for not at all. The PCI score was obtained by adding weights of responses to the problems, and therefore, for making rank order, attitude score and Problem Confrontation Index (PCI) were computed as used by Hossain and Miah (2011) and Punnoli et al. (2020).

Similarly, the PCI was computed by using the following formula:

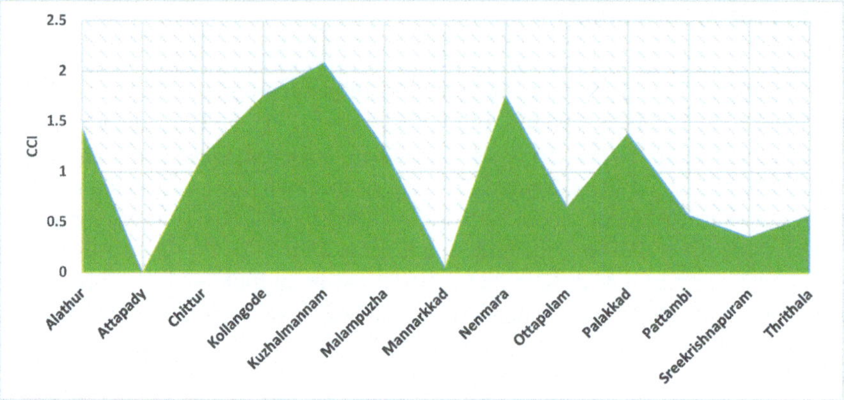


**Fig. 10.2** Block-wise rice crop concentration index in the study area. *Data Source*: Department of Economics and Statistics, Palakkad for the year 2016

$$PCI = PH * 3 + PM * 2 + PL * 1 + PN \times 0 \tag{10.1}$$

where PCI = Problem Confrontation Index; PH = Number of the respondents who expressed attitude and problem as 'high'; PM = Number of the respondents who expressed an attitude and problem as 'medium'; and PL = Number of the respondents expressed problem as 'low'; AN and PN = Number of the respondents expressed problem as 'not at all' (Hossain and Miah 2011).

Analysis of the data collected from secondary sources showed that a large concentration of rice (with an index value of 2.09) farms is located in the Kuzhalmannam block. The blocks Palakkad, Malampuzha, Alathur, Chittur, Nenmara, and Kollengode have a medium concentration of rice. There is little concentration of rice in the remaining blocks (Fig. 10.2). Hence, the survey was chiefly concentrated on the areas where there is a high rice crop concentration index.

## Results and Discussion

### *Socioeconomic Background of the Respondents*

The socioeconomic traits of the respondents are displayed in Table 10.1. Around 66% of the respondents were between the age group of 32–56 years. The respondents were specifically chosen from the individuals who had been farming in the chosen sites for at least 15 years. This explains why a sizable percentage of respondents had been farming for more than 20 years and were 40 years of age or older. Of the 136 respondents in total, around 73% were female, and 27% were male farmers. Around half of them had secondary school education (54%), 20% received primary school education, and the remaining had collegiate education (Table 10.1). The educational backgrounds of the farming community show that more than half of them have access to some kind of technology related to farming and almost 77% agreed that they were getting support from agriculture extension-related services. Enhancement of agricultural extension initiatives would help to promote greater farming community use of digital agriculture services, agromet advisories, market information, and other scientific management approaches and technology transfer. However, around 59% of them did not have any off-farm income from other diversified livelihood sources. During the survey, interactions were done with *Kudumbasree*[1] members and heads of the *Padasekhara samiti.*[2]

**Table 10.1** Socioeconomic characteristics of the respondents

| Age (years) | Frequency | Percentage |
|---|---|---|
| > 32 | 23 | 16.91 |
| 32 to 40 | 29 | 21.32 |
| 41 to 48 | 32 | 23.53 |
| 49 to 56 | 31 | 22.79 |
| 57 to 64 | 14 | 10.29 |
| 65 to 73 | 7 | 5.15 |
| **Total** | 136 | **100** |
| *Gender* | | |
| Female | **99** | 72.79 |
| Male | 37 | 27.21 |
| **Total** | 136 | **100** |
| *Education* | | |
| Primary | 28 | 20.59 |
| Secondary | 74 | 54.41 |
| Collegiate | 34 | 25.00 |
| **Total** | 136 | **100** |
| *Size of the farmland (Ha)* | | |
| Between 0.3 to 1.13 | 60 | 44.12 |
| 1.13 to 1.96 | 44 | 32.35 |
| 1.96 to2.79 | 16 | 11.76 |
| 2.79 to 3.62 | 6 | 4.41 |
| 3.62 to 4.45 | 4 | 2.94 |
| 4.45 to 5.28 | 2 | 1.47 |
| 5.28 to 6.11 | 2 | 1.47 |
| 6.11 to 6.94 | 0 | 0 |
| 6.94 to 7.77 | 2 | 1.47 |
| Total | | 100 |

*Source* Data from authors' fieldwork

## *Climate Change-Related Hazards and Their Effect on Rice Cultivation*

Most of the arable land in the study area is susceptible to a variety of climate-related hazards, including droughts, dry spells, extreme rainfalls, floods, thunder, and lightning. Moreover, farms face biotic risks like pests and disease and attacks from wild boars and peacocks. Over the past 20 years, the basin had numerous floods that have had a significant adverse impact, especially during 2018, 2019, 2021, 2023, and 2024, three of which were of appalling severity, including the devastating Kerala floods of 2018 and 2019. The floods of August 2018 and 2019 were reported to be flash floods. Several places in the state reported receiving more than 50 mm of rain in two hours. Figure 10.3 depicts farms during floods and in moderate rainfall in the study area.

'Mini cloud bursts' (MsCB) have occurred in Kerala, and high-intensity precipitation has increased in frequency over time. MsCB events are reported in this region quite often, making it more prone to seed loss and plant damage completely or partly especially when it occurs during the harvest operations. In such MsCB incidents, the amount of rainfall is so high that a region can receive the same amount of rainfall in one or two days or even a few hours that it usually receives in two or three months

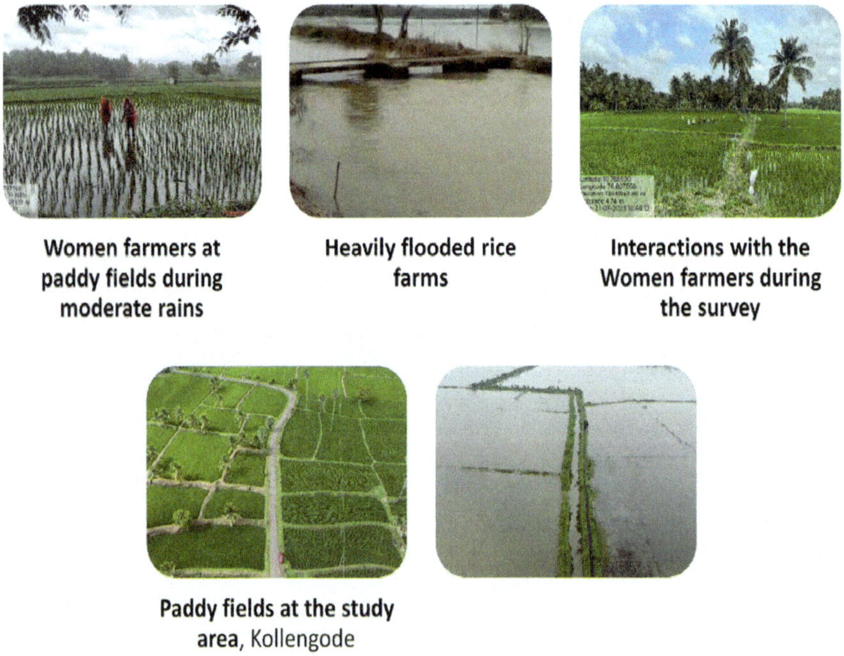


**Fig. 10.3** Pictures of the survey area in Bharathapuzha Basin. *Source* Taken by authors during field survey

on average Mishra et al. (2018). Rice productivity is affected by such heavy rainfall, which deters farmers' livelihoods and agricultural investments. There have also been numerous reports of crop damage and seed loss during these floods (KSDMA (2020), Nirmala (2023). River flow monitoring stations at sub-watersheds require a denser network for better flood monitoring.

As per the farmer's opinion, rainfall has become highly unpredictable, and Bharathapuzha river flooding has become more frequent affecting various farm activities. Interruption of rainfall also creates an irrigation crisis for Malampuzha canal irrigation systems during major crop growth stages, making them depend on groundwater even during monsoon season, which used to rarely happen in the past. Farmers have to do a weeding activity more times due to excessive rainfall or use costly herbicides to destroy weeds due to their overgrowth during seasons that receive above-normal rainfall. Even cloudy weather and excessive rainfall can be associated with stunted growth and an increase in pests and diseases in crops.

Changing rainfall patterns are also impacting the well-being of the farmers adversely. At present, fluctuating rainfall patterns and their unpredictability are making the farmers avoid either one or two of the cropping seasons. They are planting crops in one cropping season in recent times. Sometimes it is insufficient rainfall that is making farm activities and rice production impossible. Seed loss is reported by the majority of the farmers because of erratic rainfall, especially at a time just after transplantation and vegetative phase during June and July. Contrary to this, above-normal rainfall and extremities during August and September have become a frequent issue that triggers seed loss and double the farmer's labour in the initial stages of the cultivation season itself. In most cases, farmers are forced to redo the same operations because of either excessive above-normal rainfall or cessation of rainfall after the transplantation. Interruption of rainfall and/or slow onset of monsoon leads to dry spells, sporadic rains, and droughts. It causes stunted growth of plants. The farmers are compelled to carry out extra efforts to provide additional irrigation to sustain the crops. The respondents in the survey emphasized the detrimental effects of high temperatures on crop quality, plant growth, and seed germination.

The data from the study villages also shows a rising trend in day and nighttime temperatures (Fig. 10.4). Severe heat stress was experienced by them when the daytime temperature exceeded 38 °C during September, and heat waves above 42 °C during summer were reported in many parts of the study area. Similarly in November, there was very heavy rainfall recorded of 178 mm in a single day in 2023 at the farmer's field (Fig. 10.5). The farmers reported that they noticed visible physiological changes in their crops, and they confirmed that high temperatures result in a decline in the quantity and quality of rice produced.

Rice, being a labour and water-intensive crop, is sensitive to water shortages. Hence, it requires highly vigilant scheduling of irrigation to prevent water waste by irrigation in fields. Unpredictable climate patterns force the farmers to shift their farming to be more resilient by adopting cost-efficient mechanizations, resilient varieties as per the climate predictions provided by IMD, and adopt resilient crop varieties, etc. Empirical research has projected a 14% reduction in rice yield for the major rice-growing varieties in the study region under climate change threats with medium

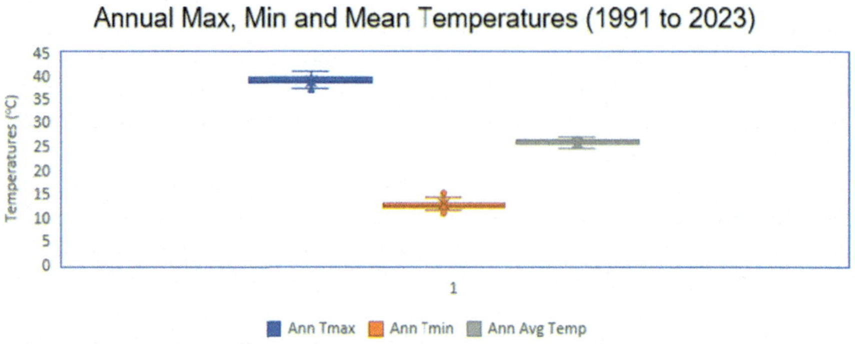


**Fig. 10.4** Temporal range in day and nighttime mean highest temperatures in the study area, 1991–2023. *Source* Author's compilation from field data

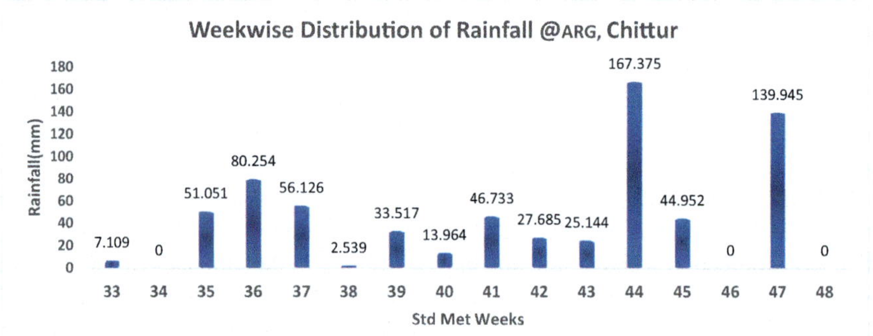


**Fig. 10.5** Rainfall recorded during the crop growing season in the study area, 2023. *Source* Author's compilation from field data

emission scenarios by the middle of the twenty-first century (Punnoli and Jayarajan 2023). This would not only put the existing farmers under pressure but also disrupt the ecosystem services that benefit all inhabitants (GoK 2017). Reviving Kerala's declining acreage under traditional rice types is crucial. It is significant because native traditional rice varieties like Kuruva/Kuruka, Njavara, and Kattamodan were shown to be resilient to flood and drought calamities, respectively.

## *Physical and Mental Health Issues Related to Extreme Climate Change Among Rice Farmers*

It was discovered that the difficulties associated with physical labour or drudgery during the three crop-producing seasons vary according to different seasons. There is a clear difference in dynamics on how various climate-induced disasters impact

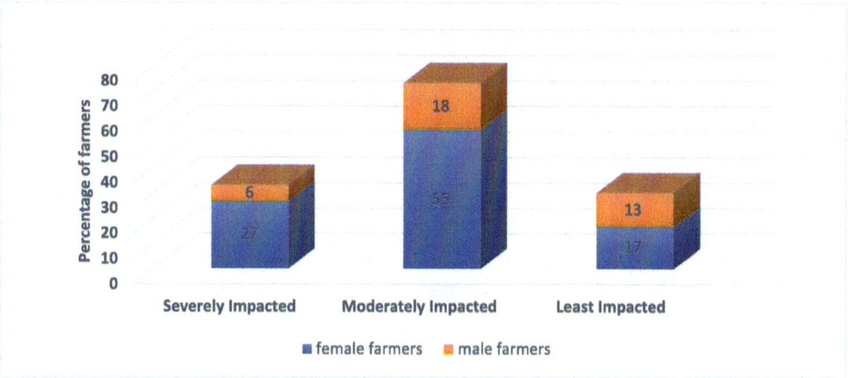


**Fig. 10.6** Severity of the climate and weather faced by surveyed farmers. *Source* Author's compilation from field data

farmers. Through a participatory survey, approximately a hundred female and forty male farmers discussed how weather aberrations have affected their everyday lives and farm operations. Out of the 99 female farmers surveyed, 27 said that they are severely impacted by climate-induced disasters (Fig. 10.6). However, only 6 out of the 36 male farmers responded that they are severely impacted by climate-related challenges in rice farming. The least impacted people were either having high adaptive capacity in terms of economic conditions, better living standards, higher educational qualifications, or less sensitivity and exposure to impacts. In most cases, their farms were located far away from the flood plains of the river.

The major struggles faced by farmers during the kharif season are during the rice nursery preparations and its transplantations. Mostly during southwest monsoon season, delays in the onset of monsoon, dry spells, and slow onset events like drought, during the initial months of cropping season, have given stress to female farmers the most. Through problem confrontation analysis (Fig. 10.7) it was understood that extreme rainfall during kharif crops and rabi crops act as the severe primary challenge and rising heat and heat waves during kharif and summer crops pose as the secondary excruciating problem faced by farmers recently during the farm operations.

The sad truth is that, compared to their male counterparts, women farmers are more negatively impacted in general by heat stress during farm operations. Several health problems like severe skin damage, irritation, and itching have been reported by the farmers due to exposure to scorching sunlight. Dehydration-related illness is the most prevalent health condition identified by the respondents, followed by problems of heat boils and prickly heat. Prolonged exposure to high daytime temperatures during harvest is making them dehydrated, especially when people do not have access to enough drinking water or shade from the trees nearby indigenous hats etc. Heat exhaustion and heatstroke are common health risks during heatwaves, especially for susceptible populations like the elderly and those with underlying medical conditions. Severe headaches and urinary tract infections (UTI) were also reported, followed by

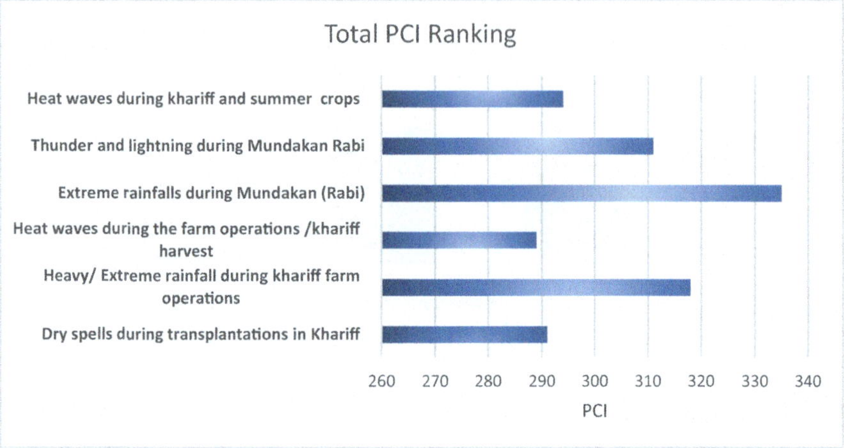


**Fig. 10.7** Problem confrontation index among the surveyed farmers. *Source* Author's compilation from field data

sunburn, fatigue, skin rashes, etc., during summer. Along with that, fetching water for drinking and washing purposes becomes an additional duty for women farmers apart from their regular tedious farm activities.

Farmers expressed their fear of severe lightning and thunder. They restrain themselves from work if there is severe lightning or thunder. Catastrophic events happen when extreme rainfall on consecutive days creates flash floods in the farmlands. Apart from physical strains, climate change, and weather extremities have intensified their fears, mental stress, and anxieties. It was very evident from the survey that the experienced farmers are somewhat mentally resilient, still, the recent incidences of havoc that floods, high-intensity rainfall, heat stress, and lightning damage have instilled serious concerns negatively impacting their mental health.

During extreme rainfall events or floods, the respondents face health challenges like frequent fever and cold, cough, dysentery, joint pains and cramps, and other airborne and waterborne diseases. During floods, they also lose their daily wages. Women become overburdened with household chores along with taking care of sick family members. The extreme rainfall events not only affect women's physical health but also stress them mentally and worry about the financial security of their families, especially after losing jobs and destruction of property and other assets.

Farmers reported that the intrusion of wild boar in the farmlands is a serious problem for farming communities in the villages. They come in groups, and they are very aggressive. Due to this menace, farmers are unable to venture into their fields during early mornings or evenings alone without fear.

Studies have found that groundwater in Kerala has a significant fluoride level, especially in Palghat (Shaji et al. 2007) and Alleppey region (Shaji et al. 2018). Secondary data from the State Groundwater Department confirms higher concentrations of fluoride in several blocks in the state. Villages consuming this water are at health risk. The spatial pattern also shows a stark correlation between population density and high irrigation usage with water shortages. Thus, the study area suffers from the availability of both quality and quantity of potable water. Thus, women farmers get overburdened by the responsibility to fetch safe drinking and cooking water for their families in addition to their labour on paddy farm operations. During the times of extreme weather events, non-cooperation from family members and unhealthy or toxic relationships within their community affect their well-being and put them under stress. Emotional attitudes and socialization of women to please and make peace with others at the cost of their mental health lead to greater emotional and physical distress.

From our survey, we could observe that most women were incapable of influencing farming choices like what crop variety to plant when to plant or harvest, or when to provide fertilizers, etc. Delay in getting the rice procurement prices also affects the smallholder's livelihood. The interviews helped a long way to understand how crucial it is to create workable plans and solutions to lessen the negative effects of excessive heat and protect the welfare of female farmers.

## Local and Global Programs to Combat the Negative Impacts of Climate Change on Women Farmers

Government, local administrations, the Department of Agriculture, as well as NGOs can help in raising awareness and equipping the farmers with appropriate adaptation strategies to either minimize the loss or avoid it altogether. Sustaining paddy farming is very critical to maintaining ecosystem services (GoK 2017). Due to the rising impacts of climate-induced disasters, the 'Rebuild Kerala Initiative' is a government program that has started to reduce the long-term effects of climate change, like water scarcity, flood, and heat wave impacts (Devi 2022; Richter et al. 2023). This may have the potential to benefit the vulnerable primary stakeholders.

Since the role of the female workforce is crucial for sustaining rice, which is a labour-intensive crop, focus from the government on any form of capacity building for women is much appreciated. Kudumbasree is one such initiative by the government that provides helping hands to the women and community as a whole during critical times. During the recent distressing floods, affected homes were cleaned and sanitized as part of a large cleaning campaign started by the Kudumbasree women's collective. Additionally, the organization provided counselling to thousands of families that experienced psychological trauma because of the tremendous damage brought on by the constant rains and floods. It has now been established as a program for women's empowerment in support of broader social goals such as women empowerment, poverty, and eradication.

To address the obstacles and difficulties faced by women farmers in agrifood systems, as well as to advance gender equality and the empowerment of all women in agriculture, FAO is planning to celebrate the International Year of the Woman

Farmer in 2026 to provide a forum for the adoption of appropriate policies and initiatives. Additionally, the observance will emphasize how important peasant women and other rural women are to their families' financial well-being and the rural and national economies. UNDP and FAO are working together to undertake the Scaling up Climate Action in Land Use and Agriculture Sectors (SCALA) program, which aims to increase knowledge and evidence on commercial potential in the land use and agriculture sectors and benefit farmers. Other initiatives focus on the overall development of the farmers and gender equality and inclusion (FAO 2023). The FAO works on the thrust areas to end poverty, hunger, and malnutrition as well as to manage natural resources sustainably are to be met through better production, better nutrition, better environment, and better life initiatives.

Although the majority of farmers in the area grow rice with traditional technologies, modern equipment like power tillers and combined harvesters are now commonly used. In recent times, mechanization of these farm operations has largely helped the women farmers to minimize the physical load and exposure to harsh climates, especially during harvesting which is majorly an activity of women farmers. Sometimes technology is a boon, but other times, it can replace the labour force and displace the marginalized women labourers. The application of heavy loads of chemical herbicides is replacing the manual weeding jobs of female farmers partly or completely in many places. Some of them even opined that in Mahatma Gandhi National Rural Employment Guarantee Mission (MGNREGS), it is providing them with secured jobs comparatively for 100 days than through farming.

## Recommendations and the Way Forward

According to case studies, when women have the power to make decisions, they are more likely to adopt climate-resilient strategies like drought preparedness measures (Grillos 2018) and crop and livelihood diversification strategies that improve household diets (De Pinto et al. 2020; Gumucio et al. 2022; Mersha and Van Laerhoven 2016; Smucker and Wangui 2016). To sustain their livelihood and overall well-being, female farmers are challenged by unpredictable weather patterns to transform their farming practices to be more resilient by implementing cost-effective mechanizations and resilient varieties. The Department of Agriculture and state agriculture universities must enhance the dissemination and reachability of contingency crop planning, setting up custom hiring centres, and alternative livelihood options by the Indian Meteorological Department's seasonal climate predictions.

From the observations from the survey, the authors propose some recommendations for improving the livelihoods and lives of women farmers. They are as follows:

- Wide gender disparity in wages in the rice farm sector is a cause of concern that has to be addressed at the local scale and must be minimized

- In the purview of extreme climate change manifestations intensifying in recent times, it is essential to treat farms as a workplace and provide essential infrastructural solutions. There should be provision for shade net rooms, rest rooms, etc., for the female farmers to take rests in between farming tasks.
- Developing proactive support systems such as enhancing drinking water portals, first aid kits, credit facilities, resilient variety seeds, and essential types of machinery before disasters should be made part of the adaptative solutions.
- Provision of shades or growing large trees in between the farms will be a solution to tackle the issues of heat stress during farm operations.
- Enhancing the dissemination and accessibility of early warnings, agronomic advisory services, and seasonal climate forecasts would help the farmers to tackle their drudgeries, crop loss, and damage to a great extent.
- Active participation in the Self-Help Groups (SHGs) can also contribute immensely to improving their income, autonomy, and decision-making ability. Apart from that, sharing responsibilities lessens tedious workloads for each member individually.
- Ensuring crop weather insurance and other safety nets is vital to enhance the quality of life and well-being of women farmers, especially under the scenario of rising climate change challenges.
- Not only the public sector, but private sector also must come forward with a gender lens and enhance their contribution towards strengthening the safety and well-being of the communities through adaptation interventions.

## Conclusion

It is seen that agrifood systems and the communities that support and depend on them are disproportionately affected by loss and damage associated with climate change. Sustainability of agrifood systems therefore is central to communities' resilience to manage climate change risks and impacts. The study identified several critical constraints experienced by smallholder farmers especially women as extreme rainfall during the rice-growing seasons, kharif and rabi. Similarly rising temperatures and heat waves throughout the kharif and summer crops were identified through problem confronting analysis as the most painful issues women farmers have been dealing with lately in their agricultural operations. The findings also showed that these challenges were found to be increasing year by year due to the increasing nature of extreme weather events in the region. Consequently, the physical and mental stress of the women farmers is also increasing with time.

Water is one of the key factors where females have a large role to play. Furthermore, a lack of critical infrastructure is making them less adaptive. The study also highlighted the need for better living conditions for women farmers and emergency healthcare facilities near their farms. Not only pay parity but their overall physical and mental well-being is always taken for granted either by the family or society. Their voices need to be heard by the policymakers and their crucial role as farmers and breadwinners must be taken seriously. By resolving these issues, women farmers' drudgery concerning climate-induced challenges may be addressed and resolved through timely interventions from the concerned authorities at all levels. The voices of the marginalized women farmers are often left out of policy discussions, and this must be changed. They also should be part of the policy discussions and dialogues, ensuring women hold space to voice their perspectives and challenges. The government needs to take extra care, support the existing women farmers to change their status, and encourage the youth to take up agriculture. Taking urgent and significant action can evade crop losses and enhance the resilience of the women farmers to climate change.

**Acknowledgements** This survey was part of the project titled 'Augmenting climate change adaptive capacity and mitigation potential among the rice farmers of Bharathapuzha River Basin in Palakkad through ICTS And Ecosystem-Based Adaptation (Eba)interventions', funded by the Kerala State Higher Education Council (KSHEC), Thiruvananthapuram. Under Chief Minister's Nava Kerala Post-Doctoral Fellowship, which was initiated as part of the Rebuild Kerala initiative after the flood, authors would like to thank the farmers of the Bharathapuzha River basin for their support throughout the project implementation and especially during the survey.

## Notes

1. Kudumbasree was established in 1998. It is a significant poverty eradication and women empowerment program in Kerala, India.
2. Padasekhara samitis are committees that are formed as part of the program's efforts to encourage collective farming and increase agricultural production among women farmers, with the ultimate aim of improving their economic status.

## References

Chanana-Nag N, Aggarwal PK (2020) Woman in agriculture, and climate risks: hotspots for development. Climatic Change 158(1):13–27. https://doi.org/10.1007/s10584-018-2233-z

Chandra A, McNamara KE, Dargusch P, Caspe AM, Dalabajan D (2017) Gendered vulnerabilities of smallholder farmers to climate change in conflict-prone areas: a case study from Mindanao, Philippines. J Rural Stud 50:45–59.https://doi.org/10.1016/j.jrurstud.2016.12.011

FAO (2023) Accessed from https://www.fao.org/newsroom/detail/women-s-equality-in-agrifood-systems-could-boost-the-global-economy-by-1-trillion-reduce-food-insecurity-by-45-million-new-fao-report/en on 22 July 2024

FAO (2023) The status of women in agrifood systems. Rome. https://doi.org/10.4060/cc5343en

FAO (2024) Accessed from https://www.fao.org/newsroom/detail/the-climate-crisis-is-unjust-for-rural-women--fao-gender-expert/en

GoK (Government of Kerala) The Kerala Conservation of Paddy Land and Wetland Act (2008) Accessed 13 Feb 2023. Available: https://ildm.kerala.gov.in/wpcontent/uploads/2017/01/PADDY-ACT-2008015.pdf

Grillos (2018) Women's participation in environmental decision-making: quasi-experimental evidence from northern Kenya World Dev 108:115–130

Gumucio T, Hansen J, Carr ER, Huyer S, Chiputwa B, Simelton E, Partey S, Schwager S (2022) Enhancing climate services design and implementation through gender-responsive evaluation. Front Clim 4. https://doi.org/10.3389/fclim.2022.908602

Hossain MS, Miah MAM (2011) Poor farmers' problem confrontation in using manure towards integrated plant nutrition system. Bangladesh J Extension Educ 23(1&2):139–147

Indira Devi P, Sam AS, Sathyan, AR (2022). Resilience to climate stresses in South India: conservation responses and exploitative reactions. In: Haque AKE, Mukhopadhyay P, Nepal M, Shammin MR (eds) Climate change and community resilience. Springer, Singapore. https://doi.org/10.1007/978-981-16-0680-9_8

IPCC (2022) Summary for policymakers. In: Pörtner HO, Roberts DC, Poloczanska ES, Mintenbeck K, Tignor M, Alegría A, Craig M, Langsdorf S, Löschke S, Möller V, Okem A (eds) Climate change 2022: impacts, adaptation, and vulnerability. Contribution of working group II to the sixth assessment report of the intergovernmental panel on climate change. Cambridge University Press, Cambridge

IPCC (2019) Accessed from https://www.ipcc.ch/srccl/cite-report/ on March 2022

KSDMA (2020) Accessed from https://sdma.kerala.gov.in/wp-content/uploads/2020/08/CWC-Report-on-Kerala-Floods.pdf, on 18 April 2024

Lipton M (2006) Can small farmers survive, prosper or be the key channel to cut mass poverty? eJ. Agric Dev Econ 3(1):58–85

Mani Murali R, Ankita M, Amrita S, Vethamony P (2013) Coastal vulnerability assessment of Puducherry coast, India, using the analytical hierarchical process. Nat Hazards Earth SystSci 13(12):3291–3311

Mersha AA, Van Laerhoven F (2016) A gender approach to understanding the differentiated impact of barriers to adaptation: responses to climate change in rural Ethiopia. Reg Environ Change 16(6):1701–1713. https://doi.org/10.1007/s10113-015-0921-z

Mishra V, Aadhar S, Shah H, Kumar R, Pattanaik D, Tiwari A (2018) The Kerala food of 2018: combined impact of extreme rainfall and reservoir storage. Hydrol Earth Syst Sci Discuss. https://doi.org/10.5194/hess-2018-48

Nirmala R (2023) Landslide and its impact on agriculture in Kottiyoor Panchayath, Kannur District, Kerala. In: Das J, Halder S (eds) Advancement of GI-science and sustainable agriculture. GIScience and Geoenvironmental Modelling. Springer, Cham. https://doi.org/10.1007/978-3-031-36825-7_17

De Pinto et al ( 2020) Alessandro De Pinto, Greg Seymour, Elizabeth Bryan, Prapti Bhandari Women's empowerment and farmland allocations in Bangladesh: evidence of a possible pathway to crop diversification. Climatic Change 163(2):1025–1043. https://doi.org/10.1007/s10584-020-02925-w

Prabhu S, Vishwas C (2024) Decoding India's changing monsoon patterns—a tehsil level assessment. New Delhi, Council on Energy, Environment and Water

Punnoli D, Jayarajan K (2023) The ecological significance of maintain rice cropping areas in the rice bowls of Kerala for sustaining food security under the purview of climate change. In: Alam A, Rukhsana (eds) Climate change, agriculture and society. Springer, Cham. https://doi.org/10.1007/978-3-031-28251-5_9

Punnoli D, Geethalakshmi V, Jagannathan R (2020) Preliminary situational analysis of impacts and risks of COVID-19 pandemic on crop production in Parambikulam Aliyar Basin, South India: need for capacity building. Disaster and Developm 9(1):112

Punnoli D, Geethalakshmi V, Ramanathan S, Senthilraja K, Sreeraj P, Pradipa C, Bhuvaneshwari, K, Vengateswari M, Dheebakaran G, Kokilavani S, et al (2022) Impacts and climate change adaptation of agrometeorological services among the maize farmers of West Tamil Nadu. AgriEngineering 4:1030–1053. https://doi.org/10.3390/agriengineering4040065

Resurrección BP, Bee BA, Dankelman I, Park, CMY, Halder M, McMullen CP (2019) Gender-transformative climate change adaptation: advancing social equity. Background paper to the 2019 report of the Global Commission on Adaptation. Rotterdam and Washington, DC

Richter BD, Ao Y, Lamsal G, Wei D, Amaya M, Marston L, Davis KF (2023) Alleviating water scarcity by optimizing crop mixes. Nat Water 1(12):1035–1047. https://doi.org/10.1038/s44221-023-00155-9

Shaji E, Bindu, Viju J, Thambi DS (2007) High fluoride in groundwater of Palghat District, Kerala. Curr Sci 240–245

Shaji E, Gómez-Alday JJ, Hussein S, Deepu TR, Anilkumar Y (2018) Salinization and deterioration of groundwater quality by nitrate and fluoride in the Chittur Block Palakkad Kerala. Abs J Geol Soc India 92(3):337–345. https://doi.org/10.1007/s12594-018-1017-4

Smucker TA, Wangui EE (2016) Gendered knowledge and adaptive practices: differentiation and change in Mwanga District, Tanzania. Ambio 45(Suppl 3):276–286. https://doi.org/10.1007/s13280-016-0828-z

Touch V, Daniel KY, Tan BR, Cook DL, Liu RC, Tran TA, Utomo A, Yous S, Grunbuhel C, Cowie A (2024) Smallholder farmers' challenges and opportunities: implications for agricultural production, environment and food security. J Environ Manage 370:122536. https://doi.org/10.1016/j.jenvman.2024.122536

WHO (2023) Accessed from https://www.who.int/data/gho/data/major-themes/health-and-well-being

World Meteorological Organization (2022) Early Warnings for All. (https://pub-lic.wmo.int/en/earlywarnings for all) Accessed from https://public.wmo.int/en/ear-lywarningsforall on 23 Feb 2023

# Chapter 11
# Hope and Despair: Towards a Gendered Emotional Geography of Water in a Changing Climate in the District of Bankura, West Bengal

**Tanusree Paul and Suranjana Karmakar**

**Abstract** An increasing number of world regions are being affected by changing climatic conditions that range from extreme events to erratic variabilities in climatic parameters. Water is an extremely crucial domain where such changes are explicitly experienced at all spatial scales. Such experiences are distinctly gendered. The patriarchal gender role stereotypes are reproduced within the waterscapes, causing men and women to have different material and emotional experiences surrounding water. Located within this theoretical understanding, the present paper explores how the relationship between gender and water is not only a matter of physical experiences surrounding fetching for domestic and other purposes but also shaped by distinct emotional responses. That said, the specific objectives of the paper are: first, to explore the perceptions of the women and men regarding the changes in climatic conditions in the study region; second, to understand how changing waterscape is differently experienced by women and men—both in terms of material realities as well as embodied emotions. Drawing on qualitative data collected through focus group discussions, key informant interviews and non-participant observation in the district of Bankura in Eastern India, this study argues that resource management practices, such as those surrounding water, are actively constituted and negotiated through embodiments, affect and emotions. Therefore, such practices and processes must not only be limited to exploring material challenges but also embodied emotions mediated through bodies, spaces and places.

**Keywords** Water · Gender · Emotional geography · Climate change · India

T. Paul (✉) · S. Karmakar
Women's Studies Centre, Institute of Humanities and Social Sciences, Visva-Bharati, Santiniketan, India
e-mail: paul.tanusree@gmail.com



A. Sarkar and N. Bandyopadhyay (eds.), *Gender-Transformative Approaches for Climate Change Adaptation*, Springer Climate,
https://doi.org/10.1007/978-3-031-90236-9_11

## Introduction

Climate change is one of the most menacing threats affecting the lives and livelihoods of people in India during contemporary times. Warmer temperatures, dwindling monsoons, untimely rainfall, and frequent tropical storms are fast reconfiguring the contours of the everyday lives of the local people, ranging from urban to rural scales. Climate change and water are intricately related, with the former affecting the latter in complex and nuanced ways. Not only do drier spells exacerbate water scarcities, but also excessive and untimely precipitation tends to cast multiple impacts, including reduction of agricultural yield. Official data indicate that due to changes in climatic parameters in India, rain-fed agriculture, especially rice farming, is projected to reduce by 20% by 2050, while the yield of wheat is projected to reduce by 19.3% by 2050 with significant spatiotemporal variations (PIB 2023). According to the 2018 Composite Water Management Index report by the National Institution for Transforming India, India is facing the worst water crisis in its history (NITI Aayog GOI 2018). A large section of the people is experiencing high to extreme water stress. India ranks 144 out of 180 countries in terms of unsafe drinking water index computed by Yale University 2024.[1]

The Ministry of Jal Shakti, Government of India, declared that the average annual per capita water availability for the years 2021 and 2031 was 1486 cubic meter and 1367 cubic meter, respectively, which indicate water-stressed condition (PIB 2024). Against this backdrop, the Government of India has introduced the *Jal Jeevan Mission* (JJM), which aims at providing clean and safe drinking water through functional tap connections (FHTCs) in every household, thereby contributing towards improving their quality of life. The Government of West Bengal also introduced the Sajal Dhara Project. Besides, several other schemes are also being introduced to promote integrated management of irrigation water, such as the Participatory Irrigation Management System (PIM), for better livelihood outcomes in the context of rural agriculture through proper management of water.

However, it is important to note that almost all these initiatives, especially those centring around water, are characterised by gross oversights at two levels. First, it focuses almost exclusively on technical and mitigation-based strategies. Second, it gives very little attention to the nuances of the gender lens in negotiations, interactions and interventions. Both these lapses can be appropriately addressed only if we recognise the critical roles individual women and men play in adapting to changing climatic conditions.

Climate change adaptation-focused studies need to take into consideration the perceptions and understandings of individuals regarding climate change in the first place. To quote Wolf and Moser (2011: 547), "What matters is their level of cognitive and emotional engagement as well as how that engagement leads to or is affected by behavioural changes and civic and political activities." This places emotional geographies of resource use, access and management at the core of climate adaptation discourses.

The people's understanding and perception are important not only in shaping their responses to climate change, determining their willingness to address the issue and rendering support to government programmes (Bord et al. 1998; Dunlap 1998; O'Connor et al. 1999; Leiserowitz 2005; Wolf and Moser 2011) but also in the context of successful management of water resources as well. The fact that gender is performed, enacted and negotiated at various scales through struggles over resources and natural environments has been amply examined and documented by feminist political ecologists (see Gururani 2002; Harris 2006; Nightingale 2006; Resurrection and Elmhirst 2008; Sultana 2009). Drawing on this scholarship as well as those in emotional geographies of resource management (e.g. Sultana 2006; Sultana 2007a 2007b 2007c; Sharp 2009; Smith et al. 2009; Pile 2010; Bondi 2017), the present article would attempt to show that adapting to changing waterscapes is not simply a function of economic models, rational choices or socio-technical issues; rather the various emotions, everyday struggles over water and experiences also shape their adaptation. Such an analysis would, to quote Sultana (2011: 164), "enrich our understanding of the ways that emotions come to matter in nature-society relations and influence how and why people use, access, control and conflict over resources the ways they do."

That said, the present paper aims to explore how the relationship between gender and water is not only a matter of physical experiences surrounding fetching for domestic and other purposes but also shaped by distinct emotional responses. The specific objectives of the chapter are as follows: first, to explore the understanding and perceptions of the women and men regarding the changes in climatic conditions in the study region; second, to understand how changing waterscape contributes towards shaping complex material realities reshaping traditional gender roles and gender identities. In doing so, the chapter vividly delves into the various emotions and meanings attached by women and men to use, access, control and conflict over water and how such emotions shape their everyday struggles and experiences of the waterscape. These are important because emotions matter not only in struggles over resources but also in how society responds to the management of resources. This chapter is organised into four sections. The introductory section is followed by a section on materials and methods, where a detailed discussion of the study site is presented, drawing on secondary literature. The third section narrates the results of the primary fieldwork, and the fourth section summarises the discussion and conclusion.

## Materials and Methods

### *The Study Site*

This study is based on primary fieldwork in the Bankura district of West Bengal, India, which is located in the south-western tracts of West Bengal. It is characterised by low precipitation, higher average annual temperature and drought-proneness. An in-depth estimation of the long-term trends of climatic parameters such as rainfall, temperature, etc., is both beyond the scope of this paper and also repetitive, primarily because changing climatic conditions in the study region have already been amply documented by government agencies, individual researchers and so forth (Goswami 2017; Singh et al. 2017; Guhathakurata et al. 2020). Hence, an attempt is made here to present the summarised findings to locate the latter discussion from a perspective.

Drawing on daily rainfall data from 1989 to 2018 from the Indian Meteorological Department, Guhathakurata et al. (2020) note significant changes in rainfall conditions over the districts of West Bengal. Their analysis shows that in terms of annual rainfall, Bankura experienced an increasing trend, though not significant. In terms of the number of rainy days, a significant increase in the stations in Bankura is also noticed. The increasing trend in annual rainfall in Bankura and elsewhere in Bengal has been noted in several other studies also and deserves some discussion here. In an extremely useful Climate Change Policy Paper, Swadesh Mishra (2012) notes that the climatic parameters in Gangetic West Bengal are exhibiting distinctive trends of change. Pal et al. (2015) analysed the rainfall pattern over the Gangetic West Bengal (GWB) during 1901–2000. They observed an increase in precipitation during specific months, especially in September. They noted that from 1901 to 2000 in Gangetic West Bengal, rainfall continued to increase monotonically at the rate of 0.73 mm per year in September. Overall, annual monsoon rainfall in the region also increased at the rate of 2.03 mm per year. Specifically, the month of September received an increasing amount of rainfall in Bankura district (along with East Midnapore, Howrah, North 24Parganas and South 24Parganas), measuring 0.84 mm per year respectively. According to Mishra (2012), this contributed to the increasing trend of annual rainfall to some extent. Bankura experienced an increase in annual monsoonal rainfall at the rate of 1.81 mm per year. Such rising amounts of precipitation during the latter part of the monsoon season resulted in several detrimental effects on agriculture- flooding, stagnation of water in agricultural fields, loss of standing crops yet to be harvested, loss of yield of kharif crops, etc. Analysing the rainfall data for 1901–2018, Hembram and Mondal (2020) note that Bankura has a decreasing trend of annual precipitation. As an average, the annual rainfall reduction is about 12.03 mm per year. However, monthly trend analysis indicated that April and August exhibited sharply decreasing trends while July and November displayed strong increasing trends.

In addition to rainfall, Mishra's analysis of the mean daily temperature from 1891 to 1970 indicated an increasing trend between 0.2 and 0.7 °C after 1970. The increasing trend in temperature is more prominent, with the daily minimum temperature rising faster than the daily maximum temperature. This resulted in the gradual reduction of the diurnal range. Winters were found to become warmer and shorter. The onset of monsoon was found to be delayed, shifting from 7th June to 13th June. Guhathakurata et al. (2020) note that there is a significant increase in dry days in Bankura district during monsoon months as well as annually. Analysis of drought conditions for a period of 100 years (1901–2000) by Mishra (2012) indicated that such conditions prevailed not only in these dry regions but also the whole of the Gangetic West Bengal for 7 years during the first 50 years period while the frequency increased to 12 during the second half of the last century. Thus, the frequency of droughts seemed to be increasing along with an increasing trend of storms and depressions.

An important point needs to be highlighted at this juncture. As per the criteria followed by the India Meteorological Department, this region does not fall in the drought-prone region. However, according to Mishra (2012), the drought-prone character of this region is not due to scarcity of rainfall but rather due to the poor moisture-holding capacity of the soil, which is a function of the undulating terrain and underlying rock structure of the region.

The preceding discussion based on secondary literature thus clearly reveals that the district of Bankura is experiencing distinct changes in climatic parameters. That said, primary fieldwork has been conducted to examine to what extent the changing climatic conditions affect the women's and men's livelihoods, the gendered adaptation strategies and the embodied emotions that shape the relationship between water and society.

According to the classification of the Central Ground Water Board, India (CGWB) and Geological Survey of India (GSI), the district can be divided into three main hydro-geological zones as follows: (a) Consolidated rocks with poor groundwater potential (Zone I); (b) Older alluvium cover with medium to good groundwater yield prospects by dug well and tubewell (Zone II); (c) Unconsolidated sediments with good groundwater prospects by shallow and deep tubewell (Zone III). For the present study, CD Blocks have been selected from each of these three zones as follows: Zone—I: Chhatna, Bankura I and Ranibandh; Zone—II: Simlapal, Sonamukhi; Zone—III: Kotulpur. Tribal villages located in these CD blocks (See Fig. 11.1 and Table 11.1) were selected in consultation with key functionaries such as Public Health Engineering department experts and Institutional Development Specialists.

## *Method, Questionnaire Design and Data Collection*

Data was collected through qualitative methods such as focus group discussions (FGD), key informant interviews (KIIs) and non-participant observation. In each CD-Block, two FGDs have been conducted, one with a group of men and one with a group of women. Thus, 12 FGDs were conducted in total. In addition, KIIs have also been conducted with key functionaries such as *Panchayat Pradhans, Panchayat* Members,[2] Block Project Managers, Community Workers, community mobilizers, etc. (see Fig. 11.4).

FGD participants have been selected using non-probabilistic, purposive sampling. In purposive sampling, samples are selected based on the judgement of the researcher (Vehovar et al. 2016). It is widely recommended in methods such as focus group discussion, which relies on the ability and capacity of participants to provide relevant information (Morgan 1996; Nyumba et al. 2018). However, utmost care was taken to ensure that the groups were socio-economically homogenous and participants were drawn from members of Water Users Associations, Village Water Sanitation

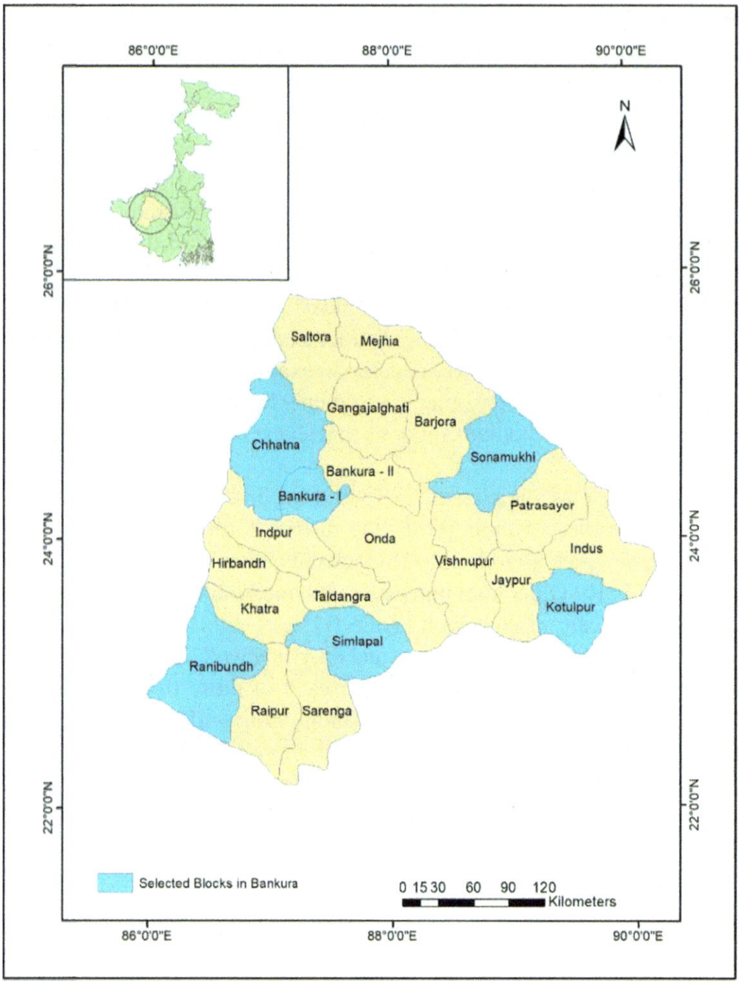


Fig. 11.1 Location of study sites. *Source* Prepared by authors

Committee (VWSC), Rural Water Supply and Sanitation (RWS) to gather perspectives on the perceptions, emotional responses and practices surrounding water in the different aspects of everyday lives. During the coding of the interview transcripts, all the names of the research participants were changed to maintain confidentiality. The names that are mentioned in the findings and discussions are imaginary.

Three concurrent aspects of climate change experiences were investigated using an unstructured interview protocol that examined three parallel domains, i.e., climate change, uncertainties in access to water resources and emotional responses surrounding water resource management through a gender lens. As the prime objective of the study is to capture the emotional responses to the water-related scenario, FGDs have been conducted to produce deep insights and information about the

**Table 11.1** Study sites

| Zone (Bankura District) | CD Block | Village |
|---|---|---|
| Consolidated rocks with poor **groundwater** potential | Chhatna | Jamdoba |
| | Bankura I | Keshiadoba |
| | Ranibandh | Haramgara, Ramdungri |
| Older alluvium **covers** with medium to good **groundwater** yield prospects by dug well and tubewell | Simlapal | Kansachora |
| | Sonamukhi | Radhamohanpur |
| Unconsolidated sediments with good **groundwater** prospects by shallow and deep tubewell | Kotulpur | Gopinathpur |

*Source* Fieldwork in Bankura, 2024

prevailing experiences and viewpoints of the participants. For this, a purposive sampling strategy has been used for the participants who are cooperative and knowledgeable. Additionally, participatory resource mapping has also been used to substantiate this information (see Fig. 11.2). In the initial stages of the fieldwork, non-participant observation was carried out in the form of attending meetings of the VWSCs at different locations, which helped in understanding the local community's concerns about water and how they responded to such concerns (Fig. 11.3). Besides in-depth conversations with the Institutional Development Specialists, the District Project Manager and Coordinator from the Public Health Engineering department of Bankura district (see Fig. 11.2) also provided sufficient insights for understanding the field sights and interpreting the data. Table 11.2 summarises the profile of the FGD participants.

## Results and Discussion

### *Changing Climate, Increasing Uncertainties*

The changing conditions of climate, especially rainfall, have already been discussed in a previous section, drawing on secondary literature. During the fieldwork, the participants were asked whether they had noticed any change in the climatic conditions in their village over the last five years. Almost all of them replied in affirmative, though they were not readily familiar with the term 'climate change' per se. Further, they described the changes that affected them the most and also tried to explain the reasons that they thought might be responsible for these changes.

When asked what they understood by climate change or what caused such changes, we received a wide variety of responses. Nirmal Tudu, a 30-year-old cultivator from the Haramgara village in Ranibandh block, observed, "*It means not having adequate rains or not having rains in time.*" Lakshmi Hansda, a 42-year-old tribal co-farmer on five bigha land from Kansachora village, Simlapal, beautifully puts it, "*The amount*

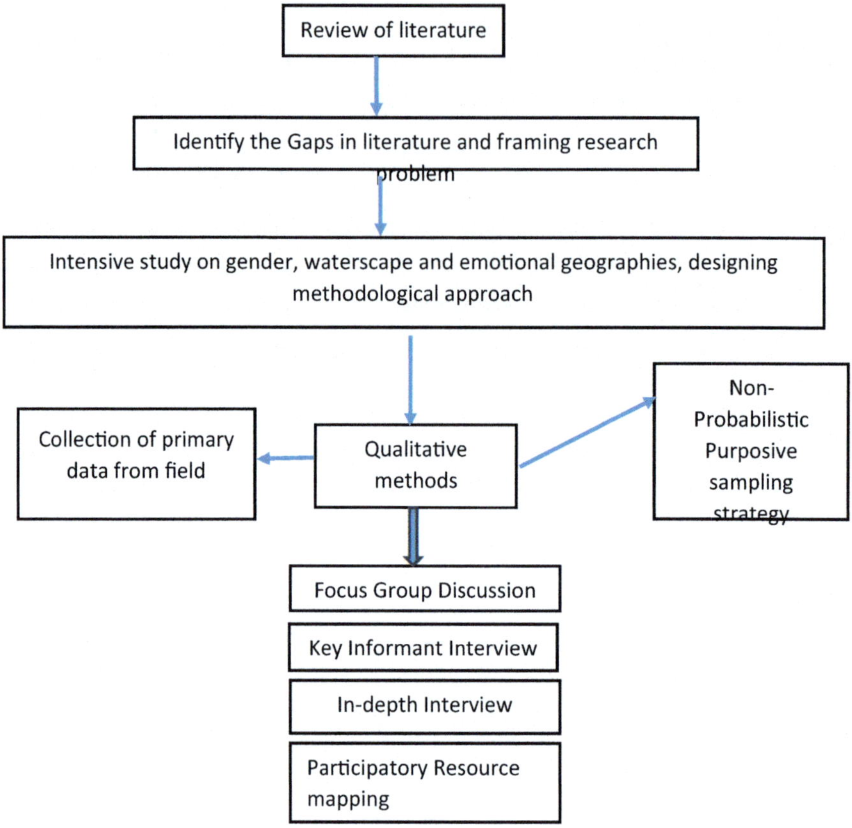


**Fig. 11.2** Methodology tree. *Source* Prepared by the authors

*of water that is sufficient to moisten our fields- we no longer receive such rains. That is the change I have noticed,*" Nanda Murmu, a 67-year-old tribal woman from Radha Mohanpur, Sonamukhi, added with exasperation, "*Earlier, the seasons were clearly distinguished from each other. Summer was followed by rain, then came autumn, followed by winter. Now, this is not happening. Round the year, it is becoming difficult to identify what is happening during which season*!".

The farmers in almost all the villages (except Kotulpur) mentioned that for the last three to four years, they have not received adequate rainfall, so it is a drought-like situation. "*It is difficult to bear the heat. Everything inside the house gets hot,*" said Renuka Sardar, a 35-year-old tribal woman from Ramdungri, Ranibundh. Durga Mandi, a 68-year-old tribal woman from the same village, also mentioned that the frequency of '*Kalbaishakhi*[4]' (also known as Nor-Wester) has reduced over the years.

Reduced rainfall during the monsoons and excessive rainfall during the harvesting season were a major concern, which was reiterated by the majority of the FGD participants across the district. Due to erratic rainfall, crop cycles are being disrupted,

**Fig. 11.3** Non-participant observation at village Water Sanitation Committee (VWSC) meeting in Sonamukhi Block, Bankura. *Source* Fieldwork in Bankura, 2024

**Fig. 11.4** Key informant interview with the officials in Gram Panchayat. *Source* Fieldwork in Bankura, 2024

resulting in large crop losses. The women farmers recollected how the rain showers were becoming more and more erratic. During the past two Bengali years,[5] there was no rain during the scheduled months, i.e. *Aashar* (June-July) and *Srabon* (July–August), while the heavy rains started in *Bhadra* (August–September). The participants mentioned they could not sow paddy in due time because of a lack of rainfall. As a result, the crop cycles were getting delayed, and the yield of paddy was declining. The yield of the seasonal horticulture crops in the region was also affected. Jagannath Murmu, a 50-year-old farmer from Kansachora village in Simlapal block mentioned,

**Table 11.2** Socio-economic profile of FGD participants

| Attributes | | Total number |
|---|---|---|
| Sex | Male | 36 |
| | Female | 36 |
| Social group | Scheduled tribe[3] | 72 |
| Age | < 35 years | 25 |
| | 36 to 50 years | 32 |
| | > 50 years | 15 |
| Education level | Illiterate | 25 |
| | Elementary | 20 |
| | Secondary | 18 |
| | Higher secondary and above | 9 |
| Size of land holding | < 2 Bigha (< 0.27 ha) | 12 |
| | 2–5 bigha (0.27–0.67 ha) | 23 |
| | > 5 bigha (> 0.67 ha) | 26 |
| House structure | Kuchha | 42 |
| | Semi-pucca | 17 |
| | Pucca | 13 |
| Grand total | | 72 |

*Source* Fieldwork in Bankura, 2024

> See, it is raining so heavily now that you cannot even move from one house to another. Such heavy rains are not good at this stage of paddy cultivation. When it is required, there is hardly any rainfall. And now, when we do not require so much water, it is continuously raining for 5–7 days. All the other crops, such as cucumber, radish, etc., are rotting in the field because of heavy Rains." In the words of Kishore Murmu from Jamdoba village in Chhatna block, "Earlier it used to rain continuously for seven to eight days. Now it does not rain for more than three days. Even the winters are becoming warmer. We do not feel cold any more during winter. The cropping seasons are getting pushed back. I have observed this. The heat has become impossible to bear.

Although all the participants spoke about changing temperature and rainfall, none of them seemed to have heard of 'climate change' or 'global warming'. However, they were familiar with various causes of environmental degradation and felt that these resulted in the observed changes in climatic conditions. At Ramdungri village in Ranibandh block, the female FGD participants reflected upon how the extensive use of chemical fertilisers was destroying the biodiversity of the soil that added to the soil fertility organically. In their words, "*Earlier, there were earthworms, frogs, snails, crabs and many other organisms. Now, we can hardly see them. The poison that we are adding is killing them. Earlier, because of those organisms, the soil particles were much looser, and we could plough easily and used to get a good harvest. Now, we are adding chemical fertilisers and pesticides. These are harmful.*"

Another major difficulty faced by the tribal villagers is the loss of non-timber forest products (NTFP), such as *bhalluk leaves, kendu leaves,* etc. that are extremely valuable to them. At Kansachora village, Machatora in Simlapal, the tribal women talked about the decreasing production of mushrooms, which is a commercially valuable product in the region. Parul Tudu, the Panchayat Pradhan of the village, observed, "*A lot is changing madam. It is raining much less these days. Earlier, in our forests, we used to get an abundant supply of Karan Chhatu.*[6] *Now, it grows in very little quantity. This is due to the weather conditions only, we think so.... Mainly because of low rainfall.*"

Decreasing quantities of *Karan Chhatu*was were also confirmed by the villagers of Kansachora at Simlapal block and those from Sonamukhi and Chhatna villages. When asked why these changes were taking place, almost all the participants promptly responded that it was due to the rampant dereliction and misappropriation caused by humankind. Suparna Ghosh, the Panchayat Pradhan of Machatora Gram Panchayat, Simlapal, remarked, "*We are only clearing forests, but not planting trees as much. The situation would have improved probably if we had planted more trees.*" Anju Soren from Kesiadoba village in Bankura I block, and many others also reiterated that rampant felling of trees was a major reason behind increasing climatic changes.

Indiscriminate felling of trees and pollution caused by vehicular emissions were also highlighted by several participants. Shyamal Mandi, a 62-year-old farmer from the Haramgara village in Ranibandh block, added, "*During Covid Lockdown, there were no vehicles, no cars, hence no pollution. We received good rain that year and also got a good harvest.*"

The awareness among the participants regarding the negative impact of plastics on the environment is especially noteworthy. Many of them articulated pollution caused by plastics to be a major reason for environmental degradation and, hence, must be addressed. Participants, for instance, Monica Tudu, explained, "*Say, there is a plastic lying in the sub-surface soil, then the sapling will not be able to grow its roots. Eventually, it will die.*"

During the fieldwork, every attempt was made to understand not only the people's perceptions about the changing climate and environmental resources but also to persuade the research participants to reflect upon their emotions and feelings to facilitate our understanding of their responses to these issues. As noted by Sultana (2011: 164), "*environmental degradation and resource crises can produce differentiated emotions that influence the very ways that resources are accessed, used, and controlled*". The FGD participants were asked how they felt about these changes in climatic conditions that they had already narrated. Dukhilal Murmu, a 42-year-old cultivator from the Haramgara village in Ranibandh block, observed, "*We feel very baddidi,*[7] the crops that are supposed to be grown in specific seasons ... we are not able to grow them during those seasons...rainfall is getting delayed. We feel bad about it." Sumitra Murmu, a 48-year-old tribal woman-cultivator from Haramgara village, Ranibandh block noted, "*It did not rain during the rainy season. You see, now it is raining so heavily in October. These showers are going to damage our crops. I am feeling so worried about this.*" In the words of Sanaka Hembrom, a 47-year-old tribal woman from Gopinathpur village of Kotulpur block, "*If it does not rain when it should be raining, then it is a matter of great concern for us. Now it is time for the paddy crop to mature, and you see, if it rains now or the wind blows strongly, then the yield goes down*".

Many of the participants also expressed concerns about the difficulties their future generations are likely to be suffering. Minoti Murmu, a 35-year-old tribal woman from Kesiadoba village, in Bankura I block, co-farmer with her husband on six bighas (0.80 ha) land, observed, "*Earlier, we used to get good rains. Our crop yields were also good. Now everything is getting affected. We do not know what our children will do! We do feel anxious about their future.*" Minoti Soren, a 35-year-old tribal woman from Haramgara village, Ranibandh, also shared her anxieties. "*We are also concerned ... we keep thinking, why the rains are behaving like this? But we do not know why*!" FGD participants from Kotulpur observed, "*We are worried about storms.*"

Perhaps an informed understanding of the climate-based uncertainties and how a poised response to such changes might facilitate the path for meaningful adaptation is reflected in the voice of Kishore Murmu, a 44-year-old farmer from Jamdoba village in Chhatna block. He reported, "*You see, didi, we are farmers. What I feel is that .... earlier, we did not have any water bodies in this region, so we could not do farming. Now, we have a big pond (Water Detention Structure- WDS). Now, we are growing crops thrice a year. Though nature is not giving us adequate time, rainfall has become uncertain, and cropping seasons are getting disrupted, we need to adjust. We have to think about growing crops that require less water. For example, we have received crops like maize, pigeon peas, mustard,* etc. *We are trying to find ways to grow these crops in less water. Then, in future, if our sons want to do farming, they will continue this. After all, we are farmers.*"

Thus, drawing on Sultana's work (2011), we argue that climate action is not only a collection of interrelated actions centred around society, ecology, economy and environment but also emotionally mediated. Several climate scientists argue that it is extremely difficult to engage people and communities, especially those living at the margins, in adaptation interventions, primarily because they are not yet aware of these changes (Akerlof et al. 2013). However, our FGDs succinctly brought out the fact that the tribal people, having very low socio-economic attainments, could distinctly make sense of the changes taking place around them. While they may not be familiar with the terms "global warming, "climate change", etc., they could vividly articulate the negative impacts of these changes—from loss of biodiversity to negative impact upon agriculture. It is worth mentioning here that the female participants in Ranibandh reported their familiarity with the concepts of greenhouse gases, carbon dioxide, CFCs, etc. Initiated anti-plastic environmental drives from the panchayat office, all the research participants were also observed to be quite environmentally aware and opposed to the use of plastics. The fact that they expressed deep concerns and anxieties about the future in the context of changing climatic conditions serves to invalidate the claim that rural communities lack the disposition to engage in climate action.

## *New Waterscapes and the Emotional Geographies of Water*

We began engaging with the research participants to understand their negotiations, struggles and everyday practices surrounding water with a participatory resource mapping exercise. The latter is expected to generate valuable insights regarding the participants' sense of acknowledgement and possession of the resources and the meanings they attached to these resources. Both male and female participants in the FGDs (in Ranibandh, Simlapal, Bankura I, and Chhatna block) were asked to draw on a blank white paper, starting with their current locations. They were provided with colour markers and were asked to choose whatever colour they liked. Willing participants were asked to identify the natural resources like different local water bodies and rivers, locations where they practised farming, collected forest products, fishing sites, grazing land, etc. and other places for monetary transactions like markets, shops, nurseries in their community, etc. They were also requested to mark locations for social interactions like schools, health centres, connecting roads, living homes, community halls, mosques, churches, temples, etc. A particular emphasis was placed on making boundaries for different types of water bodies, distance of water bodies from home, marking the principal drinking water sources, and where clean water was available. Both men and women mapped these resources, as per their knowledge and perceptions (Figs. 11.5 and 11.6). These maps were notional and did not adhere to scale or direction. Rather, these maps act as effective tools for opening up and shaping conversations.

From the analysis of the maps drawn by the participants, the following gender differences become evident:

First, the women participants mapped the landscape with far more details. They even went so far as to pinpoint the locations of the resources that were closest to them. The men mainly concentrated on outlining the locations, directions, roads, and fields.

Fig. 11.5 Resource maps drawn by female participants. *Source* Fieldwork in Bankura, 2024

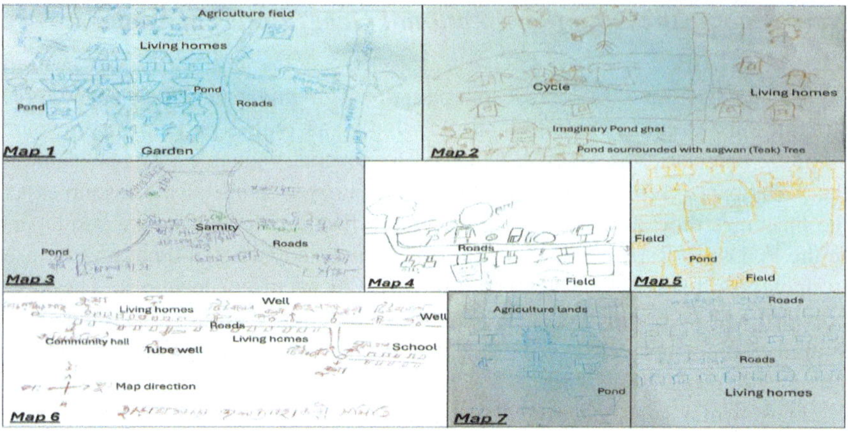


**Fig. 11.6** Resource maps drawn by male participants. *Source* Fieldwork in Bankura, 2024

Second, it is interesting to note that women outlined the resources such as water for domestic use like hand-pumps, wells and different other water bodies like ponds, dams and rivers. This reflects the gender division of labour. Women are most likely to frequently visit the water sources where water could be used for cleaning, cooking, and washing.

Third, forests are seen as a significant economic resource by both genders. However, women participants are more likely to locate nearby significant trees such as large banana trees with roots and palm trees, orchards, sugarcane gardens, and fruit gardens, paddy silos, and they differentiated the agricultural land for paddy cultivation and vegetable cultivations and others in the maps. The male participants focused more on commercial plantation teak plantation adjacent to the ponds and agroforestry farm trees agricultural land than on areas utilized for pisciculture. This difference in mapping seems to stem from their varied involvement with economic activities. Men were most likely to be involved in market-oriented primary activities, while the women tended to be engaged in paddy cultivation, kitchen gardening and maintenance of fruit orchards.

Fourth, women participants marked the ponds as fishing grounds, highlighting the 'fish' symbols to depict fish cultivation while describing its economic and cultural significance. It indicates that they were beginning to consider fishery as an important source of income. During the FGDs, almost all the participants noted that they cultivated fish in their water detention structures or ponds, partially for self-consumption and partially for sale in the market. The women mainly took care of preparing fish feed and cleaning these water bodies.

The mapping exercise facilitated subsequent conversations surrounding everyday practices and experiences with the waterscapes. For example, one of the most important observations emerging from the mapping analysis is that women participants classified the types of land based on land use and land cover. They identified lands other than agricultural land, like barren land (locally known as 'Tara'), fertile land

'Sol', and 'Baid'. 'Tara' land is infertile land which is not suitable for cultivation. 'Sol' is that low-laying agricultural lands are situated near ponds dams, where water is abundantly available for cultivation, and 'Baid' agricultural lands are water spars. In the Ramdungri village in Ranibandh block and Kesiadoba village in Bankura I block, the FGD female participants mentioned their concern about how rainfall uncertainties affected their cultivation practice in these different types of lands in other seasons. They highlighted these in the maps (Fig. 11.5).

As discussed previously in this chapter, the study sites experienced marked changes in the waterscapes over the past decade. On the one hand, rainfall had fluctuated; on the other hand, in almost all the villages, especially in Zone I and II, minor irrigation schemes have been introduced under the aegis of the West Bengal Accelerated Development of Minor Irrigation Project (WBADMIP) for integrated management of irrigation water. In addition, the installation of Functional Household Tap Connections (FHTCs) has also begun under the Jal Jeevan Mission (JJM). Most of these study sites used to be extremely water-scarce regions where women had to walk for hours to collect water for their everyday needs. During the FGDs, the participants from across the study sites narrated the immense hardship that they had once faced in meeting their daily water-based requirements such as drinking, cooking, washing, cleaning, etc. When asked about what the FGD participants considered as the most important uses of water, eight out of ten answered—drinking, cooking, bathing and agriculture- in that order. Since Bankura is located partly on a plateau region with geographical conditions not favourable for agriculture, the significance of water in their lives seemed to revolve primarily around domestic use, although at present, due to the WBADMIP schemes, some zeal towards improving agricultural yield has also been noticed. Agriculture in most of these villages transformed from being single-cropped to double or even triple, aided by the availability of irrigation water from the minor schemes constructed under the WBADMIP.

Bikram Murmu, a 64-year-old tribal farmer from the Haramgara village from Ranibandh block, narrated some indigenous methods of collecting water. He explained, "*In our small rivers or canals, say there are some depositions of sand bars in the low-lying areas… we used to take pitchers or clay pots and a hoe. We dug a hole in that sand, and water would easily sprout out of it. In our Santhali language, we call this method 'swick dadi'. Or, say, in some low-lying area, we used to dig a hole, their water would collect automatically, and then we collected water from that hole using a small leaf or a palash leaf. Now we have tubewells taps. We no longer use those methods*."

The introduction of modern technologies has changed the traditional ways of collecting domestic water. "*We used to wash our utensils using the water that dripped down from our thatched roofs*," said Deepa Murmu, a 48-year-old tribal woman from Radha Mohanpur village in Sonamukhi block. Lalita Sardar, a 45-year-old tribal woman and farmer on five bigha (0.67 ha) land, as well as the President of the Water Users' Association at Ramdungri villages in Ranibandh block, exclaimed, "*Now there is a tap connection in every household. We do not have to go elsewhere to fetch water. Earlier, we had to walk long distances to fetch water. It was troubling, and we had to suffer a lot. Earlier, we used to cycle for 2 km to the far-away dam just to take a bath. It was such immense hardship. Now we do not have to struggle so much.*"—recalled Renuka Sardar, a 35-year-old tribal woman from the same village.

Before JJM, the tubewells provided by the panchayat/block offices were the principal source of drinking water for the villagers. Although women primarily fetched water, men also participated, mostly in the absence of women. Chiranjit Murmu, a 60-year-old tribal farmer from Haramgara village from Ranibandh block cultivating about seven bighas (0.94 ha) of land, reminisced, "*In earlier days if there was any occasion in our homes, we had to fetch water from far-off wells using carrying poles fitted with two pails on both ends. This was until about 6 or 7 years ago, before the scheme. After the scheme, we are no longer required to do so.*"

The research participants mentioned that presently, their water woos reduced because the panchayat installed taps in every household. They were referring to the functional tap connections under the JJM. Yet, few places continued to suffer under the reeling water crisis even after the introduction of the various schemes. For example, in Kashachora village in Simlapal block, due to decreasing rainfall during the pre-monsoon and the early monsoon months, the local water bodies have almost dried up. Similarly, in Ranibandh region also, the farmers painstakingly narrated their story of suffering for water. In the words of Bikram Murmu, a 64-year-old tribal farmer from the village, observed, "*Our springs*[8] *are drying up. There are several places in our village neighbourhood where fountains used to form after heavy downpours. Those fountains used to have strong currents, and the water flowed into our agricultural fields as well. Now those fountains are not forming, and we are not able to use that water for farming.*"

They also mentioned that the rampant installation of tubewells with submersible pumps was causing more harm than benefits.

The participants in several study sites reported that about two to three trips were made on average to collect water for drinking and cooking purposes. Three to four buckets of water sufficed for an average household per day. In almost every *part* (local neighbourhood or locality), there is a tubewell. However, in some study sites, for instance, those in Ranibundh and Chhatna blocks, due to a dearth of adequate rainfall over the last few years, the tubewells also dried up. The Panchayat supplied water through tankers. At Jamdoba village in Chhatna block, the only source of drinking water is the High School located 1.5 km from the village.

Nevertheless, both the minor irrigation schemes, such as the check dams, water detention structures (WDS), etc., set up under the WBADMIP and the FHTCs under the JJM seem to have been considerably able to address the water woos in several study sites. In some sites, the villagers were also harvesting rainwater. For instance, Lalita Sardar from Ramdungri village from Ranibandh block mentioned, "*To conserve rainwater, we need to construct ponds and reservoirs. We make small pits of 2–3 feet in the garden/fields that can help in holding rainwater and also in groundwater harvesting. Otherwise, the rainwater simply runs off and gets wasted. Now, we also have dug a pond to hold rainwater. The depth of water is 15 feet. Earlier, that water also used to run off. We can use that water now for agriculture.*"

We observe that some operational challenges continue to hold sway. Although FHTCs have begun to be provided in households, the task of 100% installation is not yet completed. Hence, all the households are not yet receiving water. This is a major reason for the anxieties of the villagers who have been waiting for a very long time to access running water within their premises. The Block Project Manager (BPM) of the JJM at Simlapal narrated the entire process of providing drinking water connections to individual households. It involved work at three levels. At first, tap connections are installed in every household, followed by installations of pipes after all the taps have been mounted. The last stage involves making final connections with the main pipe, deciding on the village-wise distribution of water and the various modalities such as volume, speed, flow, etc. In the Machatora Gram Panchayat area of Simlapal, the installation of taps started in 2019. Since then, in many places, only taps had been mounted so far, but water connections were yet to be given.

In the words of the BPM, "*It is 2024 now. The villagers have become impatient and anxious. They have been waiting since 2019. Nobody is guiding them with proper information. Even those agencies who are coming for the installations are not giving them proper information. People are always anxious, and in every meeting, they will keep asking us two questions: first, when is water going to flow through our taps? And second, when is this work of tap and pipeline installation going to be completed? You see, Madam, there is only one tank in the village. The water supply is inadequate. In addition, since there is no water connection in the taps, there is no question of testing of water quality.*"

Such procedural delays are adding to the anxieties of the villagers and, according to the BPM, could also lead to the failure of the scheme. Additionally, another major issue is that there is limited understanding regarding the way water is supplied under the scheme. In the words of the BPM, Simlapal, "*There are two CWRs (Clear Water Reservoirs) in the Mukutmanipur region. From these CWRs, water is supplied and collected in the Overhead Reservoir (OHR) located at Khatra and from that OHR, water is supplied through pipelines to each household in the village. So, when water flows from CWR to the OHR, villages located in that region would get water. And when water is supplied from the OHR, then villages located on this side would get water. It is not possible that all villages would get water simultaneously at the same time. But the villagers do not understand this, and they get agitated.*"

Koyel Biswas, a member of the Rural Water Supply and Sanitation Department, observes, "*The women get angry, they complain about not getting adequate water. Previously, there were only three or four taps, but now the number of taps has increased phenomenally because each household has a connection. So, the flow has reduced. As a result, those who previously used to get good flow now get reduced flow; they feel agitated. They keep their taps open all the time and wastewater.*"

In Kotulpur, the BPM also noted that the women get agitated and keep asking in every meeting, "*If you cannot give us water, why have you given us a connection?*" Driven by disquieting anger, both women and men respond in manners that are detrimental to the scheme as such. They truncate the pipelines, allowing air to get into the pipe, which disrupts the flow. Sometimes, the officials and engineers are unaware of where exactly or who has truncated the pipeline and hence it remains undetected. This means that the entire village would suffer. Koyel Biswas added that sometimes, the villagers also install their taps along the pipelines laid down by the PHE. As a result, the water gets contaminated with bacteria and other germs.

Another important point worth noting is that with the formalisation of the waterscape through the implementation of the government scheme and the introduction of the FHTCs, the stress of provisioning for drinking water seems to have been shifted onto men. Now, it is no longer the women who unequivocally bear the burden of fetching and provisioning for water. Milan Ghosh, a Panchayat member and a political activist of the Machatora Gram Panchayat in Simlapal, notes, "*It is now our headache, Madam. The pressure is entirely on us. I might be having a tubewell at my home, but not everyone in the village has it. Now, if the only Panchayat-maintained tubewell in the village gets out of order, then it is our responsibility to get it fixed. We have to get in touch with the government departments to arrange for the repair work. That takes some time. Until then, we have to send tankers from the Public Health Engineering Department (PHE). Madam, until a few days back, about 8 to 10 tankers were being sent to the villagers because rainfall was inadequate and there was acute water scarcity. So, the entire burden is on us now*".

As observed by Tarun Murmu, a 40-year-old farmer from Kansachora village of Simlapal block, "Though inadequate water supply tends to concern women the most because it is their responsibility to cook food and do other household chores, then they come to us eventually, and it is we, men, who have to follow up and find ways of fixing the tap connections."

The BPM and other key functionaries complained about the fact that the villagers tended to use the clean potable water meant for drinking supplied through drinking for various other purposes such as bathing, washing, and cleaning. They primarily blamed the women for these wasteful usages of drinking water because these activities pertain to domestic chores undertaken by the women. Wasteful utilisation of water was confirmed by Koyel Biswas, too. However, she observed, "*Both women and men are involved. What the women cannot do, say illegal connections,* etc., *the men execute things. And when we try to talk to them, they would say, "everyone is doing this*!" The BPM of Simlapal observes, "*Suppose the main pipe is passing through someone's agricultural field, they will make a dent with the help of a hoe and take water out of it. Both women and men are involved.*"

These observations generate important insights regarding the fact that drinking water is no longer only a concern of women. The words of the Block Project Manager (BPM) of Sonamukhi further testify to this. Approximately 43,000 households are there in Sonamukhi Block, only 13% is covered by FHTC connection, and 5000 households have available water from FHTC. Referring to the fact that women often leave the taps open for two hours and waste the PHE water without knowing that this reduced water supply to other villages, he noted, "*Places where water is not available, mainly male members of the community protest, they try to stop the orientation programme or awareness programme that we organise for them. With so many orientation programmes being conducted day after day and no water coming, the male members of a village (in Panchal GP) got so angry and agitated that they came to beat up our community mobilizers. They tried to tear our banner and posters*".

Notwithstanding the wasteful uses of drinking water, we also found evidence of women making judicious use of the resource. For instance, at Ramdungri village of Ranibandh block, where every household gets water from 9 am to 3 pm regularly, residents showed considerable proclivity towards judicious utilisation of water. Rinita Mahato, a 26-year-old tribal woman, mentioned, "*We use this water only for drinking. We do not use it for any other purpose. For bathing, washing,* etc.*, we use water from wells, ponds* etc." Nanda Sardar from the same village added, "*If we waste this water for bathing and such activities, how will others get water for drinking? Ours is the main line. We hold the drinking water that we require for our families and then turn off the taps. Why will we waste water?*".

These excerpts indicate that women cannot be constructed as one homogenous category in terms of their response to engagement with water. The ease and extent of availability and accessibility of water determine the magnitude of struggle that women have to engage with in order to achieve the resource. This eventually shapes their emotional as well as material lives. This has a direct bearing on "*the ways water is accessed, used and struggled over in a locality*" (Sultana 2011: 165).

We also noted a heightened sense of anxiety among the research participants in some of the study sites regarding the quality of water. Sometimes, the water that comes through the taps is completely white. Participants exclaimed, "*I am not sure, but I think it's mixed with bleaching powder. Different people say different things. Some say that it is good because this water is good. After all, it helps in deworming. Some also say that consuming bleaching powder every day is not good. After a point of time, we will become paralysed. I am not sure. But this is worrying.*"

Several other research participants also seemed to have this misconception that bleaching powder was being used in the tap connections. Although this was chlorine that was periodically added to purify the drinking water made available through the tap connection, the villagers were not informed about this. In Keshiadoba, Chhatna, the women FGD participants reported the same fear of drinking water infested with bleaching powder. Hence, they avoided the tap water and used the tubewell. But since that tubewell had been operational for a long time without proper maintenance, water flow is very slow, and a lot of time is wasted in the process of fetching.

## Discussion and Conclusion

In this chapter, we highlight how the changing climatic conditions in general and waterscapes in particular are experienced and perceived by local communities. Such analyses merit academic attention because they generate insights for adaptation-focused perspectives in the context of the management of climate and natural resources. As noted by Wolf and Moser (2011: 547), in studies that focus on adaptation, "*what matters is their level of cognitive and emotional engagement as well as how that engagement leads to or is affected by behavioural changes and civic and political activities.*" Indeed, people's understanding and perception are important because these not only shape their responses to climate change but also determine their willingness to address the issue and render support to government programmes. This places emotional geographies of resource access and use at the core of management discourses. That said, the non-participant observation, focus group discussion and key informant interviews reveal the following key points:

First, contrary to a considerable volume of literature in emotional geographical studies that associate being 'angry' or 'concerned' with male responses and 'sad' or 'worried' with female responses, our study notes that women and men are rather united in the face of the onslaught of natural resources in general and water resources in particular. While most of them did agree that fetching water was primarily the responsibility of the women, men also contributed to this task. The non-availability of water equally concerned them as they struggled and suffered through their everyday embodied practices; delays and lapses were likely to make them equally resentful.

Second, we note from our fieldwork that women did resort to wasteful practices of water by leaving the taps open. Such behaviour was caused by an urge to appropriate as much water as possible in an otherwise water-scarce context. Thus, contrary to the argument developed in ecofeminist scholarship, there is a need to unpack the idea that only women have a special relationship with nature and that women's reproductive responsibilities make them inherently environmentally sensitive. Drawing on Agarwal (1992), we argue that availability, accessibility, reliability, and quality of water contribute towards shaping complex material realities, producing women as heterogenous subjects who vary in how they make sense of water crises in their lives and respond to their perceived realities.

Third, the formalisation of waterscapes in the form of taps, schemes, etc., seemed to be causing a reconfiguring of the contours of water by putting more responsibility on the men for fetching and provisioning of water. Thus, the distinctly gendered landscape of waterwhereby drinking water was considered as the domain of the women seems to be getting reconfigured.

This chapter generates significant insights towards understanding the perceptions and emotional responses among women and men surrounding climate in general and water in particular. It notes that although the research participants had limited or no exposure to factually correct knowledge about climate change, yet they were quite cognizant of the changes in rainfall and seasonal behaviour taking place around them and actively tried to make sense of these changes in the context of the rampant degradation of the environment unleashed by human action. It is their concern for the environment that made them ecologically sensitive, which was expressed through their anxieties, concerns, and willingness to plant trees, abstain from plastics and chemical fertilisers, etc. This indicates that even in the absence of adequate knowledge about climate change, they were remarkably worried about the observed trends and patterns of rising temperatures and uncertain rainfall, because of which they expressed a keen desire to adopt measures to save the environment. These findings have important policy implications from the climate change management perspective.

Specifically, concerning water, the study produces valuable insights. To start with, it notes that while the introduction of various government schemes has improved the supply of water in an otherwise drought-prone region, the intricate interplay of availability, accessibility, reliability and quality of water contribute towards shaping complex material realities producing women (and men) as heterogenous subjects, who vary in how they make sense of water crises in their lives and respond to their perceived realities. In many cases, it was found that taps were available, but a water connection was not established. Elsewhere, water might be available through the taps, but people were found to be under the misconception that it was mixed with bleaching powder hence, they abstained from drinking it and continued to suffer. There were still places where women had to travel for more than an hour to fetch drinking water. Therefore, the everyday water insecurities are produced by diverse everyday practices and make the people respond in myriad ways.

Further, the gender scape of drinking water seems to be getting reconfigured to a large extent. Men, as much as women, have been observed to be invested in the process of provisioning, collecting and management of drinking water after the introduction of taps. In those villages where women played a key role in the management of irrigation water, the latter was not only a concern of male farmers but women farmers as well. As noted by Sultana (2015: 640), in addition to sentiments caused by water crises, emotional geographies of water encompass the various aspects of water-fetching and water-sharing. Indeed, our study notes that women and men responded to these water situations in ways that were beyond the commonly documented emotions in literature, i.e. suffering and pain. They were concerned, impatient, agitated and angry—*both* women and men. As the KIIs with the BPMs reflected, these emotional sentiments largely shaped their willingness to embrace the government schemes.

Thus, this paper furthers Sultana's proposition of a feminist emotional political ecology approach (2015) whereby she demonstrates the importance of recognising the importance of emotions and meanings attached to resource access, use and conflict for a more nuanced understanding of struggles over resources. The emotional geographies of water access and use generate useful insights as to how everyday lives come to be constituted through water and how the material conditions and experience of hardship are likely to influence the management of the resource.

## Notes

1. This index is calculated using the number of age-standardized disability-adjusted life-years lost per 100,000 persons (DALY rate) due to exposure to unsafe drinking water. It is based on the Global Burden of Disease study data collected by the Institute for Health Metrics and Evaluations (IHME). A score of 100 on this index indicates a country has among the lowest DALY rates in the world while a score of 0 indicates a country is among the highest DALY rates.
2. In India, a Panchayat Pradhan is the elected head of a Gram Panchayat, which is the primary village-level governing body and an elected official of that local self-government body is known as a Panchayat member.
3. Article 342 of the Indian Constitution recognizes tribal Communities in India as Scheduled Tribes (STs).
4. Nor-Wester is a summertime, intense thunderstorm and localized rainfall phenomenon that happens in Bangladesh and India's Gangetic plains.

5. In the Bengali year, Aashar and Sraban are the two monsoon months while Bhadro is the fifth month roughly coinciding with September in the English calendar.
6. Karan Chhatu is a rare eatable mushroom wild mushroom that grows only during the autumn after the rain subsides. It is typically found in Sal woodlands during autumn in the study region.
7. *Didi* is a Bangla word used to refer to an elder sister or an elder woman.
8. These springs are kind of openings and may be referred to as an artesian springs. It occurs during the rainy season when there is a high groundwater pressure, which allows water to spontaneously rise through ground cracks or openings. When the groundwater level rises, particularly during the monsoon season, these types of natural springs can be seen in various locations, particularly near the edges or corners of the agriculture fields.

## References

Agarwal B (1992) The gender and environment debate: lessons from India. Fem Stud 18(1):119–158. https://doi.org/10.2307/3178217

Akerlof K, Maibach EW, Fitzgerald D, Cedeno AY, Neuman A (2013) Do people "personally experience" global warming, and if so, how and does it matter? Glob Environ Chang 23(1):81–91. https://doi.org/10.1016/j.gloenvcha.2012.07.006

Bondi L (2017) Making connections and thinking through emotions: between geography and psychotherapy. In: Philo C (ed) Theory and methods, 1st edn. Routledge, London, pp 563–578

Bord RJ, Fisher A, O'Connor RE (1998) Public perceptions of global warming: United States and international perspectives. Clim Res 11:75–84. https://doi.org/10.3354/cr011075

Dunlap RE (1998) Lay perceptions of global risk: public views of global warming in a cross-national context. Int Sociol 13:473–498. https://doi.org/10.1177/026858098013004004

Goswami A (2017) Analysis of meteorological drought and its trend in Bankura District of West Bengal, India. In: Recent trends in agriculture, food science, forestry, horticulture, aquaculture, animal sciences, biodiversity, ecological sciences and climate change (AFHABEC-2017), pp 6–12

Guhathakurta P, Sudeep Kumar BL, Menon P, Prasad AK, Sable ST and Advani SC (2020) Observed rainfall variability and changes over Kerala State. India Meteorological Department. Available via https://www.researchgate.net/profile/Sudeep-Kumar-6/publication/341508378_Observed_Rainfall_Variability_and_Changes_over_Kerala_State_Observed_Rainfall_Variability_and_Changes_Over_Kerala_State/links/5ec4c625458515626cb84b59/Observed-Rainfall-Variability-and-Changes-over-Kerala-State-Observed-Rainfall-Variability-and-Changes-Over-Kerala-State.pdf

Gururani S (2002) Forests of pleasure and pain: Gendered practices of labour and livelihood in the forests of the Kumaon Himalayas, India. Gend Place Cult J Fem Geogr 9(3):229–243. https://doi.org/10.1080/0966369022000003842

Harris LM (2006) Irrigation, gender, and social geographies of the changing waterscapes of south eastern Anatolia. Environ Plan D Soc Space 24(2):187–213. https://doi.org/10.1068/d03k

Hembram B, Mondal S (2020) Assessing the climate change in Bankura district of West Bengal: 117 years rainfall data analysis. CLIO Annu Interdiscip J Hist 6(14):31–35

Leiserowitz A (2005) American risk perceptions: is climate change dangerous? Risk Anal 25:1433–1442. https://doi.org/10.1111/j.1540-6261.2005.00690.x

Mishra S (2012) Climate change adaptation in arid region of West Bengal. Climate Change Policy Paper III, WWF India. Available at https://wwfin.awsassets.panda.org/downloads/climate_change_adaptation_in_arid_region_of_west_bengal.pdf. Accessed 04 Oct 2024

Morgan DL (1996) Focus groups. Ann Rev Sociol 22:129–152. https://doi.org/10.1146/annurev.soc.22.1.129

Nightingale A (2006) The nature of gender: work, gender, and environment. Environ Plan D Soc Space 24(2):165–185. https://doi.org/10.1068/d01k

NITI Aayog (2018) Composite water management index: a tool for water management. Available via https://www.niti.gov.in/sites/default/files/2023-03/CompositeWaterManagementIndex-CWMI.pdf. Accessed 01 Oct 2024

Nyumba T, Wilson K., Derrick CJ, Mukherjee N (2018) The use of focus group discussion methodology: insights from two decades of application in conservation. Methods Ecol Evol 9(1):20–32. https://doi.org/10.1111/2041-210X.12860

O'Connor RE, Bord RJ, Fisher A (1999) Risk perceptions, general environmental beliefs, and willingness to address climate change. Risk Anal 19:461–471. https://doi.org/10.1111/j.1539-6924.1999.tb00421.x

Pal I, Robertson AW, Lall U, Cane MA (2015) Modeling winter rainfall in Northwest India using a hidden Markov model: understanding occurrence of different states and their dynamical connections. Clim Dyn 44:1003–1015. https://doi.org/10.1007/s00382-014-2178-5

Pile S (2010) Emotions and affect in recent human geography. Trans Inst Br Geogr 35(1):5–20. https://doi.org/10.1111/j.1475-5661.2009.00368.x

Press Information Bureau (PIB) (2023) Impact of climate change on agriculture Available via https://pib.gov.in/PressReleaseIframePage.aspx?PRID=1909206. Accessed 01 Oct 2024

Press Information Bureau (PIB) (2024) Per capita water availability. Accessed at https://pib.gov.in/PressReleaseIframePage.aspx?PRID=2002726. Accessed 01 Oct 2024

Resurreccion BP, Elmhirst R (2008) Gender and natural resource management. Earthscan, London. https://doi.org/10.4324/9781849771436

Sharp J (2009) Geography and gender: what belongs to feminist geography? Emotion, power and change. Prog Hum Geogr 33(1):74–80

Singh PK, Rathore LS, Rao DB, Singh KK, Baxla AK, Bhan SC, Gupta A, Singh S (2017) Spatial analysis of rainfall variability and rain fed rice crop using GIS Technique in West Bengal (India). Mausam 68(2):287–98. https://doi.org/10.54302/mausam.v68i2.631

Smith M, Bondi L, Smith M, Davidson J (2009) Emotion, place and culture, vol 1. Ashgate, Farnham

Sultana F (2006) Gendered waters, poisoned wells: Political ecology of the arsenic crisis in Bangladesh. In: Lahiri-Dutt K (ed) Fluid bonds: views on gender and water. Stree Publishers, Kolkata, pp 362–386

Sultana F (2007a) Suffering for water, suffering from water: political ecologies of arsenic, water and development in Bangladesh. Ph.D. Thesis, Department of Geography, University of Minnesota, Minneapolis, MN, USA

Sultana F (2007b) Water, water everywhere but not a drop to drink: Pani Politics (water politics) in rural Bangladesh. Int Fem J Polit 9(4):1–9. https://doi.org/10.1080/14616740701607994

Sultana F (2007c) Reflexivity, positionality and participatory ethics: negotiating fieldwork dilemmas in international research. ACME Int J Crit Geogr 6(3):374–85. https://doi.org/10.14288/acme.v6i3.786

Sultana F (2009) Fluid lives: subjectivities, gender and water in rural Bangladesh. Gend Place Cult 16(4):427–444. https://doi.org/10.1080/09663690903003942

Sultana F (2011) Suffering for water, suffering from water: emotional geographies of resource access, control and conflict. Geoforum 42(2):163–172. https://doi.org/10.1016/j.geoforum.2010.12.002 Get rights and content

Sultana F (2015) Emotional political ecology. In: Bryant RL (ed) The international handbook of political ecology. Edward Elgar Publishing UK, USA, pp 633–645

Vehovar V, Toepoel V, Steinmetz S (2016) Non probability sampling. In: Wolf C, Joye D, Smith TW, Fu YC (eds) The SAGE handbook of survey methodology, vol 1. Sage, London, pp 329–245

Wolf J, Moser SC (2011) Individual understandings, perceptions, and engagement with climate change: insights from in-depth studies across the world. Wiley Interdiscip Rev Clim Chang 2(4):547–569. https://doi.org/10.1002/wcc.120

# Part III
# Women as Change-Makers: Gender Transformative Climate Adaptation

# Chapter 12
# Gender-Transformative Adaptation for Food Security, Rural Livelihoods, and Agriculture: The Case Study of Mal Pahariya Women in Eastern Coalfield Regions of India

**Aparajita Banerjee**

**Abstract** Climate extremes like prolonged droughts or excessive flooding disrupt the intertwined lives of rural women with agriculture in the Global South, especially those dependent on traditional practices like shifting cultivation or rain-fed irrigation. Agriculture is not just food production or a livelihood for these women but an extension of the care work that helps them feed their families. On the other hand, traditional hetero-patriarchal social norms continue to disempower them to make choices related to agriculture and food production as they often lack the decision-making and ownership rights on the land they toil. Women belonging to marginalized tribal communities facing compound intersectional challenges due to gender, caste, ethnicity, illiteracy, poverty, and language find it even harder to define their life goals and act upon them. This chapter shows how specific gender-transformative adaptation measures, such as promoting women's land rights, providing access to climate-resilient seeds, and offering training in climate-smart agricultural practices, can bring in, albeit small, transformative changes. Based on ethnographic research, semi-structured interviews, and focus group discussions with Mal Pahariya women and NGO workers in Jharkhand, India, this chapter shows how these measures, if designed effectively, can tackle the root causes embedded in the existing social, political, economic, and cultural context that aids gender inequality and injustice to persist. The study also shows that food and nutrition security, leadership, economic decision-making, and capacity building by learning new skills can emerge as co-benefits that can help them address other daily challenges.

**Keywords** Gender-transformative climate change adaptation · Climate-smart agriculture · Food security · Nutrition security · Marginalized communities · Mal Pahariyas

---

A. Banerjee (✉)
German Institute of Development and Sustainability (IDOS), Bonn, Germany
e-mail: Aparajita.Banerjee@idos-research.de



A. Sarkar and N. Bandyopadhyay (eds.), *Gender-Transformative Approaches for Climate Change Adaptation*, Springer Climate,
https://doi.org/10.1007/978-3-031-90236-9_12

## Introduction

Earth's climate system has always been dynamic; humans have adapted to these changes for centuries. However, the current scale at which the climate is changing is far more rapid, and adapting to these changes to match the speed and uncertainty is challenging. Limiting global warming within 2 °C will be extremely difficult. Urgent and far-reaching transformations are required in energy, food, infrastructure, and industrial systems to reduce the further accumulation of greenhouse gases (GHG) contributing to global warming (IPCC 2018). Similarly, time is limited for adaptation to climate change (Wise et al. 2014). As these systems are an integral part of human societies, any alterations in these systems will have a direct impact on human lives, from the food we eat to how we travel, or heat and cool our homes.

The risk of distributional differences in climate change mitigation and adaptation measures is possible (Ivanova et al. 2020). Distributional differences may arise in allocating resources while transitioning to sustainable food, energy, transportation, and industrial systems. Who will benefit and who will bear the burdens, for example, in the restructuring for sustainability of existing food systems and energy systems? Whose livelihood will be at risk, whose skills will become obsolete, or who will benefit from the new opportunities created? Similarly, the distribution of adverse climate change impacts is disproportionate as well. Poor people and people working in climate-dependent economic sectors, such as agriculture, will be the worst affected. Yet they will also be less capable of adapting due to limited financial means, lack of technologies, or other constraints on their adaptive capacities (Mortreux and Barnett 2017).

The climate crisis is also not gender-neutral; women will be affected more than men (Denton 2002). Even within women and other marginalized genders, the severity of adverse climate impacts on health and well-being can differ based on a host of intersectional factors like race, caste, culture, ethnicity, social class, economic stability, and geographical location, amongst other factors that marginalize some group of people more than the others (Dankelman 2012). Pre-existing social and economic inequalities can further exacerbate the vulnerability of women and minority genders in marginalized groups under climate change. For example, most rural women in the Global South, traditionally dependent on nature for food, fuel, and livelihood, will find it difficult to adjust to a rapidly changing climate compared to their counterparts in wealthy, industrialized, democratic Western countries.

Even in the Global South, women simultaneously belonging to minority groups based on caste or ethnicity, low socio-economic class status, unskilled or underskilled to engage in formal livelihoods and living in strong hetero-patriarchal norms-based society will find themselves even more affected by climate extremes than other rural women where class and caste differences are not as severe. Strong social norms create power asymmetries between men and women that limit womens ability to gather skills, knowledge, education, social networks, and access to resources like finance that would help them improve their conditions or adaptive capacities (Buchanan et al. 2016).

A range of interrelated factors exacerbates the conditions of women at the social, cultural, economic, and political margins. The meta-norm in most rural societies establishes women as primary caregivers, fuel wood collectors, and unpaid farm labourers while simultaneously being dependent on the male members of the family to make any decisions related to finance and expenditures. Male members in the family take up the traditionally assigned roles of the protector, provider, and the decision-maker. On the other hand, women in rural communities struggle with food and nutrition security. Malnutrition in women is often high (Shafique et al. 2007) and more so in women of lower socio-economic status, leading to chronic energy deficiency which further makes them physically incapable of engaging in strenuous yet often higher remunerative livelihoods.

Official records estimate that women constitute around 40% of the official agricultural labour force (Saran et al. 2024). Even then, women are often ignored in the design of climate-smart agricultural interventions (Dankelman 2012); their voices remain unheard or under-heard in the design and implementation of plans and programmes, especially those that would help them adapt to changing climate (Arora-Jonsson 2011). This is because even though women work on agricultural land, the male family members mainly own the land. To address these multiple chronic challenges, paradigmatic changes are required so that the same hetero-patriarchal norms that perpetuate conditions of inequality, oppression, and injustice that marginalize women are addressed and uprooted. Some researchers suggest that incremental changes to address gender inequalities may not be helpful in sustainable climate change adaptation, and transformative approaches are required (Fraser 2007).

In recent years, there has been an increase in the mention of gender in adaptation debates, and there is a recognition that any climate change action will be unsustainable if gender issues remain unaddressed (MacGregor 2010) and gender inequalities are perpetuated (Pearse 2017). Gender-transformative adaptation strategies explore how existing deeply rooted social systems protected by those the system empowers can be challenged (O'Brien 2012). For achieving transformative remedies, Fraser (2007) suggests three interdependent and inseparable dimensions such as redistribution, recognition, and representation can help understand and identify the leverage points for transformative changes. For example, sustainable climate change adaptation is impossible if power is not redistributed between men and women to achieve gender equality. Interrelated to that is recognizing the factors that create power imbalances between genders in a given society. To recognize the factors, it is again critical to give women the representation to express their concerns, opinions, advice, and perspectives so that climate change adaptation is designed, taking into consideration the unique perspectives of women based on the particular roles they play in families, communities, and societies. These changes are structural and deep-rooted; interventions need to be purposefully designed.

Similarly, adaptation constraints are also gendered. Adaptive capacities are often severely stunted when underlying social norms and structures continue enabling conditions that make people vulnerable (Shackleton et al. 2015), such as gender norms (Mersha and van Laerhoven 2016). Therefore, the root causes of vulnerability need to be explored, some of which are also specific to the social, economic, political, cultural, and geographical context.

This chapter explores the case of small-scale agricultural intervention projects on communal lands operated by collectives of Mal Pahariya women, one of the most marginalized people in India. Living in remote communities in the Rajmahal Hills regions of Jharkhand, India, famous for its coal-mining industry, these communities are facing severe droughts and extreme heat in recent years (Koley and Jeganathan 2023) that is negatively affecting their traditional rain-fed agriculture. Based on ethnographic research, focus group discussions, and semi-structured interviews with Mal Pahariya women, NGO workers, and government officials, this chapter contributes to a better understanding of how climate change adaptation can be designed to enable gender-transformative outcomes along with providing triple benefits of food and nutrition security, rural livelihoods, and gender empowerment. The findings draw attention to the fact that such intervention needs to be strategically designed within the realities and contexts in which marginalization thrives and in which women spend their entire lives. Traditional gender norms are often deeply entrenched in rural societies, and challenging them can create tensions, especially within families, aggravating the risks of increased gender-based violence. Therefore, gender-transformative adaptation would require finding ways to address the structural and traditional factors that create gender power asymmetries without challenging gender norms upfront for increased participation and long-term success.

## Surviving Against Heavy Odds: The Women of the Mal Pahariya Tribe of Jharkhand, India

### *Background*

The Mal Pahariya is one of the over 700 Scheduled Tribes in India, living mainly in the eastern part of the country in the states of Bihar, Jharkhand, and West Bengal. Scheduled Tribes constitute less than 10% of the Indian population, of which more than 90% live in rural areas (Mitra 2008). The term "Scheduled Tribes" is coined to denote the constitutional privileges that are provided to a specific group of historically disadvantaged people whose vulnerability continues to linger in the present time. Often stigmatized and shunted in caste-based Indian society due to their specific social norms and traditions, people belonging to Scheduled Tribes perform consistently low in many socio-economic parameters.

The Government of India previously categorized the Mal Pahariyas as Primitive Tribal Groups (PTGs) and, since 2006, replaced the term with Particularly Vulnerable Tribal Groups (PVTGs). Their population has been decimated over the years, and educational attainment in modern Western education systems continues to be low. Many Mal Pahariyas continue to practice nature-based religion or tribal animism and medicinal practices. Agriculture continues to be the main occupation, mainly rain-fed shifting cultivation. Gathering minor non-timber forest products for sustenance is common. They also sell their produce to intermediaries who visit their remote villages for additional income to buy essentials they cannot grow on their land or extract from the forests. The forests also provide them with medicinal herbs and plants that they often use in healing or fermenting to turn into alcohol, an essential part of their religious customs. Therefore, the Mal Pahariyas have a very close connection with nature, on which they depend for their food, livelihood, physical health, and spirituality.

Men are the patriarchs of Mal Pahariya households, and distinct power differentials exist between men and women. Mal Pahariyas primarily practice endogamy, marrying within the tribe while avoiding marriages with close relatives. Therefore, most of the norms, customs, and traditions have remained intact, including the norms that reinstate patriarchy. The social structure is patrilineal, and women often marry very young, often in their early or mid-teenage years, and move to their husband's families after marriage. They help the men in the family to cultivate the land, and in recent years, they have often taken up more active roles in agriculture as their husbands often migrate to local urban centres for jobs.

## *Socio-environmental Condition of Mal Pahariyas in the Study Area*

In the primarily rural Godda and Pakur districts of Jharkhand (map in Fig. 12.1), where this study is based, Mal Pahariyas lives in small and dispersed communities, mainly in the hilly highland terrains of the Rajmahal hills. Arguably, the Mal Pahariyas consider themselves the original inhabitants or indigenous people of these lands (Bandyopadhyay 2022). A large part of their land was taken away by Eastern Coalfields Limited, a subsidiary of state-owned Coal India, when rich reserves of coal were found in the area, and coal mining started in 1985. Today, active cola mining by private coal mining companies continues locally under the Rajmahal Coal Mining Ltd Company. The peak production consent for extraction for coal mining in the area is approved at 23 million tonnes per year. Most Mal Pahariyas in this area live on their traditional land on the hills, many of which are on the fringes of large open-pit coal mines. Few also have been relocated to nearby areas while their land has been blasted into rock debris by dynamites in search of coal.

Mal Pahariyas in the study area generally depend on their forest land for their agricultural practices. Mainly rice and soybeans are cultivated on hilly forest floors manually cleared before the rainy season. With fragile topsoil on the hilly slopes, crop yield is low. As rainfall has significantly dropped in recent years due to changing climate, the yield has dropped further. Agriculture primarily lacks chemical fertilizers, as most cannot afford chemical-intensive farming, especially on the hilly terrains they cultivate (Fig. 12.2).

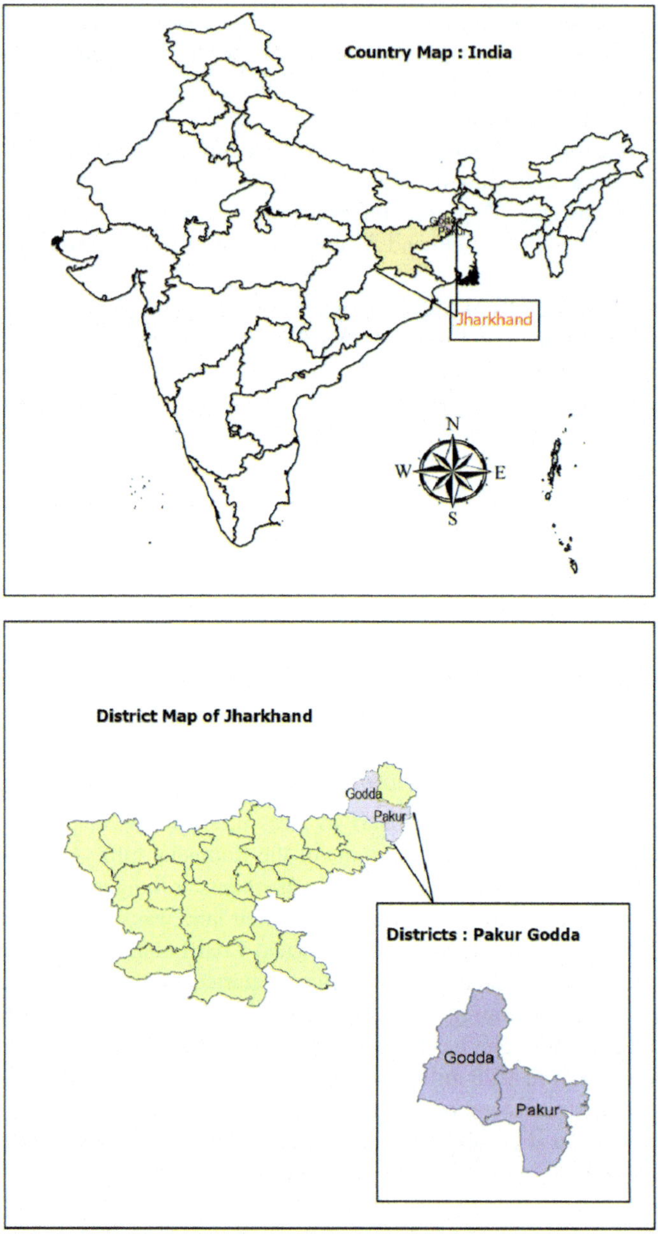


**Fig. 12.1** Location of the study area. *Source* Prepared by Adrija Roy

**Fig. 12.2** Two Mal Pahariya women overlooking coal mines near their homes. *Source* Photo by the author

Mal Pahariyas, especially women, collect *Mahua*, a locally available flower sacred to the indigenous people of the region, fermented into traditional alcohol. Dried Mahua flower is also sold to vendors who visit their communities to buy commodities they produce, along with soybeans and rice. They also keep goats and chickens, mostly donated by local NGOs. Selling agricultural produce, Mahua flowers, goats, and chicken often adds to the domestic income. The little income earned is spent buying necessary things for the family, like oil, salt, clothes, and other necessities for daily life, leaving very little to spend on nutritious food. However, the decision to spend is not their own and is often made by the husbands.

Moreover, as markets are far off from their communities in the hills, they rarely get access to fresh fruits and vegetables that would add to their dietary needs. When few families own cows, the milk is primarily fed to the calves as per traditional customs. All these factors have contributed to a high level of malnutrition in Mal Pahariyas, especially women who often have multiple children with narrow age gaps.

**Fig. 12.3** Focus group discussion. *Source* Photo by the author

Domestic abuse is common, along with alcohol addiction from traditional liquor consumption made from naturally abundant Mahua flowers, especially in men. Many men in the community migrate to distant areas, often to larger Indian cities in search of jobs, usually leaving their families in the care of their wives. However, most men who own land or have access to land owned by other male heads of the family come back home during the sowing season to help with cultivation. Due to increased rainfall scarcity in the region, this homecoming has become less common and mostly happens during specific festival times in the year.

Mal Pahariya communities in Godda and Pakur districts urgently need banks, post offices, schools, and hospitals in this region. Living in remote communities surrounded by forests in hilly areas, the available amenities often mean a challenging journey to the small towns in the plain land. With public transport, such a commute is not only time-consuming but also difficult to make regularly. As a result, children often do not go to school, and the sick avoid hospitals to seek medical care. Some remote communities avail themselves of ambulance services in extreme emergencies when ambulances are dispatched from the nearest hospitals.

Though completing school-level education in Mal Pahariya children in the study area was rare, traditional ecological knowledge, especially those related to agriculture and collecting forest products, is passed on from one generation to another orally and through hands-on training. Traditional rituals are also mostly around cultivation; each new harvest calls for a different type of ritual, in which the crop is first offered to the deity before consumption. Mal Pahariyas practice primarily nature-based religion, but over the years, many people have also converted to Christianity. They also have rules and practices that they strictly adhere to within their social groups. As a result, they are often framed as archaic, backward, and primitive, as one of their official classifications also suggests.

Living with large coal mines as neighbours has been challenging for the Mal Pahariyas in the region. Coal dust in the air is a constant phenomenon. Similarly, dynamites blasting coal shafts shake the ground multiple times a day. Jobs in the mining industry have been few for the community, especially the government jobs with benefits like health care and pensions after retirement. As private companies operate the mines, only some men from the Mal Pahariya communities have primarily secured some contractual jobs with local vendors who provide security and logistic services to the coal mines, as most extraction jobs are mechanized. Women do not get any employment in the coal mines. However, some Mal Pahariya women collect illegal coal from the mines that they sell to local vendors who collect coal from the small villages in the hills and carry it to local markets in motorized vehicles. This is, however, an extremely strenuous and dangerous activity and is done on a small scale as women suffer from chronic energy deficiency due to malnutrition.

Lack of water for drinking, cooking, and cleaning, as well as lack of sanitation facilities, add to the poor health of the Mal Pahariyas. Rainfall in the region has deteriorated over the years, and droughts have become more common. Also, the temperature has increased in this plateau region near the Tropic of Cancer. As a result, traditional water springs from the hills have mostly dried up, leaving the people with access to water through government-provided water reservoirs. These small, communal reservoirs are refilled once every few days. Every household is allotted a specific amount of water daily, severely restricting their water use. Groundwater aquifers are mostly polluted from coal mine leaching or are dried up as water is drawn from them for mining purposes. Open pits of old coal mines have turned into huge lakes, which locals avoid even in dire water stress situations due to the toxicity of the water. Many wells in the area are also abandoned due to highly polluted water from aquifer seepage and pollution from the coal dust in the air.

All the above points towards the multidimensional nature of vulnerability among Mal Pahariya women, having roots in social, economic, political, and historical factors. Most women in the communities living around the coal mines have very low education levels, with many being illiterate. They need to gain specific skills that can help them earn local livelihoods, which is very rare for women in the first place. Moreover, a range of domestic responsibilities makes it impossible for women to venture out to local towns in search of jobs, especially when public transport is non-existent and personal vehicles are unaffordable.

Women learn life skills like cooking, wood-collecting, water fetching, care work, and traditional agriculture from their elders from a young age before they are married and sent off to live with large families of their husbands. Married young women with no education or skills seldom have a few choices when making life decisions for themselves. However, their skills in traditional agriculture are beneficial as they adjust to new lives after marriage. The family patriarch has rights to land ownership, mostly on the forested hills on which they live, and the women act on the terms and conditions set by the men in the family.

Most technology training or capacity-building programmes the local state-level agricultural agencies introduce target men. Women without land tenure are rarely identified as farmers even though they engage in many agricultural activities. Additionally, with land ownership, they can get loans to buy expensive technologies, equipment, or fertilizers. Therefore, training programmes to increase local crop yield, especially under conditions of yearly droughts or any other forms of capacity building to adapt to climate change, may not reach Mal Pahariya women as the existing social, economic, and cultural systems create challenges to include them.

## *Climate-Smart Agricultural Interventions by Non-Governmental Organizations (NGOs)*

Few NGOs have taken up projects in the area to improve the physical, social, and economic health of the Mal Pahariya communities, especially given the urgent need to restrict the further shrinkage of the Mal Pahariya population. Given the gap that state-sponsored agricultural programmes failed to reach female farmers, a local NGO started a climate-smart agriculture and food production project with the Mal Pahariya women a few years back. This project was planned to establish an alternative agricultural model explicitly run by women in 20 Mal Pahariya villages in the region.

The purpose was to provide women with critical skills to produce their essential crops under a changing climate using non-synthetic fertilizers and prepare the soil in a way that was less dependent on rainwater for irrigation. Some of the women selected under the project had some access to land that the male members of the family-owned. Women with no access to land, like widows, were also selected in the program to build their capacities to find agriculture-related work locally. Widowhood was pretty high in the communities as men died young from alcohol abuse, accidents, especially in coal mines, and ill health. All women selected in the project were landless and traditionally denied land ownership or any other property rights. They also had no specific equipment or training that they could use to improve their agricultural yield. Consecutive years of droughts have added to the vulnerability of these women, receiving no compensation for lost crops and having no legal ownership of the land they toil.

The project also focussed on women farmers' capacity building, especially regarding the skills and knowledge required for climate-resilient sustainable agriculture. Every step of the project is designed so that the participants can engage in different activities to learn different steps of the alternative agricultural system from production to final marketing of their produce. For example, one of the purposes of the project is to help women find ways to reduce the input costs of seeds, fertilizers, or pesticides and help set up a supply chain to market their agricultural produce so that they can decide when, how, and to whom they can sell their products.

Creating local seed banks in specialized facilities in each selected village is also a part of the project. It is ongoing as the creation of structures and depositing seeds in the seed banks develops over time. The seed banks are essential for the long-term sustainability of these projects as women can borrow seeds from these banks before every sowing season rather than borrowing money at high interest rates from local moneylenders and entering into cumbersome debt cycles that they find difficult to escape. When fully operational, seed banks will give them certain autonomy regarding using their money rather than paying off debts.

Women from each of the selected villages are formed into groups where they are provided training on different sustainable agricultural techniques to improve the yields of their two main staple crops, soybeans and millet. Though they are part of the group, each member receives individual training in different sustainable agricultural methods. Also, to improve nutrition security, each group of women is charged with caring for a nutrition garden where they grow different types of local vegetables. Growing nutritional vegetables in community gardens also helped them understand the importance of including vegetables in their diet, which are otherwise lacking due to the lack of fresh vegetables and access to produce markets. Living in hilly regions away from local markets has created conditions of food deserts (Shaw 2006) for these women, making them rely only on crops that they grow and other produce they can collect from the forests. This helped women engage in collective action, taking turns looking after the crops they planted. The nutrition gardens are situated within each village in such a way that the run-off from the government-provided water reservoir makes use of wastewater. In communities where such central land was inaccessible due to the hilly nature of their terrain, nutritional gardens were set up in the land surrounding their houses.

## Materials and Methods

An exploratory case study approach was used to design the research to identify how small-scale agricultural initiatives for climate change adaptation can create leverage points for gender-transformative adaptation along with providing food and nutrition security. Case studies help in understanding the complex contexts, for example, the underlying social, economic, cultural, and political factors that entrench the gender asymmetries within which Mal Pahariya women exist, which are at the core of this research. Case studies help to understand 'real people in real situations' (Willis 2007, p. 239) by providing researchers with an overall understanding of the context within which research participants exist. The use of qualitative methods like semi-structured interviews and focus group discussions of women, NGO workers, and government officials responsible for agricultural development in the area helped answer specific questions requiring a deep and rich understanding of why things are the way they are.

Face-to-face interviews and focus group discussions were conducted in Pakur and Godda districts of Jharkhand. A total of ten women were interviewed face-to-face. The women were selected using snowball sampling where the interviewees were selected based on their participation and role in the projects mainly those who are responsible for steering the work of the project in each community. Around 62 women participated in the six different focus group discussions. The women selected for the face-to-face interviews and focus group discussions were part of an ongoing project of a local NGO to promote climate-smart agriculture and nutrition security in marginalized communities in Jharkhand, especially the PVTGs. This selection was appropriate, as the researcher was interested in understanding how agricultural projects can be better designed to enable gender-transformative climate change adaptation, especially in marginalized communities.

In most villages, women also participated in group discussions as they were often too shy to form or express their opinions on their own. However, within groups, they could better articulate their thoughts, especially with the support and cajoling of others. Having focus group discussions with a large group of women was also helpful in gaining a well-rounded understanding of how the projects have affected the lives of the women, which was only possible by interviewing a small number of representatives from each community. In the focus group discussion format, the women could open up and narrate their lived experiences, which helped them understand their social, economic, and cultural contexts.

## Results

### *Climate-Smart Agricultural Interventions Designed Around the Role of Women in Families*

The women's roles in the communities as primary caregivers were evident from the beginning of the research. The patriarchal system predominant in Mal Pahariya communities produced and sustained power asymmetries in gender relations, which was reflected in the social, economic, and cultural setup of the communities. Women identified themselves as a mother or the caregiver in the family, providing them with meaning and purpose in life. Agriculture was an extension of their care work, including looking after the community garden and a few animals, many of whom were raised in their homes. However, given their primary role as a caregiver, the women also mentioned that it is becoming increasingly difficult for them to provide enough food for the family at least twice a day as the agricultural yields are decreasing due to droughts and weather patterns. Some women pointed out that the food assistance they receive from the government welfare schemes differs from their staple diet, which is changing their food patterns. As a result, the project was very helpful for them as it provided them with knowledge and skills that they could use in growing their traditional food crops. Also, gaining knowledge of growing vegetables was helpful as it helped them to diversify their food choices.

Even though some women expressed that they would like to go out and work, it was evident from the conversation that the decision would be challenging. Taking care of their young children, especially when their husbands were not present throughout the day for work or had died, these women prioritized taking care of their families. The options for women to pursue any livelihood outside their communities were also limited. Lack of public transport and their lack of skills to acquire local employment further restricted their opportunities to leave home, work locally, and come back to their home at the end of the day. They also faced gender-related barriers to leaving the community to look for livelihood outside the communities, as it was uncommon in their communities for women to leave home to look for jobs. Domestic work in local towns was also limited due to caste-based challenges as women from lower castes find it difficult to work in domestic work like household help in higher caste families as caste-based untouchability-related social stigma persists in many rural and peri-urban societies in the region.

Moreover, the women in the project were at a life stage where they found it impossible to acquire new skills, as local lifelong learning facilities were non-existent. In their younger days, social norms of marrying young made it impossible for them to attain any education. As domestic violence was common in these communities, the women were also very fearful of taking action that would drastically change their life situation. Their lives were designed around care work and agriculture, which was an extension of care work, and most women were not prepared to disturb the status quo. All these factors lead to the low adaptive capacities of these women, who pursue alternative livelihoods other than working in their family fields and growing food.

For them, participating in the agriculture project was manageable as it was aligned with their predefined roles in their families and society.

## *Sense of Agency and Redistribution of Power Within Families*

Many women pointed out that as they acquired new skills and knowledge, they could share their inputs with their husbands, who sought solutions to increase agricultural yields. For a few others whose husbands and male members of the family have migrated to distant cities, it was easier to implement the newly acquired knowledge in their agricultural work. The women also pointed out that such an exchange with male family members improved their standing in the family as it changed their previous roles of simply following the advice of their husbands. The women pointed out that participating in the project helped them build their confidence in creating something that could be sold in a market and help earn money. This confidence helped them improve their negotiation skills and make decisions on how to spend their earnings. Interestingly, the finding in the study shows that increased droughts have created spaces for women to take more significant decision-making roles in daily lives as male members migrate to the cities when agriculture no longer was enough to provide them with sustenance.

The small differences that the project made to the decision-making power of the participating women did not result in significant changes like land ownership. As a result, they could not make any significant investments, for example, even buying small tools for vermicomposting. They continued to depend on the project to provide them with any necessary tools they would require to make changes in their cultivation work. However, some women also pointed out that when they attend project meetings, their husbands start caring for young children and consider their participation in the project necessary. The shift in gender norms, albeit in small measures, was evident. Not surprisingly, many women expressed the desire to send their children, especially the girls, to school if a school is set up locally. Though many women were initially shy about participating in the group discussion, it was interesting to note how, with time, most of them provided their input. The groups formed in each village provided a space for recognition; each woman found a place to express their opinions and observations, mostly about how they are progressing, which frequently progressed into other topics from the health and education of their children to domestic abuse due to alcoholism. Women in older age groups also pointed out that in these weekly meetings, they often provide advice to younger group members on how to deal with their domestic problems and how they can find help from village elders. It was evident that participation in the agricultural project helped improve their sense of agency and reduce their power asymmetries with male family members.

## Collective Action and Representation

One of the project's main focuses was to improve women's collective action. They formed small groups to look after common nutritional gardens, allowing them to make consensus-based decisions. As part of the project, the women formed informal groups. They created a fund among themselves in which each member contributed an equal amount of money at a given interval depending on the crops they could sell individually or from the community gardens they jointly operated. Group members could borrow small amounts from this fund at meagre interest rates during financial difficulties or emergencies. The whole group made decisions on lending money to a member of the group, and a consensus-based approach was used. The women found this exercise extremely helpful for multiple reasons. For many members, this was the first time they made financial decisions of their own and also collectively as a group. Women who were able to borrow from these small funds also narrated how they were able to bring money home to meet emergency expenses rather than just the male members of the family. As a result, they also found that their opinions on how the money would be spent started to matter in the family.

Moreover, the small financial independence they achieved from selling their agricultural produce from community gardens and increased agricultural yields helped them make critical household decisions. Some women shared that they could buy books for their children, and others even suggested that they were preparing to send their younger kids to boarding schools in bigger towns and cities explicitly designed to help children of vulnerable tribes attend further education.

Engagement with a group of women built relationships of trust where they could discuss their daily problems and find solutions. Forming sisterhoods was an essential part of these women's lives, as most of the gender roles they played were based on sisterhood. They collected wood for fuel in the forests and helped each other during sowing and harvesting seasons. The project developed by the local NGO recognized the role of bonding with these women and reinforced it in the project design. Many women also pointed out that after participating in the project, they could decide how to recycle and save water, use leftover food and vegetable waste in composting, and grow vegetables year-round through permaculture and small-scale greenhouse technologies. The projects focussed on helping them increase their overall agricultural yield and diversify their food choices by helping them improve access to vegetables. Moreover, it has made women more aware of how to provide healthy and nutritious food to the entire family and young children. Women learned from each other how to include the new types of vegetables they grew in their daily food and expanded their food choices.

Several women expressed the opinion that participation in the group made them feel that their voices mattered in the group's operation. They could openly represent their views and concerns, which they had never done before. They also learned from each other; one member solving a problem became a template for others. Participants were also encouraged to not just cultivate and grow crops but also participate in directly marketing their crops, as the project set up an online portal to link the producers to the vendors, reducing the role of intermediaries. Many women also pointed out that they were more aware of their roles as voters in local elections, voting for people who represented their voices, which was also an outcome of the discussions they had during their weekly meetings.

## Discussion

The Mal Pahariya women in the case study faced some of the same challenges that are also identified as global challenges even while living in their small, remote communities, like high gender inequality, extreme poverty, hunger, high malnutrition, limited access to medical facilities, water scarcity, lack of sanitation, drought and heat stress, and the lingering pollution from fossil fuel extractions. Climate change impact is already felt as unpredictable rains are causing multi-year droughts. There is an increase in summer days, extreme heat stress, and wet blub effects (Tirkey et al. 2018; Akhter et al. 2024). Interviewees were highly aware of changes in rainfall patterns and how that affected their ability to grow enough food to feed their families. Interviewees also constantly negotiated their daily lives around the daily temperatures during summer months, avoiding most of the afternoon for agricultural work in open fields. Summer days, mainly used for clearing forest undergrowth for shifting agriculture, are unbearable due to extreme heat. The temperature soars as high as 40–45 °C in April and May.

The residents of the study area also face difficulties living around open mines that are yet to be rehabilitated back to their original landscape as it was before mining. Seepage from the mines continues to pollute the rivers and lakes further locally. Rampant gender-based violence and domestic abuse give limited opportunities to the local women to challenge strong hetero-patriarchal norms that dominated their lives. In other words, there are many problems. Moreover, the Mal Pahariya population is declining, and urgent action with multipronged outcomes is needed. Given the numerous simultaneous challenges within which the Mal Pahariya women navigate their lives, it is noteworthy that the small-scale agricultural projects were strategically designed to address multiple problems.

Despite the lack of adequate resources, NGOs in the region have adopted a comprehensive approach, often targeting multiple challenges in a single project to bring about incremental progress in socioeconomically poor communities over time. On the one hand, the project aimed to improve food and nutrition security. On the other, it aimed to bring transformative changes in the women participants in ways that would be less confrontational to established social norms. This strategic approach, without openly challenging existing social norms, set forth a redistribution of gender roles to correct the gender power asymmetries. As women began to make financial decisions, contribute to domestic expenses during emergencies, and share with their husbands how they could increase their agricultural yield under droughts, a gradual shift in correcting the gender imbalance was set in motion. Not surprisingly, many participants pointed out that their husbands considered their participation in the project necessary. While the redistribution of power did not result in increased land ownership for women, the project significantly improved their wealth and well-being, enhancing their family positions regarding respect and feeling heard. Reconfiguring patriarchal social norms is a gradual process. However, the uncertainties of climate change can also create new spaces for women to take up roles they had never previously undertaken if adaptation processes are specifically designed to provide these opportunities.

The women's engagement with other group members and their regular exchange of topics facilitated critical thinking and reflection. They were increasingly aware of gender power asymmetries in their societies and found ways to tackle them. The women's role in challenging and changing societal norms is impressive and deserves respect. Some shifts in the traditional gender roles were also evident, which had previously been absent. The consciousness of how societal norms are adversely impacting their conditions and aiding the numerous injustices they navigate every day as Mal Pahariya women is essential for gender-transformative adaptation as the root of the problem is identified. With time, these spaces of critical reflection will increase as the women continue to expand the project, like building their seed backs and maintaining their online portal for direct trading of the commodities they produce. Moreover, routinized meetings under the project and the many activities where the women had to make critical decisions on their own, primarily related to finance, helped in participatory knowledge production (Hordijk and Baud 2010). As these women were in a life stage where they could not upgrade their knowledge through institutionalized education, these new avenues provided transformative learning opportunities for them.

The project recognized the specific constraints in which the women lived their lives and played specific roles critical for their identities as mothers, wives, grandmothers, and other caregiving roles women assume in their lives. It was essential to recognize the structural and relational dimensions of their inequalities that act as a barrier to finding the 'central methodological entry point' (Tschakert et al. 2013) for the project to bring transformative changes. The project's focus on creating gender-sensitive approaches in agricultural interventions was an essential aspect of its design, underlining its commitment to promoting gender equality. The climate-smart agricultural project for the Mal Pahariya women was designed so that they could align well with the roles they played in their household as caregivers while engaging in agricultural activities, which was an extension of their care work to arrange food for the family. The projects created a worthwhile pursuit for the women to learn more about food production to better feed their families. Gender-transformative changes are not possible if they fail to create spaces where women can represent their problems and find solutions. The project also created opportunities for their voices to be heard, helping them represent themselves and their views.

## Conclusion

This chapter shows how gender-transformative solutions that are specifically designed, considering the local constraints and barriers, help tackle social, economic, and cultural root causes that exacerbate gender-related vulnerabilities. They can only help evaluate whether the transformative changes will be sustained over time, challenging the firmly embedded hetero-patriarchal social norms. Nevertheless, a few key lessons can be learned on how well-designed climate-smart agriculture projects can enable transformation in food systems, shift power dynamics, and be scaled up to provide alternative livelihoods that are resilient to climate change. The findings also argue that projects aimed towards climate change adaptation should be evaluated in terms of their gender-transformative potential, especially those funded under international climate adaptation funds to change culture.

Gender-transformative climate adaptation needs time, money, patience, and long-term engagement supported by finance, research, and expertise to design projects and plans to understand the local context's complexities. Though the initial outcome of the project showed promising results, the project's success in bringing gender-transformative changes and dismantling power asymmetries would require time. Traditional norms are deeply entrenched in rural societies, especially in tribal communities where modern education and worldviews have very marginal penetration. There is also a danger of dislodging many traditional ways of life followed by these tribal groups, as most of their practices are sustainable and interlinked with what nature provides for them locally. Pushing them towards 'development' as understood in the neoliberal economic systems can set up a dangerous downhill that would dismantle the delicate balance the tribal members have with their surrounding nature.

Understanding local social, economic, cultural, and political conditions is necessary for designing projects that would bring in transformative changes. Centrally designed projects often can miss out on the small nuances in gender roles, especially in remote and marginalized communities, and may fall short of bringing effective changes. Moreover, building trust with community members is essential as tweaking agricultural activities can risk crop yield, leading to less crop production, which is the primary food source for people. Such trust building often requires years to build. The local NGO brought results quickly as it worked in the area for decades and knew the specific characteristics of the Mal Pahariya communities. However, NGOs do not build relationships of trust, but people working for them do. Funding constraints often make it difficult for NGOs to retain their workers over time.

Moreover, as the threats of climate change loom on these communities, with severe droughts becoming more and more common, a resource-intensive life may not be feasible for them in the long run, forcing them to migrate. Solutions designed with these constraints can make slow progress and require time, money, and resources often unavailable to local NGOs who depend on specific philanthropic funding. However, such funding needs to be earmarked. It seems the gains from the current project are at risk of getting lost when the project stops due to fund constraints.

**Funding** This research has been possible thanks to the financial support from the German Federal Ministry for Economic Cooperation and Development (BMZ) through the project 'Climate-resilient and nature-compatible sustainable development through socially just transformation' (Klimalog III).

## References

Akhter J, Biswas R, Das L, Midya SK (2024) Recent changes in temperature extremes and heat stress over Eastern India. Pure Appl Geophys. https://doi.org/10.1007/s00024-024-03497-x

Arora-Jonsson S (2011) Virtue and vulnerability: discourses on women, gender and climate change. Glob Environ Change. https://doi.org/10.1016/j.gloenvcha.2011.01.005

Bandyopadhyay S (2022) Mal Paharia tribe and the problem of their identity in West Bengal, India. Asian Ethn. https://doi.org/10.1080/14631369.2020.1785274

Buchanan A, Reed MG, Lidestav G (2016) What's counted as a reindeer herder? Gender and the adaptive capacity of Sami reindeer herding communities in Sweden. Ambio. https://doi.org/10.1007/s13280-016-0834-1

Dankelman I (2012) Introduction: exploring gender, environment and climate change. In: Dankelman I (ed) Gender and climate change: an introduction, 1st edn. Routledge, London, pp 29–46. https://doi.org/10.4324/9781849775274

Denton F (2002) Climate change vulnerability, impacts, and adaptation: why does gender matter? Gend Dev. https://doi.org/10.1080/13552070215903

Fraser N (2007) Feminist politics in the age of recognition: a two-dimensional approach to gender justice. Stud Soc Justice. https://doi.org/10.26522/ssj.v1i1.979

Hordijk M, Baud IS (2010) Inclusive adaptation: Linking participatory learning and knowledge management to urban resilience. In: Otto-Zimmermann K (ed) Resilient cities. Local sustainability, vol 1. Springer, Dordrecht, pp 111–121. https://doi.org/10.1007/978-94-007-0785-6_11

IPCC (2018) Special report on the impacts of global warming of 1.5 °C above pre-industrial levels and related global greenhouse gas emission pathways, in the context of strengthening the global response to the threat of climate change. In: Sustainable development, and efforts to eradicate poverty, vol 32

Ivanova A, Zia A, Ahmad P, Bastos-Lima M (2020) Climate mitigation policies and actions: access and allocation issues. Int Environ Agreem Politics Law Econ. https://doi.org/10.1007/s10784-020-09483-7

Koley S, Jeganathan C (2023) Evaluating the climatic and socio-economic influences on the agricultural drought vulnerability in Jharkhand. Environ Monit Assess. https://doi.org/10.1007/s10661-022-10557-z

MacGregor S (2010) 'Gender and climate change': from impacts to discourses. JIOR. https://doi.org/10.1080/19480881.2010.536669

Mersha AA, Van Laerhoven F (2016) A gender approach to understanding the differentiated impact of barriers to adaptation: responses to climate change in rural Ethiopia. Reg Environ Change. https://doi.org/10.1007/s10113-015-0921-z

Mitra A (2008) The status of women among the scheduled tribes in India. J Soc Econ. https://doi.org/10.1016/j.socec.2006.12.077

Mortreux C, Barnett J (2017) Adaptive capacity: exploring the research frontier. Wiley Interdiscip Rev Clim. https://doi.org/10.1002/wcc.467

O'Brien K (2012) Global environmental change II: from adaptation to deliberate transformation. Prog Hum Geogr. https://doi.org/10.1177/0309132511425767

Pearse R (2017) Gender and climate change. Wiley Interdiscip Rev Clim. https://doi.org/10.1002/wcc.451

Saran A, Singh S, Gupta N et al (2024) Interventions promoting resilience through climate-smart agricultural practices for women farmers: a systematic review. Campbell Syst Rev. https://doi.org/10.1002/cl2.1426

Shackleton S, Ziervogel G, Sallu S, Gill T, Tschakert P (2015) Why is socially-just climate change adaptation in sub-Saharan Africa so challenging? A review of barriers identified from empirical cases. Wiley Interdiscip Rev Clim. https://doi.org/10.1002/wcc.335

Shafique S, Akhter N, Stallkamp G, de Pee S, Panagides D, Bloem MW (2007) Trends of under-and overweight among rural and urban poor women indicate the double burden of malnutrition in Bangladesh. Int J Epidemiol. https://doi.org/10.1093/ije/dyl306

Shaw HJ (2006) Food deserts: towards the development of a classification. Geogr Ann B Hum Geogr. https://doi.org/10.1111/j.0435-3684.2006.00217.x

Tirkey AS, Ghosh M, Pandey AC, Shekhar S (2018) Assessment of climate extremes and its long-term spatial variability over the Jharkhand state of India. Egypt J Remote Sens. https://doi.org/10.1016/j.ejrs.2016.12.007

Tschakert P, Van Oort B, St. Clair AL, LaMadrid A (2013) Inequality and transformation analyses: a complementary lens for addressing vulnerability to climate change. Clim Dev. https://doi.org/10.1080/17565529.2013.828583

Willis, J (2007) Foundations of qualitative research: interpretive and critical approaches. Sage Publications Inc. https://doi.org/10.4135/9781452230108

Wise RM, Fazey I, Smith MS, Park SE, Eakin HC, Van Garderen EA, Campbell B (2014) Reconceptualising adaptation to climate change as part of pathways of change and response. Glob Environ Change. https://doi.org/10.1016/j.gloenvcha.2013.12.002

# Chapter 13
# Women Leading the Way: Female-Led Adaptation Initiatives in Western Himalayas

**Katyayini Sood, Abhishek Kumar, and Vipasha Negi**

**Abstract** The escalating impacts of climate change have intensified vulnerabilities across many regions of the Global South, necessitating innovative and inclusive adaptation strategies. Among these, gender-transformative approaches have gained prominence, recognizing that climate impacts are not gender-neutral and that effective adaptation must address the distinct needs and roles of different genders. In the Mandi, and Lahaul Spiti regions of Himachal Pradesh, India, the consequences of climate change, particularly the frequent flash floods and water scarcity have vividly aggravated the intersection of gender and environmental vulnerability. These floods and droughts have led to both economic and non-economic losses, severely disrupting local livelihoods and prompting male out-migration in search of alternative income sources. This migration has imposed a dual burden on women, who are left to manage household responsibilities while sustaining agricultural activities, tasks traditionally dominated by men. This chapter highlights the impacts of climate change-induced floods and droughts on women and then takes different case studies into consideration which are initiated by women residing in the region to combat climate change-induced disasters at the community level. By adopting and promoting natural farming practices, these women are at the forefront of a movement that aligns with global efforts to combat climate change and ensure sustainable food systems for future generations. Their leadership showcases the potential for community-driven, gender-inclusive approaches to foster lasting environmental and social change.

K. Sood (✉)
Department of Geography, University of Bonn, Bonn, Germany
e-mail: katyayinisood@outlook.com

A. Kumar
People for Integrated Planning and Adaptable Living (PIPAL) Foundation, New Delhi, India
e-mail: kumar.abhishek@pipalindia.org

V. Negi
Punjab State Council for Science and Technology, Chandigarh, India



A. Sarkar and N. Bandyopadhyay (eds.), *Gender-Transformative Approaches for Climate Change Adaptation*, Springer Climate,
https://doi.org/10.1007/978-3-031-90236-9_13

**Keywords** Gender · Climate action · Agriculture · Transformation · Women leaders · Himachal Pradesh

## Introduction

The impacts of climate change on human migration are well documented, with extreme weather events, rising sea levels, and environmental degradation forcing many to leave their homes in search of safer areas (Jayawardhan 2017). However, what is less frequently examined is the increased burden of responsibilities on women who remain behind in climate-affected regions. As climate change disrupts traditional livelihoods and reshapes communities, women often find themselves bearing the brunt of the social, economic, and physical demands that come with adapting to these new conditions (UN Women 2022). This study delves into this underexplored aspect, emphasizing how climate-induced migration indirectly places a heavier burden on women who stay behind.

In many communities, men are more likely to migrate in search of work when agricultural yields decrease, droughts threaten food security, or local resources are depleted. In such situations, women are often left to manage households on their own. Despite these hurdles, women in climate-impacted regions have found ways to mitigate some of the effects of climate change. Evidence from the field was analyzed in the study to show how women and community organizations are working to build resilience by adopting sustainable agricultural practices, improving water management, and diversifying income sources.

## Background and Literature Review

### *Impacts of Climate Change in the Himalayas*

Climate change is a growing concern among countries. While every country has been affected, the sensitivity and susceptibility vary based on risk transfers, mitigation measures, and adaptation to climate change (Garschagen et al. 2021). Climate change is widely acknowledged as the primary root cause of most hazards (Lizarralde et al. 2021). Over the last century, human actions have driven significant climatic shifts, notably a 1.18 °C rise in global temperatures since the late 1800s, mainly due to increased $CO_2$ emissions (IPCC 2018). The Arctic's diminishing summer sea ice (Johannessen and Olaussen 2024) and the shrinking glaciers from the Himalayas to the Alps are alarming manifestations of this change (Hock et al. 2022). Climate change, along with social, physical, and economic vulnerabilities, is increasingly exacerbating the challenges faced by highly exposed populations in the Western Himalayan Region (Malik and Ford 2024).

Himachal Pradesh, a region characterized by its stunning landscapes and rich biodiversity in the western Himalayas, has increasingly become a focal point for the impacts of climate change (Asian Development Bank 2010). The Himalayas are highly vulnerable to the effects of global climate change, and these impacts are becoming increasingly evident (Chauhan et al. 2023). A recent report by the International Centre for Integrated Mountain Development (ICIMOD) highlights that the Hindu Kush Himalayas (HKH) are experiencing a concerning trend of rising temperatures that exceed the global average. The average yearly temperature of the air near the surface in the HKH region has increased by approximately 0.1 °C per decade from 1901 to 2014 (Sabin et al. 2020). However, the rate of warming has increased to approximately 0.2 °C per decade from 1951 to 2014, which is majorly caused by anthropogenic activities (Sabin et al. 2020).

This implies that the HKH region is particularly vulnerable to the impacts of climate change and its effects on the environment and local communities. The area is now witnessing low levels of snowfall in winter months and high-intensity rainfall during the monsoon seasons (Banerjee et al. 2021). The evolved weather patterns have become a concern over the years. Most models project that this temperature increase will contribute to the precipitation, especially in the Himalayas region (Katzenberger et al. 2021). Since 2017, the state has experienced annual flash floods, resulting in devastating consequences for local communities and ecosystems. The state hosts 48 glacial lakes which can be impacted by climate change in the future and lead to flash floods (Saha 2024). These extreme weather events not only disrupt livelihoods but also pose significant threats to human health and safety (Lee et al. 2020).

Climate change is the root cause of disasters in the Himalayan range, and the Upper Beas Catchment has been one of the hardest hit over the years (Rani and Sreekesh 2019). Global warming-related rising temperatures have rebalanced hydrological processes, changing the timing, amount, and quality of available water, particularly in vulnerable areas like the Beas basin where rising temperatures are causing an acceleration of glacier melt and modifications to snowfall patterns, which may have an impact on river discharge downstream (Bolch et al. 2012). Policymakers, academics, and scientists are interested in the intricate interactions between water flow rates and climate scenarios in Himachal Pradesh because of annual flash floods (Nagamani et al. 2024).

## *Impact of Climate Change and Increasing Vulnerability of Rural Women in the Himalayas*

According to the National Health Mission (2011), 90% of the population resides in the rural areas of Himachal Pradesh in 2011. Floods and droughts have increasingly led to both economic and non-economic losses, profoundly disrupting local livelihoods (Bharadwaj and Mitchell 2023). These climate-induced changes are pushing families into hardship, with significant consequences for rural households (IOM 2024). One major response to this disruption has been an uptick in male migration, as men leave to seek alternative income sources in urban areas or other regions as returns from agriculture become uncertain in times of Climate Change. Démurger (2015) highlights that in most of the patriarchal societies, it is the men who migrate in most cases. However, this shift imposes a dual burden on the women left behind, who now bear the responsibility of both household management and sustaining agricultural activities or tasks that have traditionally been male-dominated (FAO 2023).

Many regions in the Global South are experiencing a trend, where climate-induced migration disrupts traditional gender roles, placing additional pressures on women (UN Women 2022). Often the primary caregivers and food producers, women are now responsible for tasks such as sowing, weeding, and harvesting, frequently without adequate resources or support. This shift not only increases their workload but also heightens their vulnerability to the impacts of climate change, as they have less access to technology, financial services, and decision-making processes. Gender-transformative approaches in climate change adaptation acknowledge these challenges and aim to empower women by addressing the structural inequalities that limit their resilience.

The disproportionate impacts of climate change on women stem from deeply embedded structural inequalities that exacerbate their vulnerability in multiple domains (Mbungu 2024). The gendered impacts of climate change are particularly acute in rural areas, where traditional roles and direct reliance on natural resources amplify women's vulnerability to environmental disruptions. In developing regions, rural communities often experience the immediate consequences of climate variability, such as failed rains or floods, which destabilize food security and livelihoods (Rao et al. 2017). Men, often perceived as "external actors" in these societies, are more likely to migrate to urban centers or other regions in search of economic opportunities, leaving women behind to manage households under increasingly precarious conditions (GIZ 2024). This structural imbalance is further exacerbated by systemic inequities; for instance, only 14% of landowners worldwide are women (Kaaria and Osorio 2018) leaving many without legal rights or access to essential resources like credit, agricultural inputs, or decision-making platforms. Moreover, the cascading effects of these inequalities extend beyond economic dependency to include exclusion from critical climate-related knowledge systems. For instance, in rural India, women are systematically excluded from access to weather forecasts and cropping advice, leaving them ill-prepared to anticipate or manage agricultural risks effectively.

Addressing these layered vulnerabilities requires a multi-faceted approach that targets the redistribution of economic power, dismantles discriminatory cultural practices, and ensures gender-sensitive dissemination of climate-related knowledge (Yavinsky 2012). Women's capacity to adapt to climate change is significantly hindered by their lack of autonomy and decision-making authority. They often have minimal control over household finances and assets, leaving them unable to allocate resources toward adaptive strategies (UNFCCC 2015). Additionally, in many communities, women are underrepresented in local governance and decision-making forums, limiting their ability to influence community-level adaptation plans or advocate for policies that address their specific needs and priorities. These systemic exclusions perpetuate gender disparities and restrict women's participation in shaping resilient responses to climate challenges (OECD 2024).

To overcome the impacts of climate change Sustainable Development Goal (SDG) 13, "Climate Action," focuses on urgent action to combat climate change and its impacts. It emphasizes two main strategies (i) *Climate Mitigation:* Actions aimed at reducing or preventing the emission of greenhouse gases. This includes measures to shift from fossil fuels to renewable energy sources, improve energy efficiency, reduce deforestation, and promote sustainable land and resource management. (ii) *Climate Adaptation:* Preparing for and adjusting to the current and expected impacts of climate change. Adaptation initiatives range from building resilient infrastructure to developing drought-resistant crops and implementing early warning systems for extreme weather events (Martin 2022).

While these approaches are often driven by policies and regulations at the government level, a significant impact can also be achieved through individual and community-led actions (Wannewitz and Garschagen 2023). Small, community-level changes, such as adopting sustainable agriculture, enhancing local biodiversity, or reducing waste, can aggregate into large-scale benefits (Stringer et al. 2019). This dynamic has compelled women to take on the dual responsibilities of managing household tasks and agricultural activities, while also emerging as key agents in initiating climate change adaptation strategies in the region (Antriyandarti et al. 2024). Their proactive roles in addressing environmental challenges highlight both the disproportionate burdens they face and their pivotal contributions to resilience-building efforts at the local level.

## Materials and Methods

### *Objectives and Theoretical Framework*

Previous studies in the Himalayan region have predominantly concentrated on areas such as hydrology and agriculture (Chinnasamy and Prathapar 2016). However, the current chapter shifts its focus to exploring methods for climate change adaptation, aiming to address strategies that mitigate the impacts of climatic variability and foster resilience in affected communities. This chapter focuses on two major themes in the context of climate change-induced migration. Firstly, the interconnectedness of the impacts that women face. Secondly the proactive role of women in initiating climate-sensitive practices in the western Himalayan region. Faced with the challenges of climate change, women in these communities have developed innovative, sustainable strategies to adapt to shifting environmental conditions. Their initiatives include implementing water conservation techniques, preserving biodiversity through traditional crop varieties, and adopting organic farming methods that protect soil health and reduce reliance on chemical inputs. The case studies not only focus on the adaptation strategies taken up by individual women but also on the role of women in initiating and ensuring the sustainability of such practices in the region.

This study employs a mixed-method approach. It derives its findings from analyzing both secondary and primate data. It follows the Cresswell's (2009) framework, to explore the impacts of climate change-induced disasters on women within the study region. By using a combination of primary and secondary data sources, this methodology aims to capture a nuanced understanding of women's experiences and responses to climate-related challenges, as well as the community-led initiatives they lead or participate to adapt to these impacts. The qualitative approach allows for an in-depth exploration of the subjective experiences of women facing climate-induced challenges, focusing on how they perceive, adapt to, and manage these impacts.

## *Secondary Data Collection*

To address the second research question, which focuses on identifying climate adaptation practices led by women, a rigorous literature review was conducted. Secondary data sources included academic databases such as Google Scholar, Web of Science, and LitMaps, which were used to locate relevant studies, reports, and case examples that detail women-led initiatives in the region. Additional websites and organizational reports were consulted to ensure a comprehensive review of relevant initiatives. Search terms included variations on "women-led climate adaptation," "sustainable practices," and "community resilience." From the wealth of initiatives identified in the literature, only those practices that closely matched the scope and objectives of this chapter were selected. These initiatives were chosen based on their relevance to the local context, their documented impact on climate adaptation, and their representation of diverse strategies women employ to mitigate or adapt to climate change. By synthesizing these findings from the literature, this study aims to highlight best practices that could be scaled or adapted to enhance community resilience in similar regions.

## *Primary Data Collection*

The first research question addresses the specific impacts of climate change on women in the region. To gather rich, detailed data on this topic, semi-structured in-depth interviews were conducted. This interview style allows participants to express their experiences freely, while the semi-structured format provides enough flexibility for probing further into relevant topics as they arise. The snowball sampling technique was employed to recruit participants, as this method is particularly effective in reaching individuals who may not be part of formal networks but possess valuable insights into community-level impacts of climate change. Through snowball sampling, initial participants recommended other individuals who had relevant experiences or insights into the issue, resulting in a total sample size of 15 women.

The participant group consisted of women aged 21–55, providing a range of perspectives from various stages of adulthood. The diversity within this age range allowed for capturing differences in how women of different generations perceive and experience climate change. The interview questions focused on participants' personal experiences with climate-induced events, the challenges they face, their coping mechanisms, and the role of community support in these situations. Interviews were conducted in local languages and, where necessary, were translated to ensure accuracy in representing participants' voices.

## *Impact Web Methodology*

To effectively analyze and illustrate the multidimensional impacts of climate-induced disasters on women, this study utilized the impact web methodology, as outlined by Sparkes et al. (2022) The impact web approach is rooted in systems thinking and conceptual modeling, allowing researchers to map out the cascading effects that a single disaster event can have across multiple, interconnected domains, such as agriculture, infrastructure, public health, and the local economy. Using this approach, each woman's experience was situated within a broader web of interconnected impacts, thereby providing a comprehensive view of how climate-related events disrupt different aspects of daily life and community resilience. This methodology also highlights how the impacts on one domain, such as agriculture, can ripple outward to affect others, like household income or community health, which are particularly relevant to understanding women's roles and vulnerabilities.

## *Data Analysis*

The interview data were systematically analyzed using Atlas. Ti software is a tool that enables detailed coding and thematic analysis of qualitative data. Various codes were assigned to segments of text, representing recurring themes or significant insights related to the impacts of climate change on women. Coding categories were organized to align with the domains identified in the impact web methodology, which helped in constructing an organized framework of interconnected impacts based on participants' accounts. These codes and resulting thematic patterns provided the structure for constructing the impact web, allowing for a visual and conceptual representation of how climate change affects women's lives at both individual and community levels.

## Results and Discussion

### *Impact of Climate Change on Women in the Western Himalayan Region*

The climate crisis impacts people unevenly. Women and girls, who represent a significant share of the global poor and rely heavily on nearby natural resources for survival, bear a disproportionate burden from climate change (Parikh 2007). Moreover, we also live in an increasingly interconnected world, where the impacts of climate change are not isolated but cascade across social, economic, and environmental dimensions, affecting communities in complex ways (Lawrence et al. 2020). This section highlights the interconnected nature of climate change impacts on women, particularly in vulnerable regions, using an impact web methodology to illustrate how these effects reach far beyond the immediate and visible consequences.

The impact web approach helps to represent the broad and intricate web of effects climate change has on women's lives, detailing how one change often triggers a series of additional burdens and challenges. These impact webs are based on the interviews conducted on the field and remotely (see Fig. 13.1). The impact web methodology is a valuable tool for tracing how climate change affects various facets of life and society. By following a chain of effects from a single climate-related event, we can better understand how these events amplify existing inequalities and vulnerabilities, especially for women.

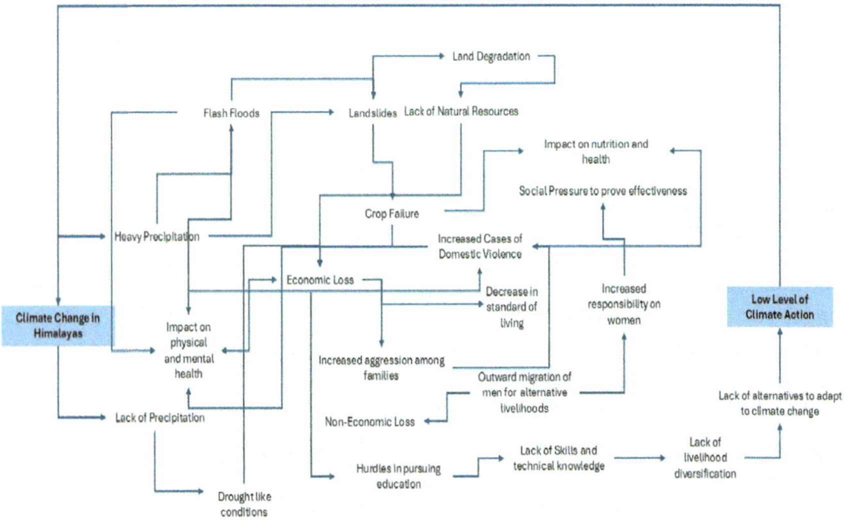


**Fig. 13.1** Impact web developed during interviews. *Source* In-depth interviews conducted during the fieldwork

Both heavy rainfall and lack of rainfall affect the people in severe ways in the villages. One respondent pointed out, "*There was not just one thing coming our way but multiple at the same time, we didn't know what to combat first.*" The cyclical nature of these events from drought-like conditions in winter to flash floods and landslides in monsoon season creates a web of challenges. The figure highlights the profound impact on land quality. Prashar (2023) described that the aftermath of flash floods leaves a thick layer of debris, which covers fertile land, causing immediate economic strain due to the costly process of land clearing. Another woman respondent commented "*It is only in the last 10 years that we have seen so many disruptions. Landslides make it even harder for us to cultivate and dry spells are also increasing. The rain in winter is so less that most of the crops did not give us the expected return. We want our kids to go to private schools, but the fee structure is so high that we might switch them to government schools. The fear of not being able to afford education is very high.*"

In the case of prolonged precipitation especially during the monsoon months leads to flash floods. As a consequence, there is decreased agricultural productivity, impacting food security and impact on local economies. One woman respondent commented, "I*t happens every year now. Over the years we have had enough losses, the response from the government is there but unfortunately, as we try to recover the region is again impacted by flash floods. If given a chance we would too move to another place where my husband works in Badi, but it is expensive there. We also have to live on rent. Here we have our land.*" In regions where agriculture is the primary livelihood, reduced crop yields force families to make difficult choices. Men often migrate to urban areas in search of work, leaving women behind to manage households, farms, and other responsibilities. This migration not only increases the physical workload for women but also places emotional and financial strain on them as they become the primary caretakers and breadwinners.

A single climate-related stressor like drought can cascade into a broader chain of impacts that affect women's physical well-being, economic opportunities, and social roles, underscoring the interconnectedness of climate effects on women's lives. Furthermore, the economic pressures and increased workload on women can have a direct impact on their children's well-being, as they are often less able to provide sufficient nutrition, healthcare, and educational support. One interviewee exclaimed, "*There are things we can do, but we have other things to spend on which are way more important. Most of the crops we farm are just for us! This year we did not have enough. The losses were very high. Sometimes the seeds are also distributed by government of cooperatives.*" Women are often the first to feel the impacts of resource scarcity, such as water and firewood shortages. With fewer resources available locally, women have to travel longer distances to secure water and fuel, which increases their vulnerability to physical exhaustion and health risks. Additionally, the time-consuming nature of these tasks limits their ability to pursue education, skill-building, or other income-generating activities.

The explanation in Fig. 13.1 offers a comprehensive view of the consequences of both low and extreme precipitation in the region. This extended impact web reflects how the burden of climate change on women has far-reaching effects, potentially leading to cycles of poverty and vulnerability across generations which in the long run influence a community's capacity to implement climate-sensitive solutions. The economic losses in the region have far-reaching consequences, particularly on access to essential services like healthcare and education. The lack of health and nutrition facilities not only causes physical health issues but also contributes to mental health problems, with women particularly affected by feelings of helplessness and aggression. The cycle of economic hardship further limits access to education, preventing individuals from benefiting from climate adaptation knowledge or technical solutions that could help mitigate future disasters.

Whenever the impacts are analyzed, the non-economic losses and damages are often overlooked. Non-economic losses and damages profoundly impact women by compounding their vulnerabilities and placing additional barriers to recovery and adaptation. The social, psychological, and cultural repercussions of these losses are significant and can have intergenerational effects, limiting not only women's well-being but also the resilience of entire communities. One woman shared in the interview, *"I come from a village which was badly affected because of a cloud burst. Being a woman I could not help my family back home. This just made me feel a little bad. Would have I had more resources I would have been in a situation to help. This impacted me deeply, I used to think about my family now and then which added to my stress."*

According to the state Agricultural Department, the lack of rain in the winter season has resulted in a massive financial loss of Rs. 130 crore (Rs. 1.3 billion or 15 million USD) in 2021 (Gupta 2021). In 2023, the agricultural sector of the state has incurred a loss of Rs 167 crore (Rs. 1.67 billion or 19 million USD) during the monsoon season (Prashar 2023). This situation exacerbates the financial challenges for farmers, especially the poor who struggle to afford the necessary resources to recover. Additionally, the degradation of land and natural resources diminishes agricultural production and the quality of resources over time. This interconnection between environmental, economic, and social factors demonstrates the broader, often compounding, impacts of climate events. As each component from land degradation to mental health challenges is intertwined, the effects ripple through the community, creating a continuous cycle of vulnerability. Figure 13.1 attempts to paint a vivid picture of how climate events, particularly extreme weather, create a chain reaction that affects not just the environment but also the livelihoods, health, and prospects of individuals in the region (Fig. 13.2).

## *Initiatives Taken to Reduce the Impact of Climate Change*

From the observations from the field, we find that women are not only impacted by climate change but also play a critical role in responding to its effects and strengthening community resilience. By analyzing the interconnected impacts (Sect. 4.1), this study emphasizes how women have organized themselves to counteract some of the challenges posed by climate change. For example, women's collectives in

*a) Women discussing their issues* *b) Respondent showing her field*

*c) Respondent working in the field*

**Fig. 13.2** Pictures from the fieldwork

some regions are working on water conservation projects, implementing rainwater harvesting systems, and teaching climate-smart agriculture techniques to adapt to changing weather patterns. These efforts highlight women's role as agents of change, using their local knowledge and community networks to mitigate climate impacts. They showcase the transformative power of women that has been uncovered due to the series of climate change adaptation efforts. Women have emerged as powerful leaders in combating climate change, particularly in rural communities where their actions directly impact environmental health and resilience. Their contributions range from developing sustainable agricultural practices to pioneering community-led conservation initiatives, creating a groundswell of transformative efforts that serve as models for climate adaptation worldwide (UN Women 2021).

In many parts of the world, women are at the forefront of implementing eco-friendly agricultural techniques that harmonize with natural ecosystems rather than depleting them (Boix-Fayos and De Vente 2023). By practicing sustainable agriculture, women are reducing reliance on intensive, chemically driven farming methods that degrade soil health, contribute to greenhouse gas emissions, and exhaust water resources (Fertő and Bojnec 2024). Instead, they are integrating indigenous knowledge with modern methods to conserve water, restore soil fertility, and increase resilience against erratic weather patterns. For example, in semi-arid regions where water scarcity is a pressing issue, women farmers are adopting techniques such as crop rotation, cover cropping, and intercropping, which improve soil moisture retention and prevent erosion, enabling crops to thrive even under challenging climate conditions (Vũ et al. 2011; Issaka et al. 2021).

A pivotal technique embraced by these women involves switching to drought-resistant seed varieties. As climate change brings increasingly unpredictable rainfall, women farmers in affected regions are leading the transition to resilient crops that require less water and can endure extreme temperatures (Aguilar et al. 2024). This shift not only ensures food security for their families but also reduces dependence on external resources like imported seeds and chemical fertilizers, empowering communities to be self-sufficient and sustainable. Moreover, by selecting native and drought-tolerant seeds, these women help preserve local biodiversity, which plays a critical role in maintaining the ecological balance and supporting wildlife, thereby enhancing the natural resilience of their surroundings (Gamage et al. 2023).

Techniques such as composting, mulching, and minimal tillage help maintain soil structure, reduce erosion, and sequester carbon, contributing to overall soil health (Ogieriakhi and Woodward 2022). By opting for organic pest control methods and reducing chemical inputs, women are protecting vital insect populations, including pollinators like bees, which are essential for crop production and ecosystem stability (FAO 2020). These sustainable practices also minimize pollution in waterways and surrounding habitats, promoting healthier ecosystems and reducing the community's carbon footprint (FAO). Beyond agriculture, women are instrumental in spearheading community-based reforestation and ecosystem restoration projects (Owen-Burge 2023). By restoring these ecosystems, women are creating a sustainable environment that serves as a buffer against climate-related disasters, supporting both human and environmental health (Bryan et al. 2023).

In this chapter, we highlight three case studies to give exemplars of climate adaptation initiatives that are led at the grassroots level by the local communities and particularly by women.

## Case Study I: Women's Leadership in Climate-Resilient Farming in Himachal Pradesh

The case of Panjayanu village in Himachal Pradesh highlights the significant role of women in promoting climate-resilient agriculture in the north-western Himalayas. Through their leadership and dedication to natural farming, the women of this village have become a driving force for community-based climate action and self-reliance (*atmanirbharta*), showcasing how local knowledge and collective effort can foster sustainable change.

In this village, the State Government's *Prakritik Kheti Khushal Kisan Yojana* (*Natural Farming Prosperous Farmer Scheme*), an ambitious scheme aimed at promoting chemical-free farming, has transformed into a women-led initiative that strengthens multiple aspects of climate resilience (Bisht 2022). By adopting natural farming practices, the women of Panjayanu are not only producing crops without chemical inputs but also working to maintain soil fertility, enhance biodiversity, and secure food supplies, advancing Himachal Pradesh as a hub for natural crop production and crop diversification. By reducing dependence on chemical fertilizers and pesticides, these women are promoting practices that benefit both the environment and public health, creating a model for others in the region to follow.

Their success has inspired approximately 30 families in the village to adopt similar practices, highlighting how community-driven initiatives led by women can expand climate resilience and food security. The success of Panjayanu's women farmers goes beyond the benefits of sustainable agriculture; it exemplifies the power of community-led climate action that complements government policies. The engagement of these women farmers demonstrates how grassroots initiatives, especially those led by women, can be vital to addressing broader environmental challenges. Their leadership showcases the effectiveness of a participatory approach, where community members actively engage in decision-making and implementation of sustainable practices, and it illustrates the broader societal shift toward sustainability that is underway in the region.

The government has provided substantial support through subsidies, training, and market access, creating an enabling environment that empowers women to lead in climate-resilient farming. This alignment between state support and community engagement allows local women to expand their knowledge, hone their skills, and access markets for their produce, which has contributed to the wider adoption of sustainable farming practices. By prioritizing women's leadership, the *Prakritik Kheti Khushal Kisan Yojana* also aligns with global calls for gender-responsive climate policies that recognize the critical role women play in building resilience and adapting to climate change. As the initiative demonstrates, gender-responsive approaches to climate adaptation not only empower women but also strengthen community resilience against climate impacts (Himachal Headlines 2022).

The success of women in Panjayanu village exemplifies how empowering women to lead in sustainable agriculture not only strengthens local food systems but also contributes to broader climate adaptation goals. Women's involvement in these initiatives strengthens food security, supports economic self-reliance, and advances environmental sustainability—all of which are critical for the climate-vulnerable regions of the Global South.

Moving forward, replicating this model across India could transform the agricultural landscape on a national scale. By promoting natural farming as a viable alternative, states can reduce agricultural dependence on chemical inputs, lower greenhouse gas emissions, and improve soil health, aligning with India's climate goals.

Replicating the Himachal model in other states could also foster more inclusive agricultural policies that elevate women's roles in climate adaptation. Such efforts would contribute to food security, protect livelihoods, and support climate-friendly farming practices, paving the way for a resilient and sustainable agricultural future in India.

## Case Study II: Community-Led Environmental and Social Initiatives in Himachal Pradesh Women's Leadership in Climate Action and Social Change

Himachal Pradesh has seen a rise in community-led environmental initiatives that are paving the way for integrated social and climate action, often spearheaded by women. These initiatives demonstrate how women's leadership in addressing climate adaptation and social issues can generate significant benefits for communities, especially in regions vulnerable to the impacts of climate change. By blending environmental conservation with social development, these initiatives are reshaping community norms and driving sustainable progress. This is one of the examples where awareness regarding climate change sensitizes communities to save the girl child.

This initiative from Mandi district in Himachal Pradesh was launched by the youngest *panchayat pradhan* (village council leader) of Tharjoon in the Seraj assembly constituency. This young woman leader has introduced a unique campaign titled *Beti Bachao, Van Bachao, Puadha Lagao*—which translates to "Save Girl, Save Forests, and Plant Trees." The campaign mandates that for every newborn daughter, 11 trees are planted, a symbolic gesture that intertwines environmental conservation with the promotion of gender equality. This initiative promotes sustainable development by encouraging locals to plant and nurture trees while addressing the deep-rooted cultural preference for sons over daughters. Parents of newborn daughters participate actively in the tree-planting drive, fostering a culture of environmental responsibility and resilience. As a result, the campaign has contributed to raising awareness about the importance of environmental health and the role of trees in mitigating the impacts of climate change.

Moreover, this initiative integrates gender advocacy with environmental conservation, bridging social issues with ecological restoration. In a region where gender inequality remains a concern, the campaign aligns with the *Save Girl Child* movement, challenging traditional norms by celebrating the birth of girls. At the same time, it addresses ecological concerns by increasing the region's green cover, an essential step for communities heavily dependent on agriculture and natural resources.

The success of *Beti Bachao, Van Bachao, and Puadha Lagao* highlights the broader potential of gender-transformative approaches in rural India, where women's involvement in climate resilience efforts is increasingly recognized as a critical factor for sustainable development. Women leaders in Himachal Pradesh, through community-led campaigns like these, are expanding the traditional roles of women and fostering greater community cohesion. By leading environmental initiatives, these women inspire a new generation of advocates for both gender equality and climate action, illustrating that social progress and environmental sustainability are intertwined goals. Community-based environmental campaigns led by women, like this one, represent a powerful model of local action that has the potential to influence broader policy discussions on sustainable development. The success of such initiatives can inspire policymakers to design and support gender-sensitive approaches to climate adaptation, recognizing that community-based and women-led movements are essential to achieving long-term sustainability goals.

These efforts are especially significant in rural Himachal Pradesh, where communities often rely on natural resources for their livelihoods and are more susceptible to climate-related hazards such as droughts, landslides, and flash floods. Women in these communities bear a considerable responsibility in managing household resources, including water and fuel, making them uniquely positioned to understand the impacts of environmental degradation. Their leadership in initiatives like *Beti Bachao, Van Bachao, and Puadha Lagao* shows how women can play a pivotal role in building climate resilience from the ground up by engaging local communities and encouraging sustainable practices that enhance environmental protection and social welfare.

The young *panchayat pradhan*'s work aligns closely with gender-transformative approaches to climate action, making it an inspiring case study for those aiming to understand the intersection of women's leadership, climate resilience, and social change. As such initiatives continue to gain recognition and impact, they serve as powerful reminders that sustainable development is most effective when led by those closest to the land and community they seek to protect.

### Case Study III: Guardians of the Enchanted Forest: Protectors of Nature's Realm

The Stingri Mahila Mandal in Lahaul, Himachal Pradesh, offers an inspiring example of women-led environmental stewardship. This women's group has developed a community-centric approach to forest conservation, empowering local women to play an instrumental role in managing natural resources and fostering climate resilience. By overseeing forest usage, enforcing regulations, and using social deterrents, these women have become guardians of the environment, combating illegal logging and ensuring sustainable practices.

In a region where harsh winters make firewood essential, the women of Lahaul have taken a firm stance against forest exploitation. The Mahila Mandal monitors forest usage closely, imposing fines or even social boycotts on those who violate conservation rules. Their commitment to forest protection is driven by the understanding that forests are vital not only for their immediate needs but also for long-term regional health, as they prevent avalanches, conserve soil, and contribute to rainfall and oxygen flow. The Stingri Mahila Mandal's conservation efforts have led to measurable environmental benefits, including an increase in wildlife sightings and a reduction in poaching. These outcomes highlight women's role in preserving biodiversity.

Their work aligns with the Dalai Lama's call to protect wildlife in the region, reinforcing the connection between local conservation efforts and broader spiritual and cultural values. Established in the 1990s by the Himachal government, Mahila Mandals in each village have been institutionalized and supported by local government and forest departments, who enforce penalties for unauthorized logging and provide guidance on forest management.

Despite their success, the women face challenges in conservation due to the slow growth rate of trees in the cold climate. To meet the community's firewood needs, they have adopted innovative solutions such as planting fast-growing willows and poplars and exploring alternative fuel sources to reduce pressure on forest resources. This adaptation reflects the resourcefulness and flexibility of the Mahila Mandal members, balancing sustenance needs with environmental preservation.

The commitment of Lahaul's women to conserving forests and maintaining ecological balance highlights how local leadership can drive sustainable practices. Their actions align with the region's cultural values of collective responsibility and community welfare, which strengthens the resilience of their conservation efforts. This model of community-based natural resource management offers an example of how grassroots movements, especially led by women, can address global environmental challenges at the local level. The success of the Mahila Mandals in Lahaul serves as a valuable lesson for scaling community-led conservation efforts in other parts of the Global South. By fostering a sense of ownership and accountability, this approach builds a foundation for sustainable conservation that is both resilient to external pressures and responsive to community needs. The story of the Stingri Mahila Mandal underscores the critical role of women in environmental stewardship and climate adaptation, proving that women's leadership in conservation not only benefits ecosystems but also strengthens social cohesion and community resilience. Empowering women in conservation is essential for fostering a sustainable, resilient future (Lenin 2016).

## Limitations of the Study

One of the major limitations of this study is the limited documentation of women-led climate initiatives and sustainable practices. Although women are at the forefront of implementing and driving community-based environmental solutions, much of their work remains undocumented, particularly in rural and remote areas where these initiatives often occur. This lack of comprehensive data makes it challenging to accurately assess the scale, impact, and diversity of their contributions to climate resilience. It also limits the ability to conduct comparative analysis or to draw on a wide range of case studies that would provide a deeper understanding of the best practices and challenges faced across different regions. Additionally, the scarcity of detailed documentation constrains the study's ability to fully explore the specific methods women employ to adapt to and mitigate the effects of climate change in varied agricultural, ecological, and socio-economic contexts. Many of these initiatives are informal, community-driven efforts that lack institutional support or visibility, and as such, they often go unrecorded in official databases and reports. This absence of formal records not only hampers academic research but also diminishes the recognition of women's vital roles in climate action, potentially hindering support and resources that could scale up these impactful practices.

## Recommendations

The Western Himalayan region is one of the most resource-scarce areas in the world, facing a unique set of challenges exacerbated by climate change. The increasing levels of precipitation and rising temperatures have created a volatile environment, leading to frequent flash floods and other extreme weather events. These climate-related hazards are having a significant impact on the local population, with both economic and non-economic damages reported every year. The increasing frequency and intensity of such events are not only eroding livelihoods but also threatening the region's already limited natural resources. In this context, the importance of involving women in climate action becomes increasingly pertinent, as they are often at the forefront of managing these challenges and leading the way toward sustainable solutions.

The case studies discussed earlier highlight how essential it is to incorporate women into climate action efforts. Women, particularly those in rural areas, have a deep understanding of local environmental issues and often possess traditional knowledge that can be invaluable in adapting to and mitigating the effects of climate change. Whether through sustainable agricultural practices, community-based conservation, or leadership in disaster preparedness, women play a critical role in strengthening the resilience of their communities. However, their contributions are often under-recognized and under-supported, which limits their potential to drive large-scale change.

The concept of "transformative climate action" is typically associated with large-scale innovations or significant shifts in societal structures that enable adaptation to climate change. Often, transformative actions are seen as those that lead to profound changes in technology, infrastructure, or policy. However, this perspective can be limiting, as it overlooks the potential for smaller, localized initiatives to trigger broader, more impactful changes. In the Western Himalayan region, even modest, community-driven actions can create transformative shifts in resilience and adaptation. For example, local women-led reforestation projects, sustainable farming techniques, or disaster preparedness programs may seem small in scope but can lead to substantial improvements in environmental health, economic security, and social cohesion.

This more inclusive understanding of transformative climate action acknowledges that change does not always have to come from large-scale innovations or sweeping societal transformations. Rather, it can emerge from the cumulative effect of many small initiatives that collectively contribute to a larger shift in climate resilience. Women, as active agents of change, are often the drivers of these smaller initiatives, demonstrating that even in resource-constrained environments, local efforts can lead to meaningful, long-term changes. These efforts can inspire other communities and policymakers to adopt similar practices, thereby catalyzing broader climate action. While the idea of transformative climate action often emphasizes large-scale innovation, this study argues that smaller, localized initiatives—particularly those led by women—can also be transformative. In regions like the Western Himalayas, where resources are scarce and the impacts of climate change are felt most acutely, even modest changes can have profound and lasting effects. Therefore, empowering women and supporting their initiatives in climate adaptation is not just a matter of social justice, but a critical component of effective and sustainable climate action in the region.

## Conclusion

This chapter highlights several initiatives within the Western Himalayan region that have begun to enhance local climate action, driven largely by the efforts of women. The case studies discussed reveal the adaptability and diversity of climate resilience strategies, underscoring a crucial insight: there is no single approach to effective climate action. Instead, climate adaptation must be flexible and responsive, shaped by local needs, cultural contexts, and available resources. Each initiative demonstrates that effective climate action can be achieved through a wide range of approaches, whether through sustainable agriculture, water conservation, or reforestation efforts.

One of the most unique aspects of the initiatives in the Western Himalayas is the shift in roles and responsibilities within the community, particularly in agriculture and environmental stewardship. Traditionally, agricultural practices and land management were predominantly the roles of men; however, as climate challenges have intensified, women have increasingly taken the lead in these areas. This transition reflects not only a shift in gender roles but also an adaptive response by communities to leverage the knowledge, resilience, and resourcefulness of women in managing climate impacts. Women's involvement in climate initiatives has proven essential in preserving biodiversity, improving soil and water conservation, and fostering community resilience in the face of recurring climate disruptions.

## References

Aguilar L, Tsegai D, Alexander S, Stoldt M, Voigt C, Bottaro M, Gerbaldo L (2024) Women-led solutions for drought resilience. https://www.unccd.int/sites/default/files/2024-10/20241005_women-led-DRAFT_V6.pdf

Antriyandarti E, Suprihatin DN, Pangesti AW, Samputra PL (2024) The dual role of women in food security and agriculture in responding to climate change: empirical evidence from Rural Java. Environ Chall 14:100852. https://doi.org/10.1016/j.envc.2024.100852

Asian Development Bank (2010) Climate change adaptation in Himachal Pradesh. https://www.adb.org/sites/default/files/publication/27999/cca-himachal-pradesh.pdf

Banerjee A, Chen R, Meadows ME, Sengupta D, Pathak S, Xia Z, Mal S (2021) Tracking 21st-century climate dynamics of the Third Pole: an analysis of topo-climate impacts on snow cover in the central Himalayas using Google Earth Engine. Int J Appl Earth Obs Geoinf 103:102490. https://doi.org/10.1016/j.jag.2021.102490

Bharadwaj R, Mitchell T (2023) International Institute for Environment and Development. Living in the shadow of loss and damage: uncovering non-economic impacts [Book]. IIED. https://www.iied.org/sites/default/files/pdfs/2023-11/21891iied.pdf

Bisht G (2022) Women of rural Himachal leading the change towards natural farming. *Hindustan Times*. https://www.hindustantimes.com/cities/chandigarh-news/women-of-rural-himachal-leading-the-change-towards-natural-farming-101650914510598.html

Boix-Fayos C, De Vente J (2023) Challenges and potential pathways towards sustainable agriculture within the European Green Deal. Agric Syst 207:103634. https://doi.org/10.1016/j.agsy.2023.103634

Bolch T, Kulkarni A, Kääb A, Huggel C, Paul F, Cogley JG, Frey H, Kargel JS, Fujita, K, Scheel M, Bajracharya S, Stoffel M (2012) The state and fate of Himalayan glaciers. Science 336(6079):310–314. https://doi.org/10.1126/science.1215828

Bryan E, Alvi M, Huyer S, Ringler C (2023) Addressing gender inequalities and strengthening women's agency to create more climate-resilient and sustainable food systems. Glob Food Sec 40:100731. https://doi.org/10.1016/j.gfs.2023.100731

Chauhan HK, Gallacher D, Bhatt A, Bisht AK (2023) The Himalayas: a climate change laboratory. Environ Develop 45:100814. https://doi.org/10.1016/j.envdev.2023.100814

Chinnasamy P, Prathapar SA (2016) Methods to investigate the hydrology of the Himalayan springs: a review. International Water Management Institute (IWMI). 28 p. (IWMI Working Paper 169), Colombo, Sri Lanka. https://doi.org/10.5337/2016.20

Creswell JW (2009) Research design: qualitative, quantitative, and mixed methods. Sage, New Delhi

Démurger S (2015) Migration and families left behind. IZA world of labor. https://doi.org/10.15185/izawol.144

FAO (2020) Towards sustainable crop pollination services. In: FAO eBooks. https://doi.org/10.4060/ca8965en

FAO (2023) The status of women in agrifood systems. In: FAO eBooks. https://doi.org/10.4060/cc5343en

Fertő I, Bojnec Š (2024) Empowering women in sustainable agriculture. Sci Rep 14(1). https://doi.org/10.1038/s41598-024-57933-y

Gamage A, Gangahagedara R, Gamage J, Jayasinghe N, Kodikara N, Suraweera P, Merah O (2023) Role of organic farming for achieving sustainability in agriculture. Farming Syst 1(1):100005. https://doi.org/10.1016/j.farsys.2023.100005

Garschagen M, Doshi D, Reith J, Hagenlocher M (2021) Global patterns of disaster and climate risk—an analysis of the consistency of leading index-based assessments and their results. Climatic Change 169(1–2). https://doi.org/10.1007/s10584-021-03209-7

GIZ (2024) The climate crisis is tougher on women. Akzente. https://akzente.giz.de/en/climate-crisis-gender-women

Gupta V (2021) Winter drought in Himachal impacts agriculture and drinking water supply. Mongabay-India. https://india.mongabay.com/2021/04/winter-drought-in-himachal-impacts-agriculture-and-drinking-water-supply/

Himachal Headlines (2022) Women of Panjayanu village of Mandi district set an example with innovative efforts in natural farming. HIMACHAL HEADLINES. https://www.himachalheadlines.com/opinion/women-of-panjayanu-village-of-mandi-district-set-an-example-with-innovative-efforts-in-natural-farming/

Hock R, Golam R, Jackson M, Kääb A, Kang S, Kutuzov S, Federation R, Milner A, Molau U, Morin S, Orlove B, Allen S, Arenson L, Banerjee S, Barr I, Bórquez R, Carey M, Cogley G, Fischlin A, Weyer N (2022) High mountain areas. In: Cambridge University Press eBooks, pp 131–202. https://doi.org/10.1017/9781009157964.00

IOM (2024) Climate-induced migration and the opportunities for youth|environmental migration portal. https://environmentalmigration.iom.int/blogs/climate-induced-migration-and-opportunities-youth

IPCC (2018) Global warming of 1.5 °C. In: IPCC. Retrieved November 11, 2024, from https://www.ipcc.ch/site/assets/uploads/sites/2/2019/06/SR15_Full_Report_High_Res.pdf

Issaka A, Makame AA, Sewwandi C, Quiacain RM, Chenevard A, Raymann KL (2021) Autonomy through Agroecology. https://swissaid.kinsta.cloud/wp-content/uploads/2021/09/2021-09-SWISSAID-Bericht_final.pdf

Jayawardhan S (2017) Vulnerability and climate change induced human displacement on JSTOR. www.jstor.org. https://www.jstor.org/stable/26188784

Johannessen OM, Olaussen TI (2024) Arctic sea-ice extent: No record minimum in 2023 or recent years. Atmos Ocean Sci Lett 100499. https://doi.org/10.1016/j.aosl.2024.100499

Katzenberger A, Schewe J, Pongratz J, Levermann A (2021) Robust increase of Indian monsoon rainfall and its variability under future warming in CMIP6 models. Earth Syst Dyn 12(2):367–386. https://doi.org/10.5194/esd-12-367-2021

Kaaria S, Osorio M (2018) The gender gap in land rights. FAO. https://openknowledge.fao.org/server/api/core/bitstreams/4966d50c-233b-43a9-8fa7-8d43263dd082/content

Lawrence J, Blackett P, Cradock-Henry NA (2020) Cascading climate change impacts and implications. Clim Risk Manag 29:100234. https://doi.org/10.1016/j.crm.2020.100234

Lee J, Perera D, Glickman T, Taing L (2020) Water-related disasters and their health impacts: a global review. Prog Disaster Sci 8:100123. https://doi.org/10.1016/j.pdisas.2020.100123

Lenin J (2016) Guardians of the forest. Mint. https://www.livemint.com/Leisure/bAoIYT3cT3P3ogf7LnuFoJ/Guardians-of-the-forest.html

Lizarralde G, Bornstein L, Robertson M, Gould K, Herazo B, Petter A, Páez H, Díaz JH, Olivera A, González G, López O, López A, Ascui H, Burdiles R, Bouchereau K (2021) Does climate change cause disasters? How citizens, academics, and leaders explain climate-related risks and disasters in Latin America and the Caribbean. Int J Disaster Risk Reduct 58:102173. https://doi.org/10.1016/j.ijdrr.2021.102173

Malik IH, Ford JD (2024) Monitoring climate change vulnerability in the Himalayas. Ambio. https://doi.org/10.1007/s13280-024-02066-9

Mbungu G (2024) Creating a just social order: the role of gender equity in climate action. APRI. https://afripoli.org/creating-a-just-social-order-the-role-of-gender-equity-in-climate-action

Martin (2022) Climate action—United Nations sustainable development. United Nations Sustainable Development. https://www.un.org/sustainabledevelopment/climate-action/

Nagamani K, Mishra AK, Meer MS, Das J (2024) Understanding flash flooding in the Himalayan Region: a case study. Sci Rep 14(1). https://doi.org/10.1038/s41598-024-53535-w

National Health Mission (2011) Demographic Profile Himachal Pradesh. https://nhm.hp.gov.in/demographic-profile

OECD (2024) Harnessing the green and digital transitions for gender equality: insights from the 2024 OECD forum on gender equality. https://www.oecd.org/content/dam/oecd/en/publications/reports/2024/10/harnessing-the-green-and-digital-transitions-for-gender-equality_c9a0091a/860d0901-en.pdf

Ogieriakhi MO, Woodward RT (2022) Understanding why farmers adopt soil conservation tillage: a systematic review. Soil Security 9:100077. https://doi.org/10.1016/j.soisec.2022.100077

Owen-Burge C (2023) Empowering women, enriching ecosystems: Laos' reforestation initiative—climate champions. Climate Champions. https://climatechampions.unfccc.int/empowering-women-enriching-ecosystems-laos-reforestation-initiative/
Parikh J (2007) Women bear a disproportionate burden of climate change consequences. In: United Nations Development Programme (UNDP), India. https://www.undp.org/sites/g/files/zskgke326/files/migration/in/is_climate_change_gender_issue.pdf
Prashar R (2023) Himachal farmers struggle with crop destruction and debris. https://www.newsclick.in/himachal-farmers-struggle-crop-destruction-debris
Rani S, Sreekesh S (2019) Evaluating the responses of streamflow under future climate change scenarios in a western Indian Himalaya watershed. Environ Processes 6(1):155–174. https://doi.org/10.1007/s40710-019-00361-2
Rao N, Lawson ET, Raditloaneng WN, Solomon D, Angula MN (2017) Gendered vulnerabilities to climate change: insights from the semi-arid regions of Africa and Asia. Climate Dev 11(1):14–26. https://doi.org/10.1080/17565529.2017.1372266
Sabin TP, Krishnan R, Vellore R, Priya P, Borgaonkar HP, Singh BB, Sagar A (2020) Climate change over the Himalayas. In: Springer eBooks, pp 207–222). https://doi.org/10.1007/978-981-15-4327-2_11
Saha B (2024) Flash flood risk in Himachal, Uttarakhand as glacial lakes in Himalayas rise: Report. https://www.preventionweb.net/news/flash-flood-risk-himachal-uttarakhand-glacial-lakes-himalayas-rise-report
Sparkes E, Hagenlocher M, Cotti D, Banerjee S, Masys A, Rana S, Shekhar H, Sodogas V, Surtiari GAK, Valdiviezo Ajila A, Werners ES (2022) Understanding and characterizing complex risks with impact webs: a guidance document. In: UNU-EHS. https://collections.unu.edu/eserv/UNU:9266/UNU-EHS_Impact_Webs_web.pdf
Stringer L, Fraser E, Harris D, Lyon C, Pereira L, Ward C, Simelton E (2019) Adaptation and development pathways for different types of farmers. Environ Sci Policy 104:174–189. https://doi.org/10.1016/j.envsci.2019.10.007
UN Women and UN Women's West and Central Africa Regional Office (2021) KEY results of UN women's flagship programme empowering women through climate-resilient agriculture in West and Central Africa [Report]. https://wrd.unwomen.org/sites/default/files/2022-04/EN%20Women%20in%20Climate%20Resilient%20Agriculture%20in%20WCA_0.pdf
UN Women (2022) Explainer: how gender inequality and climate change are interconnected|UN Women—Headquarters. UN Women—Headquarters. https://www.unwomen.org/en/news-stories/explainer/2022/02/explainer-how-gender-inequality-and-climate-change-are-interconnected
UNFCCC (2015) Expert group meeting, UN Women, UN DESA, UNFCCC, Collantes V, Newman F (2015) Implementation of gender-responsive climate action in the context of sustainable development. In: REPORT [Report]. UN Women. https://unfccc.int/files/gender_and_climate_change/application/pdf/egmreport.pdf
Vũ TH, Tebtebba, Entwicklungsdienst E (2011) Indigenous women, climate change and forests. http://ci.nii.ac.jp/ncid/BB1367119X
Wannewitz M, Garschagen M (2023) Collective adaptation to climate change. Curr Opin Environ Sustain 61:101248. https://doi.org/10.1016/j.cosust.2022.101248. What is the Sendai Framework for Disaster Risk Reduction? (2023). UNDRR. https://www.undrr.org/implementing-sendai-framework/what-sendai-framework
Yavinsky R (2012) Women more vulnerable than men to climate change. PRB. https://www.prb.org/resources/women-more-vulnerable-than-men-to-climate-change/

# Chapter 14
# Can Women's Empowerment Lead to More Effective Climate Action? A Transformative Experience from Rajasthan, India

**Kritika Mishra and Farhat Naz**

**Abstract** Climate change poses serious challenges to the environment, biodiversity, economy, agriculture, livelihood, and society. Thus, the idea of transformative approaches to tackle climate change is becoming increasingly important to implement inclusive and effective climate adaptation and mitigation strategies. The manifold challenges imposed by the negative impacts of climate change further aggravate gender inequalities that need careful consideration. This chapter provides an alternative narrative to the prevailing dominant narrative of women as victims of climate change. It presents women as a solution to climate change mitigation and adaptation. With the case study of Barefoot College, Tilonia (BCT), a small village in Rajasthan located in the desert region of western India, we show how gender transformation is possible when women get an opportunity to act as change-makers. The case study demonstrates how global Sustainable Development Goals (SDGs) on 'climate action', 'no poverty', 'gender equality', 'reduced inequalities', and 'good health and well-being' can be achieved through grassroots actions. To examine BCT's unique model, our research is grounded on the framework of Sustainable Development Theory, and it incorporates a blend of theoretical elements of gender empowerment, emphasizing the role of women's empowerment in sustainable development. To understand the dynamics of how transformative grassroots knowledge meets SDGs and climate action, we employ a qualitative approach, involving observations, interviews, surveys, and case studies. The findings highlight BCT's women as resilient warriors, battling challenges and showcasing sustainable trajectories of transformations in marginalized communities of the global south.

K. Mishra (✉)
PhD Research Scholar, School of Liberal Arts, Indian Institute of Technology, Jodhpur, Rajasthan, India
e-mail: mishra.27@iitj.ac.in

F. Naz
Associate Professor, School of Liberal Arts, Indian Institute of Technology, Jodhpur, Rajasthan, India
e-mail: farhat@iitj.ac.in

 
A. Sarkar and N. Bandyopadhyay (eds.), *Gender-Transformative Approaches for Climate Change Adaptation*, Springer Climate,
https://doi.org/10.1007/978-3-031-90236-9_14

**Keywords** Climate change · Climate action · Sustainable development goals · Gender equality · Women empowerment

## Introduction: Gender, Climate, and the Power of Transformation

Climate change has always been haunting humanity, but the momentum of recent climate change is leaving no realms untouched. Gradually but certainly, crop failures are becoming frequent, coral reefs are deteriorating rapidly, biodiversity and habitat systems are altering their seasonal patterns, water resources are fast depleting, and rising sea levels are imposing big threats to life in the larger picture (Davies et al. 2017). The effects of climate change pose serious challenges to almost every field, including the environment, biodiversity, economy, agriculture, livelihood, and social structures (Chebrolu and Dutta 2021; Davies et al. 2017; Newsome et al. 2023). However, the vagaries of climate change are disrupting the lives of men and women unequally. Owing to pre-existing social, economic, and cultural disparities, gender norms may limit their access to education, land ownership, and financial resources, further constraining their ability to adapt and hence becoming a major victim of climate change impacts (Resurrección and Elmhirst 2020). Marginalized communities, especially women, face greater risks from climate change due to social, economic, political, and cultural inequalities. Their daily lives are impacted by geophysical, agro-environmental, and economic challenges, which are often intensified by pre-existing social differences and unequal access to resources and opportunities (GGCA 2016; Rao et al. 2019).

Several studies reflect women when empowered are capable of bringing great changes. For instance, women acted as community leaders and decision-makers for relief and rehabilitation support when the Indian Ocean Tsunami hit the Andaman and Nicobar Islands of India in December 2004. Likewise, women played a major role in planning sustainable agriculture with a participatory approach in the Development of Sustainable Agriculture in the Pacific programme (DSAP), in handling new irrigation and water management technologies, and in leading several environmental movements globally (Revelo and Kan 2008). Across the globe, women have played pivotal roles as community leaders, farmers, and environmental defenders. However, there is an emerging narrative that positions women as solution-oriented change-makers in the wake of climate change (Terry 2009).

Climate change further also poses a significant threat to achieving the Sustainable Development Goals (SDG)[1] particularly those related to poverty, hunger, health, and environmental sustainability. The ever-rising crisis associated with climate change

[1] The 2030 Agenda for Sustainable Development, embraced by all United Nations Member States in 2015, serves as a collective framework for achieving peace and prosperity for both people and the planet, now and for the future, core to this agenda are the 17 Sustainable Development Goals (SDGs), which call for urgent, cooperative action across both developed and developing nations (Transforming Our World: The 2030 Agenda for Sustainable Development 2015). The SDGs are:

and the tussle of documenting the reach of SDGs, not only hinder the progress towards these global objectives in the real landscape but also disproportionately affect SDG 5 which aims to achieve Gender Equality (Panda 2019). The integration of a gender perspective into the dimension of achieving other SDGs is essential, as women play a crucial role in managing resources, their empowerment is key to fostering sustainable solutions that address both climate resilience and equitable development (Esquivel and Sweetman 2016). However, the impacts, specifically gendered implications of climate change, are distributed unevenly across various social groups (Sultana 2018). Given these disproportionate challenges, the need for transformative approaches to climate change adaptation and mitigation is becoming increasingly vital.

A critical gap becomes evident in this context as there is growing recognition of women as solution-oriented change-makers in the context of climate change, but limited empirical evidence exists on how grassroots initiatives operationalize this potential to achieve gender equality (SDG-5) and climate action (SDG-13). This chapter gives insights into the successful implementation of a unique model that exemplifies how gender empowerment can drive sustainable development and climate adaptation. It gives a detailed case study of Barefoot College in Tilonia (BCT), a small village in western India, which serves as an exemplary model where transformative knowledge aligns with the SDGs and addresses climate change through women's empowerment. It provides insights into ideas on how such programmes can be of crucial importance in other similar regions of the Global South.

The idea of transformative approaches to tackle climate change is a reality in the BCT, Rajasthan, where the inclusive implementation of SDGs has been brought to life for more than half a century (Barefoot College Social Work and Research Centre 2024). Women are at the forefront of these initiatives, which not only foster livelihood but also align with key SDGs, as shown in Fig. 14.1 where SDG-5, Gender Equality is at the core of other SDGs fulfilling more than 12 out of 17 SDGs which are directly and indirectly linked to the initiatives of BCT. Thus, the chapter aims to showcase how the BCT and its community-driven initiatives empower women and achieve multiple SDGs, with a primary focus on SDG-5, gender equality, and

---

SDG 1: No Poverty.
SDG 2: Zero Hunger.
SDG 3: Good Health and Well-Being.
SDG 4: Quality Education.
SDG 5: Gender Equality.
SDG 6: Clean Water and Sanitation.
SDG 7: Affordable and Clean Energy.
SDG 8: Decent Work and Economic Growth.
SDG 9: Industry, Innovation, and Infrastructure.
SDG 10: Reduced Inequality.
SDG 11: Sustainable Cities and Communities.
SDG 12: Responsible Consumption and Production.
SDG 13: Climate Action.
SDG 14: Life Below Water.
SDG 15: Life on Land.
SDG 16: Peace, Justice, and Strong Institutions.
SDG 17: Partnerships for the Goals.

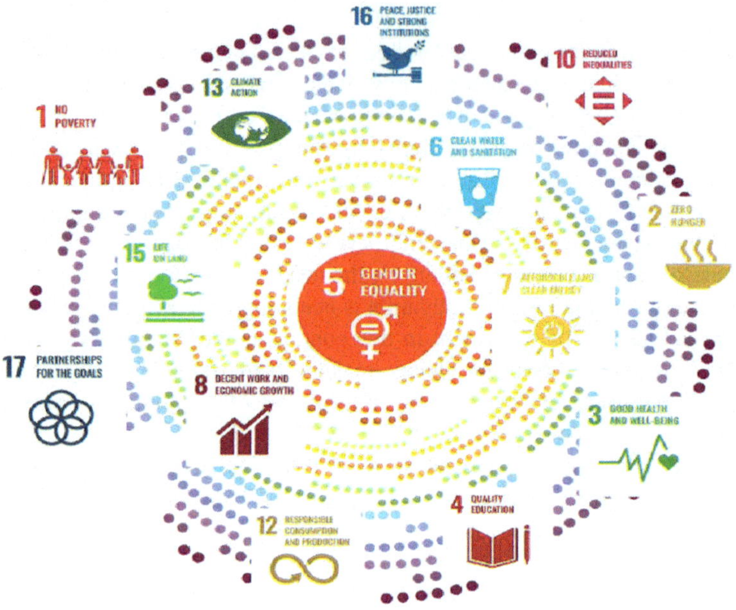


**Fig. 14.1** Showing gender equality at the core. *Source* Created by authors

SDG-13, climate action. Also, the chapter further ponders to answer the question, how does the BCT demonstrate the role of gender-based grassroots initiatives in addressing climate change and advancing multiple SDGs?

## Study Area: Barefoot College, Tilonia, Rajasthan, India

The location of the BCT is shown in Fig. 14.2. It is situated in the village named Tilonia, in the Ajmer District of Rajasthan, a state in western India. The village is surrounded by an arid plain, the typical topography of rural Rajasthan. The region faces multiple adversities, including but not limited to agricultural productivity, power supply disruptions, inadequate infrastructure, and water scarcity on top of everything. Despite these constraints, BCT has emerged as a transformative model demonstrating how SDG can be localized to benefit vulnerable communities.

BCT has become internationally known for community-based development. The roots of this non-profit organization date back to 1972, when Sanjit Bunker Roy registered at the Social Work and Research Centre (SWRC). SWRC promoted decentralization and empowerment, to empower rural communities through decentralized and environment-friendly solutions, particularly in the realm of community-based grassroots organizations working to make marginalized communities become self-sufficient with sustainable solutions. The organization has been actively working for

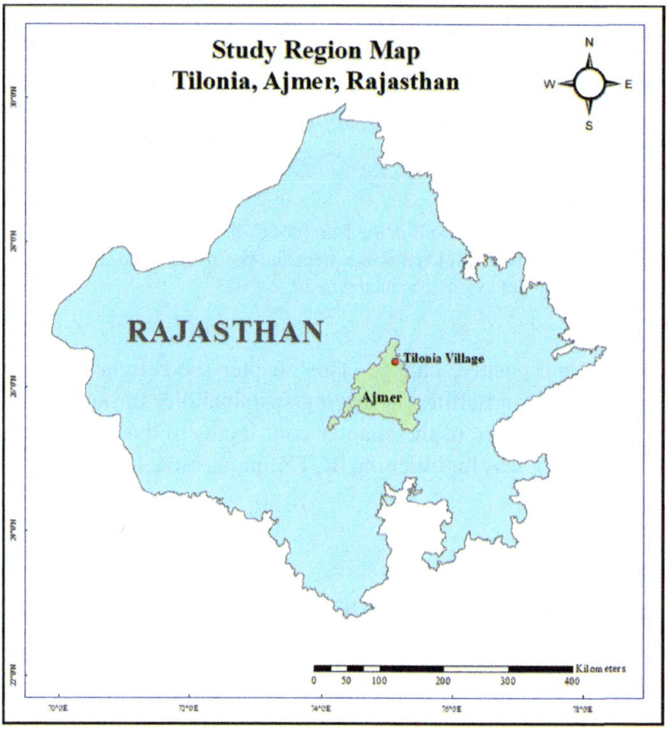


**Fig. 14.2** Location of Barefoot College in Tilonia village. *Source* Created by authors

more than five decades on areas such as renewable energy sources, water management and conservation, environment conservation, education, healthcare, livelihood generation, communication, and women's empowerment. These resources are evidently a major determinant for inclusive development. This organization conducts workshops and training beyond the national borders of India, especially in the impoverished regions of Africa. Such workshops have been conducted in more than 96 countries (Ministry of New and Renewable Energy 2022; Barefoot College Social Work and Research Centre 2024). These workshops played a pivotal role in transforming the lives of rural women by empowering them as solar engineers, architects, teachers, and community leaders. Such empowerment demonstrates how climate action can be translated into basic grassroots actions.

The BCT is the sole college in India dedicated to the poor as stated by the founder Mr. Sanjit (Bunker) Roy who conceptualized the visionary idea of immense knowledge and skills of the rural population living on less than 50 cents per day. His thoughts of actions were inspired by Mahatma Gandhi's principles of service and thoughts on sustainability (Barefoot College Social Work and Research Centre 2024). The organization runs on the learning-by-doing style of education which Bunker Roy brought to reality by leading the community. Figure 14.3 depicts a wall at the Barefoot College campus that displays inspiring messages symbolizing the institution's commitment

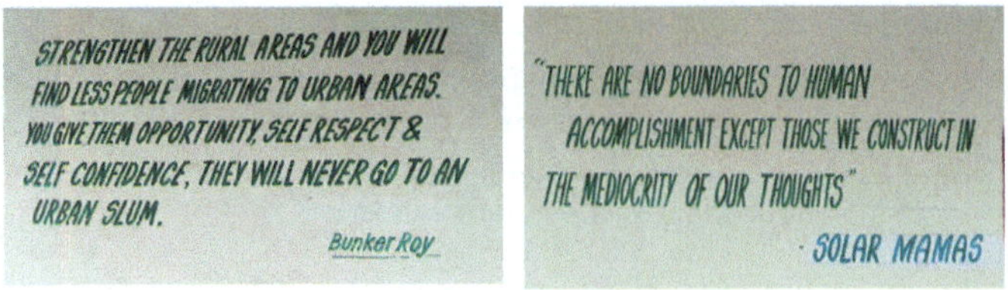


**Fig. 14.3** Images captured from a wall in the Barefoot College campus with inspiring words that reflect the institution's commitment to empowerment and positive change. *Source* Images captured by the Authors during the field visits conducted in 2022–2023

to empowerment and positive change. This chapter sheds light on the remarkable achievements of BCT and efforts in driving sustainability through community-led initiatives. The significance of the chapter contributes to the discourse on gender-based climate adaptation by highlighting BCT's unique model of empowering women as agents of change.

## Reviewing the Roots: Unpacking Gender and Sustainability Narratives

The section will provide insights into a review of pertinent literature related to sustainable development and the role of women in achieving them with a special focus on gender empowerment and grassroots innovations. The scholarly papers and reports are based on the link between SDGs from the gendered perspective, as well as the intersection of gender equality, sustainability, and climate change that shape current discourses on gender and climate action. The idea of sustainable development came to reality in 1987, the United Nations' Brundtland Commission described sustainability as meeting current needs in a way that does not hinder future generations from meeting their own needs (Our Common Future: Report of the World Commission on Environment and Development 1987).

Traditionally, gender roles have been explicitly confined to domestic spaces and excluded from formal economic, decision-making, and policy-making spheres. At the beginning of the 1970s, gender was significantly addressed in the development field owing to the pioneering work of Ester Boserup in *Women's Role in Economic Development* (1970), advocating the concept of Women in Development (WID). It argued for the inclusion of women in economic development and highlighted the potential of women's contributions to sustainable practices if given equal access to resources and opportunities. This laid the foundation for the Gender and Development (GAD) framework,), which shifted focus from integrating women into development to addressing structural inequalities that perpetuate gender disparities (Chakrabarty 2023; Okali 2011; Zeidenstein 1974). Contemporary research repeatedly underscores the insufficient representation of women in decision-making roles, and often as a significant barrier to achieving the SDGs (Esquivel and Sweetman 2016).

Women are the backbone of grassroots development, the social pillars of sustainable development with dimensions ranging from equity, awareness, participation, and social cohesion because it is rooted in the discourses of social sustainability often not understood when the social dynamics are linked to environmental dynamics (Murphy 2012). The expertise of gender and the connections it has with the technical environment and associated development lead to transformative approaches while navigating social change (Resurrección and Elmhirst 2020). Feminist perspectives have become instrumental in understanding the complexities of gender inequality within sustainable development and climate action. These views stand as critics of the traditional development frameworks that historically prioritized economic growth over social justice, sidelining the unique vulnerabilities and contributions of women in climate-vulnerable regions (Resurrección and Elmhirst 2020).

The pool of literature reflects the inadequate meaningful engagement of women in decision-making and participation (Chakrabarty 2023). Gender expertise is essential in formulating policies that are both socially and environmentally sustainable. Amartya Sen's ideas also assert that real empowerment involves expanding individuals' abilities to pursue meaningful goals, which necessitates conceiving sustainable development beyond its traditionally confined limits (Sen 2013). Across communities and generations, sustainable development aims to flourish and ensure that policies are being met by social and environmental objectives (Murphy 2012). The grassroots innovations are vital for sustainable development, offering a bottom-up approach driven by local communities and non-profit organizations, which prioritizes local needs fostering inclusive and community-centric solutions (Hossain 2016). Grassroots innovations play a vital role in addressing climate change and advancing sustainable development. By integrating social, economic, and environmental goals, grassroots innovations contribute to climate change adaptation and mitigation (Robinson and Herbert 2001). Moreover, combating climate change through renewable energy sources is recognized as a sustainable alternative to fossil fuel-based energy systems. The emphasis on the transition to renewables plays a critical role in mitigating climate change and its associated socio-environmental impacts (Owusu and Asumadu-Sarkodie 2016).

The potential of sustainable development along with gender empowerment offers the way to resilient and inclusive communities for social and environmental problems. Women's role as actors and decision-makers within sustainable development centralizes the attainment of both ecological and social long-term goals (Davies et al. 2017).

Despite being the most disproportionately affected group, women have always been portrayed as passive actors in the discourse of climate change rather than active actors in change. This narrative is increasingly contested by feminist scholars and development practitioners on gender research in the environment and sustainability practices (Figueiredo and Perkins 2013; Resurrección 2013; Yadav and Lal 2018). Gender empowerment is very crucial to sustainable development, especially through social sustainability, which appreciates equal participation of all demographics.

Empowerment of women is considered the most important aspect of attaining social sustainability because it fills the gap between economic aims and societal welfare. The studies show that women in community decision-making contribute to social cohesion and resilience, which are the necessary components of adapting to environmental changes (Murphy 2012). To add knowledge to the understanding of how women can act as change-makers, this chapter studies the exemplary work of BCT. It examines how community-based grassroots initiatives can support climate action (SDG 13) and foster sustainable development while contributing to both climate change adaptation and mitigation,

## Methodologies for Understanding Transformative Changes

A case study methodology, complemented by participant observation serves as a key approach to provide a comprehensive perspective to examine the role of BCT, as a milestone example of a women-led multi-dimensional initiative that facilitates the perspective of the interplay between gender empowerment and climate action. The methodological pluralism adopted in the chapter is rooted in the theoretical framework of Sustainable Development Theory, which has facilitated the conceptualization of the holistic impacts of sustainability and social inclusiveness (Madureira and Torre 2019). This study utilizes a qualitative and mixed methods approach to provide a holistic lens that incorporates principles of climate action and gender empowerment.

Data collection is carried out through a combination of focus group discussions (FGDs), in-depth interviews, and semi-structured interviews. Insights into outcomes and social impacts were obtained through six FGDs, each comprising 5–8 women participants involved in different programmes within BCT. The FGDs, in-depth interviews, and semi-structured interviews included solar engineers, solar mamas, health workers, water engineers, water management teams, sanitation, environment, and ecological restoration sections. Additionally, 16–18 semi-structured interviews were conducted with key figures such as programme coordinators, trainers, and community leaders. The blend of these methods provided a deeper understanding of the personal journeys of women, their roles in climate action, and the process of knowledge transfer. The utilization of first-hand accounts of the interactions between women and their environment, roles, responsibilities, and engagement in sustainability practices, participant observation method played a critical role in capturing the daily lived experiences. A purposive sampling method also enabled the understanding of flexibility in exploring perspectives related to climate action efforts and grassroots-level initiatives that contribute to SDG advancement.

The chapter focuses on how demystified knowledge is helping to combat climate change. The case study approach helped in achieving objectives along with triangulating insights from FGDs, in-depth interviews, and participant observation. Moreover, given the involvement of marginalized women, ethical considerations were given importance in the culturally sensitive rural setting, where informed consent was obtained before conducting FGDs and interviews. This qualitative, multi-method approach provides a robust foundation for analyzing the role of BCT's grassroots initiatives in addressing climate change, advancing SDGs, and promoting gender empowerment.

## Results and Discussions

The criticality of climate change, integrated with SDG, enhances the effectiveness of both domains leading to the implementation of climate action (Nerini et al. 2019). The unpredictability of climate change underscores the dire need for successful implementation of climate action. Thus, BCT plays a vital role in empowering rural marginalized communities by enabling them with access to sustainable solutions, like initiatives related to water, solar, education, health, livelihoods, and the environment. The organization works in seven core areas (Fig. 14.4) to contribute positively to using renewable solar energy, water conservation, improving health and education, environmental protection, livelihood generation, and training and communication. These interconnected initiatives demonstrate how local efforts can support broader global objectives. However, in the chapter, we highlight the transformative role that BCT is playing in fostering sustainable practices that contribute directly to climate resilience and adaptation. Hence we particularly focus on using solar power as a tool for reducing $CO_2$ emissions and other initiatives on water conservation and environmental programmes that foster climate change adaptation. The organization pioneers an integrated approach that addresses the challenges faced by the negative impacts of climate change through grassroots programmes specially designed for advancing gender empowerment.

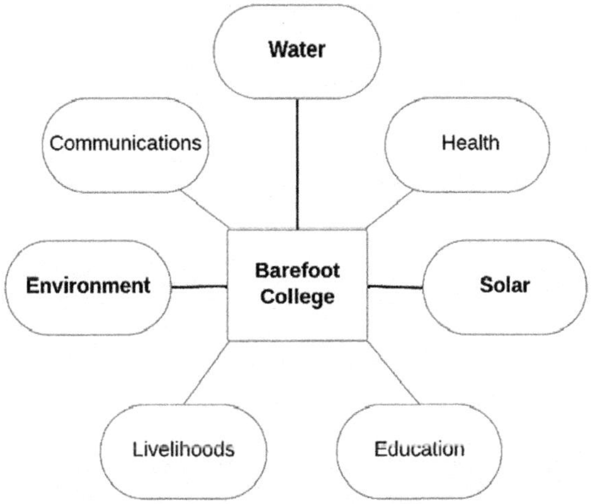


**Fig. 14.4** Seven core principles on which the various programmes are implemented at BCT. *Source* Created by authors

## *Solar Energy and Climate Change*

BCT's solar programme began as a hope for villagers for livelihood generation. This began with the women-led programmes, where semi-literate women are trained to master the skill of solar circuit installation, repair, and management. Today after decades it is a triple-win fostering development, climate change mitigation, and adaptation. As depicted in Fig. 14.5, the solar programme at BCT aims to achieve multi-faceted goals, providing a powerful example of how renewable energy can be harnessed to contribute to climate change mitigation while also contributing to various Sustainable Development Goals (SDGs).

The organization sustains its energy needs through solar modules and annually trains over more than 10 groups from various underdeveloped and developing nations at BCT. Over the past five decades, BCT has empowered over 1700 women from 96 countries to become solar engineers, known as 'Solar Mamas' (Barefoot College Social Work and Research Centre 2024; Mininni 2017). Ahead of exploring the findings, the notion, of a signature attribute of BCT is 'Solar Mamas' or 'Solar Engineers' which needs to be presented first. The question arises, who are they? Are they degree professionals or research scientists? The answer is 'no'. They are semi-literate or illiterate village women, who are trained as solar teachers, solar engineers, and solar cooker engineers. At this point, the notion comes to the fore of how this section of society, the poor vulnerable village-bound women with social and cultural constraints, are shining beyond these constraints, brighter every passing decade, and are not the victims of the climatic limitations but winning warriors. Is it possible in rural India? Yes, and it has been successful for decades.

Solar power systems lower household consumption of electricity and reliance on conventional fuels reduces transmission and distribution loss and lowers economic and environmental costs. This renewable energy-led development model leads to equitable gender development and challenges existing gender inequalities. This renewable form of energy has so far saved more than 45 billion litres of kerosene from polluting the environment. This shift reduces environmental pollution, enhances local energy security, and supports the pursuit of SDG 7 and SDG 13 (Barefoot College Social Work and Research Centre 2024).

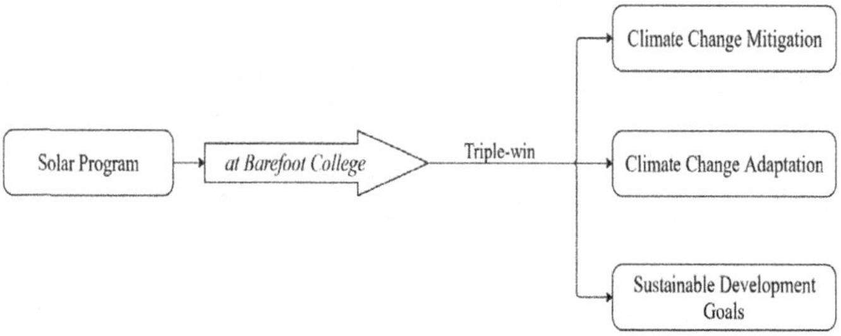


**Fig. 14.5** Triple-win with the solar programme at BCT. *Source* Created by authors

This knowledge transfer mechanism embodies a 'train-the-trainer' approach, ensuring that empowerment is not limited to a single cohort but continues to ripple across communities and countries. Many interviewees highlighted the uniqueness of BCT, because, unlike other institutes that charge fees, it offers a salary for learning, empowerment, reversing gender stereotypes, development, and livelihood in return. In a FGD with seven solar engineers, it is found that all of them started the journey of their sustainability and independence at BCT. The training programme led by solar engineers has been successfully running for the past five decades. In 2000 First Solar Mama was selected by the organization. Solar Mamas are women who return to their respective communities after training to implement solar electrification and spread this transformative knowledge.

People here understand the gravity of their work which brought changes in the longer term towards environmental sustainability, development, and sustainable livelihood. This not only reduces carbon emissions but also contributes directly to SDG 7 (Affordable and Clean Energy) and SDG 13 (Climate Action). The initiative, now a reality, has not only provided sustainable energy usage but also promoted gender equality, and empowered by offering women the technical expertise, financial independence, and leadership opportunities that coincide with the SDGs in the bigger picture. Global reach and international partnerships are achieving remarkable heights with the help of multiple collaborations. BCT's solar training programmes are supported by international collaborations such as the Indian Technical and Economic Cooperation (ITEC) initiative and partnerships with organizations such as UN Women.

This collaboration reflects a commitment to enhancing women's agency in regions where energy access remains a significant hurdle to development. Projects such as the electrification of Pacific Island communities, as well as training initiatives in African nations, showcase the global reach and adaptability of the BCT model. During the FGDs, it was stated by women that several batches of solar training have been held in the states of India like Assam, Sikkim, Manipur, Andhra Pradesh, Karnataka, MP, Ladakh, Jammu & Kashmir, Bihar, Jharkhand, and many others. The organization fosters the records of more than 1700 trained solar mamas making more than 60,000 houses electrified since its initiation (Barefoot College Social Work and Research Centre 2024). The programmes substantiate empowering women as solar engineers as a sustainable solution that promotes gender equity, energy access, economic resilience across diverse contexts, and thus meeting the agendas of SDG 5, 7, and 13.

### Challenging Gender Norms and Energy Self-sufficiency

The bonds formed between women at BCT extend beyond national borders. When women from different countries were first trained at Tilonia, they built networks that facilitated continued learning and support. This model is anchored in building connections from the bottom-up approach, particularly those related to poverty reduction, gender equality, and clean energy access. The organization itself utilizes its

**Fig. 14.6** Solar energy generation and training centre at BCT. *Source* Images captured by the authors during the field visits conducted in 2022–2023

products in the form of solar cookers, lanterns, and solar lights. For instance, the food at BCT is cooked by the solar cookers made and installed by the women. Moreover, most people have solar lanterns at their houses which are indigenously made at BCT. It's been more than decades that most of the power in the organization and nearby regions comes from the sun, and therefore they are not dependent on any aid from the government. They are self-sufficient in terms of power until there is the sun. The images below in Fig. 14.6 give a glimpse of the Solar energy section and training centre at BCT.

In conversation with a woman at Tilonia, she shared her transformative journey from a traditional handicraft worker to becoming a Solar Mama. The need for financial stability became a priority and she transitioned from handicraft-based work (learning stitching, threadwork, and garment-making) to solar engineering for a more stable income. She narrated her 36-year journey, which was not without challenges. The technical nature of solar circuits and energy systems initially intimidated some of the women, but through six months of intensive, hands-on learning, they mastered their new skills. Today, many such women with limited formal education, are adept at building, repairing, and installing solar circuits, and managing components such as resistors, CFLs, and LEDs, narrating their success with pride. Over 3000 women from India and beyond have been trained in solar technology under her mentorship, reflecting the growing global effort to build climate resilience through technology transfer. She recounted travelling to Senegal and Tanzania to teach solar skills, adapting teaching methods to overcome language barriers and ensure that knowledge could be shared regardless of cultural or linguistic differences. Her unwavering dedication resulted in praise from the Prime Minister of India in 2016 during a global showcase of Solar Mamas' skills in Tanzania and the Mahila Shakti Puraskar award in 2016 underscored the significance of such work and also emphasized cross-cultural harmony. She also described how they promote harmony and togetherness with basic changes, like when women from different countries trained at the College, their accommodation is intentionally mixed to foster unity, partnership, and mutual learning. Also, the communal dining experience, offering simple yet nutritious meals prepared with care and cleanliness, further reinforces the sense of togetherness and dignity. Through its holistic model, BCT not only equips women with life-changing skills but also fosters a global network of empowered leaders committed to climate resilience, and mitigation of carbon emissions, and fosters a global network of empowered change-makers.

The findings from several studies also support the fact that, unlike fossil fuels, the complete lifecycle of renewable energy systems results in net-zero emissions. This feature positions renewables like solar energy as vital for reducing future global greenhouse gas emissions and supporting climate action goals (Owusu and Asumadu-Sarkodie 2016). Also, solar energy is a promising solution to combat climate change by providing a clean, renewable alternative to fossil fuels, which are one of the biggest contributors to greenhouse gas emissions (Olabi and Abdelkareem 2022). Thus, by harnessing the sun's power, BCT reduces their carbon footprints and lessens the impact of climate change. The advances in solar technology, such as more efficient solar electricity, cookers, lights, solar pumps, solar water desalination and filtration systems, etc., are providing better energy storage, and have made solar energy more accessible and affordable in remote regions. This shift to solar energy is a milestone example stating how the rural remote population overcame implementation challenges and promoted widespread solar energy adoption at national and international levels. The synergy of solar energy, climate action, and sustainable development enabled the role of solar energy in building a greener, and more resilient future.

## *Sustainable Water Initiatives*

Challenges associated with climate change are not new to the regions of western India which are further intensified by low rainfall, falling groundwater levels, recurrent droughts, and salts and fluorides in the groundwater causing illnesses. Preservation and conservation of every drop of water is necessary in desert regions where water accessibility is scarce. The approach of BCT combined with modern technology has brought water security to this region. Around the 1970s, BCT started working towards water conservation efforts, motivated by the approach that every raindrop is crucial and needs to be preserved. It resulted in successful rainwater harvesting units, the number reaching more than thousands, and more than 230 community centres already working towards conservation methods (Barefoot College Social Work and Research Centre 2024).

BCT at present has adopted several measures of water conservation including the deployment of hand pumps, efficient rainwater harvesting systems, and installation of solar energy desalination systems. They also spread awareness about water quality and make people aware of the difference between potable and non-potable water. Following the footsteps of successful solar technologies and women empowerment strategies, the college also facilitates Water Engineers, who have combined their traditional wisdom and modern technology along with bringing public participation, hands-on training on skill development, and implementation, all led by women. The organization where several women have been trained as hand-pump mechanics is also shattering the global patriarchal notion of male-dominant roles. Adding to solar energy knowledge, the organization has set up a water desalination plant using solar methods and community-operated reverse osmosis plants for desalination.

In-depth interviews with the women handling the water desalination process and filtration and discussions of replenishment of groundwater tables revealed the participation of rural communities to reduce dependency on external aids and to reduce the burden on women who had to walk miles to fetch water. The decentralization of accessibility and availability of water in the organization and nearby villages are one of the benchmarks of tried and tested methods of water management. From managing, handling, and installing systems, to teaching and learning everything is led by women in this organization, which is a pioneer in achieving SDG-6 with women empowerment at its core.

During the FGDs, women stated that lower caste people are not allowed to fetch water from village wells, a prevalent caste system deeply rooted in India. Here, breaking all stereotypes, the barefoot approach dared to establish more water reservoirs and wells and taught people water conservation beyond the social demarcations of caste and creed, which gradually united people irrespective of their caste. Women from the water management and conservation section highlighted that the organization has worked so far in more than 200 villages to solve water-related issues, established nadis, tanka,[2] ponds, and wells, taught rainwater harvesting techniques, water quality testing, and many more. Here, women are knowledgeable and skilled enough to test water quality even in scientific terms. One participant narrated that she is trained to calculate the amount of fluoride, chloride, phosphates, total dissolved solids, turbidity, and sulphates through their testing systems. These women are experts in repairing and installing such water filter systems enriched with RO filters in several villages as per the needs and population size.

Despite the challenges posed by climate change, including rising summer temperatures and prolonged dry spells, BCT programmes continue to advance climate action. Their efforts are centred on promoting climate resilience and adaptation, enabling rural communities to withstand and respond to climate-related shocks. Along similar lines, rainwater harvesting methods are not new to rural regions of Rajasthan, but they have been managed efficiently to sustain the needs. It is remarkably astonishing in BCT, that the traditional water harvesting methods of rural Rajasthan are being taught to people living in the dry regions through several workshops and visits organized regularly. The management of rainwater harvesting units is being done through regular meetings at the community level. FGD with a veteran revealed that more than 1600 schools have been equipped with rainwater-harvesting systems and water filter systems. The organization also has a huge underground tank with a capacity of nearly 40,000 L which collects all the rainwater, filters it, and purifies it for various uses.

The impacts of BCT initiatives and programmes are not limited only to the geographical boundaries of Rajasthan but also to several countries where they organize their workshops, where they successfully showcase the impacts as shown in Fig. 14.7, related to rainwater harvesting, hand pumps, reviving drought-ridden regions of Rajasthan and solar energy desalination process. The interlinkages of these

[2] Nadi and Tanka are traditional rainwater harvesting structures in Rajasthan.

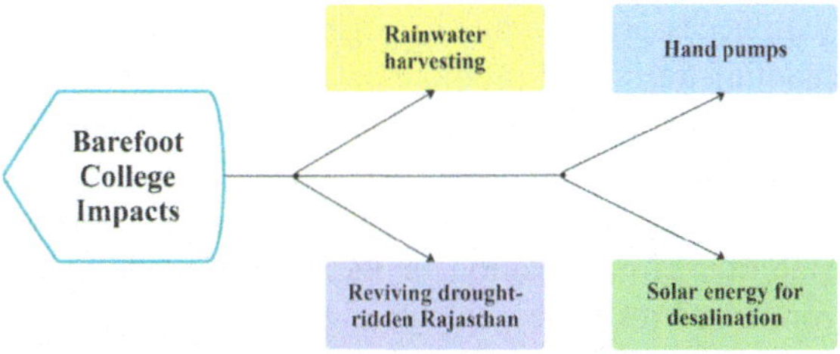


**Fig. 14.7** Impacts of BCT on water conservation and climate change adaptation. *Source* Barefoot College Social Work and Research Centre (2024)

decentralized, small-scale water conservation technologies are exemplary examples of how climate change adaptation can be done in drought-prone areas with local innovations.

These initiatives also have a profound impact on health equity, and it has reduced maternal and child mortality rates, improved access to safe and affordable health products, and contributed significantly to community resilience. In this regard, BCT, through its grassroots approach, shows how localized initiatives can be instrumental in the achievement of broader goals on climate adaptation, public health, social equity, and community empowerment for disadvantaged regions. The magnificent journey of BCT began in 1972 by addressing water issues in the drought-prone region and eventually evolved into a mighty source of social change and transformation that empowers and turns rural women into change-makers.

## *Sustainable Environmental Solutions Through Training and Practice*

BCT's education programmes and initiatives in Rajasthan have made deep social impacts that have altered the perceptions of social constraints, development, environmental management, ecological restoration, waste management, and community engagement. All of this, in turn, makes cleaner and healthier rural environments, while fostering community resilience and ecological awareness. These initiatives, spanning education, environmental management, and social transformation, align with climate adaptation goals by empowering local populations, particularly women, to engage with and lead sustainable practices.

### Tree Plantation

One of the most significant ways BCT contributes to climate adaptation is its afforestation efforts. In the last fifty years through drought-resistant plantations and regeneration of degraded lands, BCT enhanced the biodiversity by planting more than 60,000 trees. Environmental education programmes, with school tree-planting campaigns at their core, equip the students and the community members themselves with necessary information regarding the conservation of the ecological setting by planting trees. BCT promotes awareness about the benefits of tree plantation, particularly its role in soil rejuvenation and climate change mitigation by carbon absorption by plants. This large-scale greening initiative not only helps combat soil erosion and desertification but also enhances local biodiversity, thus promoting ecological stability. The plantation of indigenous plant species further contributes to environmental resilience by fostering plant varieties adapted to local conditions. The BCT also has the barefoot nursery which preserves beneficial species for plantation. Through environmental education campaigns, BCT teaches community members, particularly students, about the importance of trees in carbon sequestration, soil health, and climate change mitigation.

### Waste Management

Waste management services offered by BCT, a decentralized system appear like the solution that empowers communities to manage their methods for waste disposal. This community-led model has been successfully implemented at Chota Narena Village, Tilonia, and then replicated in other villages to have zero-waste villages, by engaging villagers in designing and operating waste management systems. Accordingly, BCT promotes source-level segregation, organic waste composting, and recycling activities, greatly reducing pollution and greenhouse gases resulting from traditional waste management practices such as open dumping and burning.

BCT has not only reduced environmental pollution through its waste management methods but also has created economic opportunities for marginalized communities. These initiatives contribute to a circular economy, where waste is minimized, resources are reused, and local livelihoods are strengthened. Moreover, the section on plastic-free sanitary pads also teaches sustainable ways to dispose of them in the old earthen pots without harming or leaving any long-term harmful impacts on the environment. Such exemplary models with sustainable approaches, if implemented on a large scale in every such village, will surely produce dignified jobs, for the economically marginalized in a manner, which in turn creates an avenue for economic empowerment towards environmental responsibility. Women's involvement in environmental awareness campaigns not only helps mitigate the effects of climate change but also elevates their role as community leaders in sustainable practices (Murphy 2012).

The waste management programme has also ignited social change of a wider magnitude by redressing the imbalances of generations-old caste-structured inequalities, and it is offering the deprived sections worthwhile livelihoods. For instance, the members of communities historically relegated to manual scavenging, placed at the bottom of all such discriminatory caste structures, are now being given an honourable livelihood with improved health and safety results, by raising community awareness about waste disposal methods. Accordingly, BCT has shifted the social perception of waste management work, granting dignity and improved living standards to those involved. The holistic environmental education, ecological restoration, and waste management practices that BCT practices, stand testimony to the impact of these initiatives on rural society. It leads not only to sustainable development through environmental conservation but also empowers the less privileged, instils environmentalism in the minds of the youth, and even breaks the social tradition and hierarchies that prevail in society.

Women's active involvement in sustainable initiatives strengthens community responses to ecological challenges, demonstrating the value of inclusive decision-making by integrating women into the planning and implementation of projects. Thus, communities not only benefit from a diverse range of perspectives but also create a more resilient social structure that is equipped to address issues such as climate-induced displacement and resource scarcity (Murphy 2012). Apart from such remarkable achievements, women here are introduced to community-to-community exchanges of knowledge and skills and collective efforts to ensure environmental sustainability. The organization directly results in reduced dependence on outside urban skills and increased livelihoods at the bottom of the pyramid, thereby reducing migration from rural to urban areas and redefining rural professionalism. BCT's comprehensive approach to environmental management, waste reduction, afforestation, and women's empowerment serves as a model for climate adaptation and climate action. By blending traditional knowledge with modern technology, BCT has equipped rural communities with the tools to face climate change while fostering resilience, social equity, and sustainable development. These programmes not only mitigate environmental degradation but also empower individuals to actively participate in the creation of a sustainable future.

## *Bridging Traditional and Modern Practices*

Apart from solar, water, and environmental initiatives, BCT has also redefined rural education by merging traditional knowledge with modern practices. BCT, through projects like the Solar Digital Night School, Children's Parliament, Day School, Residential Bridge School, Digital Village Learning Centres, and the Barefoot Teacher training programme, has come up with out-of-the-box solutions for environmental problems for more than five decades. Additionally, livelihood programmes at BCT demonstrate how developing skills and empowering local individuals, especially women, can safeguard traditional crafts and promote economic resilience in marginalized rural communities.

BCT's Hatheli Sansthan programmes focus on 'learning by doing', connecting traditional craftsmanship-aided artisans like weavers, potters, metalworkers, and carpenters by rejuvenating traditional crafts with a sustainable livelihood for rural artisans. Moreover, programmes like radio and communications for spreading awareness related to combating climate change, environmental schemes, and government programmes. BCT has also developed a comprehensive healthcare model that delivers quality diagnostic and preventive care in rural communities through the training of grassroots health workers. Barefoot doctors and workers organize programmes in many villages where they spread awareness regarding government programmes, vaccinations, insurance, and subsidies which can be beneficial for the vulnerable section of society. The miracles of BCT spread to other sections of women's hygiene and nutrition. A separate section of women is working to make low-cost sanitary napkins, and the 'Super 5' nutrition supplement developed by BCT. Conceptualizing programmes from the needs of the community and focusing on experiential accessible learning, BCT has facilitated social cohesion and gender empowerment and developed a model of education that not only serves to meet the immediate needs of the rural population but also prepares them for growth and resilience.

## *A Beacon in the Desert: Sustainability and Climate Action*

The footprints BCT is leaving behind and its impacts extend beyond their professional lives. They are lighting their villages and homes, changing perspectives, and lifting entire communities. The core values of BCT, as shown in Fig. 14.8, aim towards collective decision-making, decentralization, self-reliance, equality, and austerity in thoughts and actions showcasing consistent engagement that aligns with community practices to support climate change mitigation and adaptation.

The impact of BCT programmes in promoting climate-resilient communities is by directly addressing the vulnerability of rural regions to climate change, which includes energy self-sufficiency and sustainable livelihoods. Climate change-induced energy disruptions (like floods or storms) often affect necessities like access to electricity. By fostering community-driven solar power systems, BCT makes communities less reliant on unstable national grids. Additionally, the utilization of solar energy is a mitigation strategy resulting in reduced carbon emissions and greenhouse

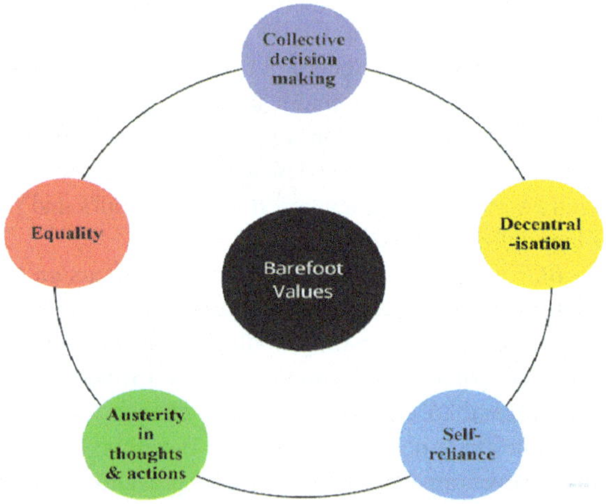


**Fig. 14.8** Values of Barefoot College. *Source* Barefoot College Social Work and Research Centre (2024)

gas (GHG) emissions, thus strengthening local climate adaptation strategies. BCT with its advancing gender-responsive climate action empowers women as engineers, decision-makers, and leaders of energy transitions—The Solar Mamas. The direct links to climate actions with local energy security also build resilience by creating a network of female climate leaders, not only in India but in other parts of the world especially in Africa. The promotion of circular economy is also deeply implemented by BCT, as it produces and maintains its own solar cookers, lanterns, and equipment, promoting the reuse and recycling of solar parts, thereby reducing waste.

Challenging gender stereotypes and local capacity building with local climate champions bring climate action knowledge back to their communities. BCT's programmes emphasize gender-inclusive employment and lifelong learning related to water management, rainwater harvesting, water quality testing, and water conservation. Such diversifying methods by training women as engineers, teachers, and trainers allow for a better ability to adapt to climate shocks. The BCT harnessing the rainwater also helps in arresting groundwater depletion as well as promoting sustainable agriculture by providing a stable water supply. Rainwater harvesting supports climate-resilient farming practices, thereby enhancing food security.

BCT's environmental programmes mitigate environmental degradation, promote climate adaptation, and foster sustainable development. Reforestation and afforestation play vital roles in mitigating the impacts of climate change. BCT with its green impacts of tree-planting drives, nursery development, and community-led afforestation efforts helps in natural carbon sequestration, as trees are also carbon sinks. Reforestation strengthens local climate resilience and healthy vegetation also regulates local weather patterns.

These programmes are deeply rooted in the principles of community-led, bottom-up approaches, ensuring that local communities, especially women, are at the forefront of environmental action. The Barefoot model is a global inspiration, showing that when women are empowered, communities are more prepared to tackle climate-related disruptions.

## Conclusion

BCT stands as a beacon of hope in the face of the multi-faceted challenges of climate change and is a great example of bottom-up solutions, its approach highlights how grassroots transformative actions can meet SDGs for the betterment of society as a whole. Can women's empowerment lead to effective climate action? The answer lies in the multi-dimensional impacts of barefoot programmes targeting environment and social well-being. The distinct programmes of the BCT, as shown in Fig. 14.9 are directly or indirectly linked with several SDGs.

The notion that gender equality stands at the core of every SDG by giving living examples of women working in renewable energy, health care, water management, waste disposal, communications, education, and sustainable livelihood at BCT justifies it. The exemplary model of climate resilience and sustainable development is

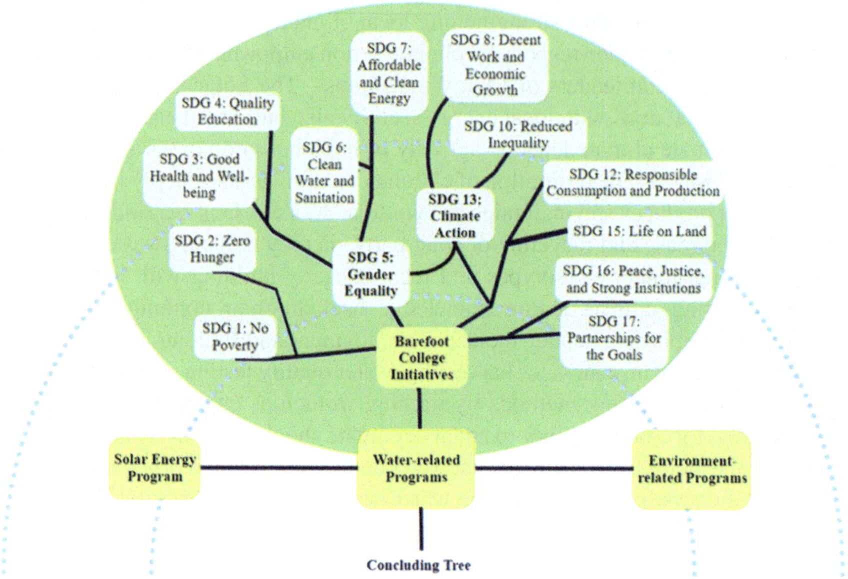


**Fig. 14.9** Linking programmes at BCT for promoting SDGs. *Source* Created by authors

anchored on women. These women in the communities are empowered through skills to become change-makers to combat and mitigate the negative impacts of climate change. The BCT model and its initiatives challenge traditional gender roles as their influence brings a positive impact that affects everyone in society. The organization following the approach where the *teacher is a learner and the learner is a teacher* for augmenting conventional education with contemporary vocational training puts it in the successful promotion of SDGs while recognizing local culture, considering the community-led nature of embeddedness and collective engagement of BCT.

Stating the founder Mr. Bunker Roy's words *'the college is for the poor'* is breaking the vicious cycle of poverty, deriving from the partnership of educated experts transforming their ways to share knowledge which has created these barefoot experts and contributed significantly to professionalizing rural development. Stated otherwise, BCT empowers individuals as it builds the framework of future sustainable societies. Its approach reflects the people-centred and integrated development model that works actively towards achieving several SDGs cumulatively by promoting equitable access to resources besides serving as an effective social catalyst in combating climate change. The BCT model shows that sustainable development, at the heart of which are people and women in particular, strengthens the potential for self-reliance, resilience, and dignified survival that promotes a sustainable and equitable future.

**Acknowledgements** The authors extend heartfelt thanks to Barefoot College Tilonia for their warm welcome and invaluable cooperation during the visits. We are especially grateful to the dedicated workers, inspiring women, and supportive villagers whose engagement and insights enriched the understanding and made the experience truly meaningful. We appreciate and thank KJ George for his helpful editorial suggestions.

## References

Barefoot College Social Work and Research Centre (2024) The Barefoot College. Available at: https://barefoot.college. Accessed 20 Oct 2024

Chakrabarty A (2023) Feminizing decentralized renewable energy programs: a study of a community renewable energy project from India. J Asian Energy Stud 7:107–120. https://doi.org/10.24112/jaes.070008

Chebrolu SP, Dutta D (2021) Managing sustainable transitions: institutional innovations from India. Sustainability 13(11):6076. https://doi.org/10.3390/su13116076

Davies K, Adelman S, Grear A, Iorns Magallanes C, Kerns T, Rajan SR (2017) The declaration on human rights and climate change: a new legal tool for global policy change. J Human Rights Environ 8(2):217–253. https://doi.org/10.4337/jhre.2017.02.03

Esquivel V, Sweetman C (2016) Gender and the sustainable development goals. Gend Dev 24(1):1–8. https://doi.org/10.1080/13552074.2016.1153318

Figueiredo P, Perkins PE (2013) Women and water management in times of climate change: participatory and inclusive processes. J Clean Prod 60:188–194. https://doi.org/10.1016/j.jclepro.2012.02.025

Global Gender and Climate Alliance (GGCA) (2016) Gender and Climate Change: a closer look at existing evidence. Women's Environment & Development Organization. Available via GGCA. https://wedo.org/gender-and-climate-change-a-closer-look-at-existing-evidence-ggca/. Accessed 12 Jun 2023

Hossain M (2016) Grassroots innovation: a systematic review of two decades of research. J Clean Prod 137:973–981. https://doi.org/10.1016/j.jclepro.2016.07.140

Madureira L, Torre A (2019) Innovation processes in rural areas. Reg Sci Policy Pract 11(2):213–218. https://doi.org/10.1111/rsp3.12215

Mininni GM (2017) The 'Barefoot Model' of economic empowerment in rural Rajasthan. J Indian Ocean Region 13(1):54–75. https://doi.org/10.1080/19480881.2016.1272813

Ministry of New and Renewable Energy (2022) Barefoot College, Tilonia. Available via MNRE. https://mnre.gov.in/en/barefoot-college-tilonia/. Accessed 20 Oct 2024

Murphy K (2012) The social pillar of sustainable development: a literature review and framework for policy analysis. Sustain Sci Pract Policy 1(8). https://doi.org/10.1080/15487733.2012.11908081

Nerini FF, Sovacool B, Hughes N, Cozzi L, Cosgrave E, Howells M, Tavoni M, Tomei J, Zerriffi H, Milligan B (2019) Connecting climate action with other sustainable development goals. Nat Sustain 2(8):674–680. https://doi.org/10.1038/s41893-019-0334-y

Newsome D, Newsome KB, Miller SA (2023) Teaching, learning, and climate change: anticipated impacts and mitigation strategies for educators. Behaviour and Social Issues 32(2):494–516. https://doi.org/10.1007/s42822-023-00129-2

Okali C (2011) Searching for new pathways towards achieving gender equity: beyond Boserup and 'Women's role in economic development.' In: ESA working papers 289012. Food and Agriculture Organization of the United Nations, Agricultural Development Economics Division (ESA). https://ideas.repec.org/p/ags/faoaes/289012.html

Olabi AG, Abdelkareem MA (2022) Renewable energy and climate change. Renew Sustain Energy Rev 158:112111. https://doi.org/10.1016/j.rser.2022.112111

Owusu PA, Asumadu-Sarkodie S (2016) A review of renewable energy sources, sustainability issues and climate change mitigation. Cogent Eng 3(1):1167990. https://doi.org/10.1080/23311916.2016.1167990

Panda PK (2019) A study on the achievement of sustainable development goals from a gender equality perspective. Glob J Nutr Food Sci 1(4). https://doi.org/10.33552/GJNFS.2019.01.000516

Rao N, Lawson ET, Raditloaneng WN, Solomon D, Angula MN (2019) Gendered vulnerabilities to climate change: insights from the semi-arid regions of Africa and Asia. Climate Dev 11(1):14–26. https://doi.org/10.1080/17565529.2017.1372266

Resurrección BP (2013) Persistent women and environment linkages in climate change and sustainable development agendas. Women's Stud Int Forum 40:33–43. https://doi.org/10.1016/j.wsif.2013.03.011

Resurrección B, Elmhirst R (2020) Negotiating gender expertise in environment and development: voices from feminist political ecology, 1st edn. Routledge, London. https://doi.org/10.4324/9781351175180

Revelo LA, Kan FM (2008) Gender perspectives: integrating disaster risk reduction into climate change adaptation, good practices and lessons learned. International Strategy for Disaster Reduction, United Nations

Robinson JB, Herbert D (2001) Integrating climate change and sustainable development. Int J Glob Environ Issues 1(2):130. https://doi.org/10.1504/IJGENVI.2001.000974

Sen A (2013) The ends and means of sustainability. J Hum Dev Capab 14(1):6–20. https://doi.org/10.1080/19452829.2012.747492

Sultana F (2018) Gender and water in a changing climate: challenges and opportunities. In: Fröhlich C, Gioli G, Cremades R, Myrttinen H (eds) Water security across the gender divide. Water security in a new world. Springer, Cham. https://doi.org/10.1007/978-3-319-64046-4_2

Terry G (2009) Climate change and gender justice. Practical Action Publishing in Association with Oxfam GB, Warwickshire, UK

World Commission on Environment and Development (1987) Our common future: report of the World Commission on Environment and Development. Brundtland Commission. Available via United Nations. https://sustainabledevelopment.un.org/. Accessed 28 Oct 2023

Yadav SS, Lal R (2018) Vulnerability of women to climate change in arid and semi-arid regions: the case of India and South Asia. J Arid Environ 149:4–17. https://doi.org/10.1016/j.jaridenv.2017.08.001

Zeidenstein SA (1974) Review of woman's role in economic development by E. Boserup. Bangladesh Econ Rev 2(1):507–510. Available via: http://www.jstor.org/stable/40795751

# Chapter 15
# Uplifting Women Farmers by Adopting Climate-Smart Practices with Climate Finance: Evidence from Central and Eastern States of India

**Tarun Katoch and Parul Gurjar**

**Abstract** Climate change impacts are acutely felt worldwide, yet women in agriculture-dependent developing nations in the Global South face disproportionate challenges. Despite their critical role, entrenched socio-cultural barriers often exclude women from financial decision-making, exacerbating their vulnerability to climate change. This study focuses on improving institutional credit access, fostering market linkages, and enhancing capacity building to promote climate-smart agriculture (CSA) among smallholder women farmers. By employing remote sensing, artificial intelligence, and machine learning, we analyzed farm activities over the past three years while accounting for localized climate change impacts. This analysis underpins the development of the *KhetScore*, an analytical farm score that serves as an alternative to traditional credit scores. This innovation addresses the critical credit gap faced by over 85% of Indian smallholder farmers. To empower women in decision-making and income diversification, tailored financial products, including crop loans, dairy loans, digital gold savings, weather, and health insurance, were introduced through financial institutions. Additionally, we used Farmer Producer Organizations (FPOs) as platforms for economies of scale, delivering customized crop advisories, climate-smart agri-inputs, weather forecasts, and market connections to women farmers. A survey of 1498 farmers demonstrated significant outcomes: borrowing costs decreased by 20–50%, input costs were reduced by 10%, and price realization improved by 2–5%. These targeted interventions are transforming smallholder agriculture, integrating women into digital ecosystems, and positioning them as central actors in climate change adaptation.

T. Katoch (✉)
Dvara E-Registry, Hyderabad, Telangana, India
e-mail: tarun.katoch@dvara.com

P. Gurjar
Bankers Institute of Rural Development, Lucknow, Uttar Pradesh, India
e-mail: Parul.g.ext@nabard.org

 
A. Sarkar and N. Bandyopadhyay (eds.), *Gender-Transformative Approaches for Climate Change Adaptation*, Springer Climate,
https://doi.org/10.1007/978-3-031-90236-9_15

**Keywords** Climate change adaptation · Women farmers · Agriculture · Gender · Financial inclusion

## Introduction: Climate Change and Its Disproportionate Impact on Women

Climate Change is currently becoming the biggest challenge for humanity (Anderson et al. 2014; Steffen 2015; Pearse 2017). It is increasingly recognized as a critical threat to global agriculture, with profound implications for food security, economic stability, and rural livelihoods (Wheeler and Braun 2013). Rising temperatures, erratic rainfall patterns, prolonged droughts, and the increased frequency of extreme weather events are already eroding agricultural productivity worldwide (IPCC 2021).

In India, where 86% of farmers are smallholders cultivating less than two hectares (Agricultural Census 2019), agriculture constitutes nearly 15% of GDP and underpins the livelihoods of over 60% of the population (World Bank 2021). Climate change could reduce crop yields by 10–20% in the country in the coming decades if robust adaptation measures are not implemented (NICRA 2023). Such disruptions may precipitate widespread food insecurity, substantial income losses for farmers, and significant rural-to-urban migration, exacerbating social and economic vulnerabilities.

The gender-differentiated impact of climate change is especially pronounced in developing economies like India, where women comprise the majority of agricultural laborers, yet they face systemic barriers to resources (UN Women 2020). Women are 68% more likely than men to suffer from climate-related health impacts due to restricted access to healthcare services (UNDRR 2024). Women farmers exposed to extreme heat are more susceptible to health issues, such as dehydration and heatstroke, resulting in reduced productivity and a decrease in daily working hours during peak summer months (Ebi et al. 2021). Climate-induced displacement also disproportionately affects women, with a higher risk of migration to urban areas in search of livelihoods (Bhardwaj et al. 2024), accompanied by an increase in reports of gender-based violence (Reuveny 2007). Women and girls displaced by climate events are more likely to experience exploitation and trafficking (Sidun and Gibbons 2023). This challenge is further intensified by caste-based disparities as women from marginalized castes are less likely to have access to climate adaptation resources than their upper-caste counterparts (Khandekar et al. 2019). Additionally, climate-induced water scarcity increases the workload of women and adversely affects their health, particularly through waterborne diseases (Mahato 2024). Gender inequalities in resource access result in a 20% higher vulnerability to climate impacts for women compared to men (UN Women 2016).

In the Indian context, rural women are disproportionately affected by limited access to resources, such as clean water, and spend around fifty minutes daily fetching water, compared to twenty minutes for men, significantly reducing time for economic and educational pursuits (Choudhuri and Desai 2021). Women farmers hold just 13% of land titles (Oxfam 2013), which limits their access to credit, technology, and agricultural extension services. Alarmingly, only 3% of climate finance supports women-led initiatives despite evidence of their effectiveness in sustainable community-based adaptation (USAID 2023). Notably, just 5 out of 28 states in India follow a gender-transformative approach (Singh et al. 2021). Therefore, women face a dual burden: heightened vulnerability to climate impacts and systematic exclusion from resource allocation processes.

Addressing these gender disparities in climate adaptation is essential for fostering sustainable and equitable outcomes. As global temperatures continue to rise, integrating gender-specific measures into climate adaptation and resilience strategies becomes critical for enhancing productivity, promoting social equity, and ensuring the holistic development of communities.

## *Gaps in Climate Finance for Farmers*

Agriculture is both a significant contributor to and a casualty of climate change, responsible for approximately 20% of global greenhouse gas (GHG) emissions, with livestock playing a major role (FAO 2020). Small and marginal farmers, who form the majority in India, are disproportionately affected, with smallholders being 40% more vulnerable to climate variability and extreme weather events compared to large landholding farmers (FAO 2015). The promotion of climate-smart agriculture (CSA) is essential for sustainability, particularly for vulnerable groups, such as women farmers.

Though very little work is done in this field, some of the isolated studies show that despite the urgency for CSA adoption, smallholder farmers refrain from implementing these practices due to limited access to finance (Mizik 2021). Investments in climate-smart inputs, technology, and infrastructure are required, but many farmers, especially women, lack the financial capacity for such expenditures. Approximately 90 million smallholder farmers in India, or three out of five farmers, lack access to institutional credit (NABARD 2017). Although efforts, such as bank nationalization, to encourage priority sector lending and the rise of NBFCs have been made, barriers persist, including the absence of credit scores, geographic isolation, and limited financial literacy.

Challenges are even more significant for women farmers, who often do not hold legal land titles, further restricting formal loan access. Only 10% of women in India have access to credit services, limiting their ability to invest in climate adaptation, contributing to a 30% reduced capacity to respond to climate impacts due to their restricted credit and agricultural extension services (Oxfam 2018).

## *Gender Gaps in Access to Agricultural Inputs and Mechanization*

India's agricultural productivity falls below global standards, affected by fragmented and smaller landholdings, inequitable access to inputs, outdated agronomic practices, and low mechanization.

Women, constituting over 60% of the agricultural workforce, face systemic disadvantages in accessing essential resources like bio-fertilizers, quality seeds, irrigation systems, and climate-resilient technologies (FAO 2019). Research by NABARD (2021) highlights gender biases in distribution channels, favoring men for government subsidies, formal credit, and private supply chains. As a result, women often rely on lower-quality or traditional inputs, limiting their productivity and resilience.

Women encounter constraints in input purchasing due to dual household and farm responsibilities. Studies show they spend 2.5 h more on unpaid care work than men each day, limiting market participation (UN Women 2019). Additionally, limited access to vehicles and the challenge of transporting heavy inputs reduce their decision-making and active involvement.

Mechanization exacerbates these disparities. While programs like Pradhan Mantri Krishi Sinchai Yojana aim to promote mechanization, only 7% of women farmers access equipment, such as tractors and drip irrigation systems (ICAR 2020). This exclusion hinders productivity and the adoption of climate-resilient practices like precision agriculture, which is vital for areas facing severe environmental degradation. These barriers to inputs and technology significantly contribute to India's low productivity, disproportionately impacting women already at higher risk from climate change.

## *Limited Access to Digital Advisories, Weather Forecasts, and Capacity-Building Programs*

Access to real-time weather forecasts and agricultural advisories is crucial for climate-resilient farming, yet rural women face significant digital barriers. GSMA (2020) and FAO (2020) report that rural women are far less likely to own mobile phones or access digital platforms, hindering their ability to receive critical updates on weather, pest outbreaks, and market conditions. In rural India, the mobile ownership gap between men and women is 28%, with a 56% gap in mobile Internet usage (GSMA 2020). This leaves women particularly vulnerable to unpredictable weather, impacting their ability to manage crops effectively.

In rain-fed regions like Rajasthan, access to weather forecasts can make or break harvest success. However, the government's digital advisory services, such as mKisan and Kisan Suvidha, are underutilized due to digital illiteracy, infrastructure gaps, and cultural barriers (Dhaliwal et al. 2021). ICRISAT (2020) found that when women accessed mobile-based weather services, climate-smart practice adoption increased by 40%, yet fewer than 20% had access. This gap restricts their ability to adapt and mitigate risks.

Limited digital access also impedes women's ability to receive guidance on pest control, fertilizer use, and crop diversification, which is critical for climate resilience. In cyclone-prone areas like Odisha and West Bengal, timely weather advisories could have prevented significant crop losses (FAO 2020). Gender-insensitive digital initiatives further exclude women, whose needs are often overlooked in platform design.

Women have less access to agricultural extension services compared to men, limiting their adaptation capabilities (Ragasa et al. 2013). Only 15–33% of women report attending agricultural training, hindered by limited transportation, cultural and social barriers to participate in male-led programs, and extensive household duties (IDH Sustainable Trade Initiative 2019). These responsibilities, along with event timing, location and duration concerns, lower literacy, and digital literacy gaps, reduce participation. Women in mixed-gender training are often hesitant to engage, underscoring the need for gender-sensitive training and tailored awareness to bolster resilience.

## *Gender Gaps in Access to Risk-Mitigating Tools*

Access to crop insurance is vital for resilience against climate uncertainties. Only 5% of rural women in India have agricultural insurance compared to 14% of men (Global Insurance Index 2020) because most schemes like *PMFBY* have land ownership as a qualifying requirement. Women are automatically excluded, as only 13% of women hold land titles in India (Agricultural Census 2015–2016). Awareness gaps, cultural norms, and restricted mobility further limit access to insurance agents and programs (Oxfam India 2020). Health insurance coverage is also low at 14% (NFHS-5 2021), leaving women vulnerable to climate-related health issues. In flood-prone regions like Odisha and Chhattisgarh, poor access to weather and accident insurance drives women toward high-interest informal credit (ILO 2019). This exclusion hampers the adoption of climate-resilient practices. In states like Madhya Pradesh and Maharashtra, women without crop insurance are 30% less likely to engage in climate-smart farming than men (ILO 2019). Addressing these gaps is essential to advancing gender equity and building long-term resilience.

## *Limited Access to Storage Services and Commodity Markets*

Limited access to storage facilities and competitive markets significantly hinders women in agriculture, preventing them from maximizing their production gains. Post-harvest losses account for 10–15% of India's annual yields, disproportionately affecting women due to inadequate storage infrastructure (NABARD 2017).

Women smallholders report higher losses from immediate post-harvest sales (Nordhagen 2021). Cultural norms, male-dominated networks, and restricted mobility confine women to informal local markets, where prices are 20–30% lower than those obtained by men (International Trade Centre 2018). In horticulture-intensive regions like Andhra Pradesh and Tamil Nadu, limited market access exacerbates economic vulnerability, forcing women to accept exploitative prices (FAO 2020). Access to formal markets remains critical; women engaged in organized marketplaces can increase their incomes substantially (FAO 2023). Limited participation in Farmer Producer Organizations (FPOs) further reduces bargaining power and access to modern storage, perpetuating income losses, reduced productivity, and reliance on exploitative intermediaries. Addressing these challenges is essential to enhancing women's economic resilience and advancing climate-resilient agricultural practices.

### *Challenges in Accessing Capital for Entrepreneurship and Income Diversification*

As climate change heightens agricultural unpredictability, income diversification becomes crucial as part of climate-smart agriculture, especially for women. Traditional agriculture, vulnerable to erratic weather and water scarcity, is no longer reliable. Activities, such as dairy, poultry, and small-scale entrepreneurship, offer more stable income streams. However, without access to capital and training, women are largely excluded from these opportunities. The Government of India's Committee on Doubling Farmers' Income emphasizes diversification, yet only 5% of rural women access formal credit (Global Findex Database 2017).

Women report challenges in securing loans for dairy and poultry sectors that could provide year-round income due to gender and social norms (Accion 2024). Women's access to training for non-agricultural enterprises is also limited. In Odisha and West Bengal, women in the fishery value chain often lack formal training in fish farming and market linkages (ICAR 2020). The Microfinance Institutions Network (MFIN 2019) states that rural women are 38% less likely than men to secure microloans, and available loans often fall short of covering start-up costs for ventures like value-added processing or agribusiness. This financial exclusion restricts women's ability to engage in profitable, climate-resilient income streams, perpetuating low agricultural earnings and vulnerability to climate risks. Without effective income diversification, rural households face compounded economic challenges, thwarting efforts to achieve sustainable livelihoods (FAO 2020).

## *Decision-Making at the Household Level and Financial Literacy*

Household decision-making power greatly influences the economic outcomes of women farmers. Despite contributing significantly to agricultural labor, women are often excluded from decisions regarding crop choices, input purchases, sales, and finances. Cultural norms uphold male authority, relegating women to unpaid or low-status roles, resulting in only 10–30% participation in decision-making. This exclusion impacts their ability to implement climate-resilient practices and invest in diversified income sources (CGIAR 2022). In patriarchal regions, women often cannot act on strategies like planting drought-resistant crops without male approval (CIMMYT 2024). This limits the household's adaptive capacity and the use of all available knowledge.

Financial literacy among rural women is also low; only 27% have basic knowledge of financial products compared to 45% of men (World Bank 2017). This deficit hampers women's ability to access services, negotiate fair terms, and engage in financial planning. In states like Uttar Pradesh, Bihar, and Rajasthan, women's mobility is severely restricted due to gender norms, making them reluctant to visit bank branches, exacerbating reliance on moneylenders who charge high interest (FAO 2024). This dependency undermines financial security and autonomy, making women more vulnerable to economic shocks from climate change.

Limited roles in household decision-making mean income allocation often prioritizes immediate needs over long-term investments in sustainable practices. Men may favor traditional methods, hindering climate adaptation. When women are excluded, the household cannot effectively address climate challenges. Studies by FAO (2020) show that empowering women in decision-making and financial literacy training boosts household productivity and climate resilience. Addressing these gender gaps is critical for sustainable rural development and effective climate response.

## Rationale of the Study

This study explores interventions aimed at helping women farmers overcome these barriers, focusing on integrated models promoting financial inclusion, market connectivity, and capacity building for climate-smart agriculture. By examining initiatives by Dvara E-Registry, the research investigates how comprehensive approaches empower women as active contributors to climate resilience. Access to financing, input supplies, market linkages, and training programs are emphasized as essential components of effective adaptation. The aim is to assess the impact of data-driven interventions and analyze lessons from successful interventions to identify best practices for creating inclusive agricultural models that enhance women's participation and decision-making.

## Methodology

This study employed a multi-method approach to evaluate the effectiveness of interventions aimed at enabling women farmers to adopt climate-resilient practices. The research emphasized holistic, data-driven solutions and integrated both qualitative and quantitative methodologies to gain a comprehensive understanding of the challenges women face and the efficacy of targeted strategies.

### *Study Area*

Six Indian states—Uttarakhand, Rajasthan, Madhya Pradesh, Maharashtra, Odisha, and Manipur—known for their varied agricultural practices and climate vulnerabilities were selected for the study, where 52 FPOs were formed (Fig. 15.1). The key methods included literature reviews, case studies, and data analysis. An aggregation model centered on FPOs has been adopted, comprising 500 to 1000 farmers.

### *Data Collection and Analysis*

The research combined primary data from interviews with women farmers and secondary data from existing literature and reports. Quantitative data from DER's digital platform Doordrishti was analyzed, engagement with FPO board members, government agencies, financial institutions, and non-governmental organizations, and an AI-based language tool was utilized to enhance the clarity and precision of the article. The various activities, interventions, services and outcomes observed in the study are shown in Fig. 15.2.

## Results and Discussion

In India, 76% of employed women work in agriculture, contributing to 63% of female labor participation (Labour Force Survey 2022). To cater for this challenge, the establishment and support of twelve women-only and women-led Farmer Producer Organizations (FPOs) have empowered women-led collectives, enabling them to independently manage agricultural processes and market engagement (Harrington et al. 2024). The presence in six states across India highlights the regional adaptability and responsiveness of the interventions to local agricultural conditions.

Women-led FPOs not only offer women specialized training, capacity-building programs, and tailored financial and market linkage services but also strengthen their resilience against the adverse effects of climate change (Jhansi et al. 2024; Bryan

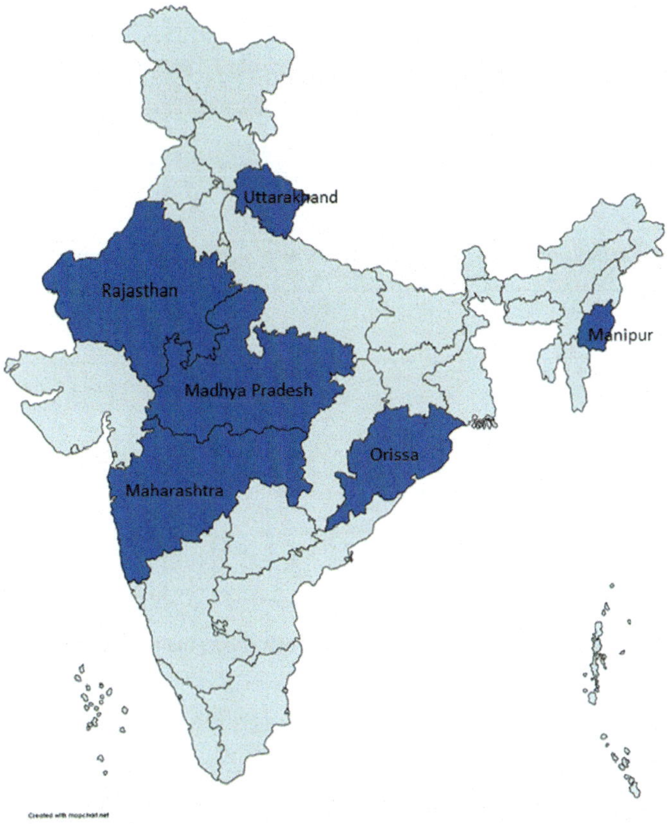


**Fig. 15.1** Map showing Indian states selected for study. *Source* Created by Authors, 2025

et al. 2023). To further support women's empowerment, women-only and women-led FPOs (Fig. 15.3) have been established, with twelve such organizations currently in operation. The implemented interventions have demonstrated significant impacts in promoting climate-resilient agricultural practices, with a strong emphasis on empowering women farmers and fostering inclusivity across diverse farmer communities. This section provides an in-depth discussion of the outcomes across various key impact areas.

## *Digital Technology and Data-Driven Services: Holistic Solution to Farmers and FPO Needs*

Data-driven platform—"Doordrishti", which delivers customized and climate-resilient solutions—collects detailed information on farmers, such as their KYC data,

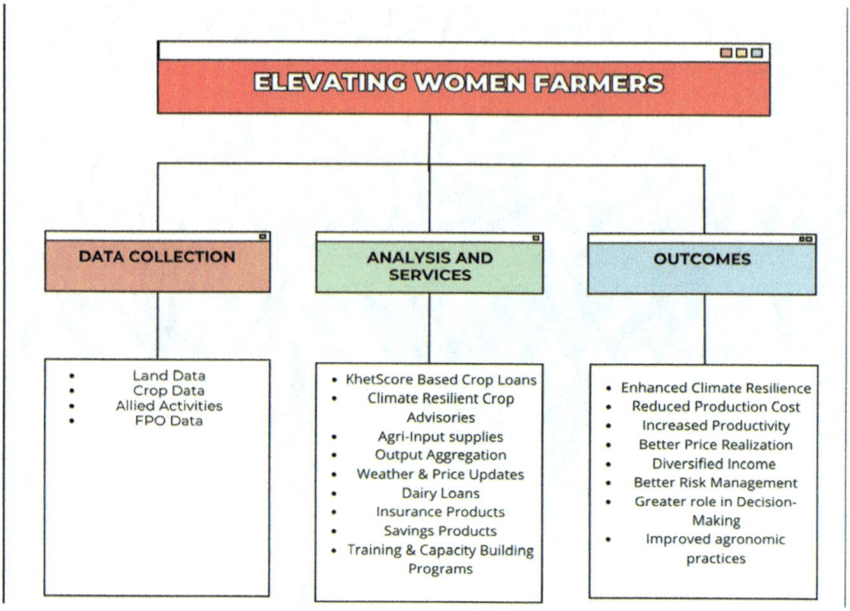


**Fig. 15.2** Conceptual analysis of study. *Source* Author's conceptualization, 2025

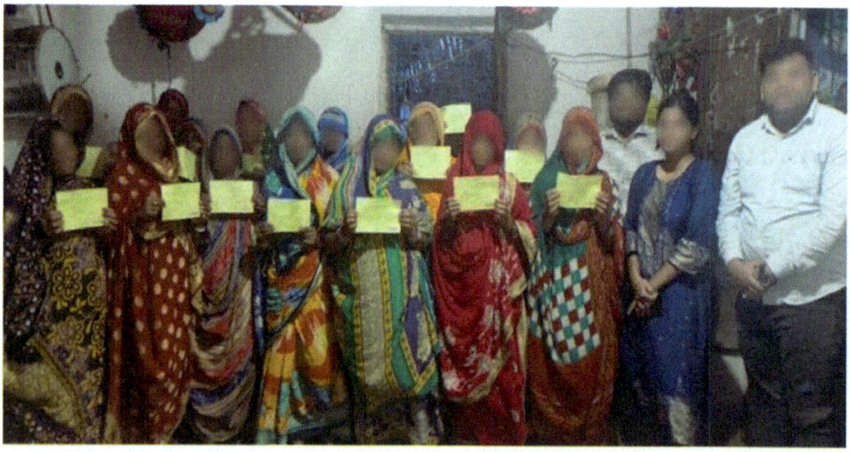

**Fig. 15.3** Membership mobilization in FPOs. *Source* Field photograph by Dvara E-Registry, 2024

land and irrigation status, crop information, and soil quality. It has three interfaces—FPO Dashboard, Farmer App, and Field Representative App (Fig. 15.4). Detailed information, such as crop type, variety, sowing date, and crop duration, is gathered from each farmer to develop customized crop advisories tailored as per soil condition, nutritional requirements, crop sowing date, and local weather forecast, which

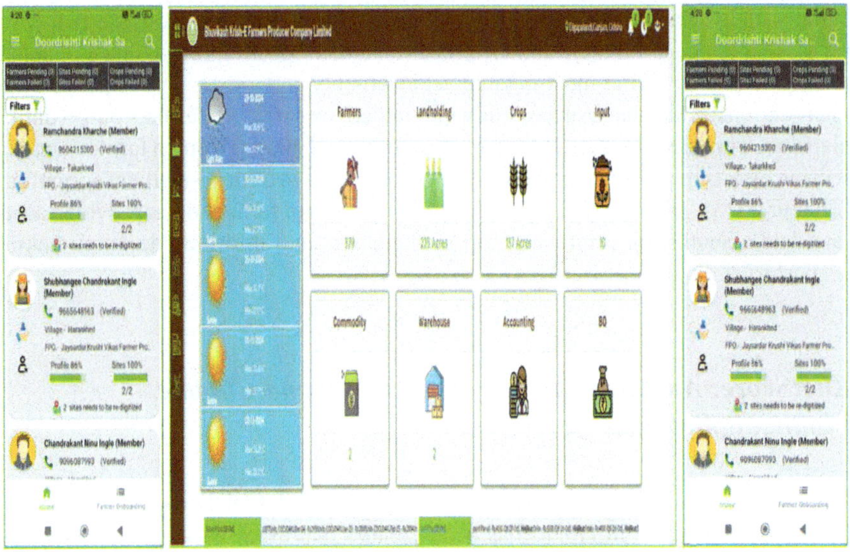


**Fig. 15.4** Interface of Krishak App (Farmer App), FPO Dashboard, and Krishak Sathi App (Farmer's friend App). *Source* Screenshot of FPO dashboard and apps by Dvara E-Registry, 2025

are delivered via SMS for feature phone users, and through notification for App users, ensuring accessibility for all and promoting climate-smart practices.

Using the Doordrishti FPO Dashboard, FPOs monitor their member farmers' field activities at various levels. Real-time data allows FPOs to coordinate backwards and forward market connections based on farmers' input requirements and their marketable surpluses. The platform provides functionality to FPOs to enter information about their agri-input products for sales and agri commodities for purchase through a virtual shop on Dashboard, through which farmers receive notifications in the Krishak mobile app and SMS alerts.

Furthermore, it also provides detailed information on Warehouse Development and Regulatory Authority (WDRA) approved warehouses in the vicinity. It shows complete information on warehouses, with geospatial mapping with latitude–longitude coordinates for precise distance calculation from the farmer location, along with data on storage capacities, addresses, and warehouse manager contact details, streamlining communication for produce storage. WDRA-approved warehouses offer benefits, such as electronic negotiable warehouse receipts (eNWRs), which maintain produce quality and help farmers avoid distress sales during low market prices. Farmers can leverage eNWRs to access commodity financing at competitive rates, improving liquidity and strengthening financial resilience among women farmers.

This model benefits women farmers who face travel constraints due to household duties, allowing them to engage in purchase and sales activities through local aggregation points. Krishak mobile app also provides spot price data for commodities and futures prices from platforms like NCDEX, empowering farmers to negotiate better with local aggregators. Access to accurate market prices is particularly helpful for women, enhancing their ability to negotiate and boosting profitability (Gautam et al. 2024). One of the most notable achievements is the extensive reach of these interventions, which have successfully onboarded over 100,000 farmers onto a digital platform. This includes active participation from over 40,000 women farmers, underscoring efforts to integrate women as key stakeholders in climate-resilient agriculture (Thakur 2023). The digitization of over 175,000 acres of farmland has provided accurate data essential for improved crop management and informed decision-making (Papaskiri et al. 2020).

## *KhetScore: An Alternative Credit Scoring and Climate Financing Solution*

To bridge the financing gap, an innovative analytical tool known as KhetScore (meaning farm score in Hindi) has been created that assesses farmers' cultivation activities in the field (Fig. 15.5). The score considers factors like the number of crop seasons taken in the last three years, farm productivity, the health of the crops, and the impact of climate change on the location. It is done by utilizing remote sensing (RS), artificial intelligence (AI), and in-house algorithms. KhetScore provides insights on farmers' cash flow through their cropping activities and evaluates risk associated with lending due to climate change in the geography by monitoring parameters like temperature, rainfall, events of floods and droughts, etc. This provides an alternate credit score option for those farmers who are without traditional credit scores.

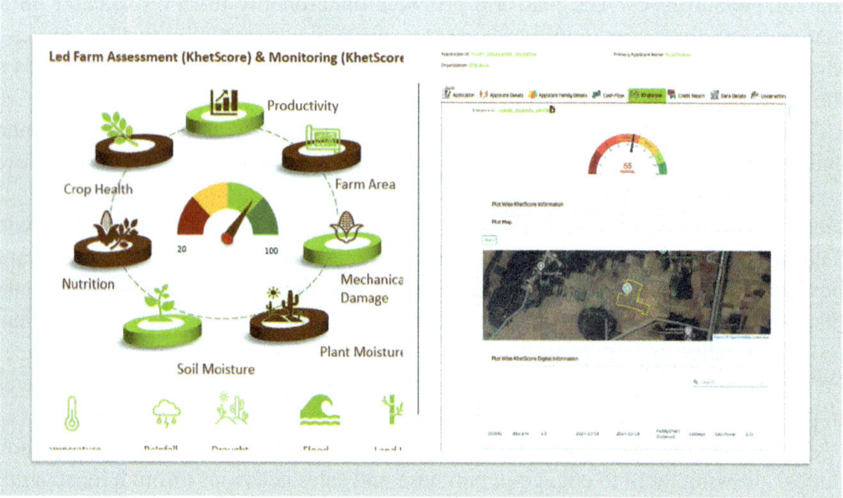


Fig. 15.5 KhetScore parameters and sample score report. *Source* Screenshot by Dvara E-Registry, 2025

The KhetScore analytics uses the following key data points as input:

i. Geo-referenced plot details for land verification,
ii. Cross-checking with government survey records (in select states),
iii. Land-use classification to confirm agricultural activity,
iv. Historical temperature, rainfall, drought, and flood impacts on land and crops,
v. Seasonal productivity assessments covering crop health, soil and plant moisture, nutrient status, and mechanical damage,
vi. Area under cultivation.

KhetScore integrates over 200 data points across eleven parameters having different weightages in the score, employing advanced remote sensing indices, such as NDVI, GNDVI, NDMI, NMDI, and VHVV. Score leverages multispectral and other data to assess agricultural conditions and, consequently, the creditworthiness of farmers.

KhetScore ratings range from 20 to 100, with different score range indicating different risk levels like: (i) 20–40: very high risk; (ii) 40–60: high to moderate risk; (iii) 60–80: moderate to low risk, and (iv) 80–100: very low risk. Farmers scoring above 40 are eligible for credit, while those scoring lower receive guidance and advisory services and targeted inputs aimed at enhancing farm productivity and, subsequently, farmer creditworthiness. KhetScore enables financial institutions to extend credit for climate-resilient investments even when conventional credit scores are unavailable. Once a farmer is considered to be creditworthy by the use of KhetScore, documentation is hassle-free, paperless, and conducted at the doorstep, leading to women farmers gaining convenient credit facilities. The generation of more than 50,000 KhetScore has been pivotal in providing farmers with critical financial insights, improving their credit access and fostering financial inclusion (Kramer et al. 2024).

A study by IFPRI found that KhetScore loans, bundled with insurance, have a wide range of significant impacts that benefit participants, and it increased overall utilization of credit, especially among women. Much of this overall increase came from an expansion in formal credit uptake by about 20% on average and by roughly 26% among women and not merely from a shift from informal to formal credit. Farmers' revenues and profits increased per acre by more than eight thousand rupees (USD 110) and nine thousand hundred rupees (USD 120), respectively. Farmers increased their cultivated area by roughly 1.3 acres, while the total cost of production per acre fell by nine thousand six hundred rupees (USD 130), perhaps as farmers generated economies of scale. As a result, farmers had significantly higher agricultural profits across both agricultural seasons, with Kharif profits nearly doubling those in the control group and Rabi profits offsetting losses suffered by farmers in the control group (Kramer et al. 2024).

To promote equitable financial inclusion, women in loanee families are required to serve as either the primary borrower or a co-borrower on all loans, ensuring their involvement in financial decision-making. Access to climate-resilient financing through KhetScore has been substantially enhanced, with more than a hundred forty crore rupees (USD 16 Million) disbursed through 30,000+ collateral-free crop loans.

Notably, 36% of these loans have been directed to women as primary borrowers and the rest as co-borrowers, promoting gender equity in financial access (Pal et al. 2022). The inclusivity of these efforts is evident, with 95% of loan beneficiaries being small and marginal farmers and 85%+ belonging to SC/ST and OBC communities. This inclusive financial strategy supports the most vulnerable groups, expanding their opportunities to invest in climate-resilient practices. Additionally, reductions in borrowing costs by 20–50% and the onboarding of 45% of new-to-credit farmers have further expanded financial inclusion, positioning these interventions as transformative for rural financial systems. It was observed that 74% of farmers who accessed credit demonstrated an improvement in their KhetScore. This indicates that the credit was utilized to purchase higher-quality agricultural inputs, which enhanced climate resilience, leading to increased productivity and higher profits.

## *Promoting Savings and Mitigating Risks*

### Digital Gold Savings Products

To encourage savings among women farmers, digital gold savings products have been introduced, recognizing gold's cultural and economic importance in India. Partnerships with financial agencies allow women to invest in digital gold from just a hundred rupees (USD 1.1), only fostering saving habits without large capital. This investment builds stable savings and serves as collateral for low-interest loans, reducing debt risk.

### Weather Insurance Products

By collaborating with insurance providers, weather insurance products are introduced to address increasing temperature variability and erratic rainfall, which provide up to 20 times the premium in coverage, supporting women farmers against climate uncertainties.

### Health Insurance Policies

To mitigate health risks exacerbated by climate change, affordable health insurance policies have been developed through partnerships. Starting at forty rupees only (USD 0.45), these policies cover hospitalization costs, ensuring women can receive necessary medical health care. Coverage includes illness and accidents, protecting families from financial burdens that could deepen poverty and limit investments in farming.

## Allied Activities and Value-Added Products: Diversifying Income for Resilience

To diversify income sources and lessen reliance on crop farming, other agricultural allied activities like livestock rearing, fisheries, poultry, and goat farming are promoted, focusing on empowering women who often oversee these activities.

In Manipur, women received support for poultry and piggery farming, including high-quality feed, day-old chicks, and training in sustainable practices. Model farms provide climate-smart livestock management training. In Maharashtra, partnerships with a state university and agriculture department train cattle farmers to boost dairy productivity. The establishment of five dairy FPOs in Amravati provides financing, training, input supplies, and market connections to women in dairy farming.

In Odisha, a project promotes sustainable fisheries for women managing farm ponds. An Aqua Lab monitors pond conditions by testing dissolved oxygen, ammonia, alkalinity, plankton, and pH levels. Customized advisories based on these results help optimize fish health, reduce mortality, and increase income.

## Market Connectivity and Value Addition

Small and marginal farmers encounter substantial challenges in accessing profitable markets (Mandala et al. 2021). Limited surplus production often restricts their ability to engage in value addition, which could otherwise yield higher returns. The integration of market connectivity strategies has effectively improved price realization and profitability for farmers. Over one hundred output buyers have been connected with FPOs, facilitating better market access and ensuring competitive pricing. This has led to a 2–5% increase in price returns for farmers participating in FPO-led sales. The price dissemination service has been particularly beneficial, with more than 70% of farmers acknowledging its role in enhancing local negotiations.

Women-led value-addition activities, including pickling, papad making (thin Indian wafer), oil extraction, and seed production (Fig. 15.6), have created supplementary income opportunities by converting raw produce into higher-value products and contributed to more sustainable, diversified agricultural business models. As registered entities, FPOs sell the produce in major APMC markets that offer better prices, and they also list their value-added products on online platforms, such as ONDC and Amazon, providing access to customers nationwide—a reach that individual farmers rarely achieve independently. Around 51 FPOs have listed their value-added products on online platforms, such as ONDC and Amazon, expanding their market reach and income potential.

**Fig. 15.6** Value-added products. *Source* Field photograph by Dvara E-Registry, 2025

## *Leveraging Efficient and Effective Agricultural Input Supply*

Through the digitization of land parcels, detailed data on farm operations, including crop types, varieties, sowing dates, and crop duration, is systematically collected. This data enables precise estimation of agricultural input demand at farm, village, and FPO levels. By aggregating input demand, bulk purchases are negotiated at wholesale prices, ensuring affordable access to climate-resilient agricultural inputs for smallholder farmers. Currently, over two hundred agricultural input manufacturers have partnered with FPOs, which serve as licensed local hubs (Fig. 15.7) equipped with credit facilities, ensuring the timely delivery of high-quality inputs. Daily dissemination of commodity prices from 4000+ APMC markets has enabled farmers to make informed marketing decisions, boosting their negotiating power and profitability (Aliyi et al. 2021).

Through the Krishak mobile application, FPOs operate virtual shops where products and services are listed for easy access. Women farmers can browse these options from home, overcoming distance barriers to buy the products and services that best meet their needs (Fig. 15.8). Additionally, women can express interest in items not listed, and the FPO arranges delivery, expanding their purchasing choices beyond limited local retail options.

The strategic connection between FPOs and Agri-input suppliers has yielded business of more than thirty-seven crores rupees (USD 4.2 Million) and resulted in a 10%

**Fig. 15.7** Agricultural input Shop of FPO. *Source* Field photograph by Dvara E-Registry, 2025

**Fig. 15.8** Seed distribution at Farmer's doorstep. *Source* Field photograph by Dvara E-Registry, 2025

reduction in input costs, easing the financial burden on smallholder farmers (Gurung et al. 2023). With partnerships with agri-input suppliers and bio-product companies, a reliable supply chain for quality and eco-friendly inputs has been established. This network supports sustainable farming practices and bolsters the resilience of agricultural systems to environmental stressors.

To mitigate methane emissions from livestock management, FPOs also provide access to high-quality feed and veterinary products, enhancing livestock health and productivity, increasing household income, and reducing environmental impact. FPOs help overcome barriers to mechanization by offering pay-per-use access to equipment, such as tractors, land levelers, harvesters, and threshers, through village-based common service centers (CSCs), giving women farmers benefit from affordable, localized services that decrease labor demands and improve crop yields.

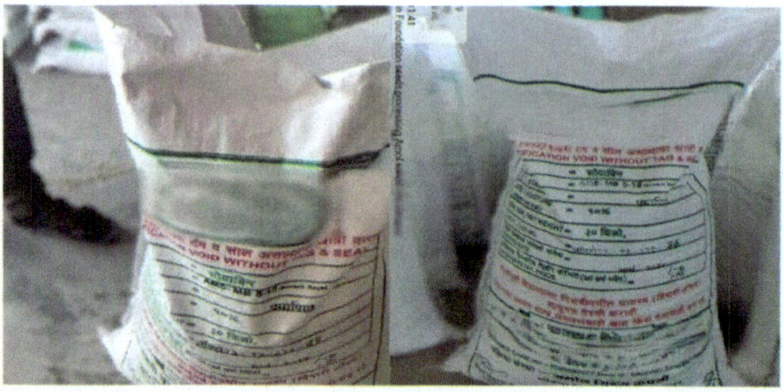

**Fig. 15.9** Soybean Seed bags produced by FPO. *Source* Field photograph by Dvara E-Registry, 2025

## *Incentivizing Climate-Smart Practices*

### Seed Multiplication Program for Climate Resilience

Low seed replacement rates and limited use of climate-resilient seeds hinder farm productivity. To address this, seed multiplication programs have been developed for FPOs in partnership with the International Rice Research Institute (IRRI), State Agricultural Universities, and private seed companies across various states to benefit their member farmers, especially women farmers who have less access to quality inputs. These programs focus on essential crops like *paddy, wheat, groundnut, soybean, and Bengal gram,* promoting the use of breeder and foundation seeds engineered to endure extreme temperatures, water scarcity, and saline soils (Fig. 15.9). The Seed Multiplication Program has involved five major crops and benefited 13 FPOs, facilitating access to high-quality, climate-resilient seeds (Vernooy et al. 2022).

### Alternative Transplanting Method

Adoption of climate-smart practices, such as direct-seeded rice (DSR), an alternative to the traditional transplanting method, is introduced to conserve more water and reduce methane emissions. Since women contribute more than 50–80% of the labor in paddy cultivation (SRI-2030), training programs are tailored to equip them with the necessary skills to implement these practices, empowering women as active participants in sustainable agriculture. Later, these farmers can be incentivized by the creation and sale of carbon credits. This kind of monetary benefit can help bring behavioral change, encouraging them to continue with climate-smart practices. Direct-seeded rice (DSR) has been adopted across 20,000+ acres, saving approximately 100 billion liters of water and reducing $CO_2$ emissions by approximately 20,000 tons every year.

**Fig. 15.10** The "Seed Cafeteria", where 21 Varieties of rice are grown on 1 acre of farmland. *Source* Field photograph by Dvara E-Registry, 2025

**Promoting Biodiversity for Climate Resilience**

Biodiversity in agriculture enhances resilience to climate change by reducing risks and improving ecosystem health. In collaboration with the International Rice Research Institute (IRRI), bio-diversification initiatives were taken by the adoption of nitrogen-fixing crops, such as Dhaincha, pulses, and oilseeds. Demonstration plots, referred to as "Seed Cafeterias," showcase 20–25 crop varieties (Fig. 15.10), allowing farmers to evaluate the performance of each crop under local conditions. To facilitate adoption among women farmers, local women leaders at the Gram Panchayat level coordinate these activities, educating their communities on the benefits of bio-diversification and empowering them in agricultural decision-making (Fig. 15.10).

**Soil Health Card**

To support soil health, water conservation, and climate resilience, soil testing services are provided through FPOs using specialized testing machines that analyze twelve nutrient profiles of soil. These assessments yield precise recommendations for fertilizer use, preventing overuse, protecting soil quality, reducing nutrient imbalances, and lowering greenhouse gas emissions. Biological products that control pests and diseases while enhancing soil health are also promoted (Fig. 15.11).

**Fig. 15.11** Demonstrating bio-agricultural inputs for plant root treatment. *Source* Field photograph by Dvara E-Registry, 2023

**Weather Forecasting Services**

Localized weather forecasting services have covered 300 + blocks, equipping farmers to better prepare for climate variability, while personalized advisories have reached over 7,500 women farmers, enhancing their agronomic knowledge and productivity (Schwalbert et al. 2020).

## *Gender-Inclusive Capacity Building and Training*

Capacity-building efforts have been instrumental in promoting the adoption of sustainable practices (Piñeiro et al. 2020). The establishment of 11 village-level soil testing labs has enabled scientific soil management and tailored fertilizer recommendations, resulting in optimized input use and enhanced yields (Havlin 2020). Digital tools increase women's awareness of climate risks, reduce crop losses, and improve productivity through timely interventions (Balasundram et al. 2023). Over 4 lakh customized advisories have been issued to encourage the adoption of climate-smart practices tailored to specific local conditions (O'Grady et al. 2021). More than 60% of farmers have reported benefits from customized advisories, citing reduced cultivation costs and improved productivity.

Limited access to training and agronomic resources has hindered the adoption of climate-resilient practices, especially among women. To address this, targeted advisories and capacity-building programs are offered to men and women in FPOs. These include training sessions, field demonstrations, vernacular videos, and exposure visits to boost modern agronomic knowledge. Women farmers are trained to use

**Fig. 15.12** A digital training session for women farmers. *Source* Field photograph by Dvara E-Registry, 2025

mobile apps for market prices, weather forecasts, and storage locations, improving interactions with FPOs (Fig. 15.12).

## *Comprehensive Impact on Women Farmers and Marginalized Groups*

These integrated interventions have emphasized gender equity, promoting women as central players in climate-resilient farming and decision-making. With significant representation in both FPO participation and financial access, women farmers have seen notable increases in their agronomic knowledge, productivity, and financial autonomy (Jhansi et al. 2023). The targeted efforts toward marginalized communities, particularly Schedule Caste (SC)/Schedule Tribe (ST) and Other Backward Class (OBC)[1] group demonstrate an inclusive approach that supports the most vulnerable sections of the rural society.

It is observed that female farmers are 30% more likely to report making contributions to household decisions regarding borrowing and about 25% more likely to report making contributions to household decisions about how to use borrowed money. What is encouraging is to see that the beneficial effects on women's empowerment have spilt over to other female household members (Kramer et al. 2024).

The results illustrate that a comprehensive, multi-faceted approach combining digital platforms, financial innovations, capacity building, and market integration can significantly enhance the resilience and profitability of smallholder farmers (Atanga

[1] Scheduled Castes (SC), Scheduled Tribes (ST), and Other Backward Classes (OBC) are historically disadvantaged social groups in India, recognized by the government for affirmative action in education, employment, and political representation to promote social equality.

2020). The active engagement of women farmers and marginalized groups underscores the importance of inclusivity in climate-resilient agricultural interventions. The widespread adoption of these practices not only strengthens economic stability but also contributes to broader goals of gender equity and sustainable development.

Through these concerted efforts, India's agricultural sector is better positioned to meet the challenges posed by climate change, with women farmers and disadvantaged communities playing a central role in this transformation. Continued investments in digital infrastructure, targeted financial products, and comprehensive advisory services are essential for sustaining these positive outcomes and promoting further advancements in climate-resilient agriculture.

## Conclusion

The interventions discussed in this chapter represent a transformative approach to enhancing climate resilience in agriculture, particularly focusing on empowering women and marginalized groups. Comprehensive efforts integrating digital tools, financial inclusion, and capacity building have demonstrated significant positive impacts. The Doordrishti digital platform and initiatives like KhetScore have revolutionized financial access, while targeted training programs have improved knowledge and adoption of sustainable practices. KhetScore credit-scoring methodology, notably its ability to reduce transaction costs for both borrowers and lenders, serves to mitigate credit rationing while simultaneously broadening and enhancing the inclusivity of rural finance (Kramer et al. 2024) in addition to opening up the scope for climate financing in agriculture.

Women's empowerment is central to all interventions and outcomes. Through gender-specific FPOs and inclusive policies, women have gained agency, contributing to increased productivity, financial stability, and broader community resilience. Data-driven advisories and improved market connectivity have further bolstered women's participation, addressing traditional barriers to resource access and decision-making.

These findings underscore multi-faceted strategies, encompassing technology, education, and policy support, crucial for sustainable agricultural development. Scaling these successful models can strengthen resilience across broader regions, enhancing economic security and gender equity. Continued investment and adaptive policy frameworks which adopt and promote holistic models like these are vital to maintaining momentum and ensuring the agricultural sector's stability in an era of climate uncertainty.

## References


Accion (2024) Growing women's role in dairy value chains. Available from https://www.accion.org/growing-womens-role-in-dairy-value-chains/

Agricultural Census Division, Department of Agriculture, Cooperation & Farmers Welfare, Government of India (2019) Agricultural census 2015–16: All India Report on number and area of operational holdings. Available from: https://www.fao.org/fileadmin/templates/ess/ess_test_folder/World_Census_Agriculture/WCA_2020/WCA_2020_new_doc/IND_REP_ENG_2015_2016.pdf

Aliyi I, Faris A, Ayele A, Oljirra A, Bayessa M (2021) Profitability and market performance of smallholder vegetable production: evidence from Ethiopia. Heliyon 7(9):e08008. https://doi.org/10.1016/j.heliyon.2021.e08008

Anderson B, Borgonovo E, Galeotti M, Roson R (2014) Uncertainty in climate change modeling: can global sensitivity analysis be of help? Risk Anal 34(2):271–293. https://doi.org/10.1111/risa.12117

Atanga SN (2020) Digitalization of agriculture: how digital technology is transforming small-scale farming in Ghana. Agrarian, Food and Environmental Studies (AFES). Available from: https://thesis.eur.nl/pub/55705/Atanga-Sam-Nicholas_MA_2019_20_AFES.pdf

Balasundram SK, Shamshiri RR, Sridhara S, Jeganathan N, Thiruchelvam S, Teh CBS et al (2023) The role of digital agriculture in mitigating climate change and ensuring food security: an overview. Sustainability 15(6):5325. https://doi.org/10.3390/su15065325

Bhardwaj LK, Rath P, Jain H, Purohit S, Yadav P, Singh V (2024) The impact of climate-induced livelihood, health, and migration on women and girls. In: Marketing the green school. IGI Global, pp 100–118. https://doi.org/10.4018/979-8-3693-2806-4.ch006

Bryan E, Alvi M, Huyer S, Ringler C (2023) Addressing gender inequalities and strengthening women's agency in food systems. Glob Food Secur 34:100731. https://doi.org/10.1016/j.gfs.2023.100731

CGIAR (2022) Six practical steps to support women farmers to drive climate resilience. Available from: https://gender.cgiar.org/news/six-practical-steps-support-women-farmers-drive-climate-resilience

Choudhuri P, Desai S (2021) Lack of access to clean fuel and piped water and children's educational outcomes in rural India. World Dev 145:105535. https://doi.org/10.1016/j.worlddev.2021.105535

CIMMYT (2024) Bargaining for better: how gender roles in household decision-making can impact crop disease resilience. Available from: https://www.cimmyt.org/news/bargaining-for-better-how-gender-roles-in-household-decision-making-can-impact-crop-disease-resilience/

Dhaliwal A, Singh G, Singh S (2021) Digital inclusion of farmers and rural hinterland: the case of Digital India. In: The role of e-business during the time of grand challenges. Lecture notes in business information processing, vol 414. Springer, Cham, pp 91–110. https://doi.org/10.1007/978-3-030-79454-5_9

Ebi KL, Capon A, Berry P, Broderick C, de Dear R, Havenith G, Honda Y, Kovats RS, Ma W, Malik A, Nybo L, Seneviratne SI, Vanos J, Jay O (2021) Hot weather and heat extremes: health risks. Lancet 398(10301):698–708. https://doi.org/10.1016/S0140-6736(21)01208-3

Food and Agriculture Organization of the United Nations (FAO) (2015) Climate change and food security: risks and responses. Available from: https://openknowledge.fao.org/handle/10986/22977

Food and Agriculture Organization of the United Nations (FAO) (2019) Good practices for integrating gender equality and women's empowerment in climate-smart agriculture programmes. Rome. Available from: https://openknowledge.fao.org/handle/10986/22977

Food and Agriculture Organization of the United Nations (FAO) (2020) Climate-smart practices and the role of women in productivity gains. Available from: https://www.fao.org/climate-smart-agriculture-sourcebook/enabling-frameworks/module-c6-gender/c6-overview/en/

Food and Agriculture Organization of the United Nations (FAO) (2023) The status of women in agrifood systems. Rome. https://doi.org/10.4060/cc5070en
Food and Agriculture Organization of the United Nations (FAO) (2024) Rural women and financial inclusion—technical guidance note. Rome. https://doi.org/10.4060/cc9522en
Gautam S, Rai IM (2024) Smartphones in the hands of rural women: embracing informal learning for digital literacy. Adult Learn. https://doi.org/10.1177/10451595241263142
Global Findex Database (2017) Measuring financial inclusion and the Fintech revolution. World Bank, Washington, DC. https://doi.org/10.1596/978-1-4648-1259-0
GSMA (2020) Connected women: the mobile gender gap report 2020. Available from: https://www.gsma.com/mobilefordevelopment/wp-content/uploads/2020/05/GSMA-The-Mobile-Gender-Gap-Report-2020.pdf
Gurung R, Choubey M, Rai R (2023) Economic impact of farmer producer organisation (FPO) membership: empirical evidence from India. Int J Soc Econ 51(8):1015–1028. https://doi.org/10.1108/IJSE-06-2023-0451
Harrington T, Narain N, Rao N, Rengalakshmi R, Sogani R, Chakraborty S, Upadhyay A (2024) A needs-based approach to promoting gender equity and inclusivity: insights from participatory research with farmer-producer organisations (FPOs). J Soc Econ Dev 26(2):409–434. https://doi.org/10.1007/s40847-023-00280-x
Havlin J, Heiniger R (2020) Soil fertility management for better crop production. Agronomy 10(9):1349. https://doi.org/10.3390/agronomy10091349
IDH Sustainable Trade Initiative (2019) Participation of women in agricultural training. Available from: https://www.idhsustainabletrade.com
Indian Council of Agricultural Research (ICAR) (2020) Mechanization and soil health programs in India. Available from: https://www.icar.org.in
Intergovernmental Panel on Climate Change (IPCC) (2021) Climate change 2021: the physical science basis. Contribution of Working Group I to the Sixth Assessment Report of the Intergovernmental Panel on Climate Change. Cambridge University Press, Cambridge. https://doi.org/10.1017/9781009157896
International Crops Research Institute for the Semi-Arid Tropics (ICRISAT) (2020) Gender barriers in rural finance for climate adaptation. Available from: https://www.icrisat.org
International Labour Organization (ILO) (2019) The impact of financial exclusion on rural women. Available from: https://www.ilo.org
International Trade Centre (2018) Annual report 2018. Available from: https://www.intracen.org/sites/default/files/inline-files/Annual%20Report%202018_English.pdf
Jhansi B, Kalal AN, Rajeshwari N (2023) Empowering rural women through farmer producer organizations (FPOs). Arch Agric Allied Sci 8(2):203–210. Available from: https://www.researchgate.net/profile/Dr-K-Vignesh/publication/373927302_Archives_of_Agriculture_and_Allied_Sciences/links/65042b45a69a4e63180c9a32/Archives-of-Agriculture-and-Allied-Sciences.pdf#page=203
Jhansi B, Kalal AN, Rajeshwari N, Maraddi GN, Chitagubbi G (2024) Training need assessment of women members in selected farmer producer organizations (FPOs) of Karnataka. J Farm Sci 37(1):95–99. https://doi.org/10.61475/JFS.2024.v37i1.22
Khandekar S, Sahoo S, Mukherjee A et al (2019) Climate adaptation and caste disparities in India. Environ Justice 12(3):210–218. https://doi.org/10.1089/env.2019.0023
Kramer B, Pattnaik S, Ward PS, Xu Y (2024) Impacts of a digital credit-insurance bundle for landless farmers: evidence from a cluster randomized trial in Odisha, India. Paper presented at the 2024 Allied Social Sciences Association (ASSA) Annual Meeting, January 5–7, 2024, San Antonio, Texas
Labour Force Survey (2022) Annual report: Periodic Labour Force Survey (PLFS) 2021–2022. Available from: https://www.mospi.gov.in/sites/default/files/publication_reports/AnnualReportPLFS2021-22F1.pdf

Mahato N (2024) Advancing clean water and sanitation in India: aligning sustainable development goals for improved outcomes. AIJRA 9(1):23–30. Available from: https://www.ijcms2015.co/file/2024/aijra-vol-9-issue-1/aijra-vol-9-issue-1-23.pdf

Mandala GN, Sangode PB, Devi SA, Gandreti VRR (2021) Problems and constraints faced by farmers in financing and marketing of agricultural produce in India. Univ J Account Finance 9(2):139–144

Microfinance Institutions Network (MFIN) (2019) Annual report 2018–2019. Available from: https://mfinindia.org/assets/upload_image/publications/AnnualReports/Annual%20Report%20%202018-19.pdf

Mizik T (2021) Climate-smart agriculture on small-scale farms: a systematic literature review. Agronomy 11(6):1096. https://doi.org/10.3390/agronomy11061096

National Bank for Agriculture and Rural Development (NABARD) (2017) Financial inclusion for smallholder farmers in India. Available from: https://www.nabard.org

National Bank for Agriculture and Rural Development (NABARD) (2021) Agricultural finance and gender impact report. Available from: https://www.nabard.org

National Family Health Survey (NFHS-5), 2019-21 (2021): compendium of fact sheets. Available from: https://mohfw.gov.in/sites/default/files/NFHS-5_Phase-II_0.pdf

NICRA (2023) Impact of climate change on agriculture. Available from: https://pib.gov.in/PressReleaseIframePage.aspx?PRID=1909206

Nordhagen S (2021) Gender equity and reduction of post-harvest losses in agricultural value chains. GAIN working paper series 20. Global Alliance for Improved Nutrition, Geneva. https://doi.org/10.36072/wp.20

O'Grady MJ, Langton D, Salinari F, Daly P, O'Hare GM (2021) Service design for climate-smart agriculture. Inf Process Agric 8(2):328–340. https://doi.org/10.1016/j.inpa.2020.07.003

Oxfam India (2013) When women farm India's land: how to increase ownership? Oxfam India Policy Brief No. 8, October 2013. Available from: https://oxfamilibrary.openrepository.com/bitstream/handle/10546/304918/bn-when-women-farm-india%27s-land-increasing-ownership-301013-en.pdf

Oxfam India (2018) India inequality report 2018: widening gaps. Available from: https://www.oxfamindia.org/sites/default/files/WideningGaps_IndiaInequalityReport2018.pdf

Oxfam India (2020) India inequality report 2020: on women's backs. Available from: https://www.oxfamindia.org/sites/default/files/2020-01/Oxfam_Inequality%20Report%202020%20single%20lo-res%20%281%29.pdf

Pal M, Gupta H, Joshi YC (2022) Social and economic empowerment of women through financial inclusion: empirical evidence from India. Equal Divers Incl 41(2):294–305. https://doi.org/10.1108/EDI-04-2021-0113

Papaskiri TV, Semochkin VN, Ananicheva EP, Popova ES, Dronina DA (2020) Digital land management and land resource data generation. IOP Conf Ser Earth Environ Sci 579:012131. https://doi.org/10.1088/1755-1315/579/1/012131

Pearse R (2017) Gender and climate change. Wiley Interdiscip Rev Clim Change 8(2):e451. https://doi.org/10.1002/wcc.451

Piñeiro V, Arias J, Dürr J, Elverdin P, Ibáñez AM, Kinengyere A, Morales Opazo C, Owoo N, Page JR, Prager SD, Torero M (2020) A scoping review on incentives for adoption of sustainable agricultural practices and their outcomes. Nat Sustain 3(10):809–820. https://doi.org/10.1038/s41893-020-00617-y

Ragasa C, Berhane G, Tadesse F, Taffesse AS (2013) Gender differences in access to extension services and agricultural productivity. J Agric Educ Ext 19(5):437–468. https://doi.org/10.1080/1389224X.2013.817343

Reuveny R (2007) Climate change-induced migration and violent conflict. Polit Geogr 26(6):656–673. https://doi.org/10.1016/j.polgeo.2007.05.001

Schwalbert RA, Amado T, Corassa G, Pott LP, Prasad PV, Ciampitti IA (2020) Satellite-based soybean yield forecast: integrating machine learning and weather data for improving crop yield prediction in southern Brazil. Agric for Meteorol 284:107886. https://doi.org/10.1016/j.agrformet.2019.107886

Sidun NM, Gibbons JL (2023) Women, girls, and climate change: human rights, vulnerabilities, and opportunities. Int J Psychol 58(1):17–25. https://doi.org/10.1002/ijop.12942

Singh C, Solomon D, Rao N (2021) How does climate change adaptation policy in India consider gender? An analysis of 28 state action plans. Clim Policy 21(7):958–975. https://doi.org/10.1080/14693062.2021.1953434

Steffen W (2015) Climate change: science, risks and responses. Available from: https://search.informit.org/doi/epdf/10.3316/informit.920627660065350

Thakur N (2023) Women farmers and technologies in agriculture: a review of current practices. In: Munshi S, Singh M (eds) Women farmers: unheard being heard. Springer, Singapore, pp 169–182. https://doi.org/10.1007/978-981-19-6978-2_12

UN Women (2016) The impact of climate change on women: global and local perspectives. Available from: https://www.unwomen.org/en/digital-library/publications/2016/11/the-impact-of-climate-change-on-women-global-and-local-perspectives

UNDRR (2024) Women and climate-related health impacts: an analysis. Available from: https://www.undrr.org/publication/women-and-climate-related-health-impacts-analysis

USAID (2023) The multiplier effect: how investing in women-led climate solutions drives sustainable impact. Available from: https://www.climatelinks.org/blog/multiplier-effect-how-investing-women-led-climate-solutions-drives-sustainable-impact

Vernooy R, Rana J, Otieno G, Mbozi H, Shrestha P (2022) Farmer-led seed production: community seed banks enter the national seed market. Seeds 1(3):164–180. https://doi.org/10.3390/seeds1030015

Wheeler T, von Braun J (2013) Climate change impacts on global food security. Science 341(6145):508–513. https://doi.org/10.1126/science.1239402

UN Women (2019) Facts and figures: economic empowerment. Available from: https://www.unwomen.org/en/what-we-do/economic-empowerment/facts-and-figures

UN Women (2020) Women in agriculture: barriers and progress. Available from: https://www.unwomen.org/en/news/in-focus/rural-women-day/2020

World Bank (2017) Measuring financial inclusion and the Fintech revolution. World Bank, Washington, D.C. Available from: https://documents.worldbank.org/curated/en/332881525873182837

World Bank (2021) India overview: agriculture and GDP contributions. Available from: https://www.worldbank.org/en/country/india/overview

# Chapter 16
# Women on the Frontline: Agriculture and Climate Change in Uttar Pradesh, India

**Surabhi Singh and Farhat Naz**

**Abstract** Uttar Pradesh, the state with the highest population in India, relies significantly on agriculture. Nevertheless, agriculture in Uttar Pradesh encounters two distinct challenges: the vulnerability of women farmers and the mounting threat of climate change. Although women constitute a significant proportion of the agricultural labour force in Uttar Pradesh, they continue to face issues of unequal access to land ownership, access to resources, and decision-making power in agriculture. Women have reduced levels of labour mobility and migration in comparison with men in response to floods or droughts, which hampers their ability to adapt to the effects of climate change. This chapter explores the twin difficulties faced by women farmers in western Uttar Pradesh. Firstly, it examines the existing socio-economic gender inequalities. Secondly, the research analyses how natural climatic factors and institutional factors influence women in agriculture. The study uses feminist standpoint theory to recognize structural power relations of gender inequality by situating it inside women's lived experiences and their productive functions. It uses both quantitative and qualitative approaches. Data was collected through Participatory Rural Assessment, interview schedules, focussed group discussions, and four case studies. The study shows how the marginalization of women increases with the rising adverse effects of climate change, particularly manifesting in increasing uncertainty of rainfall and aggravating groundwater depletion. The study calls for ensuring women's empowerment in the agricultural sector as an important strategy for achieving food security. It also proposes measures and policy directions that can improve the empowerment of women farmers in Uttar Pradesh.

**Keywords** Women farmers · Climate change · Agricultural practices · Feminist standpoint

S. Singh (✉)
Ph.D. Scholar, School of Liberal Arts, Indian Institute of Technology, Jodhpur, India
e-mail: Singh.103@iitj.ac.in

F. Naz
Associate Professor of Sociology, School of Liberal Arts, Indian Institute of Technology, Jodhpur, India
e-mail: Farhat@iitj.ac.in



A. Sarkar and N. Bandyopadhyay (eds.), *Gender-Transformative Approaches for Climate Change Adaptation*, Springer Climate,
https://doi.org/10.1007/978-3-031-90236-9_16

## Introduction

Uttar Pradesh is India's most densely populated state holding significant cultural, economic, and social importance within the country, with the agrarian economy of 47% population directly persisting on agriculture (Gulati et al. 2021). Agriculture in Uttar Pradesh serves not only as a source of employment but also as the cornerstone for food security and economic stability for millions of households. The state is responsible for the production of approximately 28% of India's wheat, 12% of its rice, and sugarcane is also produced in significant proportions, comprising 44% of the nation's total production (Gulati et al. 2021). Despite agriculture's crucial role in sustaining livelihoods, numerous challenges hamper the sector's growth and resilience, including gender issues and climate change, both of which pose significant obstacles to agricultural stability in Uttar Pradesh and the well-being of its rural population.

Women are integral to agricultural systems in Uttar Pradesh. However, they are often underestimated as farmers by both male and female members of their families and communities (Galiè et al. 2013). Also, gender plays a critical role in deciding the division of work, rights on access to resources, and taking decisions related to agricultural activities. In Uttar Pradesh, women's engagement in agricultural operations is limited to 17%; however, about 85% of female workers perform in farming and related activities (Renu 2023). They are heavily involved in intensive labour tasks like weeding and animal husbandry. However, despite their substantial contributions to agricultural output, women most often are not legally and socially acknowledged as farmers. The Government of India's present definition of a farmer encompasses solely individuals with legal land ownership, hence marginalizing the significant contributions of women (Verma et al. 2017). In the absence of land ownership, women predominantly serve as labourers on land possessed by male relatives or as informal wage earners on external properties.

Climate change has adverse effects on women in Uttar Pradesh. It is increasingly acknowledged as a barrier to attaining sustainable development objectives and is profoundly impacting the lives and livelihoods of communities worldwide. As the scientific community increasingly acknowledges the fact of climate change, it is essential to comprehend its effects on the ground, especially for communities reliant on agriculture and natural resources (Sugden et al. 2014). The concept of vulnerability is commonly used in climate change debates to concentrate on the risks and propose policies and adaptation plans that address the current and future effects (Friedman et al. 2019). Vulnerability in terms of gender inequality is fundamentally influenced by predominant socio-economic issues, including gender roles, land ownership and access, social norms, livelihood diversification, limited mobility, lack of employment options, and subdued voices in decision-making (United Nations Environment Programme 2011). These variables disproportionately affect historically marginalized and highly vulnerable populations, especially women. Thus, in this chapter, the vulnerability of women in agriculture because of climate change is looked at through three parameters in Uttar Pradesh, i.e. land ownership, access to resources, and inclusion in decision-making.

To assess the vulnerability of women in Uttar Pradesh, the study uses the conceptual framework of Feminist theories, such as Haraway and Hartsock that are used to recognize structural power relations of gender inequality by employing a systemic perspective on power, situating it inside women's lived experiences and their productive functions. This chapter utilizes Hartsock's notion of power as "essentially contested", signifying that power is continuously negotiated, opposed, shared, and redefined (Hartsock 2013). The agency is intrinsically linked to power, rendering it a site of contention, occasionally aligning with authority and at other times resisting it. Feminist theories effectively emphasize women's agency and their ability, or inability, to act, acknowledging them as subjects who are both influenced by and actively engage with unequal and oppressive power structures (Liddle and Wright 2001). The concepts of situated knowledge and social position are essential for examining how women's everyday activities and social situations influence their understanding and reactions to external pressures, including economic restructuring and climate variability. This method facilitates the examination of how women's knowledge and systemic power dynamics affect or mitigate their vulnerabilities, hence influencing their ability to adapt to an unpredictable climate.

## Review of Literature

There is an availability of literature on agriculture in Uttar Pradesh. The chapter by (Gulati et al. 2021) gave an in-depth knowledge about the overview of agriculture in Uttar Pradesh. It shows a region-wise study about the nine agro-climatic zones, output, cropping pattern, livestock, etc. It also gives an entire overview of the irrigation ratio, road density, and terms of trade between agriculture and the market. Further, there has been research done on the condition of women in Uttar Pradesh based on gender roles, land ownership, rights over resources, and inclusion in decision-making. The work by (Jain et al. 2023) and (Jassal 2005) shows how women play an important role in agriculture over time but still suffer over land ownership due to deep-rooted societal norms and weak policy enforcement by the state. The work of (Verma et al. 2017) shows the dedication of women in agriculture in western Uttar Pradesh yet their work goes unnoticed and they lack access to resources. Additionally, the work of (Joshi et al. 2016) shows how women have very little say in farming decisions that affect their work. This is due to socio-economic causes like limited education and societal expectations which hold them back from making the key choices, Hence, to empower women, they must be included in decision-making for a stronger voice in agriculture.

Further, research has been done concerning climate change in Uttar Pradesh. In the research of (Deo et al. 2015), the changing trend of rainfall patterns under the influence of climate change is shown in different sectors of Uttar Pradesh. It states that rainfall has been decreasing over the past few decades in the state with western Uttar

Pradesh seeing the biggest drop. Also, there is a late onset of monsoon leading to less rainfall and a short rainy season. Hence, there is a need for the planning of crops to deal with rainfall fluctuation and groundwater depletion. Additionally, studies have been abundant on the cropping pattern and its relation to groundwater depletion in western Uttar Pradesh. In the region, water-intensive cropping pattern requires huge amounts of water for irrigation causing groundwater depletion. Groundwater depletion is compounded by uncertain rainfall that affects crop productivity aggravating both food and water security.

This chapter focuses on Uttar Pradesh due to its significance in India's food and nutrition security initiatives and its pronounced vulnerability to climate change and variability (O'Brien et al. 2004). Although climate change has gathered significant attention in India concerning research, the majority of studies concentrate on changes in climatic variables or the effects of climate change (Tripathi 2017). Few investigations evaluate vulnerability to climate change and the majority evaluate susceptibility to natural hazards, such as cyclones, in coastal areas (Tripathi 2017). However, it is essential to examine socio-economic factors alongside natural systems to comprehend the impact of climate change on vulnerable populations. While certain studies have sought to assess agriculture's vulnerability, they primarily focussed on coastal states, neglecting inland states, such as Uttar Pradesh, which also face climatic challenges, such as floods and droughts, though to a lesser extent than coastal regions. Furthermore, there exists a gap in the literature about the impact of climate change on women in agriculture in Uttar Pradesh. This chapter aims to provide new insight by examining an inland state and analysing the relationship between gender and climate change through the lens of socio-economic factors.

## Objectives

Agriculture in western Uttar Pradesh is characterized by intense cropping and major exploitation of groundwater through tube wells for irrigation. In this context, the present chapter investigates the twin issues of gender inequality and climate change in the context of agriculture in Uttar Pradesh. This chapter uses the feminist standpoint theory to look at the current issues with relation to gender and climate change in Indian agriculture at the district level in western Uttar Pradesh and identifies the possible gaps in its approach. This research has two main objectives: first, to understand the unique challenges faced by women farmers in agriculture in terms of gender roles, land ownership, access to resources, and decision-making power and second, to analyse how climate change exacerbates the issues faced by women by analysing the changing rainfall patterns and groundwater exploitation through irrigation activities and cash cropping patterns. The chapter draws attention to the great paradox of agricultural practices in western Uttar Pradesh along with the gender inequalities and climate impact. Further, there are some insights drawn from the site of the field study based on the objectives mentioned above and discussion on actionable solutions focussed on empowering women in agriculture by identifying policies and practices that can address both gender disparities and climate vulnerabilities.

## Study Area

The most populated state in India is Uttar Pradesh, which is located in the fertile plain of Indo- the Gangetic region (Fig. 16.1) drained by five major rivers named Ganga, Yamuna, Ghaghra, Gomti, and Ramganga (Singh 2020). The economic backbone of the state is agriculture. It is divided into four zones, namely Bundelkhand, Central, Eastern, and Western zone. These four zones have varying levels of rainfall, groundwater supply, and other issues depending on their environment. The land is composed of alluvial deposits which makes the state a good reservoir of aquifer systems (Deo et al. 2015). But still, it faces the increasing extent of groundwater depletion as groundwater is easily accessible to people for irrigation. The state's agriculture relies on rainfall and groundwater to a greater extent. Additionally, most of the needs for drinking water and nearly all its industrial needs are met by groundwater, leading to its ongoing increased abstraction, and diminishing water levels throughout the whole state, in both rural and urban areas. The average annual rainfall of Uttar Pradesh has decreased from 960 cm in the 1980s to 901 cm in the 1990s before sharply declining to 795 cm in 2001–2010 (Singh 2019). The state experiences less rainfall than the national average, but western Uttar Pradesh reported a greater decline in average rainfall than other parts of the state (Deo et al. 2015). Also, significant variations in a region's rainfall pattern affect the productivity, profitability, cropping pattern, and way of life of the local population (Singh 2019). Hence, Uttar Pradesh is considered an important site for studying the effects of climate change on agriculture in this chapter.

Over 80% of the state's irrigated area is supplied by groundwater, making it a significant source of irrigation (Wing MI 2017). Compared to other areas of the state, western Uttar Pradesh has high development of irrigation facilities and high cropping intensity (Rehman and Lata 2022). Additionally, compared to other zones in the state, this zone has the highest fertilizer usage per hectare and irrigation coverage. Farmers in the western region of the state largely cultivate sugarcane, and this region gives around 50% of agricultural output (Tripathi 2017). The cropping intensity is high in the state due to usage of canal water and groundwater which is resulting in groundwater depletion more rapidly in western Uttar Pradesh. This subsequently has an impact on the sustainability of the resource in many places where groundwater availability is fast declining.

Saharanpur district was purposefully selected among thirty districts in Western Uttar Pradesh for the study. The area has a system of canal networks, privately owned tube wells, sugar plants, and regulated markets. In the district of Saharanpur, groundwater irrigation governs 85.66% of the area (Central Ground Water Board). The district's most significant cash crop is sugarcane, followed by wheat, rice, and mustard. Also, it faces uneven rainfall often resulting in floods as it comes in the

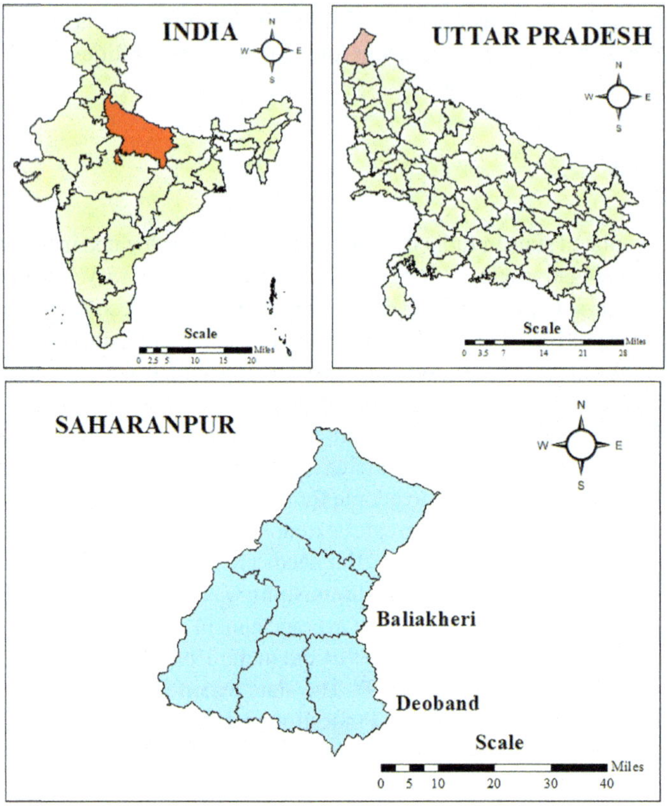


**Fig. 16.1** Location map of Saharanpur district in Uttar Pradesh. *Source* Prepared by Author

**Fig. 16.2** Sampling design of study. *Source* Sampling plan by authors

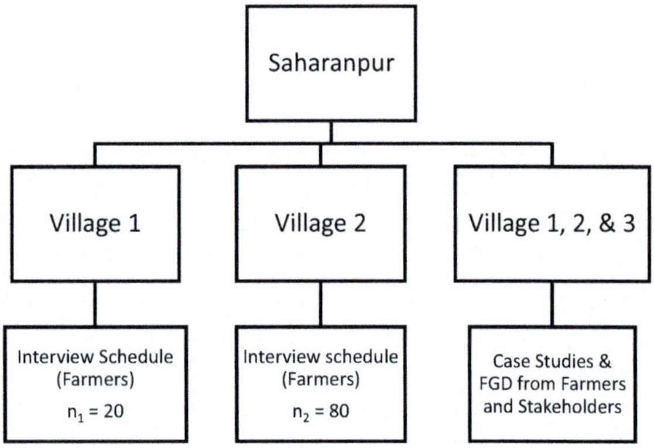

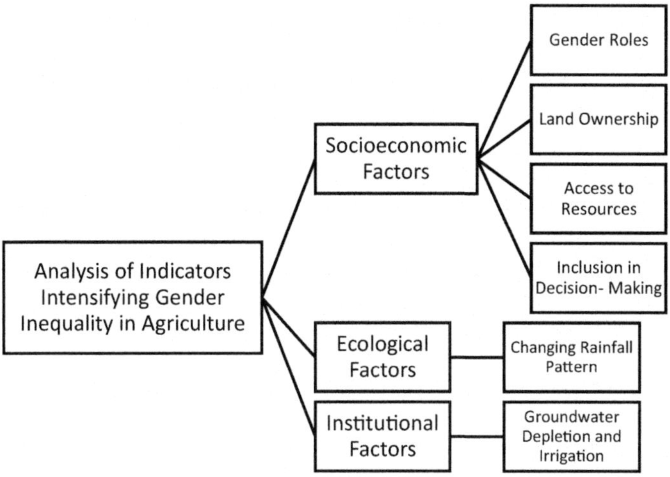


**Fig. 16.3** Indicators for analysing gender inequality in ggriculture. *Source* Conceptual diagram prepared by authors to understand gender inequality in agriculture

Terai region of the state (Joshi 2016). This affects the crops and increases the burden on farmers for livelihood. Females are more vulnerable in such situations due to restricted mobility. There is a high dependency on groundwater, but the district also faces a high level of groundwater depletion, as evidenced by the fact that out of eleven blocks, five blocks are classified as over-exploited four as critical, one as semi-critical, and only one as safe.[1] The district currently and in future will be vulnerable to negative impacts of groundwater depletion in addition to several environmental threats. A decline in the water table will lead to a reduction in the output of bore wells, failure of shallow wells, an increase in the cost of withdrawing water from wells with greater depths, and increasing power consumption for pumping water from greater depths.

## Data and Methodology

The approach adopted for data collection was Participatory Rural Assessment which considered multidimensional characteristics of the field inquiry along with the combination of quantitative and qualitative data to interpret the issues undertaken in the study. A combination of tools was used to perceive the insights of residents, their experiential knowledge, and their viewpoints. Data was collected from the 100 selected farmers through purposive sampling and snowball sampling from the villages with a mix of qualitative and quantitative techniques from two villages. 20 respondents were selected from the smaller village, and 80 were selected from the larger village as per the population. Also, data was collected from other stakeholders like village heads, Village Panchayat Development Officers, Agriculture

Development Officers, Block Development Officers, Village Employment Assistants, etc., associated directly or indirectly with agriculture and irrigation practices from government and private agencies in the region. Further, case studies, cases, and focussed group discussion (FGD) were used to make the study participatory. A cluster of 2–3 villages from two different blocks named Deoband and Baliakheri of Saharanpur district in western Uttar Pradesh were chosen to interview the stakeholders and learn about solution-oriented farming practices. A thorough survey was conducted in villages to collect data. Information about women's participation in agricultural activities was gathered through a carefully constructed interview schedule. The schedule was designed to be easy to understand and explicit to guarantee comprehensive documentation of both men's and women's engagement in agriculture.

## Findings and Discussions

The insights from the field of western Uttar Pradesh are aligned based on the objectives discussed before in the chapter. The problems that women farmers encounter were first examined using four criteria: gender roles, their rights to land, their ability to access resources, and their inclusion in decision-making when working in agriculture. Second, to analyse how ecological factors like changes in rainfall patterns and institutional factors like dependency on groundwater exacerbate the issues faced by women under climate change.

## Gendered Roles and Women's Rights

### *Gender Roles in Agriculture*

The findings show that agriculture predominantly highlights the contributions of males though women perform a significant share of the agricultural workforce. They rarely receive acknowledgement for their contributions to agriculture, which are often regarded as secondary, while policies predominantly reflect the perception that farmers are male (Pande 2007). Studies indicate that ancient agricultural methods, especially plough agriculture, have significantly influenced gender roles over time in agriculture (Giuliano 2017). This separation established persistent conventions seeing women's roles as confined to the household chores and some extent in livestock management. Their role in crop production remains undermined. Moreover, women's engagement in the agricultural workforce and their subsequent remuneration are also influenced by the socio-cultural context. Women own a very small share of agricultural land, and very often wage rates for men and women vary in agriculture.

In Uttar Pradesh, an agrarian economy is characterized by a significant rural workforce where female participation rates are markedly lower than those of male

**Fig. 16.4** Women working in sugarcane field. *Source* Author's image from the field

workers, but the majority of female workers account for 81.2% in rural regions employed in agriculture (GOI 2021). The rural non-farm sector in Uttar Pradesh has experienced remarkable growth. This growth has favoured men over women. The proportion of women in agriculture exceeds that of men, although this proportion is smaller in the non-agricultural sector (Singh and Mishra 2013). Nonetheless, over 50% of women are employed in the agricultural sector throughout all regions of Uttar Pradesh (Kumari 2018). However, a significant proportion of women are employed as casual labourers in agriculture as also observed in the field study (Fig. 16.4).

## *Gender and Land Ownership Rights*

In agriculture, women in Uttar Pradesh encounter substantial gender-based obstacles that impede their growth. The major obstacle that women are facing is restriction over land ownership which arises from inheritance laws and societal traditions mostly favouring male ownership. This lack of ownership over land is not only limiting their economic stability and growth but also restricting their ability to obtain loans or investments, as a guarantee is often required (Verma et al. 2017). Additionally, government initiatives and policies are majorly in favour of male landowners (Mugadur and Hiremath 2014). More women are working in agriculture but very few of them in the area use their rights to land. During the study, it was found that very few women had land rights, primarily because there were no male members. Even when women do have access to land, it is often very limited. It is crucial to have access to land for women because it also provides access to essential resources related to land, as highlighted by this study's focus on groundwater. Unfortunately, when access to land is restricted, it also means that access to vital resources, such as groundwater, is also constrained.

In rural India, women encounter considerable gender disparity in land ownership, notwithstanding the Indian Constitution's assurances of equality in Articles 14 and 15. Before 1956, inheritance rules adhered to personal community customs, and although the Hindu Succession Act of 1956 conferred property rights to women, it exempted agricultural land, which remained under state control (Akhtar and Saba 2014). Agrarian reforms predominantly overlooked gender equality, with women's land rights receiving consideration in government-distributed land only from the 1980s onwards. In 2005, a significant change transpired with the Hindu Succession (Amendment) Act, which granted Hindu women equal inheritance rights to agricultural land, potentially benefiting millions of rural women (Akhtar and Saba 2014). In Uttar Pradesh, women's rights to agricultural land are constrained by the U.P. Zamindari Abolition and Land Reforms Act of 1950 which privileges male heirs, granting women only contingent rights that revert to male lineage upon their demise (Jassal 2005). Daughters are barred from inheritance in the presence of sons, so reinforcing male supremacy and constraining women's economic strength. This legal gender bias obstructs women's opportunities for social and economic advancement, highlighting the necessity for more inclusive changes to achieve gender equality in land rights.

## *Access and Control Over Resources*

In conventional contexts, gender profoundly influences access to and management of resources in Uttar Pradesh. Also, cultural standards underline the restricted access women have to resources. Access denotes the opportunity to utilize resources for economic or social advantage, regulated by societal norms. In this context, women encounter restricted access relative to males, particularly concerning capital and land, frequently attributable to cultural beliefs that prioritize male land inheritance (Akhtar and Saba 2014).

It is seen that if women do not have land rights, then it becomes difficult to make decisions also related to farming. Hence, they depend on male members of the family for access to the resources. Notwithstanding established inheritance laws, women frequently abstain from asserting their rights to prevent familial dissonance, operating under the assumption that males require land and resources more to perform their roles as providers. Exerting power over resources, considered as the capacity to determine their utilization, also exposes a significant gender prejudice. Men, as heads of households, generally wield authority over the majority of resources (Jassal 2005). In occasional instances, women exert authority over agricultural implements, seeds, or livestock, mainly in the absence of a male presence. Nonetheless, women predominantly oversee household resources, indicating restricted agency in wider economic or agricultural choices. This gap highlights the influence of gender roles on resource accessibility.

### *Inclusion in Decision-Making*

It was noticed that both men and women participate in domestic and agricultural tasks, with responsibilities that are frequently complementary. Nevertheless, the participation of women obtains minimal acknowledgement or assistance in decision-making positions. Also, patriarchal systems do not encourage the participation of women in agricultural committees or village councils. As a result, they rely on male members to represent their voice, which keeps their voice unheard over important agricultural laws and practices (Kumari 2018). Therefore, in Uttar Pradesh cultural and social norms restrict women's participation in decision-making. Men are traditionally the primary decision-makers in agricultural planning and resource management. Therefore, in Uttar Pradesh, the ingrained societal and cultural norms restrict women's participation in agricultural decision-making (Kumari 2018).

This dynamic means that crucial agricultural decisions, such as which crops to plant and where to invest financially, are mostly made by men, limiting women's ability to influence these choices and respond effectively to challenges like climate change. As a result, women often find themselves sidelined perpetuating gender inequality and impacting rural communities where successful agriculture is essential for food security. Their expertise can be crucial in implementing sustainable agricultural techniques, such as organic farming and crop rotation, rather than cash cropping, which contribute to soil health and biodiversity preservation. When women have secure land rights, access to financial resources, and training in sustainable farming techniques, they become much more capable of adopting environmentally friendly practices. By blending their traditional knowledge with modern sustainable methods, these women can effectively tackle the negative effects of climate change and nurture healthy ecosystems.

## Climate Change and Disproportionate Burden on Women

Uttar Pradesh is India's leading state in terms of agricultural output due to the fertile soil, extensive river system, and robust agricultural foundation, which are essential for maintaining India's food and nutritional security (Tripathi 2017). Still, the state's agricultural sector is very susceptible to climate change, necessitating the development of climate resilience in farming techniques. Floods due to changing rainfall patterns are the most prevalent annual events in Uttar Pradesh, impacting several regions of the state, particularly the eastern and the Terai region, including Saharanpur districts. Small and marginal farmers, who depend significantly on favourable monsoon patterns and climatic stability for crop yield, are the most adversely impacted by these changes (Kaithal 2021). The prolonged dry and warm season results in drought like conditions, transforming productive land into desolate patches, and floods represent a persistent natural hazard that endangers the state's population and agricultural sector. Floods hinder agricultural production by destroying crops, elevating injury rates, and resulting in temporary homelessness. In recent decades, Uttar Pradesh has seen significant flooding, with notable events occurring in 1998, 2000, and 2008 (Kaithal 2021).

Moreover, in Uttar Pradesh, the challenges posed by climate change intensify existing gender inequities, disproportionately impacting women farmers compared to their male counterparts. Climate change makes life even harder for women in rural areas who are already facing challenges due to floods. There are significant hurdles women frequently operate on smaller land patches, dedicate fewer hours, or concentrate on cash crops, restricting their output (Singh and Singh 2019). In Uttar Pradesh, women are the backbone of households and farms, shouldering the bulk of daily chores and agricultural responsibilities. This heavy load can take a toll on their time, energy, and overall well-being. When resources become scarce due to these environmental changes, it is often the women who feel the brunt of the impact first. They face heightened risks of food insecurity, struggling to provide for their families amidst dwindling supplies. Their vital roles in both the home and the fields make them particularly vulnerable during tough times (Tripathi 2017).

In Uttar Pradesh, women farmers face significant challenges when it comes to migration and adapting to climate-related disasters like floods, droughts, or crop failures. Unlike men, who often have the option to seek livelihood in cities, women are usually tied down by family responsibilities and social expectations that limit their freedom to migrate (Mitra 2018). Climate change affects women compelling them to migrate in situations of natural hazards which contributes to food insecurity and resource scarcity. For many women, being unable to move freely poses significant challenges. Although migration serves as a crucial adaptive strategy for men, it frequently exacerbates the demands on women, who take on additional tasks in the absence of male household members, so further straining an already precarious agricultural system (Singh and Singh 2019). Financial struggles, combined with cultural expectations that often discourage women from seeking opportunities elsewhere, leave female farmers with fewer choices to face or escape the difficulties brought on by climate change. This situation underscores the need for greater support and opportunities for women in agriculture.

## Gender and Groundwater Exploitation in Agriculture

Water scarcity and excessive groundwater irrigation may precipitate a significant water crisis in the forthcoming decades, particularly in places, such as Uttar Pradesh, where water is essential for agricultural output (The World Bank 2021). Examining how climate change worsens the challenges faced by women reveals a stark reality in many districts of the Ganges floodplain in Uttar Pradesh. These areas are particularly vulnerable to the impacts of climate change, showing high levels of exposure and sensitivity while struggling to adapt. Additionally, practices like intensive groundwater irrigation and excessive groundwater extraction make these communities even more vulnerable. Western Uttar Pradesh exhibits considerable vulnerability, with districts seeing over 90% groundwater extraction, leading to a risky water crisis in the context of maladaptation to changing climatic conditions (Mohan and Sinha 2016). This excessive groundwater extraction, driven by rising water demands and limited surface water resources, threatens long-term sustainability and future availability, especially for women. Irrigation reduces dependence on precipitation. Also, it significantly deletes gr resources, which can be considered a form of maladaptation to climate change. The low adaptive capacity of Uttar Pradesh, marked by insufficient infrastructure and limited access to essential services, makes it not equipped to address the adverse impacts of climate change, thereby intensifying both ecological and socio-economic vulnerabilities.

Historically, the Green Revolution significantly transformed agricultural practices in Uttar Pradesh. The most noticeable transformation has been the increasing number of pumping tube wells and the decrease in the usage of canal water. Traditionally, agriculture relied on canals and surface water resources in Uttar Pradesh but with the advent of the green revolution, there was a shift in HYV seeds, chemical fertilizers, and groundwater pumping systems. In the post-Green Revolution Period, the number of tube wells in Uttar Pradesh grew rapidly as can be seen in Fig. 16.5. Due to the heightened need for irrigation to sustain high-yielding variety (HYV) crops, numerous farmers opted to set up their tube wells. A significant number of farmers made investments in privately owned tube wells, driven by the necessity for reliable irrigation water to sustain a high level of agricultural productivity. These agricultural activities majorly performed by the males impact the climatic conditions of the state and will affect the women adversely shortly.

Uttar Pradesh has been overlooked in the examination of climate change implications, as it is rich in resources and relatively insulated from natural disasters due to its

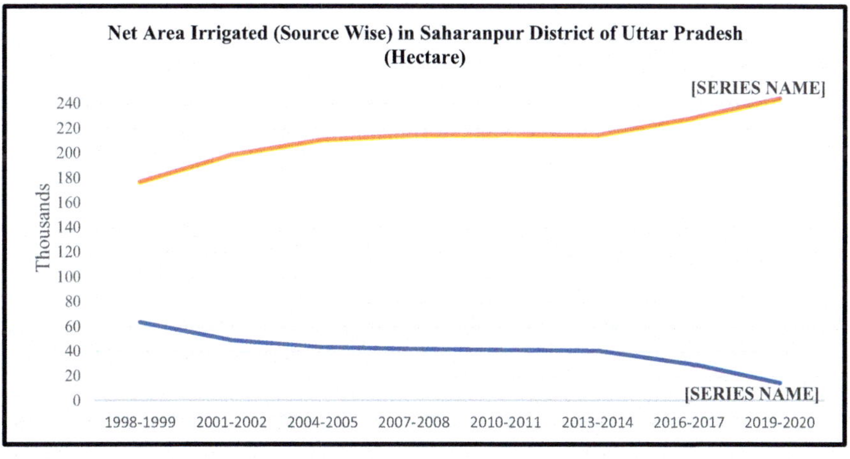


**Fig. 16.5** Shift in sources of irrigation after green revolution in Saharanpur district. *Source* Prepared by authors based on data accumulated from different reports of Land Use Statistics from the Directorate of Economics and Statistics, Department of Agriculture & Farmers Welfare, Ministry of Agriculture and Farmers Welfare, Government of India

geographical boundaries with neighbouring states. Despite possessing the greatest aquifer, it has begun to experience water scarcity, particularly when the surface water supply is inadequate, prompting farmers to utilize groundwater resources. This significantly diminishes the state's adaptive potential. The number of cultivators in the region has already reduced while the agricultural workforce, predominantly composed of women, has risen. Furthermore, as males migrate to urban regions in pursuit of employment, women are often left behind, particularly in areas affected by hazards. Consequently, when the region's groundwater can no longer meet future irrigation demands, women will be disproportionately impacted, exacerbating their workload and health issues. Therefore, an adaptive strategy is required for the state, together with the engagement of women in decision-making, to facilitate more sustainable agricultural choices.

## Solutions to Empower Women in Agriculture of Uttar Pradesh

Due to their marginalized socio-economic status, women are particularly **susceptible** to the consequences of climate change. However, because women are more sensitive to risks, more equipped to modify their behaviour, and more inclined to support policies and actions related to climate change, gender equality and women's involvement in decision-making can lessen climate change susceptibility (Tripathi 2017). During the field visit, women in Uttar Pradesh started thinking collectively during the group discussion to determine the causes of their disadvantages in terms of access to land, resources, and decision-making power. They quickly came up with their answers and formulated their demands and objectives in a way that suited their own pressing needs. To gain insight into how women view their relationship to land, the conversation centred on topics like the unique challenges faced by rural women, the gendered nature of rural labour, gender-based wage disparities, issues that exacerbate women's resentment, the types of rights that women demand, and the impact of resource depletion and climate change on women in the form of droughts and floods.

The conclusion can be categorized into three distinct segments: The primary emphasis was on the eradication or modification of the prevailing labour conditions for women in domestic settings and on agricultural land, hence accepting the existence of gender roles and trying to bring change in them; second, liberation from male subjugation and from authorities that attempt to exert power on them; and third, implementation of policies to address these identified issues and necessary actions through the introduction of various reforms in the form of demands. These encompass the rights to sufficient land for family sustenance, rights to the earnings of their employment, the right to education and knowledge, residential rights in both natal and married residences, and equal rights with husbands regarding land and labour harvests. Therefore, there is a necessity for land inheritance changes, land redistribution initiatives, and legal aid to enhance women's land ownership. Moreover, augmenting resource accessibility for women via targeted initiatives by government and NGOs might substantially improve their agricultural productivity and resilience to climate adversities (Singh and Singh 2019). Education and training in sustainable agricultural practices are equally essential.

The involvement of women in agricultural decision-making processes is a crucial aspect of empowerment. In any case, inclusive decision-making lets women give their opinions and help choose projects that are good for the whole family and community. Decision-making in agricultural communities is traditionally dominated by males, with women's perspectives frequently marginalized or absent (Kumari 2018). Enhancing women's representation in village councils, agricultural committees, and cooperatives helps rectify this disparity and bolster community resilience. Furthermore, full community engagement may encourage a cultural shift in perceptions of women's contributions to agriculture. It is imperative to motivate local authorities to implement gender-sensitive policies that acknowledge and encourage women's contributions to agriculture (Panda et al. 2014). Enhancing women's visibility in agricultural and community leadership positions enables communities to progressively deconstruct preconceptions that devalue women's labour and perspectives.

Agricultural risk management requires high attention due to the increasing effects of climate change in Uttar Pradesh. Hence, already knowing the threats induced by climate change, it is essential to establish effective systems by developing adaptive strategies to address any natural disasters that may arise from shifting climate patterns. The state must concentrate on enhancing its efforts and strategies to mitigate the risk factors associated with the threats posed by climate change, particularly for women in the agricultural sector. Therefore, a significant section should be prioritized within the comprehensive policy framework for adaptation. Unfortunately, the state experiences insufficient prioritization by the government regarding definite allocations. For instance, there is no allocation for disaster management, and the forestry sector has been assigned a low-priority (Panda et al. 2014). Agriculture and related activities require heightened prioritization, particularly considering that a significant portion of the population relies on this sector for both livelihood and food security. Thus, addressing these issues through climate-resilient farming practices, such as the use of drought-resilient seeds, drip irrigation, rejuvenating surface water bodies, and mixed and diversified farming, is necessary (Fig. 16.6). Also, sustainable water management and improved flood prevention infrastructure are necessary to lessen the effects of climate change on the state's food security and rural economy.

## Summary and Conclusion

In conclusion, it is seen that pointing out the issues of women and empowering them is essential for the agriculture sector in Uttar Pradesh specially with relation to climate change. This will be a step forward in ensuring sustainable agricultural growth and

**Fig. 16.6** Climate resilient practices in study area. *Source* Author's images from the field

increasing the resilience of the women's community towards climate change. The main issues that are addressed in this chapter in the agriculture sector of Uttar Pradesh are the deep-rooted gender inequality and the increasing threat of climate change. The socio-economic indicators of gender inequality in agriculture which are analysed in this chapter are fixed gender roles, minimal or no land ownership of women, limited access to resources, and minimal involvement of women in decision-making processes. Additionally, natural indicators like floods due to changes in rainfall and groundwater depletion are also analysed.

All these factors are responsible for reducing the adaptive capability of the state as women comprise the major workforce in agriculture, and they are devoid of equal rights. Hence, this chapter shows that women's marginalization in agriculture arises from historical, cultural, and legal factors which favour male control over land and resources resulting in unequal climatic impacts on women. Thus, there is a need for reform in legal rights and policies to resolve the issue, as equal rights will enable women to make important decisions which would help in sustainability of the agriculture in the state. The legal framework like The Hindu Succession (Amendment) Act of 2005 supports women by guaranteeing land rights but the implementation is very low in the state, especially in rural regions, thus not resulting in improvement in the condition of women.

Further, after pointing out the issues that exist in the state it becomes very important to provide access to resources, such as land, water, and new technology which will help in the sustainable development of the sector. As women are impacted more by the effects of climate change, it becomes important to prepare them for these changes. Hence, they should possess access to the requisite tools and information to survive in adverse climatic conditions. Thus, community-oriented training programmes and access to resources can provide them with sustainable and climate-resilient methodologies. Through the implementation of traditional knowledge of crop rotation, water conservation, and organic farming, women can help reduce the current issues that are caused by cash cropping and pesticides in the state responsible for the depletion of natural resources. These measures will support women in reducing their reliance on male members of the family and also position them as essential contributors in the battle against climate change.

This chapter emphasizes the inclusion and representation of women farmers in the decision-making in governance related to agricultural activities. Encouraging women's involvement in village councils and agricultural committees will help in understanding the issues that women undergo and will provide a different perspective for understanding local ecosystems and resource management. Furthermore, it is crucial to strategize for the adaptive capability to withstand different impacts of climate change like changing trends in rainfall patterns and groundwater depletion on women. Equal representation, resources, and policies are essential for maintaining resilience. Therefore, policies that increase adaptive capacity and reduce vulnerabilities to promote climate resilience are crucial.

By considering the needs and viewpoints of women, the policies can promote a more inclusive agricultural framework that addresses environmental concerns as well as gender inequality. In conclusion, empowering women in Uttar Pradesh's agriculture sector will promote climate resilience and sustainable development in addition to equity. Thus, a better future can be created if women's contributions are included and valued by encouraging inclusive governance and building adaptive capacities. Thus, this chapter promotes a holistic perspective by including efforts by communities, NGOs, and policymakers in Uttar Pradesh to promote the participation of women in agriculture by overcoming socio-economic and natural barriers.

## Notes

1. The status of groundwater extraction is generally classified by percentage. It is calculated as the ratio of annual groundwater extraction to annual extractable groundwater resource. Based on the state of extraction, the assessment units are categorized as Safe ($\leq 70\%$), Semi-critical ($> 70\%$ and $\leq 90\%$), Critical ($> 90\%$ and $\leq 100\%$), and Over-exploited ($> 100\%$).

## References

Akhtar S, Saba P (2014) Small, marginal and women farmers in Uttar Pradesh: land rights and credit flows. Int J Res Soc Sci 4(3):396–419

Deo K, Tripathi P, Kumar A, Singh KK, Mishra S, Mishra A, Singh A (2015) Trend of rainfall in different sectors of Uttar Pradesh under present scenario of climate change. Int J Environ Sci 6:303–310

Friedman R, Hirons MA, Boyd E (2019) Vulnerability of Ghanaian women cocoa farmers to climate change: a typology. Climate Dev 11(5):446–458

Galiè A, Jiggins J, Struik PC (2013) Women's identity as farmers: a case study from ten households in Syria. NJAS-Wageningen J Life Sci 64:25–33

Giuliano P (2017) Gender: an historical perspective

GOI (2021) Annual report, periodic labour force survey (PLFS), July 2019 to June 2020. National Statistical Office, Ministry of Statistics and Programme Implementation, July

Gulati A, Terway P, Hussain S (2021) Performance of agriculture in Uttar Pradesh. Revitalizing Indian Agriculture and Boosting Farmer Incomes, pp 175–210

Hartsock N (2013) Foucault on power: a theory for women? In: Feminism/postmodernism. Routledge, pp 157–175

Jain C, Saxena D, Sen S, Sanan D (2023) Women's land ownership in India: evidence from digital land records. Land Use Policy 133:106835. https://doi.org/10.1016/j.landusepol.2023.106835

Jassal ST (2005) Gendering agrarian issues: the Uttar Pradesh experience. women of India: Colonial and Post-Colonial Periods, pp 259–282

Joshi K, Dash H, Gangwar B (2016) Exploring gender involvement in agriculture decision-making: a case study of Meerut district in Uttar Pradesh. J Hum Ecol 54(1):41–48

Joshi S (2016) Farmers' perception about climate change and strategies to cope with climate change in Uttar Pradesh [PhD Thesis]. GB Pant University of Agriculture and Technology, Pantnagar-263145 (Uttarakhand)

Kaithal AK (2021) Climate change and Uttar Pradesh: Issues, challenges and way forward. J Emerging Technol Emerg Res 8(10):783–789

Kumari AR (2018) Role of farm women in agriculture and their involvement in decision making-A study in Deoria district of Uttar Pradesh. J Pharmacognosy Phytochem 7(1S):1249–1253

Liddle J, Wright C (2001) Introduction—understanding local/global identities: towards a materialist discursive feminism. Women's Stud Int Forum 24(3–4):273–281

Mitra A (2018) Male migrants and women farmers in Gorakhpur-climate adaptation and changing gender relations

Mohan D, Sinha S (2016) Assessing vulnerability to climate change in the Ganges Basin using a combined macro-and micro-level approach. Climate Dev 8(4):312–323

Mugadur N, Hiremath R (2014) Socio-economic condition of agriculture women labour: a case study. Ind J. Res. 3(12):50–52

O'Brien K, Leichenko R, Kelkar U, Venema H, Aandahl G, Tompkins H, Javed A, Bhadwal S, Barg S, Nygaard L et al (2004) Mapping vulnerability to multiple stressors: climate change and globalization in India. Glob Environ Chang 14(4):303–313

Panda GR, Shrivastava S, Kapoor A (2014) Climate change and gender: Study of adaptation expenditure in select states of India. Handbook of Climate Change Adaptation. Springer, Berlin

Pande R (2007) Role of women in Indian agriculture: a historical perspective. Indian Women in development perspective. Anmol Publication Pvt. Ltd., New Delhi, pp 66–96

Rehman H, Lata S (2022) Comparative analysis of irrigation and cropping intensity in Uttar Pradesh, India. Nat Geogr J India 58(3):25–34

Renu (2023) Determinants of female workforce participation in agriculture of Uttar Pradesh, India. Current Agriculture Research 11(1)

Singh S (2019) Determinants of agriculture production in Uttar Pradesh, India: a regional analysis. Res Rev Int J Multidisc 4(1):1–14

Singh A (2020) Sustainable management of groundwater resources in Uttar Pradesh. Adm J LBSNAA 60(1):58–80

Singh U, Mishra NK (2013) Women work participation in rural Uttar Pradesh: a regional analysis. Int J Soc Sci Interdisc Res 2(8):47–57

Singh S, Singh A (2019) Farmers' perception of climate change and livelihood vulnerability in rainfed regions of India: a Gender-environment perspective. Int J Environ Clim Change 9(12):878–889

Sugden F, Maskey N, Clement F, Ramesh V, Philip A, Rai A (2014) Agrarian stress and climate change in the Eastern Gangetic Plains: Gendered vulnerability in a stratified social formation. Glob Environ Chang 29:258–269

The World Bank (2021) World Bank Indicators. https://data.worldbank.org/indicator/SP.RUR.TOTL.ZS?locations=IN

Tripathi A (2017) Socioeconomic backwardness and vulnerability to climate change: evidence from Uttar Pradesh state in India. J Environ Planning Manage 60(2):328–350
United Nations Environment Programme (2011) Women at the frontline of climate change: gender risks and hopes. https://wedocs.unep.org/20.500.11822/7985
Verma N, Kashyap P, Arya M, Singh M, Nath A, Singh S et al (2017) Economic quantification of women's work in farming system: a case study of the western plain zone of Uttar Pradesh. Indian J Agric Sci 87(3):355–362
Wing MI (2017) Report of 5th census of minor irrigation schemes. River Development and Ganga Rejuvenation, Ministry of Water Resources, Government of India

# Part IV
# Addressing Gender In Climate Change Adaptation: Policies And Programmes

# Chapter 17
# Gender Integration in Climate-Resilient Agriculture: Exploring Policies, Programmes, and Transformative Pathways

**Mamta Mehar and Twinkle Jaspal**

**Abstract** This chapter examines the evolution of gender integration in climate-resilient agriculture, focussing on the transition from gender-neutral to gender-transformative approaches. Drawing on literature from the Consultative Group on International Agricultural Research (CGIAR) and other key studies, this chapter highlights the limitations of gender-neutral policies that neglect gendered disparities and fail to address the systemic barriers faced by women in agriculture. It categorizes approaches into gender-neutral, gender-sensitive, gender-accommodative, and gender-transformative frameworks, emphasizing their implications for equitable resource distribution and women's empowerment. The analysis underscores the potential of gender-transformative strategies to challenge entrenched power dynamics and foster systemic change, while also exploring challenges in implementation, such as limited access to resources, training, and decision-making processes. By integrating a gender lens in agricultural and climate adaptation policies, this chapter contributes to the discourse on inclusive and equitable development, offering actionable insights for policymakers and practitioners.

**Keywords** Gender integration · Climate-resilient agriculture · Adaptation · Gender-transformative approaches

M. Mehar (✉)
Evidence Synthesis Consultant, Climate Change Adaptation and Mitigation Impact Action Platform of the CGIAR, Delhi, India
e-mail: mamta.dse@gmail.com

T. Jaspal
Assistant Manager, MicroSave Consulting, Delhi, India



A. Sarkar and N. Bandyopadhyay (eds.), *Gender-Transformative Approaches for Climate Change Adaptation*, Springer Climate,
https://doi.org/10.1007/978-3-031-90236-9_17

## Introduction

The imperative of integrating gender considerations into climate policy and practice is increasingly recognized as a fundamental element for attaining sustainable development and facilitating effective adaptation to climate change. The discourse on climate resilience advocacy and policies is now looking at gender justice and climate justice as critical concepts, as emphasized in the 2024 United Nations Climate Change Conference (more commonly referred to as CoP 29). These interconnected concepts highlight the need to address the disproportionate impacts of climate change on marginalized groups, including women, and to ensure equitable access to resources and decision-making power. Gender justice, in this case, calls for addressing the systemic inequalities that limit marginalized groups, especially women's access to resources, information, and decision-making power, hindering their ability to adapt to and mitigate climate change. In turn, this highlights the importance of recognizing and valuing women's unique knowledge and contributions to climate-resilient agriculture, while actively ensuring their full participation in the planning and implementation of climate-related initiatives.

Policies, programmes, research, and governance frameworks play a crucial role in promoting gender justice and climate justice at macro-, meso-, and micro-levels. At the macro-level, national policies and international agreements can set the framework for gender-responsive climate action. This includes supporting research initiatives that investigate the gendered impacts of climate change and inform policy development. At the meso-level, regional and local governments can implement programmes and initiatives that address the specific needs of women and other marginalized groups. At the micro-level, community-based organizations and grassroots movements can empower women and advocate for gender justice in climate adaptation strategies.

It would not be an overstatement to assert that gender diversity and inclusion considerations within climate change research have emerged as a focal point in recent years (Mehar and Prasad 2022). In the past few decades, several studies and initiatives have elucidated the need to address gendered roles in decision-making, the influence of intersecting identities, intra-household dynamics, and the importance of empowering women through access to information and resources (Acosta et al. 2021; Birachi et al. 2020; Gutierrez-Montes et al. 2020; Mutenje et al. 2019; Mehar et al. 2016; Ravera et al. 2016). More recently, there has also been a growing rise for implementing a gender lens in climate financing that specifically seeks to build locally led solutions and empowering women as actors of change (Tye et al. 2021).

This chapter explores the evolution of gender integration within strategies for climate-resilient agriculture, moving from gender-neutral to gender-transformative approaches. It highlights the limitations of gender-neutral policies, which often fail to consider differences in gendered experiences, leading to the exclusion of women from decision-making processes and access to resources. This exclusion is particularly critical in regions reliant on agriculture, where climate change disproportionately affects women (Mehar and Prasad 2022). Therefore, understanding how climate-resilient approaches and programming integrate gender inclusivity and equity is crucial. To explore this issue, this chapter utilizes literature from the Consultative Group on International Agricultural Research (CGIAR), a global organization that acknowledges the significance of gender in tackling climate change and has embedded gender considerations into its research and development endeavours. This literature is publicly available through their online directory, CGSpace (https://cgspace.cgiar.org/). This inquiry seeks to enrich the ongoing discourse surrounding gender and climate change and offers valuable insights for policymakers and practitioners endeavouring to foster gender-inclusive and equitable climate-resilient agricultural development.

## Conceptualizing Gender Inclusivity and Equity

This chapter conceptualizes gender inclusivity and equity through a spectrum of approaches, including gender-neutral, gender-aware categorized as gender-sensitive, gender-accommodative or gender-responsive, and gender-transformative approaches. These terms are primarily derived from the gender equality continuum model developed by the Inter-Agency Gender Working Group (IGWG) and USAID (IGWG and USAID 2017), as well as the more recently developed UNICEF gender continuum (UNICEF 2019). These frameworks hold pertinence across various sectors, including agriculture, education, water, sanitation and hygiene (WASH), and health, where gender dynamics significantly impact outcomes. Considering that literature about CGIAR traverses these varied sectors, the adoption of these frameworks affords a holistic perspective for analytical inquiry.

Each of the approaches in these frameworks provides distinct perspectives on the integration of gender-related issues, exhibiting differing levels of efficacy in the advancement of gender equity. The progression from gender-blind to gender-transformative approaches signifies a transition from the neglect of gender disparities to the proactive reconfiguration of gender norms and power relations. An elaboration on these approaches is presented below:

**Gender-neutral approach**: A gender-neutral approach aims to avoid distinguishing roles or making assumptions based on people's sex or gender. This approach strives to promote equality and inclusivity by creating an environment where individuals are treated with fairness and respect, regardless of their gender identity. Gender norms, roles, and relations are not affected (worsened or improved) (Mehar and Prasad 2022; UN Women n.d.).

**Gender-aware approach**: This approach recognizes gender disparities and the imperative to address these differences in policy and practice. Gender-aware policies and programmes examine and address the set of economic, social, and political roles, responsibilities, rights, entitlements, obligations, and power relations associated with being female and male and the dynamics between and among men and women, and boys and girls (IGWG and USAID 2017). Based on the degree of equity consideration, the approach can be further categorized into three types:

**Gender-sensitive**[1]: Acknowledges gender inequalities but does not robustly address them. It often involves interventions that unintentionally or intentionally discriminate against women or reinforces harmful gender norms.

**Gender accommodative or gender-responsive approach**: Identifies and addresses the different needs of girls, boys, women, and men to promote equal outcomes within the existing institutional frameworks. This approach may involve the implementation of policies that support women or other marginalized genders without interrogating the fundamental causes of inequality (IGWG and USAID 2017).

While both gender-sensitive and gender-accommodative approaches aim to address gender disparities, they differ significantly in their scope and depth. For instance, a gender-aware or sensitive approach might involve a climate strategy that provides general training on climate-resilient agricultural practices to both men and women, without considering their specific needs or constraints (Resurrección 2017). In contrast, a gender-responsive approach would tailor the training to address the unique challenges faced by women and men, such as providing separate sessions for women to discuss their specific concerns and needs related to accessing resources, decision-making, and labour-saving technologies (Arora-Jonsson 2011). However, both of these approaches commonly focus on identifying and responding to practical needs through the provision of resources but do so without challenging entrenched gender norms and systemic inequalities.

**Gender-Transformative Approach**: This approach actively seeks to alter gender norms and power dynamics to attain gender equity. It encompasses the reconfiguration of societal norms, the redistribution of resources, and the promotion of collective initiatives to confront gender inequalities (IGWG and USAID 2017). While gender-transformative approaches are increasingly acknowledged for their capacity to realize authentic gender equity, they necessitate a thorough comprehension of intersectionality and the multifaceted determinants influencing gender dynamics. It contests conventional gender roles and aspires to engender systemic change. However, its implementation may prove challenging due to deeply embedded societal norms and resistance to transformation that often call for a longer cycle of interventions, training, and/or programmatic efforts as well as transforming the policy prescriptions within 'the local' context (Acosta et al. 2021).

While these frameworks provide a structured approach to addressing gender issues, it is important to recognize that the effectiveness of these approaches can vary based on context and implementation. Gender-transformative approaches, although promising, require careful consideration of local contexts and engagement with power holders to achieve sustainable change (Njuki et al. 2022). Moving from gender-blind to gender-transformative approaches, there is an evolving need to aim

---

[1] We have replaced the 'gender-exploitative approach' proposed by the IGWG framework with the 'gender-aware sensitive approach' proposed in the UNICEF gender continuum. The latter is available at https://www.unicef.org/evaluation/media/1221/file/UNICEF%20Guidance%20on%20Gender.pdf, Accessed on November 10, 2024.

for systemic change rather than individual-level adjustments (Backman-Levy and Greene 2024).

## Research Methodology

This exploratory study employs a narrative review methodology instead of a systematic review due to budget and time constraints. The narrative review allows for a time-flexible in-depth synthesis of studies relevant to understanding the focus on gender integration in climate-resilient agricultural strategies. The review focuses on journal articles sourced from the CGSpace portal, published between 2012 and 2023, addressing the integration of gender considerations in climate-resilient agricultural strategies and exploring the intersection of climate stress and gender.

This delimited timeframe is justified as it demarcates a pivotal juncture in the discourse surrounding gender integration within development initiatives. The year 2012 holds particular significance due to the 'World Development Report' (World Bank 2012), which was exclusively dedicated to gender equality under the title 'Gender Equality and Development'. This seminal report provided compelling evidence illustrating the detrimental impact of gender disparities on economic growth and development, thereby underscoring the imperative for targeted interventions. Furthermore, the period following 2012 witnessed a discernible paradigm shift from gender-neutral to gender-transformative approaches. Furthermore, the post-2012 period marks a critical juncture as global frameworks, such as the UNICEF Gender Continuum (2019) and the IGWG Gender Equality Continuum Model (2017), began gaining prominence, emphasizing a shift from gender-neutral to gender-transformative strategies.

A rigorous qualitative content analysis of over 70 papers employing diverse methodologies (like mixed-methods, case studies, qualitative, and experimental) identified key themes related to gender and climate-resilient agriculture. The themes included are differential access to resources, shaped by intersecting social identities (Ravera et al. 2016); the influence of social norms on women's decision-making (Mutenje et al. 2019); and the efficacy of gender-responsive policies in fostering women's empowerment (Acosta et al. 2021; Gutierrez-Montes et al. 2020; Birachi et al. 2020). The existing research emphasizes the investigation of the gendered impacts of climate change, highlighting the influence of gendered roles in shaping actions and decision-making processes (Mehar et al. 2016). Through a critical examination of these themes, the study assessed the extent of gender inclusivity employed and the typologies of gender approaches adopted across the reviewed studies.

While the heterogeneity of the studies limits direct comparisons, the findings, contextualized by regional and cultural factors, inform recommendations for enhancing gender equality and empowering women in agriculture. This study emphasizes the need for systemic changes that promote inclusive decision-making processes.

## Gender Integration in Climate-Resilient Strategies

In this section, we discuss the extent of gender integration in the design, implementation, and adaptation of climate-resilient strategies through three key areas of inquiry as seen emerging from the content analysis. These include gender considerations in climate or agricultural policies, direct or indirect impact on women's participation, and gender-focussed empowerment and challenges in technology adoptions. These areas of inquiry have been assessed under each gender approach to showcase comparisons across the approaches. These comparisons are backed with examples from the various studies under the research in CGIAR to map experiences and outcomes of varying levels of gender integration within agricultural and climate programmes. These evidence-backed arguments seek to advise and support similar programmes for adequately integrating gender within their design and implementation.

### *Gender-Neutral Approaches*

As mentioned earlier, gender-neutral approaches within programme and policy-making often neglect the gendered roles and responsibilities as well as their capacities, resulting in interventions that are insufficient and inequitable in responding to individuals' specific needs. Moreover, these approaches can instead perpetuate systemic inequalities, especially regarding global issues where gender roles guide the disproportionate impact that individuals experience (Mehar and Prasad 2022). This aligns with the concept of 'gender nescience' documented by Acosta et al. (2019), which refers to the lack of understanding of gender issues within climate change policymaking. Their research found that this narrative was present in 11% of the stories analysed, indicating a significant gap in knowledge and awareness. Such gaps often result in superficial gender inclusion, focussing on quotas rather than addressing the root causes of inequality.

Approximately 90% of reviewed studies surpass the stage of gender neutrality, moving towards gender-aware approaches. However, 19% still exhibit gender-neutral characteristics by neglecting gendered differences. For instance, zero-tillage technologies, while improving soil conditions and reducing atmospheric carbon (Mondal et al. 2020), have traditionally been less accessible to women due to barriers like limited equipment access, lack of training, and exclusion from decision-making (Sapkota et al. 2021; Kumar et al. 2018). Although advancements, such as ergonomic designs, are making these technologies more accessible, further research is needed to explore their gendered implications and ensure inclusive design and implementation.

Gender-neutral approaches also often lack an ecosystem lens and equitable support services, overlooking factors like digital literacy and disproportionate access to climate-smart technologies. For example, the Agrometeorological Advisory Programme (AAP) in Mali has been criticized as 'gender-insensitive' due to its neglect of equitable farmer welfare (Carr and Onzere 2017). Senior male farmers benefited the most, while women and junior males had limited engagement, exacerbating disparities in access to climate-smart agricultural practices (Jost et al. 2015). Bundled climate insurance services, while promising, must be gender-inclusive and accessible to all, regardless of digital literacy levels.

Examples from REDD+ initiatives further illustrate how gender-neutral policies restrict women's participation in decision-making. In Nepal, only two out of five climate policies acknowledged gender issues, with even fewer addressing women-specific adaptation measures, reflecting a gender-neutral approach to climate-smart agriculture (CSA) (Paudyal et al. 2019).

In summary, gender-neutral approaches fail to fully integrate gender considerations into programme and policy work, contributing to inequities in resource access and decision-making. These approaches highlight the need for policies that address diverse gendered experiences and embed inclusivity at all levels of design and implementation.

## *Gender-Aware Approaches*

Gender aware policies and programmes are crucial in addressing the unique challenges faced by women in the context of climate change. These programmes and policies can be categorized based on the degree of gender integration, ranging from gender-sensitive to accommodative to transformative approaches. Understanding these categories and the contexts they undertake helps in designing interventions as well as understanding the consequences of viewing women as either as vulnerable or virtuous and/or active agents in adaptive measures to climate solutions. This section explores the nuances of these gender-aware approaches and their implications in climate change adaptation, as evidenced by reviewed studies.

### Gender Sensitive

Gender-sensitive programmes aim to address the practical needs of women and men by providing resources tailored to their distinct experiences and challenges, recognizing gender disparities in access to resources and decision-making power. These programmes often focus on practical solutions, such as introducing technologies like biogas stoves and maize shellers, which alleviate physical burdens and improve time management for women (Bernier et al. 2013; Ragasa 2012). Similarly, technologies, such as direct-seeded rice (DSR) and machine-transplanted rice (MTR), help reduce women's labour burdens and greenhouse gas emissions, particularly in agriculture where women play a crucial role (Gartaula et al. 2020). However, while addressing immediate needs, these programmes often fail to challenge systemic gender inequalities or address underlying social norms.

Despite their benefits, many gender-sensitive approaches treat women as a homogenous group, neglecting intersectional factors, such as age, class, and caste that shape their experiences. For instance, women are frequently underrepresented in policymaking processes, leading to limited consideration of their diverse roles and needs (Howland et al. 2021). This lack of inclusion can be seen in programmes like forestry and REDD+ decision-making in Vietnam, where women's rights and empowerment are often overlooked (Pham et al. 2018; Larson et al. 2018). Extension services also frequently target male farmers, excluding women from accessing vital information and resources (Farnworth et al. 2016). For example, training sessions are often held during times when women are occupied with household responsibilities, or they use materials that are not relevant to women's specific farming contexts. Participatory initiatives, such as Radio Listeners' Clubs (RLCs) and Integrated Climate Services for Agriculture (PICSA), have shown promise in improving women's access to information and reducing gender disparities (Dayamba et al. 2018). However, the success of these initiatives often hinges on localized implementation and sustained institutional support. Addressing these systemic gaps requires a shift in how programmes are designed and delivered, ensuring they meet the diverse and context-specific needs of women.

While gender-sensitive approaches increase the participation of women in training and provide them access to resources to directly benefit them, they do not fully utilize the potential of women in agriculture, which could otherwise contribute to increased productivity and food security (Nchanji et al. 2022; Diarra et al. 2021). Training programmes and resources are frequently designed without considering the specific needs and constraints of women and often treat them as a homogenous group, failing to consider their diverse needs and vulnerabilities. These approaches also do not challenge social norms and gendered roles, further entrenching gender disparities in agricultural practices (Farnworth et al. 2016). Socially, these efforts can perpetuate cycles of poverty and inequality, as women remain marginalized in both household and community decision-making processes (Paudyal et al. 2019).

In summary, gender-sensitive approaches address the practical needs of women and men, enhancing participation, empowerment, and technology adoption in climate and agricultural policies. However, their lack of a systemic and intersectional perspective limits their potential for long-term change. Future efforts must integrate intersectional gender analyses to tackle diverse challenges, address root inequalities, and develop inclusive, transformative policies and programmes.

## Gender Accommodative

Gender-accommodating or gender-responsive programmes can be quite similar to gender-sensitive programmes, in the way that they recognize and address gender inequalities by providing resources and support to both men and women for climate adaptation. These resources encompass information, training, equipment, and financial support pertinent to the execution of climate-adaptive strategies. However, gender-accommodating programmes focus on working within existing frameworks rather than implementing remedial or transformative actions to alter these structures. Gender accommodative programmes comprise climate-smart and women-friendly equipment and provision of equipment that is conducive to women's participation, such as the Conoweeder, which aims to alleviate the labour-intensive nature of wedding activities. Roof-rainwater harvesting or other small harvesting structures for kitchen gardens or small farms are another example (Gonda et al. 2016). Further, gender accommodative approaches also aim for targeted support for women. For instance, several initiatives have created women-friendly groups where they directly or indirectly receive support for agricultural activities. For example, microfinance saving groups are quite popular in Bangladesh, especially for fisherwomen in coastal regions. Results from the evaluation of the project Rwanda Climate Services for Agriculture (RCSA) from 2016 to 2019 project show that inclusion in participatory communication processes improved women's and men's awareness and access to climate information (Birachi et al. 2020). The programme has eliminated the disparities in awareness, access, and use of climate information that otherwise exist between women and men smallholder farmers and that female Radio Listeners clubs' participants (RLC) access climate information through a greater diversity of channels than non-participants female or male RLC participants (Ingabire 2021). However, an important point to note here is that while these programmes have made strides in addressing practical needs for both genders, they have not necessarily challenged power structures that give rise to these gendered inequalities. As a result, these programmes instead focus on accommodating women within existing systems rather than instilling measures to transform them (Barooah et al. 2023).

Another consequence of this approach is the characterization of women as 'virtuous' within these strategies, which may serve to reinforce conventional roles rather than foster genuine gender equality (Paudyal et al. 2019). This framing can limit the full participation and empowerment of women in climate adaptation strategies. As a broader framework, gender mainstreaming builds on the principles of gender integration by systematically embedding gender considerations across all levels of climate policies and programmes. Efforts, such as the establishment of gender focal points within ministries and the development of sector-specific gender strategies, exemplify gender mainstreaming in action, aiming to create more inclusive and equitable climate responses (Acosta et al. 2019).

To summarize, gender accommodative programmes play a crucial role in the equitable distribution of resources as well as capacity building for climate adaptation but often fall short in challenging existing structures that give rise to gender norms and gender inequality. There is a need for initiatives to shift their perspective from seeing women as merely 'virtuous' or 'vulnerable' and realize their role as active agents in agricultural and climate adaptation strategies. This shift could lead to more sustainable and equitable outcomes in the face of climate change. Gender accommodating programmes, mostly embedded as part of other larger multidisciplinary projects, while beneficial in providing short-term support, cannot ingrain long-term benefits and positive cycles of change if they do not challenge the structural norms and roles that perpetuate gender disparities. By merely adapting to existing gender norms and roles, these programmes fail to address the root causes of gender inequality. This highlights a critical need for standalone gender projects focussed on climate resilience.

## Gender-Transformative Approaches

Gender-transformative initiatives endeavour to eliminate structural barriers or systematic challenges to gender equality, by compelling reconsideration of the roles and opportunities provided to both men and women, thereby moving beyond deeply rooted social and cultural norms (Mehar and Prasad 2022). Commonly referred to as gender redistributive, this approach acknowledges women as proactive participants in the adoption of climate-responsive solutions. Policies, research, and intervention frameworks under this paradigm prioritize enhancing women's skills and access to critical resources while addressing strategic needs across genders.

Our conceptualization of empowerment draws upon Naila Kabeer's framework: 'the expansion of the capacity to make strategic and meaningful choices by those who have previously been denied this capacity, but in ways that do not merely reproduce, and may indeed actively challenge, the structures of inequality in their society' (Kabeer 1999). Empowerment involves significant transformations in individual awareness, together with institutional strategies that challenge deeply rooted normative beliefs concerning gender dynamics and power relations (Cornwall 2016).

Programmes like the Agroclimate Information in Southeast Asia (ACIS) project demonstrate the transformative potential of gender-responsive interventions. By integrating gender awareness training, this initiative aimed to dismantle restrictive gender norms and amplify women's agency in agricultural decision-making (Gumucio et al. 2022; Mulema et al. 2022). Through participatory methods like Participatory Scenario Planning (PSP), ACIS empowered women and ethnic minorities to collaborate with scientists and service providers, tailoring climate services to diverse needs. Additionally, the programme's collaboration with Village Savings and Loan Associations (VSLAs) in Cambodia enabled women to access financial resources, improve economic livelihoods, and challenge traditional gender roles.

In contrast to gender accommodative approaches, which work within existing norms, gender-transformative initiatives like ACIS aim to fundamentally reshape these norms. By enhancing women's participation in decision-making and control over resources, such programmes foster equitable and sustainable community development.

Structural barriers, however, pose significant challenges to gender mainstreaming in climate policies. For instance, in Uganda, institutional and political inertia often hinder the effective integration of gender considerations, despite policy commitments (Acosta et al. 2021). Gender-transformative programmes provide a promising alternative by reshaping power dynamics, centring marginalized voices, and influencing policymaking to foster equitable resource distribution and social conditions (Singh et al. 2021; Mersha & Laerhoven 2018).

Empowering women through gender-sensitive climate initiatives can lead to substantial improvements in poverty reduction, family health, and education outcomes. For example, the climate-smart village (CSV) approach in India has enhanced women's participation in decision-making, contributing to their empowerment and adoption of climate-smart agricultural practices (Hariharan et al.

2018). Such methodologies demonstrate the potential for long-term benefits when gender-transformative strategies are embedded in programme design.

In summary, the adoption of gender-transformative approaches is critical for building community resilience and addressing systemic inequalities in climate adaptation strategies. These approaches allow for deeply entrenched systemic inequalities to be challenged and reconfigured so that individuals within the community can be empowered. Women, especially, not only stand to gain long-term equity and support against the impacts of climate change but also experience better social conditions and enhanced agency over their livelihoods, a benefit that is generally short-lived in interventions otherwise. However, the realization and incorporation of gender-transformative measures in one's policy or programme require an in-depth understanding and interrogation of existing systems and inequalities that give rise to varying vulnerabilities of gendered groups, thus allowing for a programme for moving from gender-blind to gender-sensitive to gender-transformative.

## Discussion

The reviewed studies explicitly acknowledge the differential perception, experience, and responses of men and women as well as the influence of adaptation strategies on men and women's resilience (Mehar et al. 2016; Twyman et al. 2014). It highlights how gender norms and cultural factors shape the varying vulnerabilities and adaptive capacities of men and women to climate change impacts (Khatri-Chhetri et al. 2017; Twyman et al. 2014). Additionally, they inform the influence of gender roles on women's capacities to adapt, as well as the importance of channels, such as education (Aryal et al. 2020), training, access to resources (Peterman et al. 2014), community and organizational support, and policy and institutional support in improving adaptation for both men and women in agriculture. The findings also highlight the importance of considering gender in research and development. This recognition of gender as a relevant factor demonstrates a move beyond a gender-blind or merely gender-sensitive approach. The majority of the studies have also discussed the need for transformative approaches and emphasize participatory research methods that thoroughly include women's voices and perspectives.

### *Implications of Gender Integration in Climate-Resilient Strategies*

The intent to adapt as well as the impact of climate-resilient strategies on men and women depends largely on how they are designed and implemented as well as on individuals' access to resources, their intersecting identities, and the prevailing social norms that shape their agency and decision-making power. While these technologies hold the potential to enhance agricultural productivity and resilience, their gendered implications can be both positive and negative. For instance, zero-tillage machinery, as discussed in Sect. 4.1, is often considered gender-neutral. Traditionally, this machinery, by design, was physically demanding and not conducive to use by women. Additionally, it requires financial investment, and women face structural barriers to accessing these funds (Sapkota et al. 2021; Kumar et al. 2018). It is important to acknowledge that zero-tillage technology is evolving, and newer models may be less physically demanding. Another aspect to consider is the rise in weed growth associated with this technology, which can increase the workload of women. This is particularly significant as women are primarily responsible for weeding and other post-harvest activities, which might become more labour-intensive with these technologies. In this way, zero tillage as well as direct-seeded rice with potentially reducing labour burden can be categorized as either gender accommodative or gender-sensitive, depending on their implementation. To achieve gender-accommodative outcomes, these technologies must address women's specific needs through inclusive design and equitable resource allocation. Participatory approaches, such as microfinance and community-based programmes, show promise in enhancing women's decision-making authority and access to financial resources, though their impact remains contingent on local contexts and gender dynamics.

Participatory approaches, such as microfinance and community-based programmes, show promise in enhancing women's decision-making authority and access to financial resources by addressing gendered barriers to credit and resource access. Microfinance and savings groups can promote gender responsiveness and transformative change by increasing women's financial access and recognition in households and communities. However, their impact may be limited if women lack autonomy over loan use. Additionally, access to information and bargaining power significantly influence gender dynamics in adopting climate-smart agriculture practices, as shown in a cost–benefit analysis in Southern Africa (Mutenje et al. 2019).

Gender-transformative approaches build on these participatory models to address the structural inequalities that shape the varied experiences of men and women in climate adaptation. Treating women as agents of change, rather than as vulnerable individuals, fosters equitable and sustainable outcomes (Sultana 2017). These approaches not only enhance women's agency but also improve the effectiveness and longevity of climate adaptation efforts.

## *Challenges in Implementing Gender-Inclusive Climate Strategies and Policies*

Despite growing attention to gender-transformative measures, significant challenges hinder their integration into climate programmes. Many organizations lack gender-disaggregated data and context-specific research, limiting their ability to design tailored interventions. Insufficient understanding of gender dynamics often reduces gender to quotas, with little consideration of deeper issues, and can lead to community resistance if gender initiatives are mis-framed (Acosta et al. 2021). Furthermore, inadequate monitoring and evaluation systems with few gender-sensitive indicators hinder the assessment and refinement of policies (Larson et al. 2015).

Structural and institutional barriers, such as bureaucratic inertia, inadequate budgeting, and limited staff capacity, further impede gender-responsive approaches. Competition for funding and entrenched bureaucratic practices often deprioritize gender-specific initiatives (Bryan et al. 2017). Cultural and social norms frequently conflict with formal gender mandates, complicating implementation.

Policies often include gender language without actionable strategies, resulting in a gap between policy and practice. For instance, in Uganda, while gender mainstreaming is a policy goal, structural and cultural barriers limit its implementation, with policymakers showing superficial engagement and scepticism (Acosta et al. 2021).

Projects like REDD+ , a global initiative aimed at mitigating climate change by incentivizing conservation and sustainable management in developing countries, highlight these limitations. While aiming for gender transformation, REDD+ initiatives often fall short, focussing on accommodative measures like including women in meetings or distributing benefits without challenging power dynamics. This underscores the need for structural reforms and robust guidance to ensure genuine gender equity and empowerment in climate programmes.

## Conclusion

Overall, the paper contributes to the existing literature by providing evidence-based recommendations to enhance gender equality and empower women in agricultural contexts. The arguments highlight the necessity of gender-transformative strategies that reconfigure societal norms, address systemic disparities, and empower women as agents of change. By embedding gender inclusivity across policy and programme design and emphasizing participatory frameworks, the chapter advocates for equitable resource distribution and the inclusion of women's voices in decision-making. These strategies are essential for fostering community resilience, improving adaptive capacities, and achieving sustainable development in climate-resilient agriculture.

## References

Acosta M, van Wessel M, van Bommel S, Ampaire EL, Jassogne L, Feindt PH (2019) The power of narratives: explaining inaction on gender mainstreaming in Uganda's climate change policy. Dev Policy Rev 37:631–651. https://doi.org/10.1111/dpr.12396

Acosta M, Riley S, Bonilla-Findji O, Martínez-Barón M et al (2021) Exploring women's differentiated access to climate-smart agricultural interventions in selected climate-smart villages of Latin America. Sustainability 13:10951. https://doi.org/10.3390/su131910951

Arora-Jonsson S (2011) Virtue and vulnerability: discourses on women, gender and climate change. Glob Environ Change 21:744–751. https://doi.org/10.1016/j.gloenvcha.2011.01.005

Aryal JP, Farnworth CR, Khurana R, Ray S, Sapkota TB, Rahut DB (2020) Does women's participation in agricultural technology adoption decisions affect the adoption of climate-smart agriculture? Insights from Indo-Gangetic Plains of India. Rev Dev Econ 24:973–990. https://doi.org/10.1111/rode.12679

Backman-Levy JK, Greene ME (2024) Gender-transformative programmes: a framework for demonstrating evidence of social impact. BMJ Glob Health 9:e014203. https://doi.org/10.1136/bmjgh-2023-014203

Barooah P, Alvi M, Ringler C, Pathak V (2023) Gender, agriculture policies, and climate-smart agriculture in India. Agric Syst 212:103751. https://doi.org/10.1016/j.agsy.2022.103751

Bernier Q, Franks P, Kristjanson P, Neufeldt H, Otzelberger A, Foster K (2013) Addressing gender in climate-smart smallholder agriculture. ICRAF Policy Brief 14. World Agroforestry Centre (ICRAF), Nairobi, Kenya

Birachi E, Hansen J, Radeny M, Mutua M, Mbugua MW, Munyangeri Y et al (2020) Rwanda climate services for agriculture: evaluation of farmers' awareness, use, and impacts. CCAFS Working Paper No. 304. CGIAR Research Program on Climate Change, Agriculture and Food Security (CCAFS), Wageningen, the Netherlands

Bryan E, Bernier Q, Espinal M, Ringler C (2017) Making climate change adaptation programmes in sub-Saharan Africa more gender-responsive: Insights from implementing organizations on the barriers and opportunities. Clim Dev 9:393–405

Carr ER, Onzere SN (2017) Effective (for 15% of the men): lessons in understanding and addressing user needs in climate services from Mali. Clim Risk Manag 22:1–14

Cornwall A (2016) Women's empowerment: what works. J Int Dev 28:342–359

Dayamba DS, Ky-Dembele C, Bayala J et al (2018) Assessment of the use of participatory integrated climate services for agriculture (PICSA) approach by farmers to manage climate risk in Mali and Senegal. Clim Serv 12:27–35

Diarra FB, Ouédraogo M, Zougmoré RB et al (2021) Are perception and adaptation to climate variability and change of cowpea growers in Mali gender differentiated? Environ Dev Sustain 23:13854–13870

Farnworth CR, Baudron F, Andersson JA, Misiko M, Stirling CM (2016) Gender and conservation agriculture in East and Southern Africa: towards a research agenda. Int J Agric Sustain 14:142–165

Gartaula H, Sapkota T, Khatri-Chhetri A, Prasad G, Badstue L (2020) Gendered impacts of greenhouse gas mitigation options for rice cultivation in India. Clim Change 163:1045–1063

Gonda N (2016) Climate change, "technology" and gender: "Adapting women" to climate change with cooking stoves and water reservoirs. Gend Technol Dev 20:149–168

Gumucio T (2022) Enhancing climate services design and implementation through gender-responsive evaluation. Front Clim 4

Gutierrez-Montes I, Arguedas M, Ramirez-Aguero F et al (2020) Contributing to the construction of a framework for improved gender integration into climate-smart agriculture projects monitoring and evaluation: MAP-Norway experience. Clim Change 158:93–106

Hariharan VK et al (2018) Does the climate-smart village approach include gender equality in farming households? A case of two contrasting ecologies in India. Clim Change

Howland F, Acosta M, Muriel J, Le Coq JF (2021) Examining the barriers to gender integration in agriculture, climate change, food security, and nutrition policies: Guatemalan and Honduran perspectives. Front Sustain Food Syst 5:664253. https://doi.org/10.3389/fsufs.2021.664253
IGWG and USAID (2017) Gender integration continuum. Available via: https://www.igwg.org/wp-content/uploads/2017/05/FG_GendrIntegrContinuum.pdf. Accessed 8 Nov 2024
Ingabire C (2021) Closing gender gaps in farmers' access to climate information: The case of radio listeners clubs (RLCs) in Rwanda, preliminary results of a mixed-method analysis. CCAFS Info Note. CGIAR Research Program on Climate Change, Agriculture and Food Security (CCAFS), Wageningen, the Netherlands
Jost C, Kyazze FB, Naab JB, Neelormi S, Kinyangi J, Zougmoré RB, Aggarwal P, Bhatta GD, Chaudhury M, Tapio-Biström M, Nelson S, Kristjanson P (2015) Understanding gender dimensions of agriculture and climate change in smallholder farming communities. Taylor & Francis 8:133–144. https://doi.org/10.1080/21582041.2015.1069922
Kabeer N (1999) Resources, agency, achievements: reflections on the measurement of women's empowerment. Dev Change 30:435–464. https://doi.org/10.1111/1467-7660.00150
Khatri-Chhetri A, Aggarwal PK, Joshi PK, Vyas S (2017) Farmers' prioritization of climate-smart agriculture (CSA) technologies. Agric Syst 151:184–191. https://doi.org/10.1016/j.agsy.2017.10.002
Kumar U, Bharati RC, Kant R, Chaubey KK et al (2018) Gender perspective of conservation agriculture. Indian J Agric Sci 88:1202–1207
Larson AM, Dokeen T, Duchelle AE et al (2015) The role of women in early REDD+ implementation: Lessons for future engagement. Environ Sci Policy 17:43–62. https://doi.org/10.1016/j.envsci.2011.07.006
Larson AM, Solis D, Duchelle AE et al (2018) Gender lessons for climate initiatives: a comparative study of REDD+ impacts on subjective wellbeing. World Dev 108:86–102. https://doi.org/10.1016/j.worlddev.2018.04.019
Mehar M, Mittal S, Prasad N (2016) Farmers coping strategies for climate shock: is it differentiated by gender? J Rural Stud 44:123–131. https://doi.org/10.1016/j.jrurstud.2015.12.008
Mehar M, Prasad N (2022) Climate change, adaptation and gender: Methodological underpinnings. CABI-Climate Change Series. CAB International, Wallingford, Oxfordshire, UK; Boston, MA, USA
Mersha AA, Laerhoven FV (2018) Gender and climate policy: a discursive institutional analysis of Ethiopia's climate resilient strategy. Reg Environ Change 19:429–440. https://doi.org/10.1007/s10113-018-1345-4
Mondal S, Chakraborty D, Bandyopadhyay K, Aggarwal P, Rana DS (2020) A global analysis of the impact of zero-tillage on soil physical condition, organic carbon content, and plant root response. Land Degrad Dev 31:1–16. https://doi.org/10.1002/ldr.3390
Mulema AA, Cramer L, Huyer S (2022) Stakeholder engagement in gender and climate change policy processes: lessons from the climate change, agriculture and food security research program. Front Sustain Food Syst 6:862654. https://doi.org/10.3389/fsufs.2022.862654
Mutenje MJ, Farnworth CR, Stirling C, Thierfelder C, Mupangwa W, Nyagumbo B (2019) A cost-benefit analysis of climate-smart agriculture options in Southern Africa: balancing gender and technology. Ecol Econ 163:126–137. https://doi.org/10.1016/j.ecolecon.2019.04.027
Nchanji EB, Kabuli H, Onyango N, Cosmas L, Chisale V, Matumba A (2022) Gender differences in climate-smart adaptation practices amongst bean-producing farmers in Malawi: The case of Linthipe Extension Planning Area. Front Sustain Food Syst 6:1001152. https://doi.org/10.3389/fsufs.2022.1001152
Njuki J, Melesse M, Sinha C, Seward R, Renaud M et al (2022) Meeting the challenge of gender inequality through gender transformative research: lessons from research in Africa, Asia, and Latin America. Can J Dev Stud 44:206–228. https://doi.org/10.1080/02255189.2022.2081131
Paudyal BR, Chanana N, Khatri-Chhetri A, Sherpa L, Kadariya I, Aggarwal P (2019) Gender integration in climate change and agricultural policies: the case of Nepal. Front Sustain Food Syst 3:66. https://doi.org/10.3389/fsufs.2019.00066

Peterman A, Behrman JA, Quisumbing AR (2014) A review of empirical evidence on gender differences in nonland agricultural inputs, technology, and services in developing countries. In: Quisumbing A, Meinzen-Dick R, Raney T, Croppenstedt A, Behrman J, Peterman A (eds) Gender in agriculture. Springer, Dordrecht, pp 145–186

Pham TT, Angelsen A, Hoang MH, Brockhaus M (2018) Women's participation in REDD+ decision making in Vietnam. Int for Rev 20:358–370. https://doi.org/10.1505/146554818823405701

Ragasa C (2012) Gender and institutional dimensions of agricultural technology adoption: a review of literature and synthesis of 35 case studies. In: Int Assoc Agric Econ Triennial Conf, Foz do Iguaçu, Brazil, 18–24 Aug 2012

Ravera F, Iniesta-Arandia I, Martín-López B, Pascual U, Bose P (2016) Gender perspectives in resilience, vulnerability, and adaptation to global environmental change. Ambio 45(Suppl 3):235–247. https://doi.org/10.1007/s13280-016-0842-1

Resurrección BP (2017) Gender and environment from "women, environment and development" to feminist political ecology. In: MacGregor S (ed) Routledge handbook of gender and environment. Routledge, United Kingdom, pp 71–85

Sapkota S, Shrestha SR, Adhikari PS, Jha SJ, Gairhe S (2021) Zero till in wheat from the gender perspective in Nepal. J Agric Nat Resour 4(1):106–117. https://doi.org/10.3126/janr.v4i1.33588

Singh C, Solomon D, Rao N (2021) How does climate change adaptation policy in India consider gender? An analysis of 28 state action plans. Taylor & Francis 21(7):958–975. https://doi.org/10.1080/2157930X.2021.1946517

Sultana F (2017) Gender and water in a changing climate: challenges and opportunities. Adv Water Resour Manag 110:45–52. https://doi.org/10.1016/j.advwatres.2017.09.009

Twyman J, Green M, Quinn S, Bernier Q, Kristjanson P, Russo S, Tall A (2014) Adaptation actions in Africa: Evidence from working paper adaptation actions in Africa. CCAFS Working Paper 83. CCAFS, Wageningen, Netherlands

Tye S, Mukasa H, Kratzer B, Businge G, Atieno R, Wamalwa R, O'Connor R, Trivedi A, Coger T, Elwell N (2021) Strengthening gender equity in locally led adaptation processes in Africa. Environ Sci Policy 45(3):123–140

UNICEF (2019) Guidance on gender integration in evaluation. Retrieved 22 Jan 2025, https://www.unicef.org/evaluation/media/1221/file/UNICEF%20Guidance%20on%20Gender.pdf. Accessed on 10 Nov 2024

UN Women (n.d.) Gender equality glossary. Retrieved from: https://trainingcentre.unwomen.org. (Accessed 20 Dec 2024)

World Bank (2012) World Development Report 2012: Gender equality and development. World Bank, Washington, DC

# Chapter 18
# Addressing the Crisis of Climate-Induced Internally Displaced Women in Bangladesh

**Zelina Sultana**

**Abstract** One of the climate change ramifications is forced displacement from the place of origin to a new place due to floods, storm surges and cyclones within the State, permanently or temporarily. Women face greater vulnerability than men during climate crises. Yet, their issues are inadequately addressed, limiting effective solutions for the harm they endure after displacement due to the negative impacts of climate change. Besides, the intersection of climate change vulnerability and existing gender inequalities undermines the principles of climate justice and the norms of Sustainable Development Goals (SDGs), particularly the commitment to leave 'no one behind'. This chapter focuses on the unique vulnerabilities of climate-induced internally displaced women. It identifies gaps in different terminology to define these groups of women that hinder their legal protection. The study finds that most of the time, the displacement issues are merged with the discussions of climate migrants and refugees in the name of environmental refugees and environmental migrants. Using various terminologies creates a vacuum in the extant literature and makes it difficult to take effective legal measures. This increases the existing vulnerabilities and keeps them unaddressed. This confusion also reduces the chances of accessing climate justice and achieving SDGs. Thus, the paper calls for addressing their vulnerability in all phases of climate displacement by using proper terminology, initiating effective legal protection and monitoring the Climate-Induced Internally Displaced Women situation. These are essential to ensure the legal protection of Climate-Induced Internally Displaced Women and ensure achieving climate justice as well as the related SDGs.

**Keywords** Climate change · Crisis · Displacement · Women · Legal · Protection · Vulnerabilities

Z. Sultana (✉)
Faculty of Business and Law, Anglia Ruskin University, Cambridge, UK
e-mail: zelina2011@gmail.com

Department of Law, Jagannath University, Dhaka, Bangladesh



A. Sarkar and N. Bandyopadhyay (eds.), *Gender-Transformative Approaches for Climate Change Adaptation*, Springer Climate,
https://doi.org/10.1007/978-3-031-90236-9_18

## Introduction

Bangladesh is one of the countries most affected by climate change and disasters often prompt displacement of people. People mostly migrate to cities or urban areas as observed in many regions in the global South. Bangladesh experiences both extreme events and slow onset events like cyclones, riverbank erosion, droughts, floods, saline water intrusion, sea level rise, etc. (Murshid and Yasmeen 2004). It is being called as a region of the climate crisis. Evertsen (2023) calls it a '*riskscape*' which is defined as a geographical space covered with estimated risks that must be managed.

The livelihoods of Bangladeshi people are diverse, but most residents depend on primary economic activities that are more dependent on nature like agriculture, fishery and forestry. Therefore, their livelihoods are at a greater risk of climate change disasters and events and require greater attention (Murshid and Yasmeen 2004). People who are displaced temporarily first try to adjust their livelihood strategies. It is difficult and, in many cases, causes maladaptation becoming a concerning issue on a national and global scale (Magnan et al. 2015). For example, some infrastructures like embankments have been installed in Koyra Khulna to prevent flood water and cyclone water. However, it has proved to be detrimental to the local people by raising their vulnerabilities due to waterlogging of saline water, decreasing the freshwater, the decline in native fishes and biodiversity, and blocking navigation by sedimentation of river channels (Thomas 2020).

It is projected that the susceptibilities of Climate-Induced Internally Displaced Women (CIIDW) will increase by 2050 and beyond 2100 because of the deteriorating impacts of climate change (Kirezci 2020). Considering the unequal effects of climate change and the degree of vulnerabilities of displaced women, they need more concentration, aid and facilities to adjust to the current situation. It is observed that women in women-headed households face more challenges in climate emergencies than housewives. Moreover, the climate crisis is not the same for all women of the same locality and varies with their age, social status, educational background, displaced location, socioeconomic and demographic conditions, etc. It demands intersectionality to understand the degree of oppression the women encounter based on their different variables such as education, race, ethnicity, language, age, and geographic location (IDLO 2021). This intersectionality lens will help to devise solutions needed for their protection by the policymakers. It will equip individual researchers to study the nuances of climate change impacts on vulnerable communities and their needs (Jackson Humphrey 2022).

## Material and Methods

The unequal crises or issues of displaced women cannot be treated with blanket policies as they can also initiate injustice towards the communities that suffer most (Kane 1996). The failure to provide distinct measures for the most vulnerable people will increase their vulnerability and inequality, which will again cause injustice. Based on several policy gaps that do not address the specific vulnerabilities of CIIDW, this chapter analyses the climate-induced internally displaced women's different vulnerabilities to ensure effective protection for them. It highlights that their crises affect their entire lives and livelihoods and exacerbate gender-based violence, trafficking and their well-being in general. The findings of this chapter contribute to providing an inclusive policy direction ensuring 'no one is left behind'.

The findings of this chapter are based on existing published literature, policy documents and secondary data. Reports and scholarly articles on climate change impacts on women in Bangladesh particularly in its coastal areas, are reviewed. As the term 'climate-induced internally displaced' is used with different terminology in the extant literature, all the literature related to the impact of climate change on climate migrants and environmental migrants of Bangladesh is also reviewed. However, some general literature relating to climate justice and SDGs has also been reviewed to highlight the necessity of addressing the vulnerabilities of the CIIDW. After careful review, the related information on vulnerabilities of CIIDW is used to highlight their distinct susceptibilities. Above all, a policy analysis method is applied to highlight briefly the legal and policy gaps to ensure the future protection of the CIIDW.

## Defining Climate-Induced Internally Displaced Women

In Bangladesh, most of the published literature particularly since the 1990s has frequently been using the term 'migrant' in place of 'displaced' without awareness of a concrete definition. It seems not using definite terminology is a result of the failure to understand the difference between the climate-displaced crisis and the climate migrant crisis in the current climate change context (Chowdhury and Uddin 2021). At present the term 'climate-induced internally displaced person' is not defined in any law or academic literature of Bangladesh but originates from the definition of Internally Displaced People (IDP) provided by UN Guiding Principles on Internal Displacement 1998 (GPID). According to these principles, IDP is defined as:

> Persons or groups of persons who have been forced or obliged to flee or to leave their homes or places of habitual residence, in particular as a result of or to avoid the effects of armed conflict, situations of generalized violence, violations of human rights or natural or human-made disasters, and who have not crossed an internationally recognized State border (Art. 2, Page 1 GPID).

This definition shapes the international protection of IDPs and encompasses three core elements: the involuntary nature of movement, movement due to natural disasters and the fact that such movement occurs within national borders. However, it is purposively descriptive rather than normative because it offers a list of characteristics of IDPs that can identify them as vulnerable. Still, this definition fails to distinguish/ or define the people who are displaced by climate change's slow onset effects such as salinity, sea level rise, or increase in temperatures. This legal definitional gap also paves the way for the use of different terms for climate-induced displaced, such as environmental refugees, environmental migrants, climate migrants, climate refugees, etc., deliberately. The frequently used term in existing literature is 'migrant' instead of 'displaced'. However, the submerging of their study in Bangladesh creates a lack of focus on the distinct crisis of climate change-displaced women. Moreover, it is difficult from the reviewed extant literature to find the data relating to the exact number of women deaths, displaced, impact on livelihoods, loss of livelihoods, oppression, violence, health hazard based on their age, ethnicity, wealth and education.

The gap in using definite terminology in existing literature creates a vacuum in the policies of Bangladesh. Table 18.1 shows how policies do not address the issue of displacement and CIIDW. For example, policies like Delta Plan 2100, Mujib Climate Prosperity Plan (2022–2041), and Nationally Determined Contributions (NDCs) lack addressing specific vulnerabilities of displaced communities, specifically women. Bangladesh's notable climate change policies include the Bangladesh Climate Change Strategy and Action Plan (hereafter BCCSAP, 2009), and the Climate Change Gender Action Plan (hereinafter CCGAP 2013). These plans addressed the strategies for disaster risk reduction and to some extent, included the international guidelines of UNFCCC related to the climate change issue but ignored the matter of climate-induced internal displacement (see Table 18.1 for more detail). These legal documents are general protection and development strategies, which also need specifications based on gender, age, ethnicity and their necessities. The gaps trigger injustices concerning gender, social, and climate and negatively impact attaining the goals of climate actions/ policies, including the development strategies that should prioritize the safeguarding of displaced women's rights in times of disasters, particularly to save their lives and livelihood (Shehu and Abba 2020).

Finally, the National Strategies on Internal Displaced Management, 2021, was developed to cover all internally displaced populations. However, this plan lacks a specific focus on the distinct vulnerabilities of displaced women, their relocation strategies and livelihood rehabilitation. This plan provides no solution for CIIDW who are displaced permanently. This less cognizant of their crisis is the direct cause of the legal protection gap, which is a constraint to ensuring climate justice (Jayawardhan 2017).

## CIIDWs and Their Vulnerabilities

The susceptibilities of CIIDWs vary with their status in society determined by their level of education, empowerment, socio-cultural background, and even the geography of their residence. For example, women of coastal areas in Bangladesh face multiple inequalities that are exacerbated after displacement (Murshid and Yasmeen 2004). Though there are few studies on CIIDWs and their issues of disproportionate vulnerabilities in Bangladesh, it still lacks in capturing the vulnerabilities of CIIDW's livelihoods (Garai 2016; Rahman 2013).

**Table 18.1** Reviewed legal documents with provisions for climate change migrant/displaced population

| Laws/policies with year | Addresses issues of displacement | Addresses issues of CIIDW |
|---|---|---|
| Bangladesh climate change strategy and action plan 2009 | Yes | No |
| Bangladesh climate change trust Act 2010 | Yes | No |
| 6th five-year plan (2011–2015) 2011 | Yes | No |
| Disaster management act 2012 | Yes | No |
| Bangladesh climate change and gender action plan 2013 | No | No |
| National sustainable strategy (2010–2021) 2013 | Yes | No |
| Bangladesh climate fiscal framework 2014 | Yes | No |
| Nationally determined contribution 2015 | Yes | No |
| Plan of action to implement sendai framework for disaster risk reduction 2015–2030, 2015 | No | No |
| Bangladesh delta plan 2100, 2018 | No | No |
| Standing orders on disaster 2019 | Yes | No |
| National strategy on internal displacement management 2021 | Yes | Yes |
| National plan for disaster management 2021–2025, 2021 | Yes | No |
| Mujib climate prosperity plan 2022–2041, 2022 | Yes | No |
| National Adaptation plan (NAP), 2022 | Yes | No |

*Source* Author's creation

Studies have reported that climate change is responsible for re-creating gender-based violence like rape, sexual exploitation, etc. for which women suffer the most. Violence can happen both during the disaster and after displacement. Women remain susceptible to such crimes and injustices due to a lack of protective measures in existing legal regimes and loopholes in their executions (Shehu and Abba 2020). As a result, affected women cannot recover their loss due to legal protection gaps and their issues remain unaddressed.

In 2016, CARE Denmark defined 'climate-induced internal displacement' and 'climate-induced internal migration', which also confirms that 'climate-induced internal displacement' people have faced more vulnerabilities than 'climate-induced internal migrants' due to the nature of the situation (CARE 2016) (Table 18.2). In

**Table 18.2** Distinguishing features of climate-induced displacement and climate-induced migration

| | Climate-induced displacement | Climate-induced migration |
|---|---|---|
| Decision | Unplanned/not decided | Pre-decided/Planned |
| Movement | Involuntary movement | Voluntary movement |
| Destination | Not Known | Mostly Known |
| Causes | Climate disasters like cyclones, floods, storm surges, tornadoes, earthquakes, river erosion, etc | Slow onset climate events like sea level rise, salinization, desertification, droughts, etc. that cause shortages of food, water, and economic opportunities |
| Nature | Temporary or permanent | Form of survival or erosive coping migration |
| Mental shock | Sudden and affected shocked mentally | Pre-planned, mental shocks can be avoided |
| Crisis | Uncertainty of destination increased the humanitarian crisis | Crisis faced comparatively lower |
| Reproductive health/illness | Illness increased both physical and psychological | Less illness is found as it is pre-planned |
| Violence | Gender-based violence increased | Comparatively less Gender-based violence |
| Effect on girl children | Increased child marriage and school dropouts | Comparatively less for pre-planning activities |

*Source* Prepared by the author based on the definition and secondary literature

displacement, people cannot foresee the severity of its hazard and immediate impacts, so they cannot plan accordingly, making them more vulnerable than in similar situations. This matter is also ignored in the literature. Thus, their vulnerabilities need a specific focus to understand the level of vulnerability and protective measures they need.

## *Vulnerabilities Before and After Disaster Displacement of Women*

Bangladesh features both temporary and permanent climate-induced internal displacement. It is home to different climate disasters and events, different sociopolitical consequences, and perpetuated risk and vulnerabilities (Murshid and Yasmeen 2004; Shamsuddoha and Chowdhury 2007). In addition to this, uncontrolled population growth is creating pressure on the agricultural sector and available resources reducing its availability. For instance, during the inter-census period between 1960

**Table 18.3** Pre- and post-displacement issues of women

| Pre-disaster issues | Post-disaster/displaced issues |
|---|---|
| — | Mental trauma |
| — | Post-traumatic stress disorder (PTSD) and suicide |
| — | Conflict over natural resources and domestic insecurity |
| Less access to properties | Less access to properties and disputes |
| Less participation in decision-making | Less participation in decision-making and marginalization |
| Gender discrimination<br>Gender-based violence<br>Victims of child marriage<br>Less awareness regarding rights | Gender discrimination<br>Increased gender-based violence<br>Dropout from school and education<br>Increased child marriage<br>Less awareness regarding rights |
| Wage discrimination | Fewer opportunities for work<br>Wage discrimination<br>Increased unemployment and anti-social activities |
| Human trafficking | Increased human trafficking |

*Source* Prepared by the author based on the definition and secondary literature

and 1996, it was discovered that the agricultural dependency of the coastal households changed significantly due to less arable land, extreme weather patterns and shifting agro farms into fish farms (Shamsuddoha and Chowdhury 2007). However, the living standard of coastal people, per capita income, farmland, education, health access and social security are not up to the mark with other parts of the country (Murshid and Yasmeen 2004). Presently, disasters have become frequent, severe, and more devastating, making people more vulnerable affecting their lives and livelihood even to a greater extent. Thus, the existing vulnerable situation of women generates some unique crises when they move from their residents due to climatic hazards to other places (Table 18.3). They face the following challenges in both temporary and permanent displacement situations.

## *Mental Trauma and Shock*

There is a strong connection between climate change disasters and mental disorders. Climate-induced displacement triggers social tensions and psychological damage with the experience of post-traumatic stress, anxiety, depression, and sleep disorders (Goudet et al. 2023). Some disasters like cyclones and flash floods affect instantly, creating fear of losing lives and homes and this fear creates mental shock and trauma. After disasters, women become more anxious and mentally stressed than men due to the loss of livelihoods and scarcity of food and water. Married women show more stress to maintain the family expenditures and well-being of their children due to the loss of livelihoods of their husbands.

Goudet et al. (2023) found a relationship between climate change disasters and the increase in mental stress of men and women based on an empirical study. This study reports women's social and family responsibility with the climate change situations make them more affected by mental health issues than men. The study notes that men become vulnerable mentally due to livelihood and economic security, while women are more stressed and have responsibilities of caregiving and increased workload due to water scarcity and food insecurity. Though the study mentions the mental stress of women whose husbands migrate to cities for income, it does not mention how climate change may induce poor mental health among women when their livelihood and income are in distress (Goudet et al. 2023). This is particularly relevant for women-headed households where women bear the same vulnerability regarding income and livelihoods and mental pressure as men being the primary breadwinners of their families. Goudet's research fails to cover this crucial issue.

## *Post-Traumatic Stress Disorder (PTSD) and Suicide*

The impact on mental health in the absence of mental well-being is extremely severe which induces one to commit suicide due to gender discrimination and subordinate culture (Palinkas and Wong 2020). Besides, when the PTSD symptom persists for a long time, it becomes a concern for family and community members even affecting their mental well-being. A study in Bangladesh showed that 20% of rural women were diagnosed with a major depression disorder (Sparling et al. 2020). Unfortunately, the mental health infrastructure is inadequate in Bangladesh with the presence of 260 psychiatrists for 162 million people (WHO 2020). In an empirical study on the people of Koyra in 2016, it was stated that *women fainted and became unstable mentally by just seeing the damage affected by cyclone Aila even though they are trying to recover from the shock (*Kabir et al. 2016*).* The country is now in a dual crisis with an increase in mental health crises and the climate crisis.

The catastrophic occurrences also have a disruptive influence on the physical as well as mental well-being of children. The frequency of post-disaster impact on youngsters and children ranges from 3.6 to 37% (Lee et al. 2020). Moreover, depression affects differently in boys and girls. Hayward and Ayeb-Karlsson (2021) have found girls to be more susceptible than boys to having a psychological disorder. Thus, children demand special attention and extra support in cases of climate change migrations and displacements.

## *Increased Gender-Based Violence*

Climate-induced gender-based violence (GBV) is exacerbated due to resource scarcity, unpaid household work, existing gender inequality, and social disparity (Islam 2010). Desai and Mandal (2021) explain both disasters and a lack of protective systems trigger the violence. This is true for disaster-affected areas of Bangladesh, due to the disruption of protective mechanisms and existing religious and cultural norms. GBV is often not reported in police stations due to social taboos. The violence of climate-induced displaced women is markedly different from the violence of climate-induced displaced men as women become victims of rape and sexual harassment (Le Masson 2022), which is further exuberated in communities with social and religious taboos.

Disasters are also responsible for increasing other forms of violence such as domestic violence, spouse abuse and intimate partner violence (Desai and Mandal 2021). However, during and after displacement both times require the change of places, increased tension and anxiety of male members and sometimes become reasons for physical violence (Desai and Mandal 2021; Hossain and Rahman 2021). GBV is also associated with climate change effects (Secretariat 2020). In climate-induced disaster displacement, women are more prone to sexual violence, and it also rises when families are displaced and bound to live in camps, slums or open spaces that lack the privacy of space (Hossain and Rahman 2021; Le Masson 2022). A study conducted in the coastal area of Satkhira Upazila of Bangladesh confirmed that harassment, including GBV violence, is common in this area which women mostly face while collecting fresh drinking water (Secretariat 2020). This is because the coastal areas lack fresh drinking water during and aftermath of disasters and women are forced to walk long distances to collect water.

The incidents of GBV are also more common among poor and marginalized women in coastal areas due to lack of access to resources such as food, water, and firewood. During climate displacement, women are responsible for managing and fetching these household resources making them exposed to longer times in unexplored territories increasing the risk of violence and abuse. In most cases, perpetrators go unpunished as either women lack access to protective legal measures or there are loopholes in the system. Sexual violence like rape, attempt to rape, sexual harassment, torturing of women and children and teasing and lewd comments on unmarried women and girls are reported in cases of displacement (Shahjahan 2023, Hossain and Rahman 2021).

Evidence also suggests that displaced women face greater domestic violence like being beaten by their husbands and verbal abuse in private and public spaces. Instances of domestic violence also include harassment related to the issue of dowry. Torture by the husband and his family is reported to increase in households facing displacement. To summarize, women are the victims of different violence during disasters and after displacement which needs to be noticed to initiate effective protective measures.

### *Child Marriage and Trafficking*

The most egregious form of crime perpetrated because of ineffective disaster response strategies in climate change-affected areas include human trafficking, smuggling, and other unlawful activities targeting women and children. Climate change, which is a major cause of catastrophe, is also to blame for an increase in underage marriage, particularly among girls. Bangladesh experiences the topmost incidence of child marriage in Asia and throughout the world (Nusrat 2019). Education, on the flip side, has consistently been proven to be a strong preventive strategy for children against early marriage, trafficking, and criminal activities (Ezeibe et al. 2021).

Climate change is now considered a significant cause of the trafficking of women (Ezeibe et al. 2021). Centre for International Climate and Environmental Research (CICERO) reported that the trafficking of girls and women and other forms of GBV have significantly grown due to disasters in South Asia (Desai and Mandal 2021). The coastal area mainly is treated as the hub due to the presence of both sea routes and land routes for the trafficking of girls both for forced labour and forced prostitution (Vives and Murga 2022). There is no data to accurately estimate the exact number of trafficking events from the coastal areas of Bangladesh each year. Approximately, in total, 20,000 to 50,000 Bangladeshi girls are trafficked via India every year (Vives and Murga 2022).

In an empirical study by Rahman et al. (2023), in adjacent areas of Koyra and Khulna, the participants in the research reported that girls and women frequently experience GBV, both sexual harassment and eve teasing. They argued that GBV, human trafficking, polygamy, and early marriage are increasing with climate change, exposing girls and women to further atrocities. Climate-change-induced displaced households to suffer from joblessness and economic challenges. Many times, such affected families get duped by unlawful practices and women, girls and children get lured by the promise of good life and trafficked to other countries.

## Women's Lack of Access to Resources and Insecurity

Evacuation from a disaster zone is one of the challenges for the affected people. However, 75% of displaced persons are women, including children (Rahman 2013) who face challenges during departure. In extreme disaster situations, displacement is harder for women than men due to less access to climate/disasters-related news and training and less access to land, resources and decision-making processes (Garai 2016; Rahman 2013). In the context of Bangladesh, mass displacement is a cause of domestic security threats and social conflicts over resources (Saul 2012). Besides, displaced people often settle their new homes in government *Khas* land which creates tension between local leaders or powerful persons (who already captured the land and rented it to others) and the displaced (Saul 2012). The risk of conflict is increased in Bangladesh. In such cases, women are at a higher risk of deprivation and violence than men in all situations after displacement. However, the vulnerability of women is more than men due to pre-existing social and cultural norms and legal systems.

## Less Awareness Regarding Rights

Women in rural and coastal areas in Bangladesh are mostly uneducated (Murshid and Yasmeen 2004) and not empowered (Mazzeo and Rashed 2003). In most cases, they are unaware of their rights to their assets and even other special provisions available to them (UNICEF UK n. d.). This unawareness of rights creates a gap in tackling climate change disasters and events. It becomes a common challenge to make appropriate decisions at the time of family displacement (UNICEF UK n. d.). After displacement, thus women also become more vulnerable. In most cases, girls and women have to live on the roadsides and survive by doing household work. They are sometimes deprived of basic human rights like rights to privacy and education without any voice (UNICEF UK n. d.).

The lives of women and girls in open spaces and roadsides after displacement are difficult. Especially they struggle for their privacy and lack of privacy increases the risk of sexual violence and exploitation (Henry and Adams 2018). Even women face disruption of critical services, including sexual and reproductive healthcare, life-saving assistance and social protection during disasters and after displacement. Many times, they are not aware of their rights, and it leaves them in critically vulnerable situations (Muttarak and Lutz 2014).

Traditionally, women in rural areas have to adjust to all situations, even after displacement. Women in Bangladesh are socialized to adjust to difficult situations and are taught to accept hardships as their 'fate' (Murshid and Yasmeen 2004). Due to such socio-cultural upbringing, women learn to adapt to social taboos and remain silent while suffering from hardships even when they require care and support; especially during menstruation, pregnancy, and while experiencing other health challenges. They accept poverty, unhealthy procedures and harassment as a way of life (Murshid and Yasmeen 2004). In contrast to such socializing, evidence suggests that educating girls can easily increase community resilience, support adaptation capacity, and enable girls to engage in the new green economy and contribute to green solutions (ILO 2016). Therefore, educating girls can promote sustainable practices and principles, empowering them to act for climate justice and ensuring their voices are included in climate-related discussions.

## Culture and Gender Discrimination

Persistent discrimination against women in the society of Bangladesh is the root of all evil. Despite the country's guarantee to comply with gender equality (The Constitution of People's Republic Bangladesh 1972) and gender-sensitive climate policies (ccGAP 2013), disparity indeed endures. Ferdous and Mallik (2019) highlighted that women face heightened vulnerability due to discriminatory social norms and practices. These include limited access to property ownership, inadequate educational opportunities, early marriages, a dowry system, and societal acceptance of domestic violence against women. Such discrimination obstructs women's empowerment, leaving them among the poorest and most powerless within marginalized groups. As stated, this discrimination is a structural problem that starts from their early childhood and continues throughout their lives.

The vulnerability to climate change has increased by 83.01% in Bangladesh between 2008 to 2015, which is the highest among the 20 most climate-affected countries in the world (Wheeler 2011). For instance, in Bangladesh sea level rise will cause 13.2 million people to be vulnerable, which is projected to increase two times by 2050 (Wheeler 2011). This increased vulnerability will inevitably result in an escalation of vulnerability among women. More climate-related disasters will increase climate-induced displacement disproportionately affecting women (Murshid and Yasmeen 2004). For example, the floods of 1988, 1997, and 1998 and cyclones Side and Aila were some recent devastating calamities that negatively impacted livelihood in Bangladesh and led to mass displacement of people. According to a survey conducted by Action Aid in 2009, among the total populations affected by Cyclone Aila, women in the age group of 18–60 years were affected the most followed by girls in the age group of 5–18 years as compared to the same age group of men and boys (Islam 2010). The ratio of vulnerabilities of women is high due to gender disparity, social norms, and lack of access to assets, such as property, credit, decision-making bodies, farming inputs, technology, etc. (Islam 2010).

As a Muslim-majority country, women are restricted from going outside without *pardha* (additional veil/clothes covering their body and face). This limits their freedom of mobility and especially restricts their opportunity to pursue activities outside the.

home (Gomes et al. 2022). They are not allowed to go far from home as they have domestic responsibilities at home added to a fear of sexual harassment (Mazzeo and Rashed 2003). For religious and cultural reasons, most women do not go out to learn swimming. Research has found more women than men in Bangladesh die in floods and cyclones because they cannot swim (Mazzeo and Rashed 2003).

## Lack of Education and Economic Opportunities

Education is the least prioritized and seldom restored promptly during as well as after a disaster. Common ways to curtail access to education in catastrophic emergencies for the displaced include school closures for a longer period which may be due to flooding, collapse of institutional buildings, unavailability of school buildings when used as disaster shelter centres, and so on. It has been reported that in such cases there are large numbers of dropouts, especially among female students. In most cases, the girls are not allowed to complete their academic curriculum and are forced into early marriages. The consequence is an increase in child marriages, child labour, child trafficking, prostitution and juvenile delinquency (Murshid and Yasmeen 2004; Hossain and Rahman 2021; Desai and Mandal 2021).

Social norms and practices support investing more money in boys for their education so that they are gainfully employed and can support their parents in their old age. On the contrary, since girls migrate to live with their in-laws, they are not expected to live with their parents in their old age and economically support them (Ferdaush and Rahman 2011). The social norm in both urban and rural societies of Bangladesh thus mandates girls to be taught household chores so that they can fulfil their domestic duties more effectively. Rather than being gainfully employed women are restricted to fulfilling roles of childbearing, caregiving and all domestic chores after marriage (Ferdaush and Rahman 2011). Such social expectations are manifested in the form of low gross enrolment of girls in primary schools than boys that becomes further skewed at secondary, higher secondary and eventually to university levels (Murshid and Yasmeen 2004).

Lower levels of education restrict women's choices for employment outside homes and empowerment in general impacting their future lives as women, mothers and household decision-makers. With a lack of formal education, women are either employed in the informal sector or in primary economic activities. Traditionally, women have predominantly been engaged in unpaid or non-monetized sectors, participating in subsistence activities like household chores, cooking, caring for children and the elderly, and tending to livestock such as cattle, ducks, and hens. Climate-induced disasters threaten the livelihoods of women due to their heavy dependence on natural resources. For example, women living in coastal areas of Bangladesh generally rely on natural resources like agriculture and fishing, which are directly affected by climate hazards (Garai 2016). There is however, evidence to suggest that education helps to reduce the vulnerability of women to climate change by enhancing their resilience, ability to adapt, and preparedness for disasters (Muttarak and Lutz 2014).

## *Wage Discrimination and Poor Working Conditions*

Wages vary from place to place. However, evidence suggests wage rates of women are 50% less than that of men at all levels (Murshid and Yasmeen 2004). The wage rate varies over the year depending on the harvesting seasons of crops, and if disasters damage crops, these wage rates fall, and the people, including women, lose their livelihood and wages due to disasters (Murshid and Yasmeen 2004). The draft final report of 'Vulnerability Analysis of Major Livelihood Groups in Coastal Zone of Bangladesh ' (CEGIS) recommends that women's low-income ratio is a major reason for their vulnerability to disasters (Murshid and Yasmeen 2004).

The vulnerable condition of displaced women increases by two dimensions – first being displaced and second being a woman (Chowdhury and Uddin 2021). In a study on migrant workers' wages in Dhaka city, Chowdhury and Uddin (2021) argued that displaced people are not able to promptly find jobs at their new places of residence. It is not only hard for them to negotiate and wait for opportunities and are even unable to bargain for wages. Displaced women, most of the time do not get work as they are newcomers and are unknown to all the employees, particularly in the informal sector. As a result, they mostly start to work as daily wage labourers (construction workers, agricultural workers) or domestic workers. However, women from women-headed households are bound to engage in outdoor labour because of their lack of choice irrespective of their religion (Mazzeo and Rashed 2003). It has also been observed that in many cases even middle-class and wealthy women, had to take up low-wage and (culturally) low-status jobs for survival. Due to a lack of formal education and training, most women in dire need of income support engaged in jobs of domestic servants, cleaners and daily wage labourers.

Women often face discrimination concerning payment, work environment, low wages and lack of social security net (Chowdhury and Uddin 2021). Women in most cases face discrimination at the workplace by typically getting lower wages as compared to their male counterparts (Chowdhury and Uddin 2021). However, the nature of discrimination varies depending on the nature of work and profession. Women who received less payment than their male counterparts from their employers accept such wage discrimination. Being uneducated and unaware of their rights as workers/employees, most of the women do not even know or understand this aspect of discrimination (Chowdhury and Uddin 2021). Most importantly, all forms of such workplace discrimination increase when women are displaced and struggle to settle in a new place.

## *Crime and Anti-Social Activities*

The nature and characteristics of disasters, geographical location, available resources, etc., are the determinants of the anti-social activities in a specific area. Throughout the year, crime activities may change, but research shows climate change disasters have an increased correlation with the increase in unlawful activities in coastal areas of Bangladesh (Alam and Islam 2024). In Bangladesh, climate change disasters affect the lives and livelihoods of coastal areas and after disasters and displacement theft, extortion and violence increase (Shahjahan 2023). The authors observed increasing evidence of stealing, harassment, exploitation of relief supports, smuggling, and hijacking during and after disasters like in the case of Cyclone Sidr and Aila.

People try to adopt illegal ways to meet their daily needs. Alam and Islam (2024) established a link between catastrophes and crimes. Women of coastal areas engage in some unique illegal activities, such as using illegal nets for fishing and going illegally to restricted areas (Sundarbans) or collecting natural resources like fish, crabs, timber, and honey from restricted areas. Displaced people, particularly women when moving from their local areas face challenges in finding suitable jobs in urban areas, which also triggers their anti-social activities such as drug dealing and smuggling.

## Land Rights and Access to Properties

Research suggests that women possess less ownership and control over land in rural areas of Bangladesh due to societal, cultural, and gender norms (Sultana et al. 2024). Access to property is one of the determinants of the positions of women in family and society. It makes them poorer and after displacement makes them the poorest. It is also associated with lower social and economic status. It hampers their chances of finding new work to support their families in new places. It is especially concerning when women are the breadwinner of their families. According to the inheritance law of the land, women can inherit half of their male counterparts in all situations such as daughter, wife, widow, mother and sister (Sections 29 to 54 of The Succession Act, 1925). This discriminating land rights and lack of land tenure for women exacerbate their vulnerability to climate change impact (Wickramaratne and Silva 2023).

There is also gender disparity in accessing various kinds of assets. While men usually own fixed assets women are generally endowed with movable assets (Anon 2002). For example, men have access to wealth like land, and water, whereas women have access to furniture, poultry, household utilities, jewellery, sewing machines and so on (Murshid and Yasmeen 2004). Women from wealthy families may inherit farmland or have some access to land bought by their husbands. In women-headed households, women have ownership of both homestead and farmland (Anon 2002). However, even if they own some land, most women cannot sell or transfer their title to others without the permission of the male members of their family (Mazzeo and Rashed 2003). Hence, regional differences, religious values, limited mobility, and discriminatory inheritance laws all affect women's approach to wealth and assets. Thus, women with less property and mobility experience two-fold challenges during and after disaster displacement, such as facing mobility crises and associated challenges for new employment and livelihoods.

## Lack of Participation in Decision-Making in Families

Women, being poor and having less access to property and resources can participate less in decision-making both within families and in society (Sultana et al. 2024). The lack of participation in decision-making processes has consequences on response to disasters increasing their vulnerability further in climate change context. Two decades ago, the decision-making power of women was limited to small purchases (Murshid and Yasmeen 2004). Slowly, this trend is changing to mutual decision-making in all aspects of family matters due to the economic sovereignty of women members and political empowerment. However, still today decisions regarding leaving home to migrate to other places are left to the male members of the family with some exceptions in cases of women-headed households (Mazzeo and Rashed 2003). Thus, in the presence of males, women have very little say in any type of major decision, even in times of climate emergency and during displacement.

## Major Findings

This chapter focuses on the multiple climate-induced crises of displaced women in Bangladesh and highlights that climate-induced internal displacement issues are distinct from climate-induced migration. There is increased gender-based violence, mental trauma, school dropouts, security threats, and increased displacement-related tensions as a consequence of climate change-induced disaster displacements. However ambiguous use of the terms leads to confusion mixing climate-induced displacements with other forms of displacements. As a result, policies are unable to address the unique challenges faced by the CIIDW and lack implementation of protective measures.

At present, the issues of climate displacement are merged with economic migrants. This is a conceptual problem because climate/ environment migrants and refugees are displaced or are forced to leave their place of origin. In cases of extreme events, they do not even have time to plan for the migration. On the contrary, in cases of labour migration, people can plan and then migrate in search of better economic opportunities. Due to several social, cultural and religious reasons, when women are forced to migrate, they become extremely vulnerable. Thus, studies on migrants and refugees are inadequate to evaluate and analyze the vulnerabilities of the CIIDW who face increased gender-based violations, gender disparity, less access to property, mental stress/trauma, post-traumatic disorder, etc.

In Bangladesh, climate change laws and policies have been enacted since 2009, but none of them focuses on CIIDW issues or their protections. Most of the laws and policies are general, and Bangladesh needs to implement and enact special and robust plans and policies for the effective protection of CIIDW. It is important to address issues of CIIDW to ensure legal protection, climate justice, social equity and achieving SDGs. However, climate change displacement is overlooked within policy frameworks, as current national policies have yet to establish a critical connection between climate change impacts and the worsening of associated challenges (Wilkinson et al. 2016).

## Policy Recommendations

Guaranteeing access to rights as citizens and humans is essential in incorporating CIIDW's issues at the national and local levels (IDLO 2021). Displacement caused by climate change deprives them of human rights such as the right to life, health, privacy, a safe environment, and freedom from gender discrimination. This also acts as a barrier to achieving climate justice. Addressing their crises in disaster and displacement situations will help to take proper protective measures for them and future generations, including assisting to empower and reduce disparity. According to UNDP Bangladesh Gender Equality Strategy 2023–2026 (UNDP n.d.), some crucial approaches to mainstreaming gender are empowering them and reducing gender inequality. It is based on core principles such as guaranteeing human rights, leaving no one behind, transforming gender and power relations, engaging men and boys, contextualizing challenges and monitoring. Based on the findings from the study, some broader policy goals and suggestions are suggested (Table 18.4).

Challenges of women's displacement are essential to address differently to avoid injustice and ensure climate justice. Although climate justice (CJ) is a comb of interdisciplinary research, it is related to gender justice, social equity and environmental movement. CJ should recognize that the impacts of climate change affect women unequally and disproportionately increasing their vulnerabilities. It also triggers several forms of injustice like recognizing, distributing and procedural injustice (Sultana 2022). Therefore, ensuring CJ will not be possible for Bangladesh if we leave behind the CIIDW. Their susceptibilities need to be addressed on a priority basis and checked accordingly. The government can ensure, social and climate justice and protection of the women by adopting necessary steps like monitoring the places where the displaced are settled and looking after supplying sufficient shelter, livelihood, food, water and health care facilities (UNHCR UK n.d.).

CIIDW's crisis should be addressed in the policies and plans relating to development policies highlighting the pre-existing inequalities and vulnerabilities of women, including social protection. Social protection policy is a policy instrument that the government uses to help individuals and households manage income and livelihood risks or to reduce poverty and inequality (Costella 2023). It can play a vital role in addressing the socioeconomic implications of climate change by managing poverty and income risks. Social protection policies which are known as a social safety net (SSN) such as old age aid, widow aid, disabled aid, etc. are provided by the Government of Bangladesh to a limited extent and are not yet connected with climate-related plans and policies (Costella 2023). SSN can help to establish equity and enhance social justice by protecting the climate-displaced livelihood and work as a part of individual rights protection.

As the displaced people remain inside the territory of the country of origin, it is the responsibility of the States to protect CIIDW from the disaster displacement crisis (UNHCR UK n.d.). Now, the responsibility of the Government of Bangladesh is to save its CIIDW from the current displacement crisis and disaster impacts by

**Table 18.4** Findings and suggestions for CIIDW

| Findings | Broader policy goals | Suggestions |
|---|---|---|
| • Increased and unique vulnerabilities of CIIDW<br>• Uncertain terminology<br>• Lack of legal protection | • Protect affected women<br>• Ensure climate justice, social equity<br>• Achieve SDGs 5 and 13<br>• Ensure robust legal protection | • Definite terminology uses<br>• More research focusing on CIIDW issues<br>• Promoting gender justice and social equity<br>• Protection of individual rights by implementing laws<br>• Monitoring CIIDW situation regularly |

*Source* Author's creation

implementing robust legal protection. Addressing the distinct vulnerabilities faced by the CIIDWs will help to determine the protective measures they need. Focusing on the extant definition of 'internally displaced person' (IDP) and 'climate-induced internal Displacement' (CIID) provided by the document and report, the displaced persons should be addressed with specific terminology to identify their hurdles in the current climate change context and displacement situation. However, empirical research on CIIDW vulnerability is essential to understand the degree of their vulnerabilities and to determine the degree of their protection.

Considering the crises of climate-induced displaced women, it is worthwhile for Bangladesh to minimize gender discrimination and associated vulnerabilities by implementing the norms of climate justice and social equity. Thus, for the protection of the human rights of the CIIDW in the future, their crisis and challenges should be considered in embracing new policies and plans, including women-centred legislations and policies, as well as promoting a strong and effective justice system. For this, embracing gender-responsive strategies (focused on eliminating gender discrimination) is crucial to enhancing socioeconomic incorporation and providing access to health, education, livelihoods, decent work and SSN.

Ensuring CIIDW rights and safety by securing the rule of law and ensuring climate justice is crucial (Murshid and Yasmeen 2004). Considering the greater gender disparity in the country, the combined efforts of the government, donors, and NGOs should be employed to promote gender development and achieve the shared goal of improving the gender development index in Bangladesh. The human rights protection approach can be an effective tool for undertaking policies and programs to minimize climate-induced susceptibilities (Hossain et al. 2021). Considering the approach is an effective way of overcoming the challenges faced by CIIDW, the Government should initiate a monitoring system for monitoring CIIDW conditions after their displacement, increase the budget on climate risk and adaptation research, increase social protections and develop effective policies.

Bangladesh should take initiatives to mitigate gender inequality, encourage decision-making, strengthen the regulatory framework, enhance the governing system of land and resources and embrace social justice, which will also pave the way to achieve some SDGs (like 3, 4, 5, 8, 10, 16) (IDLO 2021). In addition to this, the individual rights protection of CIIDW will contribute to achieving climate justice. If the Government of Bangladesh wants to take SDGs delivery plans, it should include the CIIDW issues there. Implementing and enacting policies and plans will be the first step to addressing their vulnerabilities.

## References

Alam N, Islam MM (2024) Understanding the nexus: causes, hotspots, and impacts of crimes amid climate Change in the Southern region of Bangladesh. Int J Res Innovat Soc Sci 8:1646–1656. Accessed 10 Sept 2024

Anon (2002) Perceptions of direct stakeholders on coastal livelihoods. working paper WP004, Program Development Office for Integrated Coastal Zone Management Plan Project; Bangladesh Dhaka

International Development Law Organization (IDLO). Creating a Culture of Justice' (2021) Available via IDLO https://www.idlo.int/sites/default/files/pdfs/publications/climate_justice_policy_paper_climate_action_final.pdf. Accessed on 23 Sept 2024
ccGAP (2013) Bangladesh climate change and gender action plan. Available via the ministry of environment and forest (MoEF) Government of the People's Republic of Bangladesh. SUBMITTED_Bangladesh_climate_change_gender_action_plan.pdf. Accessed 11 November 2024
Chowdhury SA, Uddin MM (2021) Discrimination against the internally migrated women workers: a study on Dhaka city. Soc Sci Rev 38:135–157. https://doi.org/10.3329/ssr.v38i2.64466
Costella C et al (2023) Can social protection tackle emerging risks from climate change, and how? A framework and a critical review. Clim Risk Manag 40:100501. https://doi.org/10.1016/j.crm.2023.100501
CARE Denmark (2016) Fleeing climate change: impacts on migration and displacement. Available via https://careclimatechange.org/fleeing-climate-change-impacts-migration-displacement. Accessed 12 Apr 2024
Desai BH, Mandal M (2021) Role of climate change in exacerbating sexual and gender-based violence against women: a new challenge for international law. Environ Pol Law 51:137–157. https://doi.org/10.3233/EPL-210055
Evertsen KF (2023) Women as subjects of risk in Bangladesh's coastal riskscapes. The Sociological Rev 71:154–1171. https://doi.org/10.1177/00380261231175223
Ezeibe C, Oguonu C, Ajaero CK, Osadebe N, Agbo H, Uwaechia O (2021) From vulnerability to sustainability: implementation of free education programmes and reversal of child trafficking in Nigeria. J Human Trafficking 7:104–118. https://doi.org/10.1080/23322705.2019.1647506
Ferdaush J, Rahman KMM (2011) Gender inequality in Bangladesh. Available via Unnayan Onneshan, Dhaka Bangladesh https://archive.nyu.edu/bitstream/2451/33900/2/gender_inequality_in_bangladesh.pdf. Accessed 17 Sept 2024
Ferdous J, Mallick D (2019) Norms, practices, and gendered vulnerabilities in the lower Teesta basin, Bangladesh. Environ Dev 31:88–96. https://doi.org/10.1016/j.envdev.2018.10.003
Garai J (2016) Gender-specific vulnerability in climate change and possible sustainable livelihoods of coastal people: a case from Bangladesh. J Integrated Coastal Zone Manage 16:79–88. https://doi.org/10.5894/rgci656
Gomes Md AC, Dias JM, Cerdà A (2022) Exploring gender and climate change nexus, and empowering women in the southwestern coastal region of Bangladesh for adaptation and mitigation. Climate 10:172.https://doi.org/10.3390/cli10110172
Goudet JM, Arif FB, Owais H, Ahmed HU, Ridde V (2023) Intersecting vulnerabilities: unravelling the complex relationship between women, climate change, and mental health in Bangladeshi communities. medRxiv 10:2023–06. https://doi.org/10.1101/2023.06.06.23290721
GPID (1998) United nations guiding principles on internal displacement available via https://www.internal-displacement.org/sites/default/files/publications/documents/199808-training-OCHA-guiding-principles-Eng2.pdf. Accessed 10 May 2023
Hayward G, Ayeb-Karlsson S (2021) 'Seeing with empty eyes': a systems approach to understanding climate change and mental health in Bangladesh. Clim Change 165:29
Henry C, Adams J (2018) Spotlight on sexual violence and harassment in commercial agriculture lower and middle-income countries. International Labour Organization, Geneva
Hossain T, Rabbani A, Aktar M, Rifat MTH (2021) Addressing climate change induced displacement and migration in Bangladesh: taking a human rights-based approach (2021) ActionAid Bangladesh
Hossain S, Arifur Rahman A (2021) Nexus between climate-induced displacement and violence against women and girls: evidence from disaster Prone Areas of Bangladesh. Share-net Bangladesh
Islam MR (2010) Vulnerability and coping strategies of women in disaster: a study on coastal areas of Bangladesh. Arts Faculty J 4:147–169. https://doi.org/10.62296/afj.v4i1.22

Jackson S, Humphrey C (2022) Yale experts explain intersectionality and climate change. Available via Sustainability https://sustainability.yale.edu/explainers/yale-experts-explain-intersectionality-and-climate-change. Accessed on 22 Sept 2024

Jayawardhan S (2017) Vulnerability and climate change induced human displacement. Consilience 17:103–142. https://www.jstor.org/stable/26188784

Kabir R, Khan HT, Ball E, Caldwell K (2016) Climate change impact: the experience of the coastal areas of Bangladesh affected by cyclones Sidr and Aila. J Environ Public Health 1:9654753. https://doi.org/10.1155/2016/9654753

Kane J (1996) Justice, impartiality, and equality: Why the concept of justice does not presume equality. Polit Theory 24:375–393. https://doi.org/10.1177/0090591796024003003

Kirezci E et al (2020) Projections of global-scale extreme sea levels and resulting episodic coastal flooding over the 21st Century. Sci Rep 10:1–12. https://doi.org/10.1038/s41598-020-67736-6

Lee JY, Kim SW, Kim JM (2020) The impact of community disaster trauma: a focus on emerging research of PTSD and other mental health outcomes. Chonnam Med J 56:99. https://doi.org/10.4068/cmj.2020.56.2.99

Magnan A, Ribera T, Treyer S (2015) National adaptation is also a global concern. Institute for Sustainable Development and International Relations (IDDRI)

Le Masson V (2022) Disasters, climate change, and violence against women and girls. In: Oxford Research Encyclopaedia of Natural Hazard Science

Mazzeo J, Rashed A (2003) Livelihood Monitoring for Southeast Bangladesh. CARE, TANGO International Dhaka

Murshid S, Yasmeen A (2004) Women of the coast: a gender status paper on the coastal zone. WP 027 Program Development Office for Integrated Coastal Zone Management Plan Project. Bangladesh. Dhaka

Muttarak R, Lutz W (2014) Is education a key to reducing vulnerability to natural disasters and hence unavoidable climate change? Ecol Soc 19:42. https://doi.org/10.5751/ES-06476-190142

Nusrat N (2019) Children risk early marriage—climate change is one of the factors. Available via RELIEFWEB https://reliefweb.int/report/world/children-risk-early-marriage-climate-change-one-factors Accessed on 11 Aug 2024

Palinkas LA, Wong M (2020) Global climate change and mental health. Curr Opin Psychol 32:12–16. https://doi.org/10.1016/j.copsyc.2019.06.023

Rahaman MA, Saba Z, Quader MA (2023) Climate change and social protection: threats for contemporary human security issues in Bangladesh. South Asian Journal of Social Sciences and Humanities 4:142–163 https://doi.org/10.48165/sajssh.2023.4408

Rahman MS (2013) Climate change, disaster and gender vulnerability: a study on two divisions of Bangladesh. Am J Human Ecol 2:72–82 https://doi.org/10.11634/216796221504315

Saul B (2012) The security risks of climate change displacement in Bangladesh. J Human Sec Forthcoming, Sydney Law School Research Paper 12:58 http://ssrn.com/abstract=2138006

Secretariat CNB (2020) Climate change induced gender-based violence against women during water collection: a case study in Shatkhira Upazilla, Bangladesh

Shahjahan M (2023) Disaster and crime in the coastal area of Bangladesh- Study on police activities. Biomed J Sci Tech Res 52:43794–43798 https://doi.org/10.26717/BJSTR.2023.52.008264

Shamsuddoha M, Chowdhury RK (2007) Climate change impact and disaster vulnerabilities in the coastal areas of Bangladesh. COAST Trust, Dhaka 40–48

Shehu M, Abba A (2020) Humanitarian crisis and sustainable development: perspectives and preferences of internally displaced persons in northeastern Nigeria. J Int Human Action 5:17. https://doi.org/10.1186/s41018-020-00084-2

Sparling TM, Waid JL, Wendt AS, Gabrysch S (2020) Depression among women of reproductive age in rural Bangladesh is linked to food security, diets and nutrition. Public Health Nutr 23:660–673. https://doi.org/10.1017/S1368980019003495

National Strategies on Internal Displaced Management 2021 Available via GOV.BD https://modmr.portal.gov.bd/sites/default/files/files/modmr.portal.gov.bd/publications/d05820e8_b72f_43a1_8cfe_4b4bd34a2560/NSIDM%20Publication(Size%206.5%20X%209.5%20in).pdf. Accessed on 21 Aug 2024

Sultana F (2022) Critical climate justice. Geogr J 188:118–124. https://doi.org/10.1111/geoj.12417
Sultana T, Mahmud KT, Moniruzzaman M, Tareque M (2024) Impact of access to land on women's economic well-being: an empirical evidence from rural Bangladesh. SAGE Open 14:21582440241227704. https://doi.org/10.1177/21582440241227705
The International Labour Organization (ILO) 'What is a green Job?' (2016) Available via ILO https://www.ilo.org/resource/article/what-green-job. Accessed on 19 Sept 2024
The Succession Act,1925. Available via Laws of Bangladesh http://bdlaws.minlaw.gov.bd/act-138.html . Accessed on 28 September 2024
Thomas KA (2020) Shifting baselines of disaster mitigation. Climate Dev 12:147–150
UNDP Bangladesh Gender Equality Strategy 2023–2026' Available via UNDP https://www.undp.org/sites/g/files/zskgke326/files/2023-03/GES%202023-2026%20Final%20Draft.pdf. Accessed on 18 Sept 2024
UNHCR UK Internally Displaced People. Available via UNHCR https://www.unhcr.org/uk/about-unhcr/who-we-protect/internally-displaced-people Accessed on 18 Sept 2024
UNICEF UK, Futures at Risk' How the UK can Support Education for Girls on the Move in a Changing Climate, Working Paper, Available Via UNICEF https://www.unicef.org.uk/wp-content/uploads/2021/05/Climate-Related-Displacement-and-Migration-and-Girls-Education-WP_FINAL-compressed.pdf. Accessed on 18 Sept 2024
Vives M, Murga JDO (2022) Report on human trafficking in India, Nepal and Bangladesh. International Alliances Department Mumbai Smiles Available via MUMBAISMILES https://www.mumbaismiles.org/wp-content/uploads/2022/07/human-trafficking-in-asian-region_en-1.pdf. Accessed on 17 Sept 2024
Wheeler D (2011) Quantifying vulnerability to climate change: implications for adaptation assistance. Centre for Global Development Working Paper 240
WHO (2020) World Health Organization. Bangladesh WHO special initiative for mental health situational assessment. Available via WHO https://www.who.int/docs/default-source/mental-health/special-initiative/who-special-initiative-country-report---bangladesh---2020.pdf?. Accessed on 17 July 2024
Wickramaratne P, de Silva R, (2023) Loss and damage to land: voices from Asia https://doi.org/10.21201/2023.621531
Wilkinson E, Kirbyshire A, Mayhew L, Batra P, Milan A (2016) Climate-induced migration and displacement: closing policy gap. Available via IOM https://environmentalmigration.iom.int/sites/g/files/tmzbdl1411/files/documents/2023-11/10996.pdf. Accessed on 18 Sept 2024

# Chapter 19
# Women's Resilience in Adapting to Climate Change on the North Coast of Java Island, Indonesia

**Hari Harjanto Setiawan, Badrun Susantyo, Nyi Raden Irmayani, Habibullah, Mery Ganti, Abdullah, Luthfi Setiadiwibawa, Euis Setiawati, Ineu Maryani, Suhandy Siswoyo, Djaenudin, Mahdi Ibrahim Tanjung, Karno, Ruslan Halifu, and Rizal Akbar Aldyan**

**Abstract** Climate change starts from global warming, which causes environmental and ecosystem damage. Handling this problem is one of the goals of the Sustainable Development Agenda. The north coast of Java Island is the area with the most vulnerability. One of the vulnerable groups in coastal communities affected is women. Extreme weather and tidal flooding are challenges faced by coastal communities. Women face twice as many problems as men do in daily chores. Women must be more

H. H. Setiawan (✉) · B. Susantyo · N. R. Irmayani · Habibullah · M. Ganti · Djaenudin · R. Halifu · R. A. Aldyan
National Research and Innovation Agency (BRIN), Jakarta, Indonesia
e-mail: hari.harjanto.setiawan@brin.go.id

B. Susantyo
e-mail: badr004@brin.go.id

N. R. Irmayani
e-mail: nyir001@brin.go.id

Habibullah
e-mail: habi003@brin.go.id

M. Ganti
e-mail: mery004@brin.go.id

Djaenudin
e-mail: djae002@brin.go.id

R. Halifu
e-mail: rusl010@brin.go.id

R. A. Aldyan
e-mail: riza014@brin.go.id

Abdullah
Telkom University, Bandung, Indonesia
e-mail: sangaji@telkomuniversity.ac.id

 
A. Sarkar and N. Bandyopadhyay (eds.), *Gender-Transformative Approaches for Climate Change Adaptation*, Springer Climate,
https://doi.org/10.1007/978-3-031-90236-9_19

active in thinking and working to ensure the sustainability of their families' lives. This is inseparable from the gender roles still socially attached to women. Women on the north coast of Java must adapt to this vulnerability to build resilience and face all problems. This paper aims to describe women's resilience in facing climate change. Data were obtained using qualitative methods focused on coastal women's ideas and experiences related to climate change adaptation. Data were collected through interviews, observations, and focus group discussions with affected women. The study's results describe the vulnerability of coastal women, including environmental exposure, health risks, economy, gender equality, family social responsibility, and access to clean water. The role of coastal women in addressing climate change is to maintain family health, manage the environment, encourage economic empowerment, and advocate for the environment. Programs to improve women's resilience are household management, financial strategies, health, and environmental conservation. This chapter of the book will contribute to addressing issues of vulnerable groups facing climate change on the coast.

**Keywords** Global warming · Woman resilience · Climate change · Vulnerability · Adaptation

---

L. Setiadiwibawa
Winaya Mukti University, Bandung, Indonesia
e-mail: lutfi@unwim.ac.id

E. Setiawati
Religious Training Center Bandung, Indonesia's Ministry of Religious Affairs (MORA), Bandung, Indonesia
e-mail: euissetiawati@kemenag.go.id

I. Maryani
Guidance and Counseling Study Program, Faculty of Educational Sciences, Indonesia University of Education, Bandung, Indonesia
e-mail: ineumaryani@upi.edu

S. Siswoyo
Department of Architectural Engineering Education, Indonesia University of Education, Bandung, Indonesia
e-mail: suhandysiswoyo@upi.edu

M. I. Tanjung
Ministry of Public Works, Bandung, Indonesia
e-mail: mahdi.ibrahim@pu.go.id

Karno
Institute of Home Affairs Government, Jakarta, Indonesia
e-mail: karno@ipdn.ac.id

## Introduction

The north coast of Java Island in Indonesia is highly vulnerable to tidal flooding and abrasion, which significantly impacts the community, especially women (Setiawan 2023). Climate change poses challenges for women in this region, who are at greater risk than men. Various vulnerabilities impact women, including food crises, health, clean water sanitation, migration, social and economic conflicts, and gender-based violence (Sorensen et al. 2018). The vulnerabilities caused by climate change significantly affect women on the coast, especially those living below the poverty line and women who act as heads of families (Karmakar 2021). In this condition, the problems faced are very complex. The crucial role of women that must be considered is access to health, education, and welfare. Gender equality is essential in every decision-making at various levels to address these impacts.

Given this problem's complexity, women's role is crucial in reducing its impact (Marlina et al. 2021). This implies considering women as natural resource managers, guaranteeing women access to education, health, and training, and promoting equity at all levels of decision-making within communities right up to the international level. Women's traditional knowledge in decision-making related to community natural resource management can provide excellent solutions for adaptation to climate change. Moreover, educating and developing women's skills would enable them to lead sustainable development activities in their region. Traditional ecological knowledge of women and their perspectives on natural resource management may provide original solutions for climate change adaptation by considering the local context (Akinsemolu and Olukoya 2020).

The study of the role of women is a study that is constantly developing in the context of life. Women who used to be seen as weak and worthless are now the opposite. Not a few women have advantages and hold strategic positions as well as policymakers. The presence of women should not be denied and underestimated, let alone discriminated against. Women also have potential and responsibilities that are almost the same as men, although there is room for differences in certain conditions. The role of women can encourage global climate resilience. In situations like today, climate change is an inevitability. This article explains women's resilience in facing the global climate. In substance, the role of women on the coast has almost the same characteristics. Women have policies, demands, and responsibilities in preserving, maintaining, and protecting this universe according to local geographical conditions in adapting to climate change.

Likewise, more inclusive mitigation policies and programs increasingly address women's demands by empowering women in climate-related decision-making processes at all levels. We can create empowering dynamics to offer women valuable participation in changing the socio-economic consequences of global climate shifts. This study will look at various aspects by involving researchers from multiple disciplines. Specifically, this study will explain: (1) How vulnerable are coastal women to climate change? (2) What are the roles of women in dealing with these vulnerabilities? (3) What are the forms of adaptation of women living in coastal areas in dealing with these vulnerabilities?

## The Rationale of the Study

### *Climate Change*

Long-term changes in weather patterns are called climate change (Moazami et al. 2019). Natural causes of these changes include changes in the solar system cycle. However, since the 1800s, human activity in the form of burning fossils has become the leading cause of climate change (Ma et al. 2023). These result from burning coal to heat houses or petrol to power automobiles. Forest clearing causes global warming because it can no longer consume additional carbon dioxide in the atmosphere. Waste disposal is also a source of methane emissions. The main emitters of Cos come from transportation and industry.

Changes in the physical parameters of the Earth's atmosphere, such as temperature and rainfall patterns, affect many aspects of human life (Popovski and Mundy 2012). These physical alterations take place gradually rather than all at once. Climate change that affects the entire Earth causes global climate change. The average fluctuation of a location's climate or its variability over some time causes the environment to change significantly. Apart from that, it is also evident that human actions that alter the atmosphere's composition constantly, as well as internal natural processes or external forces, maybe the cause of climate change.

Social, demographic, and cultural shifts are impacted by global warming, which leads to climate change (Gómez-Casillas and Gómez Márquez 2023). Numerous social studies have discovered a strong correlation between social relations patterns and climate patterns. According to the findings of the IPCC research from 2007, there have been 12 of the warmest years on record based on global surface temperature data since 1850. The past 12 years have seen 11 of the 12 warmest years. Overall, temperature was a 0.76° C increase between 1850–1899 and 2001–2005. Between 1961 and 2003, the global average sea level rose at an average annual pace of 1.8 mm. An estimated 0.17 m has been added to the sea level throughout the twentieth century. Since the mid-twentieth century, the IPCC has assessed that human activity is the cause of global warming. If not stopped, it will continue to increase in the twenty-first century (Kabir et al. 2023).

### *Women's Resilience in Climate Change Adaptation*

Adaptation is a natural process that humans modify to respond to climate influences, reduce risks, or offer beneficial opportunities (Pallubinsky et al. 2023). Another way to describe adaptation is as an attempt by humans or nature to reduce past impacts. Indonesia is committed to adapting to climate change to reduce its adverse impacts on various sectors that can reduce global gross domestic product. Given Indonesia's tropical location, topography variances, and extensive land and sea areas, implementing climate change adaptation is challenging (Suyarso et al. 2023). Adaptation actions must be planned with a comprehensive strategy to meet basic human needs. Saving allows them to manage risks and increase their families' economic resilience in the face of challenges, such as natural disasters and changing weather patterns that affect fish catches and other resources (Shaffril et al. 2024). Coastal women are often the leading managers of household finances and resources (Stacey et al. 2019).

To prevent maladaptation—things that negatively impact meeting one of the fundamental needs—planned climate change adaptation measures must consider the connectedness between action programs to meet these needs (Ray Biswas and Rahman 2023). Consideration of alternative energy is required since, for instance, irrigation to satisfy food needs may increase the demand for fossil-based energy. Therefore, adaptation measures must adhere to fundamental efficiency and effectiveness criteria to use ecosystem services resources sustainably. Adaptation and mitigation strategies cannot be separated; they are complementary to being sustainable (Rej et al. 2023). This strategy was developed in research which can be seen in Fig. 19.1.

## Material and Methods

This chapter explains how women on the coast can become resilient and adapt. This research design is based on qualitative data collection and analysis through interviews, observations, and documentation studies. The qualitative approach describes the characteristics of women when exploring experiences. This approach is expected to gain appreciation, knowledge, and perception of understanding and giving meaning to life (Neuman 2006). The qualitative approach describes the characteristics of women's resilience in dealing with climate change because it emphasizes deepening experiences and theoretical observations (Rubin and Babby 2007). The approach chosen to build an understanding of this phenomenon is to develop a new theory (Alston and Bowles 1998). Another opinion is that 'the theoretical basis of qualitative research is a phenomenological approach, symbolic interaction, culture, and ethnomethodology' (Moleong 2018). Researchers make observations and interact with subjects, so researchers are data collection instruments (Creswell 1994).

The informant categories in this study are women's groups, local community leaders, and stakeholders. Women affected by climate change were interviewed by

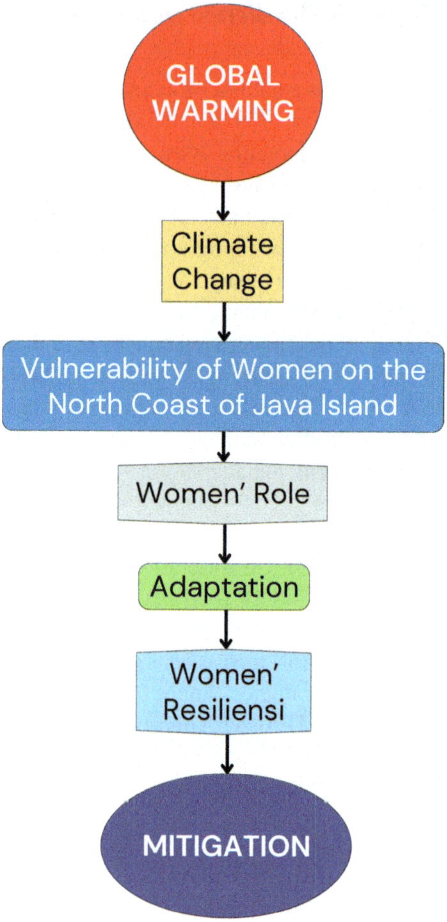


**Fig. 19.1** Framework for Women's Resilience in Adapting. *Source* Adapted from Rej et al. (2023)

as many as 25 people from five areas. This study aims to collect detailed information about the characteristics of these communities and their experiences in dealing with climate change. This includes understanding their daily challenges, adaptation measures, and resilience to changing environmental conditions. Local Community Leaders include significant leaders in collective efforts to address women affected by climate change. These informants provide insights into women's collective actions and strategies to adapt. This study aims to collect information from 25 local community leaders. Stakeholders involved in the Local Government's Climate Change Mitigation Program. These informants provide valuable information on government initiatives and programs implemented to manage the impacts.

Data were collected through face-to-face interviews conducted by visiting the informant's residence. Focus group discussions (FGD) were conducted by inviting representatives of groups that care about women. Interviews were conducted naturally to obtain factual information. FGD is a focused discussion with a predetermined topic (Pahwa et al. 2023). One group consists of seven people. Documentation study, namely studying all documents related to climate change on the north coast of Java was used. Documents are data sources that are used to complete the research (Stringer 2007).

The location of the study was determined intentionally based on specific characteristics related to the existence of the informant. This selection is a case study representing the coastal communities of the north coast of Java. The locations of this research are Tangerang Regency, DKI Jakarta, Cirebon Regency, Demak Regency, and Pasuruan Regency.

The first step in processing qualitative data is reviewing all the data already collected from different sources, such as observations, interviews, and literature reviews (Sardana et al. 2023). These findings are documented in various ways, including field notes, official and personal records, drawings, photographs, and more. Data reduction, accomplished by developing abstractions, comes next after reading it. The goal of abstraction is to produce a synopsis of the essential elements, procedures, and assertions that must be upheld to be included. Putting them in units is the next stage. These units are then grouped based on their codes into three phases (Neuman 2006).

In the first stage of coding, stories or tales from observations or interviews are described according to themes or classifications. These categories guide future iterations of the queries and observations. The second step is axial coding, where themes and categories are linked, and their connections are further examined while gathering data. The purpose of data gathering, analysis, and sampling in this second step is to generate hypotheses. The third step in creating theoretical assertions is selective coding. The nature of the connections between themes and categories is identified and included in a thorough statement. At this point, the data collecting, analysis, and sampling processes aim to identify key categories that will be the focus of integration. The observation data is presented and analyzed qualitatively using verbatim participant quotes (Miles et al. 2018). The conceptual framework of the data and methodology used can be seen in Fig. 19.2.

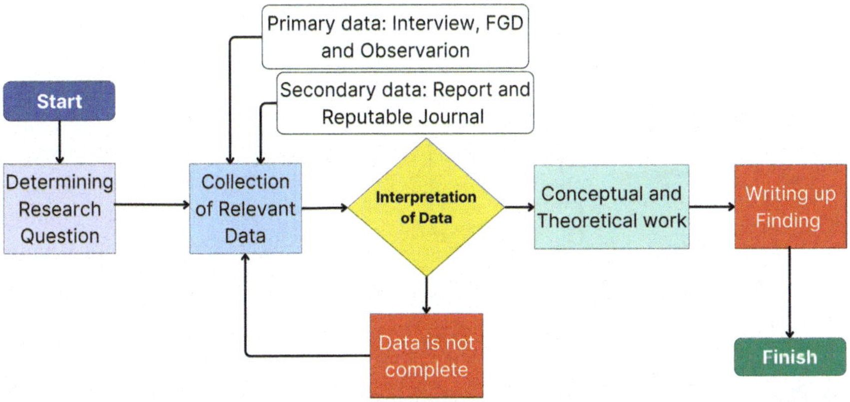


**Fig. 19.2** Conceptual framework of data and methodology used. *Source* Created by authors

## Results and Discussion

The impacts most felt by coastal communities on the north coast of Java are high sea waves or large waves and strong currents, which can cause coastal abrasion and damage settlements and fishponds. Abrasion and large waves have resulted in several places initially built on the coast now filled with seawater. Abrasion has also caused many coastal areas to be submerged in the ocean. This condition certainly needs to be considered by various parties in managing the northern coastal area of Java. Protection from coastal abrasion must also guarantee the safety of coastal residents and their settlements. This study focuses on women as a vulnerable group affected more than men.

At the national level, the Government of Indonesia is firmly committed to addressing climate challenges and has incorporated gender mainstreaming into its medium-term development strategy. In addition, gender equality is also one of the long-term development targets until 2025. General and technical guidelines on gender responsiveness in climate change adaptation were prepared in 2015. Its function is to guide the government at both the central and regional levels in development program planning, leading to gender responsiveness in mitigation and adaptation programs. This is focused on six priority sectors to reduce and adapt to climate change: food security, energy security, health, public housing, infrastructure and small islands, and coastal areas. Answering this research question, the challenges facing women on the coast can be seen in Fig. 19.3.

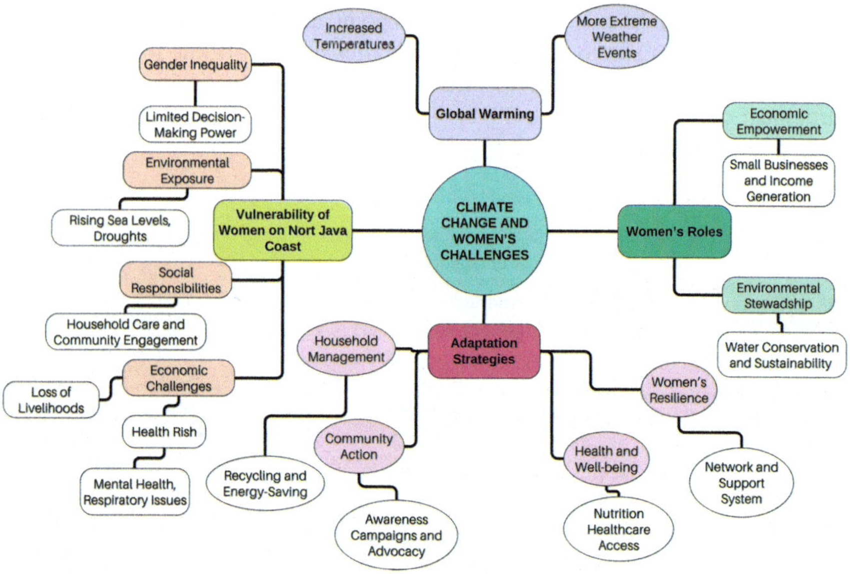


**Fig. 19.3** Research results framework for women in adapting to climate change. *Source* Created by authors

## *Vulnerability of Women on the Coast*

### Vulnerability to Environmental Exposure

Women are often the most vulnerable group in the face of environmental change. Women's vulnerability to environmental exposure is an impact caused by climate change. Due to climate change, women's vulnerability to coastal environments in northern Java is significantly impacting their lives (Gravitiani et al. 2018). Exposure to environmental changes, such as rising sea levels, puts women at greater risk (Marlina et al. 2020). This is because women are often responsible for managing households and natural resources, making them more vulnerable when these resources are threatened.

### Vulnerability to Health Risks

Climate change significantly impacts women's health, such as dry skin, itching, flu, cough, fever, and colds. Weather uncertainty and natural disasters can cause prolonged psychological stress and anxiety, which impacts women's mental health (Suciana et al. 2024). Women involved in agricultural, and fisheries activities are at high risk of exposure to malaria; this is because they often work outdoors and interact directly with the environment affected. Policies that consider gender roles and support women's adaptive capacities are essential to increasing their resilience to these impacts (Arwan et al. 2021; Jatayu et al. 2024).

Extreme temperature changes have substantial effects on expectant moms and their offspring. Expectant mothers who are exposed to severe temperature changes may suffer illnesses, and some may give birth prematurely. Such extremes in temperature can also affect an infant's size, so many are smaller than otherwise would be expected. The newborn's weight is also affected, resulting in low birth weight. Exposure to extreme temperatures also increases the risk of infant mortality and puts babies at risk of neonatal stress, which can affect the development of the baby's growth when they are born (Kuehn and McCormick 2017; Veenema et al. 2023).

### Economic Vulnerability

The economic vulnerability of coastal women is a pressing issue in creating sustainable development (Babacan 2021). This condition significantly impacts women's economy, especially in coastal areas highly dependent on natural resources (Ajani et al. 2013). Changing weather patterns affect the availability of fishery resources, which directly impacts the income of fishing families (Mustika et al. 2023; Nelayan et al. 2022). This results in increased poverty and economic instability, which have a more significant impact on women because they are often responsible for managing family finances (Ulfa 2018).

**Fig. 19.4** A woman is working looking for shrimp and crabs on the beach. *Source* By the authors

Women directly involved in fishing usually struggle harder to maintain their income (Fig. 19.4). Sometimes, women take over economic responsibilities when their husbands are forced to seek work outside the unstable fishing sector. Climate change affects women's financial vulnerability, and by developing policies that support women's empowerment, we can create more responsible and sustainable societies.

#### Vulnerability to Gender Equality

Women's social roles in accessing various resources will influence their capacity to adapt (Setiawan and Wismayanti 2023). The occurrence of gender inequality causes increased vulnerability among women because they do not have adequate capacity. Women's adaptive capacity to climate change is minimal due to their limited participation in formulating policies related to climate change. Women have a very decisive capacity to take adaptation measures. The lower the adaptive capacity, the more vulnerable they are to impacts. These things cause climate change to be gender-neutral. Some gender-biased social roles ultimately result in minimal access for women to resources, including natural resources, economic resources, and political resources.

### *Social and Family Responsibility*

Climate change poses significant vulnerabilities to coastal communities, especially women, who often bear the brunt of its impacts (Handayani et al. 2018). Climate change affects Women in coastal areas because they bear greater social and family burdens, exacerbating their vulnerability (Yadav and Lal 2018). The vulnerability of women in coastal areas is multifaceted, stemming from their social and cultural roles in society (Akinsemolu and Olukoya 2020). Women are the leading providers of food and water. However, they face more significant challenges in adapting, such as reduced fisheries productivity, difficulty accessing clean water, and abrasion.

#### Accessing Clean Water and Resource for Their Families

Women in coastal areas often bear the brunt of climate change impacts, as their social roles and environmental interactions make them particularly vulnerable to disruptions in access to clean water and other vital resources (Figueiredo and Perkins 2013) Their social roles and environmental interactions place coastal women at risk of natural resources, such as water sources, which can exacerbate their vulnerability when they become scarce. Women in coastal areas are responsible for tasks such as fetching water, and water availability disruptions can significantly impact their daily lives. As a result of climate change, women in coastal areas are experiencing increasing difficulties in accessing safe and potable water. Saltwater intrusion due to land subsidence also deteriorates groundwater quality, making it saltier and unsuitable for consumption (Khan et al. 2011). This, in turn, forces women to travel longer distances, sometimes several kilometres, to find alternative water sources, which can be highly time-consuming, physically demanding, and even dangerous. The added burden of securing water for household use exacerbates coastal women's challenges.

#### Health and Family Care

In the health sector, coastal women are critical in coping with and adapting (Hiremath et al. 2021). They are responsible for the well-being of their families, including providing food and health care. Empowering coastal women in the health sector can improve their access to critical information to address the health impacts of climate change, such as diseases caused by environmental change and degradation of natural resources. Coastal women are often at the forefront of adapting to climate change, developing strategies that focus not only on the economy but also on the health and well-being of their families (Akinsemolu and Olukoya 2020). In addition, they also often act as community mobilizers to raise awareness of health issues related, such as airborne diseases and the impact of natural disasters.

### *Environmental Stewardship*

Coastal women are often involved in managing mangrove ecosystems, which are one of the natural barriers against climate change and natural disasters. Mangrove conservation involving women can improve disaster mitigation and protect coastal environments from negative impacts (Rudianto et al. 2022). Women's participation in mangrove management also contributes to increasing public awareness of the importance of this ecosystem. The involvement of women in conservation and natural resource management activities shows that they have the knowledge and skills needed to address the environmental challenges coastal communities face (Roba and Kikwatha 2021). Women also play an essential role in addressing pollution problems, primarily concerning plastic waste, a global issue in coastal areas. Women in coastal communities are involved in beach cleanups and waste management activities, increasing environmental awareness and strengthening social solidarity among community members. Coastal women are drivers in ecological conservation efforts and sustainable resource management (Purba et al. 2023).

## *Economic Empowerment*

The role of coastal women in economic empowerment is significant, especially in facing the increasingly real impacts. Women in coastal areas not only contribute to the financial aspects of the family but also play a strategic role in the social and economic resilience of the community. Climate change brings significant challenges to coastal communities, including women. Coastal women develop adaptation strategies to deal with the impacts of climate change, such as changing business patterns and diversifying sources of income (Setiawan and Wismayanti 2023).

Women on the north coast of Java utilize mangrove leaves for micro, small, and medium enterprises (SMEs) to increase their income. Using mangrove leaves in micro, small, and medium enterprises has significant economic and environmental sustainability potential. Mangrove leaves, especially from species such as *Rhizophora mucronata* and *Avicennia marina*, have been shown to have various benefits that SMEs can utilize. Mangrove leaves from the *Rhizophora mucronata* and *Avicennia marina* species have significant nutritional potential and can be processed into multiple innovative food products.

The research found that mangrove leaves can be used as raw materials in making ice cream (Alindra et al. 2023). This activity involves socialization and training for Family Empowerment and Welfare groups. This aims to increase knowledge and skills in processing mangrove leaves into attractive food products with sale value. The research found that mangrove leaf powder can be used as an additional raw material in making chocolate. Mangrove leaves have great potential to be used in the food industry, which can support local SMEs. Research showed that mangrove fruit and leaves can be processed into various cakes and syrups, increasing their economic value and introducing people to healthier food alternatives. Using mangrove fruit and leaves provides financial benefits and contributes to the sustainability of the mangrove forest ecosystem by increasing public awareness of its importance.

Mangrove fruit from the *Avicennia marina* species can also be processed into chips, a culinary speciality in several regions. This shows that mangroves can be utilized sustainably to improve people's welfare. The nutritional content of mangrove leaves is diverse, including antioxidant compounds that can provide health benefits (Istriningsih et al. 2023). Mangrove leaf extract has significant antioxidant activity. Developing mangrove leaves into food products can increase SMES's income and positively affect public health. Using mangrove leaves and fruits in food products to increase SMES's income provides a vast opportunity for innovation and local economic development.

## *Leadership and Advocacy*

Women living in coastal areas of northern Java face disproportionate vulnerability to the impacts. This inequality stems from complex cultural, social, and economic dynamics that shape gender roles and perceptions in communities. Women, as an integral part of coastal communities, have the potential to become leaders in efforts to mitigate and adapt to climate change. The role of women in leadership is their ability to participate in decision-making related to natural resources. Women are critical in developing local adaptation strategies, leveraging their deep understanding of ecosystems and community needs (Marlina et al. 2021). Furthermore, women have emerged as influential advocates, leveraging their unique perspectives and experiences to influence climate change policies and promote more equitable and inclusive decision-making processes.

They have been at the forefront of grassroots movements, raising awareness, mobilizing their communities, and demanding accountability from policymakers (Croak and Walker 2024). Through their leadership and advocacy efforts, coastal women have played a critical role in shaping the climate change adaptation narrative and ensuring that marginalized groups' needs and voices are well represented and addressed (Reilly-Moman et al. 2023). Women also have an essential role in climate change education and outreach. Therefore, increasing women's capacity through education and training can strengthen their role in climate change advocacy. For example, information and communication technology can be used to improve coastal women's knowledge and skills to be more effective in participating in discussions and decision-making related to climate change.

Women often act as a bridge between local communities and Government policies, allowing them to voice the needs and challenges coastal communities face (Rusmadi 2017). In this context, it is essential to create space for women to speak and participate in forums that discuss environmental and climate change issues so that their voices are heard and recognized in decision-making (Adriana et al. 2017). Coastal women have been critical in driving transformative change, advocating for policies and interventions prioritizing their communities and the environment's well-being (Gill et al. 2023). Their unwavering commitment and impactful work are critical to fostering more inclusive and sustainable approaches to climate change adaptation.

## *Women's Resilience in Adapting*

The adaptation of coastal women is an increasingly pressing issue, given the enormous impacts on their lives. Women in coastal areas are often at the forefront of dealing with these impacts.

### Household Management

Coastal women play an essential role in managing and organizing household activities, which is an important responsibility in the household (Handayani et al. 2018). If empowered and supported, women's skills in household management can be an essential foundation for building more sustainable and adaptive societies (Figueiredo and Perkins 2013). Leveraging their deep knowledge of household dynamics and resource needs, coastal women in northern Java can design innovative strategies to enhance the resilience of their families and communities. These include implementing water conservation techniques, diversifying food sources, and adopting energy-saving practices.

Women's expertise in water collection, food preparation, and energy use is becoming increasingly valuable due to increasing environmental pressures. Through these efforts, coastal women play a central role in managing and organizing household activities (Shrestha et al. 2020). By leveraging household management skills, women can contribute to developing more sustainable and equitable societies (Ajani et al. 2013). By empowering and supporting household management capabilities, women can be an essential foundation for building more sustainable and adaptive communities that are better prepared for the complexities of climate change.

### Economic Strategies

Women on the north coast of Java significantly contribute to their family income through various adaptation strategies, both traditional and those that arise from socio-economic dynamics. They are involved in diversifying income sources, such as SMEs or activities involving local resources, to reduce dependence on marine catches, which are increasingly under threat due to climate change. One adaptation strategy women carry out on the north coast of Java is saving. They have an essential role in maintaining family economic resilience, especially in maintaining economic resilience due to climate change. Overall, the adaptation of coastal women on the north coast of Java through saving and economic diversification is an essential step in increasing their resilience to the impacts of climate change. Through training, community support, and sustainable resource management, women can contribute significantly to mitigating the risks coastal communities face.

### Health and Well-Being

Adaptation of women on the north coast of Java to climate change in health and family welfare is an increasingly important issue, considering the significant impacts of climate change on the lives of coastal communities. Coastal women are additional breadwinners and their family's health and welfare managers. Climate change, such as increasing temperatures and changing weather patterns, directly impacts the health of coastal communities. The results of interviews with health workers were that; 'coastal women are vulnerable to disease, especially environmental-based diseases caused by environmental conditions, sanitation, and climate. The diseases that are commonly suffered are malaria, diarrhoea, tuberculosis, skin diseases, dengue fever, nutritional problems, typhoid or typhus, upper respiratory tract infections, and digestive problems.' The increasing frequency of storms and rising sea levels can damage health infrastructure and reduce access to health services. In this context, coastal women must adapt creatively to maintain their families' health. They use traditional health practices or seek alternative, more sustainable health resources (Choirunnisa et al. 2022).

In climate change adaptation, coastal women must also improve their knowledge and skills. Therefore, education and training programs focusing on health and climate change coping are critical to increasing coastal women's capacity to deal with climate change's impacts (Ramdani et al. 2022). Health management is essential for adapting women on the north coast of Java to climate change in health and family welfare. This effort improves individual welfare and strengthens coastal communities' resilience to increasingly complex climate change.

### Environmental Conservation

Women on the north coast of Java have a crucial role in resource management and adaptation to climate change. Coastal women not only play a role in the domestic context but are also actively involved in environmental management and adaptation strategies needed to face the challenges of climate change. One of the main adaptation strategies adopted by coastal women is ecological conservation. This includes efforts such as replanting mangroves and sustainable fisheries management. By protecting and restoring local ecosystems, coastal women help to improve natural defences against climate change impacts such as sea level rise and extreme weather events (Gracia et al. 2018). The results of interviews with community leaders stated that; "Women here are active in planting mangrove trees because they have many benefits. Women's groups utilize mangroves to increase family income. Mangrove leaves are used to make chips, seeds are used to make drinks and wood is used for handicrafts. Their actions also support biodiversity and food security, which are critical to community resilience. In addition, environmental conservation initiatives led by coastal women often incorporate traditional ecological knowledge passed down from generation to generation".

These initiatives also involve community-based decision-making, where coastal women are crucial in planning and implementation. In addition, coastal women leverage their local knowledge and networks to mobilize communities for environmental management and climate change adaptation. Coastal women have developed a multifaceted approach to ecological conservation, leveraging their deep understanding of local ecosystems and their ability to engage communities (Stoffle et al. 2020). Through mangrove replanting and sustainable fisheries management, they actively strengthen natural defences and promote food security in the face of climate change. Their initiatives also incorporate traditional ecological knowledge and prioritize community-based decision-making, empowering coastal women as critical agents of adaptation and resilience.

## Limitations of the Study

This study only focuses on the research topic of women's resilience in adapting to climate change. This study cannot be used universally because each country has different policies and characteristics. However, the study's conclusions can be used to strengthen women's ability to adapt to climate change. As a result, small groups from different countries with different characteristics need to be assessed for their conditions. The results of this study cannot be made to several scenarios because of the limited observation period. This study cannot be used as a standard of measurement because the conditions of women worldwide are very different.

## Recommendations

The Sustainable Development Goals have a mandate to ensure 'no one is left behind' in the development process. Women living in coastal areas are one of the vulnerable groups to climate change that must be considered. The vulnerability of coastal women is more significant than men because they have a double burden. They are vulnerable to environmental exposure, health risks, lack of economic opportunities, gender inequality, additional social responsibility, and the burden of ensuring access to clean water and resources for their families. The issue of climate change is important to understand, and coastal women must be considered in every policy-making. The policies of Indonesia and several other countries still pay little attention to gender so that the central and regional governments can make policies to prevent and deal with the climate crisis and be fairer for all, especially women.

The vulnerable conditions of women in coastal areas require women's participation in the development process. The roles that women can play include health and family care, environmental management, economic empowerment, leadership and advocacy. The role of women in climate change adaptation is an important thing that needs to be considered. Extreme weather and abrasion have an impact on sources of income that have been managed by women. Disruption of sources of income will disrupt childcare so some women are forced to work to meet their needs. Women's empowerment programs and access to health facilities must be prioritized to improve women's lives.

Women have an important role in adapting to climate change, both as natural resource managers, food security guards, and key actors in local communities. However, women are also a group that is vulnerable to the impacts of climate change. Increasing women's resilience requires adaptation strategies to address vulnerability. Adaptation strategies carried out by women are household management, economic strategies, health and welfare. Adaptation at the community level must provide space to increase women's strategic roles in decision-making at the public level so that the principles of access, participation, control and balanced benefits can be met.

## Conclusion

Women make significant contributions to their family income through various adaptation strategies, both traditional and those emerging from socio-economic dynamics like the gender division of labour resulting in different gender roles between men and women. Women have roles and responsibilities for productive and reproductive work in the family and society, maintaining community unity and social networks. The vulnerability and non-neutrality of the Indonesian government's policies towards women's rights make climate change adaptation education essential to implement, considering the vital role of women in life. Through awareness and empowerment of women, it is hoped that they will be able to contribute to improving the quality of the environment, especially in reducing impacts through women's hands that can ultimately influence other hands to play an active role in efforts to deal with climate change. Suggestions for policymakers are as follows: First, in various sectors of life, the government needs to pay more attention to and prioritize protection for women in formulating policies. Women should also be adequately placed as stakeholders, especially in environmental policies. Second, in various sectors of life, the government needs to provide women with wider opportunities and access to obtain information, make decisions, and participate fully in supporting development in Indonesia.

**Acknowledgements** Thank you to LPDP and BRIN, who have funded this research through RIIM batch SK NUMBER 37/II.7/HK/2023.

## References

Adriana G, Pandjaitan NK, Dharmawan AH (2017) Cohesivity fisheries community in the face of climate change on the coast of Western Java. Sodality: Jurnal Sosiologi Pedesaan 5. https://doi.org/10.22500/SODALITY.V5I1.16275

Ajani E, Onwubuya E, Mgbenka R (2013) Approaches to economic empowerment of rural women for climate change mitigation and adaptation: implications for policy. J Agric Extens 17:23–34. https://doi.org/10.4314/JAE.V17I1.3

Akinsemolu AA, Olukoya OAP (2020) The vulnerability of women to climate change in coastal regions of Nigeria: a case of the Ilaje community in Ondo State. J Clean Prod 246:119015. https://doi.org/10.1016/J.JCLEPRO.2019.119015

Alindra PSA, Rohmah AN, Sabam et al (2023) Pembuatan Es Krim Mangrove sebagai Upaya Pengembangan Ide Inovasi Pangan UMKM Desa Kaliwlingi. Jurnal Pusat Inovasi Masyarakat 5:202–209. https://doi.org/10.29244/JPIM.5.2.202-209

Alston M, Bowles W (1998) Research for social workers: an introduction to methods. Allen and Unwin, Australia

Arwan JF, Dewi L, Wahyudin D (2021) The urgency of climate change-based education for sustainable development. Jurnal Ilmiah Pendidikan Lingkungan dan Pembangunan 22:23–38. https://doi.org/10.21009/PLPB.222.03

Babacan H (2021) Women and economic dimensions of climate change. economic effects of natural disasters: theoretical foundations, methods, and tools, pp 77–95. https://doi.org/10.1016/B978-0-12-817465-4.00006-6

Choirunnisa LAD, Purwaningsih Y, Prasetyani D (2022) Adaptasi Nelayan Pesisir Kabupaten Pacitan Akibat Perubahan Iklim. Jurnal Wilayah dan Lingkungan 10:166–181. https://doi.org/10.14710/JWL.10.2.166-181

Creswell JW (1994) Research design qualitative & quantitative approaches. International Educational and Professional Publisher, Thousand Oaks, London, New Delhi

Croak A, Walker GJ (2024) Hands-on climate engagement: principles for effective hands-on activities and demonstrations. J Sci Commun 23:N03. https://doi.org/10.22323/2.23030803

Figueiredo P, Perkins PE (2013) Women and water management in times of climate change: participatory and inclusive processes. J Clean Prod 60:188–194. https://doi.org/10.1016/J.JCLEPRO.2012.02.025

Gill DA, Blythe J, Bennett N et al (2023) Triple exposure: Reducing negative impacts of climate change, blue growth, and conservation on coastal communities. One Earth 6:118–130. https://doi.org/10.1016/J.ONEEAR.2023.01.010

Gómez-Casillas A, Gómez Márquez V (2023) The effect of social network site usage in climate change awareness in Latin America. Popul Environ 45:1–28. https://doi.org/10.1007/S11111-023-00417-4/TABLES/2

Gracia A, Rangel-Buitrago N, Oakley JA, Williams AT (2018) Use of ecosystems in coastal erosion management. Ocean Coast Manag 156:277–289. https://doi.org/10.1016/J.OCECOAMAN.2017.07.009

Gravitiani E, Fitriana SN, Suryanto (2018) Community livelihood vulnerability level in the northern and southern coastal area of Java, Indonesia. IOP Conf Ser Earth Environ Sci 202:012050. https://doi.org/10.1088/1755-1315/202/1/012050

Handayani W, Ananda MR, Esariti L, Anggraeni M (2018) Climate change adaptation in Tanjung Mas—Semarang: a comparison between male- and female-headed households. IOP Conf Ser Earth Environ Sci 129:012025. https://doi.org/10.1088/1755-1315/129/1/012025

Hiremath RB, Joseph V, Kumar B, Yadav R (2021) Is climate change affecting women's health more? J Clim Change 7:73–75. https://doi.org/10.3233/JCC210013

Istriningsih E, Sri Rejeki D, Anggraeni S et al (2023) Aktivitas Antioksidan Daun Mangrove (Rhizophora mucronata Lamk) Dengan Metode β-Carotene Bleaching. Kunir: Jurnal Farmasi Indonesia 1:16–25. https://doi.org/10.36308/KJFI.V1I1.525

Jatayu A, Idajati H, Umilia E et al (2024) Penguatan Kapasitas dan Strategi Adaptasi Sektor Pariwisata Gili Ketapang dalam mendukung Pengembangan Wisata Berbasis Ketahanan Iklim. Sewagati 8:1620–1632. https://doi.org/10.12962/J26139960.V8I3.920

Kabir M, Habiba UE, Khan W et al (2023) Climate change due to increasing concentration of carbon dioxide and its impacts on the environment in 21st century; a mini-review. J King Saud Univ Sci 35:102693. https://doi.org/10.1016/J.JKSUS.2023.102693

Karmakar S (2021) Gender and climate change: the condition of women in the Indian Sundarbans. Sch J Arts Humanit Soc Sci 9(1):1–5. https://doi.org/10.36347/sjahss.2021.v09i01.001

Khan AE, Ireson A, Kovats S et al (2011) Drinking water salinity and maternal health in Coastal Bangladesh: implications of climate change. Environ Health Perspect 119:1328–1332. https://doi.org/10.1289/EHP.1002804

Kuehn L, McCormick S (2017) Heat exposure and maternal health in the face of climate change. Int J Environ Res Publ Health 14:853. https://doi.org/10.3390/IJERPH14080853

Ma M, Wang Q, Liu R et al (2023) Effects of climate change and human activities on vegetation coverage change in northern China considering extreme climate and time-lag and -accumulation effects. Sci Total Environ 860:160527. https://doi.org/10.1016/J.SCITOTENV.2022.160527

Marlina S, Lautt BS, Usup A, Sunaryati R (2021) Gender role in climate change adaptation on the peat swamp ecosystem in Pulang Pisau Regency Central Kalimantan. IOP Conf Ser Earth Environ Sci 716:012090. https://doi.org/10.1088/1755-1315/716/1/012090

Marlina S, Supriyono Lautt B, Usup A, Sunaryati R (2020) The impact of climate change on community, culture, and gender in Central Kalimantan. In: E3S web of conferences, vol 211, pp 01001. https://doi.org/10.1051/E3SCONF/202021101001

Miles MB, Huberman AM, Saldana J (2018) Qualitative data analysis, p 408

Moazami A, Nik VM, Carlucci S, Geving S (2019) Impacts of future weather data typology on building energy performance—investigating long-term patterns of climate change and extreme weather conditions. Appl Energy 238:696–720. https://doi.org/10.1016/J.APENERGY.2019.01.085

Moleong LJ (2018) Metodologi penelitian kualitatif, 38th edn. PT Remaja Rosdakarya, Bandung

Mustika I, Herawati T, Muflikhati I (2023) Strategi Penghidupan, Manajemen Keuangan, Kerentanan Ekonomi, dan Ketahanan Keluarga Nelayan. Jurnal Ilmu Sosial dan Humaniora 12:269–279. https://doi.org/10.23887/JISH.V12I2.62017

Nelayan A, Kabupaten P, Akibat P et al (2022) Adaptasi Nelayan Pesisir Kabupaten Pacitan Akibat Perubahan Iklim. Jurnal Wilayah dan Lingkungan 10:166–181. https://doi.org/10.14710/JWL.10.2.166-181

Neuman WL (2006) Social research methods: qualitative and quantitative approach. Boston

Pahwa M, Cavanagh A, Vanstone M (2023) Key informants in applied qualitative health research. Qual Health Res 33:1251–1261. https://doi.org/10.1177/10497323231198796

Pallubinsky H, Kramer RP, van Marken Lichtenbelt WD (2023) Establishing resilience in times of climate change—a perspective on humans and buildings. Clim Change 176:1–19. https://doi.org/10.1007/S10584-023-03614-0/FIGURES/1

Popovski V, Mundy KG (2012) Defining climate-change victims. Sustain Sci 7:5–16. https://doi.org/10.1007/S11625-011-0138-0/METRICS

Purba NP, Pasaribu B, Faizal I et al (2023) Coastal clean-up in Southeast Asia: lessons learned, challenges, and future strategies. Front Mar Sci 10:1250736. https://doi.org/10.3389/FMARS.2023.1250736/BIBTEX

Ramdani J, Gunawan B, Darwis RS (2022) Peranan Green Social Workers Pada Program Kampung Iklim Di Wilayah Pesisir Indonesia. EMPATI: Jurnal Ilmu Kesejahteraan Sosial 11:36–43. https://doi.org/10.15408/empati.v11i1.26424

Ray Biswas R, Rahman A (2023) Adaptation to climate change: a study on regional climate change adaptation policy and practice framework. J Environ Manage 336:117666. https://doi.org/10.1016/J.JENVMAN.2023.117666

Reilly-Moman J, Jacobs K, Lough G, Moss R (2023) Understanding the effectiveness of coastal nature-based solutions: practitioner-based learning. https://doi.org/10.69925/QFMT6975

Rej S, Bandyopadhyay A, Das N et al (2023) The asymmetric influence of environmental-related technological innovation on climate change mitigation: what role do FDI and renewable energy play? Environ Sci Pollut Res 30:14916–14931. https://doi.org/10.1007/S11356-022-23182-7/FIGURES/4

Roba R, Kikwatha RW (2021) Influence of women participation on sustainability of the community conservation project implemented by Jaldesa community conservancy in Marsabit County, Kenya. Eur J Bus Manage Res 6:188–194. https://doi.org/10.24018/EJBMR.2021.6.3.898

Rubin A, Babby E (2007) Research methods for social work. Brooks/Cole Cengage Learning

Rudianto R, Darmawan V, Isdianto A, Bintoro G (2022) Restoration of coastal ecosystems as an approach to the integrated mangrove ecosystem management and mitigation and adaptation to climate changes in the north coast of East Java. J Coast Conserv 26:1–17. https://doi.org/10.1007/S11852-022-00865-4/TABLES/10

Rusmadi (2017) Pengarusutamaan Gender Dalam Kebijakan Perubahan Iklim Di Indonesia. Sawwa: Jurnal Studi Gender 12:91–110. https://doi.org/10.21580/SA.V12I1.1470

Sardana N, Shekoohi S, Cornett EM, Kaye AD (2023) Qualitative and quantitative research methods. In: Substance use and addiction research: methodology, mechanisms, and therapeutics, pp 65–69. https://doi.org/10.1016/B978-0-323-98814-8.00008-1

Setiawan HH (2023) Social resilience of local communities due to tidal flooding on the North Coast of Semarang City, Indonesia, pp 137–150. https://doi.org/10.1007/978-3-031-44397-8_8

Setiawan HH, Wismayanti YF (2023) The green economy to support women's empowerment: social work approach for climate change adaptation toward sustainability development. Clim Change Commun Respons Resil 225–240. https://doi.org/10.1016/B978-0-443-18707-0.00012-6

Shaffril HAM, Samah AA, Mazuki R (2024) A systematic literature review on the adaptation of women in fisheries-based families on climate change impacts. J Environ Stud Sci 1–18. https://doi.org/10.1007/S13412-024-00963-9/TABLES/3

Shrestha B, Tiwari S, Bajracharya S, keitsch M (2020) Role of gender participation in household energy technology to achieve sustainability: a case of Kathmandu. https://doi.org/10.21203/RS.3.RS-113822/V1

Sorensen C, Murray V, Lemery J, Balbus J (2018) Climate change and women's health: Impacts and policy directions. PLoS Med 15:e1002603. https://doi.org/10.1371/JOURNAL.PMED.1002603

Stacey N, Gibson E, Loneragan NR et al (2019) Enhancing coastal livelihoods in Indonesia: an evaluation of recent initiatives on gender, women and sustainable livelihoods in small-scale fisheries. Maritime Stud 18:359–371. https://doi.org/10.1007/S40152-019-00142-5/TABLES/3

Stoffle B, Stoffle R, Van Vlack K (2020) Sustainable use of the littoral by traditional people of Barbados and Bahamas. Sustainability 12:4764. https://doi.org/10.3390/SU12114764

Stringer ET (2007) Action research third edition. SAGE Publications

Suciana F, Yanuvianti M, Psikologi P, Psikologi F (2024) Gambaran Kecemasan Perubahan Iklim pada Mahasiswa di Kota Bandung. Bandung Conf Ser Psychol Sci 4:317–332. https://doi.org/10.29313/BCSPS.V4I1.10050

Suyarso S, Setiawati MD, Supriyadi IH, Prayudha B (2023) Climate change indicator, impact, adaptation, and innovation at the local level: learn from the people's experience of the coastal plain of Probolinggo, East Java, Indonesia. In: Climate Change, Community Response and Resilience: Insight for Socio-Ecological Sustainability, Pp 93–118. https://doi.org/10.1016/B978-0-443-18707-0.00005-9

Ulfa M (2018) Persepsi Masyarakat Nelayan dalam Menghadapi Perubahan Iklim (Ditinjau dalam Aspek Sosial Ekonomi). Jurnal Pendidikan Geografi: Kajian, Teori, dan Praktek dalam Bidang Pendidikan dan Ilmu Geografi 23:41–49. https://doi.org/10.17977/UM017V23I12018P41-49

Veenema RJ, Hoepner LA, Geer LA (2023) Climate change-related environmental exposures and perinatal and maternal health outcomes in the U.S. Int J Environ Res Publ Health 20:1662. https://doi.org/10.3390/IJERPH20031662

Yadav SS, Lal R (2018) Vulnerability of women to climate change in arid and semi-arid regions: the case of India and South Asia. J Arid Environ 149:4–17. https://doi.org/10.1016/J.JARIDENV.2017.08.001

# Chapter 20
# Climate Crisis and "*City Citizens*": A Look into the Gender-Inclusive Disaster Management Policies of Indian Smart Cities

**Rituparna Bhattacharya and Sramana Mukherjee**

**Abstract** Urban activities contribute significantly to greenhouse gas emissions exacerbating climate change, increasing global temperatures, and leading to frequent extreme weather events like heatwaves, droughts, storms, and floods. These changes affect urban infrastructure, services, livelihoods, and public health, necessitating robust disaster risk reduction (DRR) strategies for resilient urban planning. International frameworks—Sendai Framework, 2030 Agenda, and New Urban Agenda—emphasize the importance of urban resilience, urging local and subnational governments to foster inclusive, safe, resilient, and sustainable cities. Smart cities present opportunities to enhance disaster resilience through flexible land-use policies, targeted investments, and community-driven decision-making. However, gender inclusivity remains a critical yet underexplored aspect, as women and marginalized groups often face heightened vulnerabilities, especially during disasters. This paper explores the role of Indian smart cities in building and supporting disaster resilience, with a focus on gender inclusiveness. Employing content and policy analysis, it examines a hundred smart cities developed under the Smart Cities Mission (SCM) of the Government of India in 2015. Findings highlight that while some smart cities have progressive DRR, the integration of gender-responsive approaches remains inconsistent. Therefore, smart cities were categorized into—comprehensively planned (Indore, Coimbatore, Chennai); moderately planned (Pune, Jaipur, Bhubaneswar); and inadequately planned (Shillong, Agra, Guwahati). However, the reliance on secondary data restricts precise analysis. Additionally, varying levels of policy detail and definitions used across cities may introduce inconsistencies in comparative analysis. Collection of gender-disaggregated data, expanding women's

R. Bhattacharya (✉)
Institut für Geographie, Friedrich-Alexander-Universität Erlangen-Nürnberg, Erlangen, Germany
e-mail: ritu.bhattacharya@fau.de

S. Mukherjee
Department of Humanities and Social Sciences, National Institute of Technology Rourkela, Rourkela, Odisha, India

 
A. Sarkar and N. Bandyopadhyay (eds.), *Gender-Transformative Approaches for Climate Change Adaptation*, Springer Climate,
https://doi.org/10.1007/978-3-031-90236-9_20

roles in urban planning, gender-responsive training and budgeting, etc., can be helpful measures in gendering DRR strategies.

**Keywords** Urban planning · Gender-inclusive · Climate change · Disaster management · Smart cities · India

## Introduction

### *Urbanization and Climate Change*

Cities as the key driver of socio-economic changes, house over half of the world's population, even though they occupy less than 1% of the planet's area. Global projections show that by 2050, 68% of the population will live in urban areas (World Urbanization Prospects 2019). Urbanization, thus, has emerged as one of the most prominent and irreversible human interventions that are rapidly modifying land use and land cover. However, such uncontrolled and unprecedented urbanization, occurring more so in developing nations, gives rise to various socio-economic challenges, endangers the well-being of urban citizens, and entails a significant environmental risk (Sethi and Vinoj 2024).

Urban regions are particularly susceptible to the combined impacts of urbanization and climate change. Cities are often seen as contributors to environmental degradation and represent immense ecological burdens (World Bank 2010). As urbanization creates altered urban landscapes through modified land use and land cover, factors like increased anthropogenic activity, altered convection efficiency and surface albedo, and surfaces with higher thermal inertia (such as concrete and asphalt) diminish the evaporative cooling in cities. This makes them retain the heat longer and triggers the formation of urban heat islands. Consequently, this disrupts natural climatic processes like rainfall, carbon cycle, pollution levels, etc., (Oke et al. 2017). Added to that, climate change is exacerbated by the generation of greenhouse gases and other emissions brought on by increased energy needs and from activities stemming particularly from construction and transportation sectors in urban areas. Dense population and infrastructural (over)development usher in climate change impacts in the form of disasters—heatwaves, extreme weather events, flooding, drought, and so on (Dodman et al. 2022; Hari et al. 2021).

Disasters occur when hazards interact with human populations amid socio-economic, physical, and environmental vulnerabilities, resulting in destructive outcomes (Benali et al. 2018). These events adversely impact economies and living conditions by damaging physical capital such as housing, infrastructure, and fixed assets, as well as human capital through fatalities and injuries (Yun and Kim 2021; Khan et al. 2023). Urban areas, marked by sprawling developments, weak infrastructure, low disaster resilience, and limited coping mechanisms, are increasingly experiencing the escalating frequency and intensity of disasters (Bansal et al. 2017).

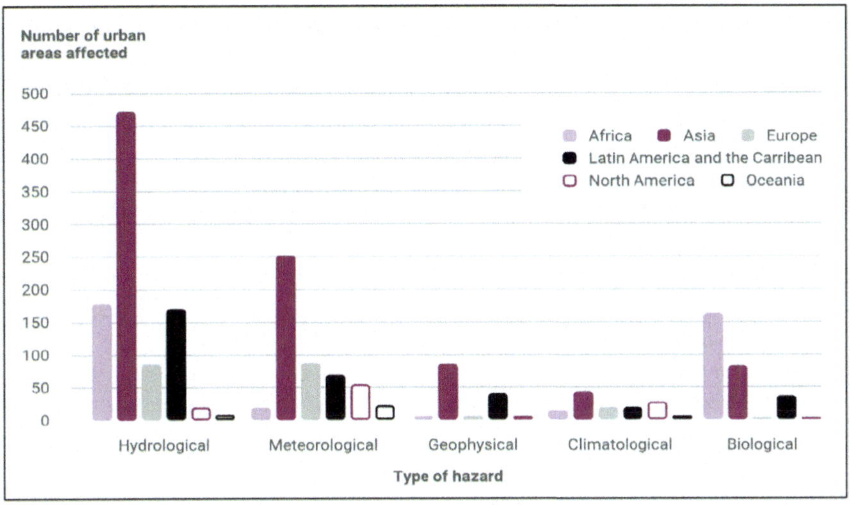


**Fig. 20.1** Number of urban areas with population over 7,50,000 affected by disasters. *Source* Gencer and UNDRR (2017)

The absence of robust early warning systems exacerbates disaster-related destruction's financial and social toll (Yun and Kim 2021; Khan et al. 2023). Moreover, rising global temperatures are intensifying extreme weather phenomena like heatwaves, storms, droughts, and floods, while also contributing to the proliferation of tropical diseases, thus endangering public health. Approximately 890 million people—roughly 60% of the urban population—reside in cities exposed to significant risks from disasters such as floods, droughts, cyclones, and earthquakes. Projections suggest that natural hazards could claim 100,000 lives annually, while the economic toll from these disasters is expected to reach around $300 billion annually by 2050 (UNISDR 2002; WSSD 2002). Consequently, the recurrence of disasters poses a persistent threat to people and development, potentially erasing decades of progress and leading to fatalities, injuries, economic setbacks, and the deterioration of environmental and urban systems (see Fig. 20.1) (Bansal et al. 2017).

## Background and Context

### *Urban Development and Disaster Risk Reduction (DRR) Strategies*

In the last 20 years, the idea of resilience has gained popularity within scientific and policy circles as many global communities, especially in urban areas, are dealing with a rising number of hazards and vulnerabilities (see Fig. 20.2). With the anticipated rise in frequency and severity of climate-related disasters, pandemics, and geopolitical conflicts, such resilience-building initiatives are yet to receive more attention soon (Sharifi 2020). Thus, the contemporary urban development framework highlights an urgent need for resilient urban planning. Cities act as a larger interconnected system, comprising individual, intricate urban systems like transportation, water supply, sanitation, housing, and other infrastructures. Significant disruptions to these urban infrastructures, services, livelihoods, and public health underscore the importance of implementing robust disaster risk reduction strategies to foster urban resilience. A resilient city minimizes risks from disasters by strengthening its capacity to address climate challenges, disasters, and economic shocks, thereby reducing its vulnerabilities.

In this context, cities have the potential to become models for environmental efficiency. Despite challenges posed by uneven and poorly planned urbanization, emerging cities possess distinct advantages. They are well-positioned to tackle global challenges like climate change at a localized and actionable level, facilitating immediate and effective communication between communities and decision-makers. By capitalizing on the co-benefits of climate change mitigation and adaptation, such cities can transform into "resilient" and "sustainable cities" (World Bank 2010).

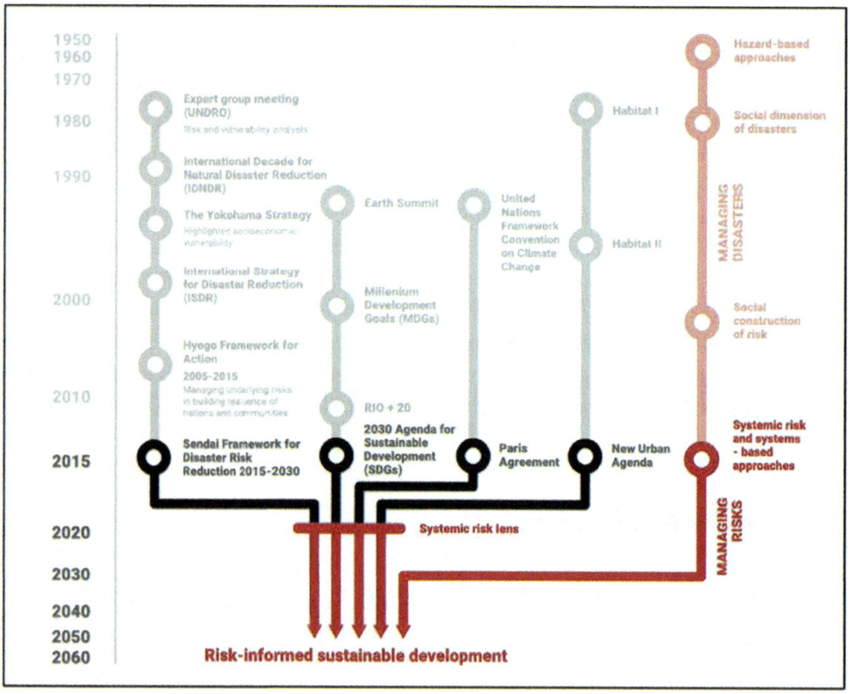


**Fig. 20.2** Spatio-temporal depiction of risk reduction strategies. *Source* United Nations Office for Disaster Risk Reduction (2019)

## *The Rise of Sustainable 'Smart' Cities*

Over the past decade, the concept of the "smart city" has emerged as a major leitmotif in the discourse of urban development, symbolizing a vision of a technologically advanced global urban future. Building on the legacy of the sustainable (or eco) city, the smart city has gained prominence as a digitalized approach to addressing urban challenges (Crivello 2015; Joss et al. 2019). Its widespread appeal is evident in the transnational exchange of smart city ideas and policies, as well as the rapid implementation of initiatives across regions in both the Global North and South (Caragliu et al. 2014; Datta 2015; Watson 2015; Karvonen et al. 2018).

The smart city is conceptually defined as "an ultramodern urban area that addresses the needs of businesses, institutions, and especially citizens through the usage of ICT," and it operates as an "urban development model" that leverages human and technological capital within urban settings (Angelidou 2015; Khatoun and Zeadaly 2016; Ahvenniemi et al. 2017). In practical terms, the smart city is associated with the strategic use of advanced ICT-enabled tools and applications that emphasize the interconnectedness of various aspects of urban life, including governance, economy, energy, mobility, security, living standards, well-being, and environmental management (Visvizi and Lytras 2019).

Joss et al. (2019) highlight several themes and critical perspectives from academic literature that frame key debates surrounding smart urban development. First, the literature critiques neoliberal urban policies, focusing on how smart city initiatives promote new forms of technocratic governance and corporatization of city management (Townsend 2013; Söderström et al. 2014; Vanolo 2014; Angelidou 2015; Calzada and Cobo 2015; Hollands 2015; Kitchin 2015; Przeybilovicz et al. 2018). Second, it conceptualizes smart cities as examples of "experimental urbanism," characterized by the creation of novel urban spaces and practices (Evans et al. 2016; Scholl and Kemp 2016; Caprotti and Cowley 2017; Raven et al. 2017). Third, the literature examines how smart cities influence norms of citizenship, public engagement, and the public sphere (Gabrys 2014; Saunders and Baeck 2015; Cowley et al. 2017; Joss et al. 2018; Cardullo and Kitchin 2018a, b). Finally, it scrutinizes the relationship between smart cities and sustainable development (Haarstad 2017; Kudva and Ye 2017; Trindade et al. 2017; Chang et al. 2018; Lytras et al. 2019).

The concept of "sustainable smart cities" gains additional relevance through the United Nations Sustainable Development Goals, particularly Goal 11: "Sustainable Cities and Communities" (United Nations 2015), which emphasizes the need for holistic, resilient, and sustainable[1] urban development (Klopp and Petretta 2017; Lützkendorf and Balouktsi 2017). Smart cities, through ICT integration, offer significant potential to enhance disaster risk resilience. IT-based systems collect environmental data and establish interconnected networks for real-time data transmission

[1] I Sustainability, here, goes beyond the notion of environmental sustainability to include socio-economic equality among individuals and societies, fostering the growth and improvement of current and future generations of city citizens, exercising civil liberties, and promoting a balanced and secure standard of living (Visvizi and Lytras, 2019).

during disaster management (Sakhardande et al. 2016; Ray et al. 2017). This data is processed and analyzed using big data and data mining techniques to improve decision-making (Rahman et al. 2017). Additionally, cloud technology facilitates data sharing, security, and backup for disaster management departments (Habiba and Akhter 2013). AI applications process UAV and satellite imagery to extract actionable insights (Amit and Aoki 2017; Ivić 2019), while remote sensing and GIS tools support disaster monitoring and risk reduction (Westen 2013; Kaku 2018).

City governments can further reduce disaster risks by employing building monitoring technologies (Cheng et al. 2017), UAVs, and sensors tailored to urban infrastructure (Erdelj et al. 2017a, 2017b). ICT plays a critical role across all four phases of disaster management: The pre-disaster phase, the Disaster response phase, the Post-disaster phase, and the Policy phase (Bansal et al. 2017; Kaur et al. 2022). By adopting flexible land-use policies, targeted investments, and community-based decision-making, smart cities can create sustainable, resilient, and inclusive urban environments, aligning with the Sustainable Development Goals.

## DRR Strategies and Smart Cities: Perspectives from a 'Gendered' Lens

Rapid urbanization, environmental degradation, climate change, poverty, and weak governance continue to be the primary drivers of global disaster risks in this century. These factors intersect across the economic, social, and environmental dimensions of sustainable development (UNDRR 2019). Addressing these challenges requires a shift from merely managing disasters to adopting a holistic approach to managing risks, which has led to the global emphasis on disaster risk reduction (DRR) strategies. The Sendai Framework for Disaster Risk Reduction 2015–2030 was followed by several global agreements incorporating DRR and resilience into their frameworks. It includes the 2030 Agenda for Sustainable Development, the Paris Agreement on Climate Change, the New Urban Agenda (NUA), the Addis Ababa Action Agenda (AAAA), and the Agenda for Humanity (UNFCCC 2016; United Nations General Assembly 2015b, 2016a, 2017b; Peters et al. 2016; Garschagen et al. 2018; Murray et al. 2017). The disaster risk management approach, therefore, enhances resilience and adaptive capacity, addressing climate-related hazards and natural disasters while supporting sustainable development objectives.

In both adaptation and disaster risk reduction (DRR) domains, there is growing recognition of the need for gender-responsive actions to address climatic and disaster risks. Vulnerability varies among groups due to factors such as socio-economic conditions, geography, demographics, culture, politics, and the environment. Gender, alongside other socio-cultural characteristics such as age, social status, health, wealth or poverty, and disability, plays a critical role in shaping an individual's vulnerability (IPCC 2012).

Women often face heightened impacts of climate change due to structural disadvantages, including lower income, limited education, lack of political representation, and exclusion from decision-making at household and community levels (Valero and Kaul 2023). These socio-economic inequities leave women with fewer assets and fewer resources to cope with adverse effects. Their increased reliance on natural resources for livelihoods further amplifies their vulnerability. For example, climatic stress on water and forest resources forces women and girls to travel longer distances to collect water or firewood, exposing them to health risks (WHO 2011) and reducing their opportunities for education, political engagement, and economic participation (UNDP 2011; FAO 2011; Habtezion 2016). These challenges highlight the importance of integrating gender-sensitive approaches into climate and disaster risk responses.

Similarly, disasters tend to disproportionately impact women more severely than men. Research indicates that disaster fatality rates are significantly higher for women, primarily due to gendered differences in coping capacities and limited access to information and early warning systems (Ikeda 1995; Oxfam 2005; Neumayer and Plümper 2007). Women often live and work near natural resources and areas most vulnerable to disasters, such as marginal lands and informal settlements.

Cultural norms and gender roles further compound these challenges, sometimes restricting women's ability to make quick decisions in disaster scenarios. In certain contexts, factors such as traditional clothing, unpaid caregiving responsibilities for children, the elderly, or the sick, and household obligations can limit women's mobility and capacity to escape disaster zones, particularly in water-related hazards (Habtezion 2016). These compounded vulnerabilities underscore the urgency of implementing gender-inclusive disaster risk reduction (DRR) strategies. Such approaches are essential for fostering sustainable and smart urban development in contemporary times.

As a result, gender inclusivity has become a pivotal element in building resilient cities, intersecting with various dimensions of identity, including sex, age, class, caste, (dis)ability, and ethnicity. This emphasis is reflected in the success of major post-2015 multilateral frameworks such as the 2030 Agenda, the Paris Agreement, and the Sendai Framework, all of which prioritize gender equality and the empowerment of women (Habtezion 2016).

## *Gender-Inclusive DRR Strategies in Smart Cities: The Indian Context*

India is currently undergoing rapid urbanization, with 36.6% of its population—approximately 520 million people—living in urban areas. This accounts for 11% of the global urban population as of 2023 (World Bank 2024). As the world's second-largest urban system, India boasts 7935 urban settlements, including 53 cities with populations exceeding one million (UN-Habitat 2023). Urban areas contribute two-thirds of the nation's GDP, cementing their role as crucial economic hubs. Recognizing the transformative potential of urbanization, the country has launched several flagship urban initiatives, including Housing for All, the Clean India Mission, the Smart Cities Mission (SCM), and the Atal Mission for Rejuvenation and Urban Transformation (AMRUT). These programs aim to improve ease of living, foster responsible governance, and promote sustainable environments while driving economic growth and employment opportunities. Together, they have reshaped over 4800 cities and towns by implementing inclusive, participatory, and data-driven approaches to urban development.

Among these, the SCM stands out with its goal to create a hundred "Smart Cities" by 2020 (see Fig. 20.3). It focuses on Area-Based Development (ABD), including redevelopment (50 acres), retrofitting (500 acres), and Greenfield development (250 acres), alongside pan-city innovations to improve infrastructure. A "Smart City" is characterized by adequate water supply, reliable electricity, sanitation, efficient public transport, affordable housing, governance, IT connectivity, citizen safety, and a sustainable environment (NIUA 2023).

While India's urban initiatives focus on environmental sustainability, the gender dimension of disaster risk reduction (DRR) remains inadequately addressed. An analysis of 28 State Action Plans on Climate Change (SAPCCs) highlighted that, despite recognizing the gendered impacts of climate change, most plans lacked concrete gender-specific measures (Singh et al. 2021). This gap underscores the urgent need for inclusive strategies to build equitable resilience for women and marginalized communities, Sustainability is central to the SCM, with projects addressing river and water body restoration, rainwater harvesting, solar plants, waste-to-energy plants, and air quality monitoring. The Climate Smart Cities Assessment Framework (2019) (Govt. of India 2020) evaluates climate-relevant parameters to align urban development with the National Mission on Sustainable Habitat, one of the eight climate missions under the National Action Plan on Climate Change (NAPCC).

Women bear disproportionate vulnerabilities during urban disasters. For DRR strategies within Smart Cities to be truly gender-inclusive, several actions are essential: designing gender-sensitive infrastructure such as safe public spaces, shelters, and accessible transportation systems; incorporating women's voices in urban planning and governance to reflect diverse perspectives; ensuring equitable access to resources like housing, healthcare, and sanitation during disaster recovery; and integrating gender-sensitive communication and safety protocols into early warning systems. As aspirational hubs of innovation, Smart Cities have the potential to lead by example, fostering inclusive, resilient, and sustainable urban environments by addressing gender disparities in disaster preparedness and response.

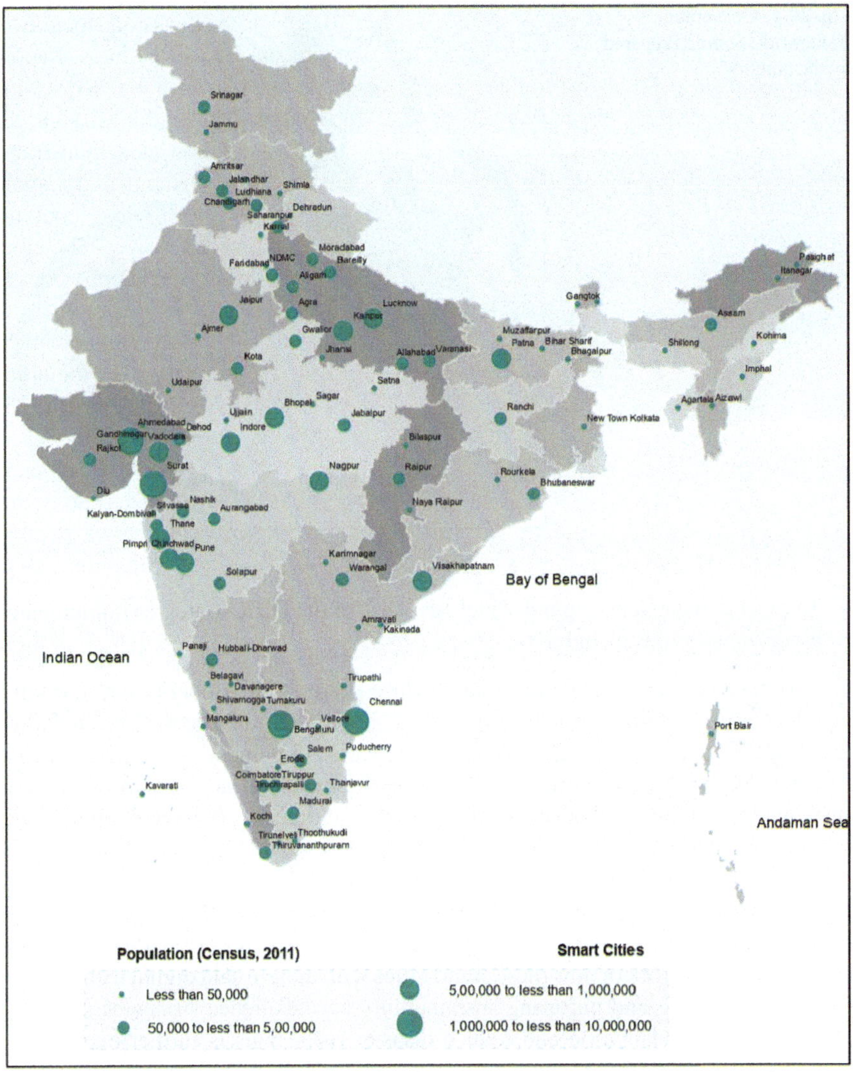


**Fig. 20.3** Map depicting the hundred smart cities of India. *Source* UN-Habitat (2023)

## Objectives and Conceptual Framework of Analysis

This paper examines how smart cities in India contribute to disaster resilience, with a particular focus on gender inclusiveness. The study is guided by two primary objectives:

- To identify the approach and strategies of Disaster Risk Resilience (DRR) in ICT-driven smart cities of India

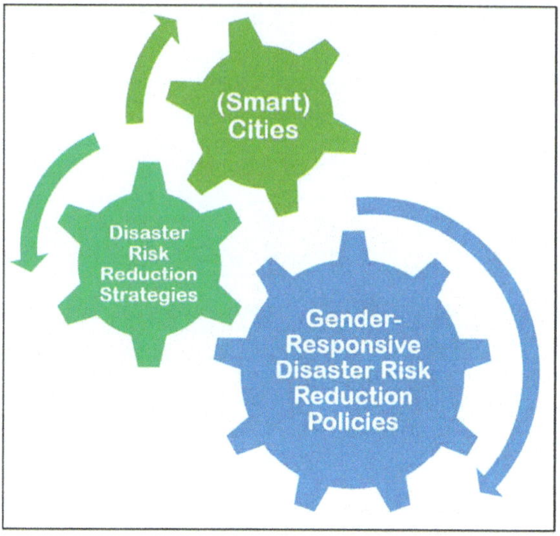


**Fig. 20.4** Conceptual framework. *Source* Prepared by the authors

- To explore the extent of gender inclusiveness of the DRR strategies implemented by the smart cities of India

Fostering urban resilience has taken centre stage at the global policy level and is reflected in the formulation of multifarious international frameworks—including the Sendai Framework, the 2030 Agenda, the Paris Agreement, and the New Urban Agenda (NUA). All these frameworks are built on the central objective of creating sustainable, resilient, inclusive, and safe, human settlements (United Nations 2015) that further reiterates the Sustainable Development Goal (SDG) number 11 which states—'to make cities and communities inclusive, safe, resilient, and sustainable.' In contemporary society, such needs for sustainable settlements have been reimagined using the newfound notion of 'smart' cities. The current paper operates under the conceptual framework of how these smart cities with their in-laid digital technologies (ICT) are fostering and nurturing sustainability and resilience strategies to further enhance climate adaptation and reduce disaster risks. This is further explored by adopting a gendered lens to focus on the vulnerable communities, in this case, women, who face a higher risk from climate change-induced disasters in their everyday lives and livelihoods (see Fig. 20.4).

## Research Design and Methodology

This study adopts a qualitative research approach to analyze the integration of gender-inclusive disaster resilience strategies in India's Smart Cities Mission, focusing particularly on the role of information and communications technology (ICT) and other climate adaptation techniques. By analyzing secondary data sources, this study evaluates disaster resilience planning through a gendered lens, identifying key patterns, strengths, and weaknesses in current urban planning frameworks. This section details the research approach, data sources, and analytical framework employed in this study.

### *Research Approach*

The research employs a qualitative research design primarily focusing on content and policy analysis approach to explore the nuances of disaster risk reduction (DRR) strategies in urban planning and ICT applications across selected Indian smart cities. This method is particularly suitable for analyzing complex policy documents and evaluating how these policies reflect international frameworks, such as the Sendai Framework for Disaster Risk Reduction (UNDRR 2015) and the Sustainable Development Goals (United Nations 2015).

### *Data Sources*

The study relies exclusively on secondary data sources, including:

- Policy Documents: Official plans, Smart city challenge proposals, self-assessment frameworks, and city-level policies developed under the Government of India's Smart Cities Mission (2015). These documents provide insights into how each city envisions and implements disaster resilience measures, ICT applications, and gender inclusivity.
- International Frameworks: Key international policies, such as the Sendai Framework (UNDRR 2015), the New Urban Agenda (UN-Habitat 2016), and the Sustainable Development Goals (United Nations 2015), are used as benchmarks for evaluating city resilience and inclusivity.
- Academic Literature: Peer-reviewed articles and reports from organizations like the World Bank (2010) and UN-Habitat are reviewed to frame the discourse around urban resilience, ICT, and gender inclusivity. Works by Ahvenniemi et al. (2017) and Khatoun and Zeadaly (2016) on smart cities and ICT applications provide a conceptual basis for assessing the technological integration within the cities.

### *Sampling*

A categorical sampling method is employed to systematically assess urban resilience across various levels of DRR planning and the extensive number of cities within the Smart Cities Mission to ensure a representative cross-section of resilience planning quality. This involves categorizing the 100 smart cities into three groups based on preliminary evaluations:

- **Comprehensive Plans**: Cities such as Indore, New Delhi, Coimbatore, and Chennai exhibit comprehensive DRR frameworks with notable ICT and gender inclusivity efforts.
- **Average Plans**: Cities like Pune, Jaipur, and Bhubaneswar demonstrate moderate ICT integration and gender-focused strategies, with room for improvement.
- **Inadequate Plans**: Cities including Shillong, Agra, and Guwahati show limited ICT application and minimal gender considerations within DRR strategies.

This categorization helps systematically analyze the variation in disaster resilience strategies, allowing for targeted analysis and comparison across the groups.

### *Methodology*

The analysis is conducted through a three-pronged approach involving content and policy analysis, thematic coding, and comparative analysis, enabling a comprehensive evaluation of policy documents. (1) A thematic content analysis approach codes and categorizes key themes within DRR policies. (2) Policy documents are examined for explicit references to ICT applications, gender considerations, and disaster resilience strategies. The cities within each category are compared based on the degree of ICT integration and the inclusivity of DRR strategies using resilience and gender inclusivity metrics to determine best practices and areas for improvement. (3) Comparative analysis reveals best practices, common challenges, and areas needing improvement, aligning with global resilience standards (UNDRR 2019).

## Major Findings

### *Climate Vulnerabilities and Responses in Indian Smart Cities*

Climate smart development addresses the challenges posed by changing climatic conditions through sustainable actions that enhance urban living standards (Sharifi and Yamagata 2022). In India, the Smart Cities Mission integrates sustainability and resilience into urban infrastructure projects to mitigate climate-related disasters, emphasizing the need for disaster risk reduction (DRR) strategies that ensure the holistic and inclusive well-being of all city residents (UN-Habitat and Ministry of Housing and Urban Affairs 2023). Across a hundred Smart Cities, impacting over a hundred million people, efforts have focused on low-carbon development, rapid adoption of energy-efficient technologies, and investments in climate-resilient infrastructure at the local level (Ministry of Housing and Urban Affairs 2020). Key areas of intervention include energy and green buildings, urban planning, green cover and biodiversity, mobility and air quality, water management, and waste management (Ministry of Housing and Urban Affairs 2020).

This section identifies specific Indian smart cities that have implemented climate-related initiatives and examines their contributions to DRR strategies and community well-being. The analysis draws from three major policy documents: "Smart Cities and Academia Towards Action and Research—Part B Climate Change and Resilient Cities" by the National Institute of Urban Affairs (2023), "Climate Smart Cities Assessment Framework 2.0" by the Ministry of Housing and Urban Affairs (2020), and "Smart Cities Mission, India: Localizing Sustainable Development Goals" by UN-Habitat and the Ministry of Housing and Urban Affairs (2023).

Most of the Indian smart cities face their unique climate vulnerabilities—ranging from flooding and urban heat islands to air pollution and water scarcity. To combat these challenges, responses have been developed that reflect a mix of technological, infrastructural, and community-driven strategies. Upon analyzing the government policy documents and/or reports, eight major smart cities—Surat, Coimbatore, Bhubaneswar, Indore, Delhi NCR, Pune, Jabalpur, and Ahmedabad—were identified that faced different types of disaster risks. The intensity of their risks ranged from moderate to high (see Table 20.1).

Flooding remains a major problem in the cities of Surat, Coimbatore, and Bhubaneswar during heavy monsoons. The city of Surat deals with excessive flooding from river Tapi. Similarly, Coimbatore is also at a high risk of flooding which gets worse due to waterlogging with clogged drains. Bhubaneswar surrounded by river Mahanadi and its distributaries like Kuakhai, Kathajodi, etc., and being exposed to severe tropical cyclones remains highly susceptible to flooding. Urban heat islands are predominantly found in the cities of Thanjavur, Surat, Delhi, Indore, Bhubaneswar, and Ahmedabad, where summer temperatures rise to extreme levels. Cities like Indore and Pune face high risks of air pollution and solid waste accumulation, although Delhi remains one of the cities to be worst affected by air pollution and smog formations. Pune and Bhubaneswar further face conditions of water scarcity leading to drought-like situations, especially during summers when temperatures become extremely high. The problem of solid waste accumulation is predominantly seen in the cities of Indore, Jabalpur, and Surat which can be directly accounted to the high density of population.

To counter these challenges, under the Smart Cities Mission, each city has formulated and implemented projects tailor-made to curb the rising deficits and restore the ease of living. Interventions like Canal Corridor Development with greenbelts, wetland restoration, flood risk mapping, drainage upgrades, water recycling, etc., became the major flood-resistant infrastructure to be implemented in the cities. Increased urban greenery, rooftop solar panels, green buildings, heat action plans, vertical gardens, and green public spaces arose as the major responses to combat urban heat islands. While air pollution remains a major concern across numerous cities, technological infrastructures like waste-to-energy plants, non-motorized transport systems (NMT), digital display boards for monitoring and information, Bio-CNG projects, phased electrification of public transport, and reduction of methane

**Table 20.1** Identification and management of disaster risks in different smart cities

| Hazard/disaster | Exposure level | Vulnerable smart cities | Shared vulnerabilities | Interventions | Key case studies |
|---|---|---|---|---|---|
| Flooding | High | Surat, Coimbatore, Bhubaneswar | • Urban flooding due to inadequate drainage<br>• Loss of green cover and natural buffers<br>• High urbanization pressure | • Flood risk mapping, *drainage upgrades*, *water recycling*<br>• Restoration of water bodies and Rainwater harvesting and retrofitting of drainage systems (Coimbatore, Bhubaneswar)<br>• Canal corridor development (Surat) | **Surat**: Canal Corridor *Development—Flood-resistant* infrastructure with greenbelts<br>**Coimbatore**: Wetland restoration as natural flood buffers |
| Urban Heat Islands | High–moderate | Delhi NCR, Indore, Ahmedabad | • High density and reduced green cover<br>• Heat stress affects vulnerable populations | • Increased urban greenery, rooftop solar, and green buildings<br>• Urban greening initiatives (Ahmedabad, Indore)<br>• Energy-efficient buildings and Rooftop solar plant installation (Delhi NCR, Thanjavur) | **Ahmedabad**: Heat Action Plan–First in South Asia, focused on vulnerable groups<br>**Indore**: Vertical gardens and green public spaces |

(continued)

**Table 20.1** (continued)

| Hazard/disaster | Exposure level | Vulnerable smart cities | Shared vulnerabilities | Interventions | Key case studies |
|---|---|---|---|---|---|
| Air pollution | High | Delhi NCR, Indore, Pune, Surat | • High vehicular emissions<br>• Industrial pollutants<br>• Waste burning | • Waste-to-energy plants, non-motorized transport (NMT)<br>• Adoption of clean energy (Bio-CNG in Indore)<br>• Implementation of Clean Air Action Plans (Delhi NCR, Pune)<br>• Greenbelt development along industrial zones for lowered industrial air pollution | **Indore**: Bio-CNG Project—Converts organic waste into clean energy, reducing methane emissions<br>**Delhi NCR**: Phased electrification of public transport |
| Drought/water Scarcity | Moderate | Pune, Bhubaneswar | • Depleting groundwater levels<br>• Over-reliance on monsoon | • Water-efficient systems<br>• Rainwater harvesting (Bhubaneswar)<br>• Community-led water management initiatives<br>• Recycling and reuse (Pune) | **Pune**: Decentralized water management—Reuse and recycling initiatives<br>**Bhubaneswar**: Women-led community water conservation projects |
| Solid Waste Accumulation | High | Indore, Jabalpur, Surat | • Limited landfill space<br>• Methane emissions from unmanaged waste | • Bio-CNG conversion, scientific landfill management<br>• Waste-to-energy plants (Jabalpur)<br>• Bio-CNG projects (Indore) | **Indore**: Bio-CNG—transforming organic waste into clean energy<br>**Jabalpur**: Scientific landfill remediation and energy recovery |

*Source* Authors findings

emissions were implemented as countermeasures in the major cities of Delhi, Pune, Indore, etc. Along similar lines, bio-CNG conversion and scientific landfill management were also used to resolve the challenge of solid waste accumulation.

As drought-like conditions remain rampant throughout the country, especially during summers, major cities fell back on methods of rainwater harvesting, water recycling and reuse, and community-led water conservation techniques under the aegis of technology-driven solutions of smart cities.

Therefore, cities like Surat, Indore, Bhubaneswar, Coimbatore, Delhi NCR, Ahmedabad, Thanjavur, Bhopal and so on, have already implemented interventions that directly counter the urban climate challenges and build resilience strategies. However, despite these efforts, there is still a need for greater attention to how these plans impact women and marginalized communities, who often face disproportionate climate risks.

## *Gender-Driven Climate Strategies and Indian Smart Cities: A Comparative Analysis*

The DRR plans implemented by the different smart cities often focus on infrastructural improvements using technological aids, such as enhancing urban drainage, managing waste, expanding green cover, and transitioning to renewable energy sources (see Fig. 20.5). Yet. At the same time, these initiatives are essential for mitigating environmental risks, and the integration of gender-responsive actions remains a major area for improvement. This section analyzes the operationalisation of gender in climate response plans of smart cities in India. There remains a significant gap, with policies often failing to translate gender-aware rhetoric into concrete, gender-responsive actions and resource allocations.

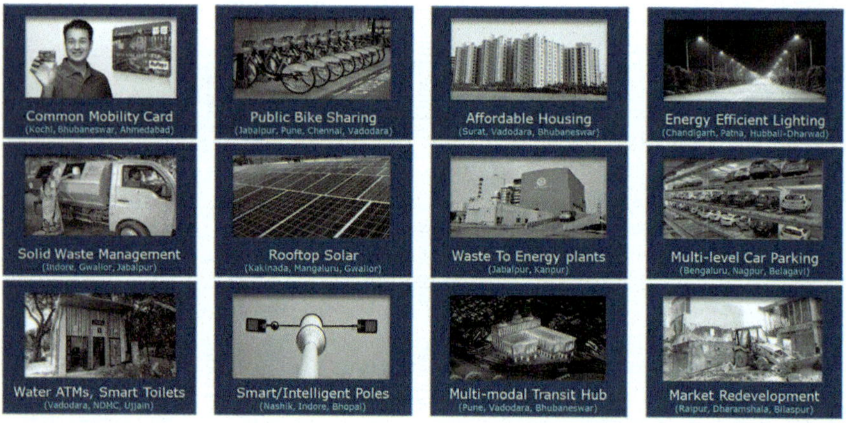


**Fig. 20.5** ICT interventions for climate resilience in smart cities. *Source* UN-Habitat (2023)

Most of the smart cities have not even considered the gender element in their plans, or discussions on the vulnerability of certain population groups or their adaptation actions. Only a few plans come close to acknowledging women as the marginal and vulnerable groups and this has been found in only a handful of major smart cities like Surat, Coimbatore, Indore, Bhubaneswar, Delhi NCR, Ahmedabad, Bhopal, Thanjavur, Kochi, Varanasi, and Pune based on the analysis of policy documents of the government. Tables 20.2, 20.3 and 20.4 highlight the major smart cities which have incorporated gender-inclusive initiatives in their smart urban planning. Cities like Delhi NCR, Indore, Ahmedabad, and Kochi showcase extensive DRR strategies with high ICT usage and strong gender incorporation. Following this, cities like Bhopal, Coimbatore, Pune, and Varanasi have average plan layouts with basic DRR strategies comprising moderate ICT usage and meagre gender inclusivity. The rest of most smart cities, specifically Shillong and Guwahati, have inadequate DRR strategies, coupled with negligible ICT usage and lack of gender inclusivity (see Table 20.2 and Fig. 20.6).

The eight major smart cities of Surat, Indore, Bhubaneswar, Coimbatore, Delhi NCR, Bhopal, Ahmedabad, and Thanjavur have implemented key projects of DRR with elements of gender inclusiveness that have been identified as case studies for this study (see Tables 20.3 and 20.4). Out of 100 smart cities formulated in India, these eight cities are the only ones where the needle has moved in the positive direction with their gender-inclusive conceptualization of climate-related and DRR interventions. Some of the case studies are discussed in further detail below (NIUA 2023, MoHUA 2020; UN-Habitat and MoHUA 2023).

The Surat Municipal Corporation undertook the Canal Corridor Development project to address unhealthy conditions along the city's canals and maximize road utilization. Spanning over 2500 m, this initiative features a 100-foot-wide greenbelt

**Table 20.2** Types of disaster risk reduction strategies in different smart cities

| Plan type | Disaster resilience strategies | ICT integration | Gender inclusivity | Smart cities |
|---|---|---|---|---|
| Comprehensive plans | Extensive DRR frameworks, early warning systems | High ICT use: IoT, GIS, big data analytics for resilience | Strong gender policies, including women in DRR plans | New Delhi, Indore, Ahmedabad, Kochi |
| Average plans | Basic DRR strategies, limited public involvement | Moderate ICT use, primarily for data collection and alerts | Some gender-focused initiatives, but are not comprehensive | Bhopal, Coimbatore, Pune, Varanasi |
| Inadequate plans | Minimal DRR initiatives, lacking clear resilience goals | Limited or no ICT applications in DRR | Few or no policies addressing gender inclusivity | Shillong, Guwahati |

*Source* Authors findings

**Table 20.3** Smart cities with gender-inclusive disaster resilience projects

| Smart cities | Disaster resilience measures | ICT integration | Gender inclusivity |
|---|---|---|---|
| Ahmedabad | • Flood control measures<br>• Disaster preparedness (e.g., emergency response systems)<br>• Public safety infrastructure | • Real-time monitoring systems<br>• GIS-based urban planning<br>• Smart traffic management | • Equal access to public spaces<br>• Women's safety measures (lighting, transport options) |
| Coimbatore | • Emergency response systems<br>• Enhanced healthcare access<br>• Improved public safety | • High-speed internet<br>• Wi-Fi-enabled zones<br>• E-governance platforms for citizen engagement | • Accessible public infrastructure<br>• Promoting gender-responsive governance<br>• Education and employment for women |
| Indore | • Expanded emergency health services<br>• Integrated ambulance systems<br>• Community disaster awareness programs | • ICT tools for smart governance<br>• Citizen engagement platforms<br>• Citywide Wi-Fi network | • Gender-sensitive urban planning<br>• Affordable housing for all<br>• Accessible public spaces for women and marginalized groups |
| Bhubaneswar | • Focus on emergency response systems<br>• Strengthened health infrastructure<br>• Redevelopment of parks for disaster refugees | • Citywide Wi-Fi project<br>• Digital grievance redressal platforms<br>• Integrated command control centre | • Improved safety measures for women in public spaces<br>• Equal access to upgraded housing initiatives |
| Pune | • Strengthened public transport and e-rickshaws for disaster recovery<br>• Slum rehabilitation programs | • High-speed internet and ICT-enabled services<br>• Centralized data management for governance | • Affordable housing development for marginalized groups<br>• Safer streets and public infrastructure for women |

*Source* Authors findings

comprising local, flowering, and medicinal plants, creating a healthier and more sustainable urban environment. The project integrates tree plantations to enhance air quality, reduce urban heat, and mitigate floods while incorporating recreational facilities like play areas and gathering spaces for the community. Additionally, it promotes gender-inclusive urban spaces by ensuring well-lit areas, safe walkways, and improved accessibility for women, children, and the elderly, fostering a more inclusive and resilient urban landscape.

**Table 20.4** Key case studies of gender-inclusive disaster resilience projects

| Smart cities | Project | Climate objective | Gender-responsive element | Impact |
|---|---|---|---|---|
| Surat | Canal Corridor Development | • Reduce urban heat<br>• Improve air quality<br>• Mitigate flooding | • Gender-inclusive public spaces (well-lit areas, safe walkways)<br>• Accessibility for women, children, elderly | • Enhanced safety and usability of public spaces<br>• Encouraged participation of diverse groups |
| Indore | Bio-CNG Project | • Sustainable waste management<br>• Reduce GHG emissions | • Inclusion of women in operational roles and awareness campaigns | • Empowered women through skill-building and employment |
| Bhubaneswar | Rainwater Harvesting and Community Water Management | • Improve water security<br>• Mitigate water scarcity | • Community-driven approach with significant participation of women<br>• Training for women in water conservation techniques | • Increased community resilience to water scarcity<br>• Women-led decision-making in water resource management |
| Coimbatore | Restoration of Urban Water Bodies | • Enhance urban biodiversity<br>• Manage flooding | • Women's participation in designing and maintaining public spaces<br>• Livelihood opportunities in urban farming along water bodies | • Improved access to safe and green spaces for women<br>• Economic benefits for women in local communities |
| Delhi NCR | Clean Air Action Plan | • Mitigate air pollution | • Women are employed to monitor air quality through local NGOs | • Promote gender equality in environmental monitoring roles |
| Ahmedabad | Energy-Efficient Street Lighting | • Reduce urban carbon footprint<br>• Improve safety | • Focus on improving safety for women and children with better-lit public areas | • Reduced gender disparity in public space usage during evenings |

(continued)

**Table 20.4** (continued)

| Smart cities | Project | Climate objective | Gender-responsive element | Impact |
|---|---|---|---|---|
| Thanjavur | Rooftop Solar Initiatives | • Promote renewable energy adoption | • Training for women in solar panel installation and maintenance | • Enhance women's technical skills and job opportunities |
| Bhopal | Smart Mobility with Non-Motorized Transport (NMT) | • Reduce emissions<br>• Improve mobility options | • Create gender-inclusive NMT pathways to ensure safety and accessibility | • Increased mobility for women and children<br>• Reduced transport-related gender barriers |

*Source* Authors findings

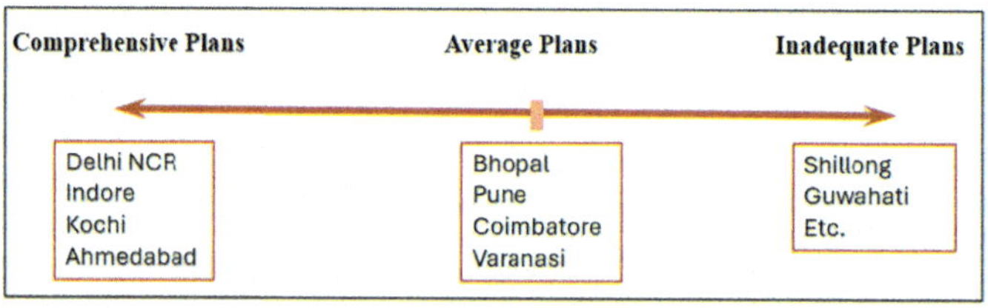


**Fig. 20.6** Type of disaster risk reduction strategies in different smart cities. *Source* Authors findings

Coimbatore's smart city initiative focuses on revitalizing its eight-lake network through an Eco-Restoration Master Plan, which includes connecting greenways to restore environmentally sensitive areas and create sustainable public spaces around the lakefronts. The concept of lake restoration under the smart city plan came up as an opportunity for the city to improve its water security and become more climate-resilient while also increasing the quality of its public realm for both its residents and tourists. While it enhanced urban biodiversity and managed the urban floods, it also facilitated women's participation in designing and maintaining public spaces, improved their livelihood opportunities in urban farming along water bodies, and improved their access to safe and green spaces.

Thanjavur City Municipal Corporation, under the smart city program, has proposed a 3 MW solar park at Sirajudeen Nagar WTP to advance non-conventional energy sources. This initiative highlights the critical role of renewable energy—particularly solar and wind in ensuring future sustainability and addressing global climate mitigation goals. By focusing on clean, zero-emission energy production, the project not only contributes to building energy security but also supports local economic development, creating opportunities for growth and community well-being. They further provide training for women in solar panel installation and maintenance and enhance women's technical skills and job opportunities.

Indore Smart City Development Limited implemented the Carbon Credit Aggregator Model, incorporating technologies such as composting, bio-methanation, energy-efficient lighting, sewage treatment plants, bioremediation of legacy waste, and decentralized composting. These green initiatives not only advanced environmental sustainability and monetization but also emphasized gender inclusivity. Women were actively included in operational roles and awareness campaigns, empowering them through targeted skill-building and employment opportunities, thereby fostering both ecological and social resilience.

Similarly, the Kochi Smart City plan includes initiatives to empower women through skill development and entrepreneurship programs within broader climate-resilient projects, as well as the creation of safe public spaces. Bhubaneswar Smart City also focuses on improving access to public services and infrastructure for marginalized communities, including women. Bhubaneswar has successfully integrated gender into water conservation programs with its women-led rainwater harvesting initiatives. It demonstrates how local women's knowledge can be harnessed to improve climate resilience.

## Discussion

The current research found that within the Indian smart city project, several cities have recognized the need for climate adaptation plans and addressed gender-specific vulnerabilities as well. The majority of Indian smart cities face unique climate vulnerabilities. Flooding remains a major problem in the cities of Surat, Coimbatore, and Bhubaneswar, while Thanjavur, Surat, Delhi, Indore, Bhubaneswar, and Ahmedabad, deal with urban heat formations. Cities like Delhi, Indore, and Pune face air pollution and solid waste accumulation. Pune and Bhubaneswar face water scarcity. To curb their rising deficits, DRR plans have been implemented by different smart cities focusing on infrastructural improvements using technological aids. Canal Corridor Development, wetland restoration, flood risk mapping, drainage upgrades, water recycling, etc., became the major flood-resistant infrastructure. Increased urban greenery, rooftop solar panels, green buildings, heat action plans, vertical gardens, and green public spaces arose to combat urban heat islands. Technological infrastructures like waste-to-energy plants, non-motorized transport systems (NMT), digital display boards, bio-CNG projects, phased electrification of public transport, and reduction of methane emissions were implemented as countermeasures to air pollution. Bio-CNG conversion and scientific landfill management were used for solid waste management.

Methods of rainwater harvesting, water recycling and reuse, and community-led water conservation techniques are popular to counter water scarcity. Of these, only a handful—Surat, Coimbatore, Indore, Bhubaneswar, Delhi NCR, Ahmedabad,

Bhopal, Thanjavur, Kochi, Varanasi, and Pune—showcase extensive DRR strategies with high ICT usage and strong gender incorporation. Thus, a comprehensive analysis of these city-level climate plans in this study reveals that the integration of gender-responsive strategies and resource allocation remains limited across the majority of the Indian smart cities. Bridging this gap is crucial for building resilient and equitable urban communities in the face of the climate crisis.

Furthermore, resource allocation remains a significant constraint. Even when gender-responsive strategies are outlined, they are rarely backed by adequate funding. Budgets often prioritize infrastructure-heavy projects, such as roadways or digital connectivity, over interventions that could address social vulnerabilities, such as childcare facilities or safe public spaces. This skewed allocation reflects a broader societal undervaluation of care work and gender-specific needs, perpetuating structural inequalities. Thus, several barriers contribute to the limited integration of gender-responsive strategies in DRR across Indian smart cities. A few of these are noted here:

*Lack of Gender-Disaggregated Data*[2]—Many cities do not collect or utilize gender-disaggregated data when planning climate adaptation measures. This absence of data limits the ability of urban planners to understand and address the specific vulnerabilities that women face, such as their disproportionate exposure to heat or lack of access to disaster relief resources. Moreover, the lack of such data undermines efforts to track women's participation in climate resilience initiatives or quantify the benefits they receive from these programs.

*Insufficient Institutional Capacity*—There is a widespread lack of capacity within urban governance systems to mainstream gender considerations into climate adaptation projects. While gender-sensitive policies may be drafted, their implementation is often hindered by a lack of technical knowledge, resources, and dedicated personnel to ensure that gendered perspectives are fully integrated into urban planning.

*Gender Gaps in Decision-Making*—Women's representation in urban planning and decision-making bodies remains low, leading to climate adaptation measures being implemented that often fail to reflect their needs and priorities. This lack of representation in decision-making exacerbates the marginalization of women in climate-resilient urban development.

*Inadequate Resource Allocation*—Even when gender-responsive actions are outlined, they often lack the necessary financial backing. Gender-focused initiatives are often sidelined in favour of larger infrastructure projects that prioritize technological innovation or physical infrastructure over social inclusivity. This reflects a systemic bias in urban development priorities, where tangible outputs are valued

[2] Gender-disaggregated data refers to the collection and analysis of information separated by gender, highlighting differences in experiences, vulnerabilities, and access to resources between men and women. In the context of urban climate adaptation, such data is critical for understanding how climatic and disaster risks uniquely impact women.

more highly than addressing underlying social inequalities. The absence of gender-inclusive climate finance further exacerbates this issue. Most climate finance frameworks often fail to integrate gender as a core consideration, leaving women and marginalized groups underrepresented in funding allocation.

## Limitations of the Study

While this study provides valuable insights, there are several limitations. The reliance on secondary data means that the analysis is limited to information available in public documents, which may not capture all aspects of DRR planning and implementation. Additionally, varying levels of policy detail across cities may introduce inconsistencies in comparative analysis. The evolving nature of ICT in disaster resilience and the variation in gender inclusivity definitions may also affect the study's findings.

## Policy Implications and Conclusion

As climate change continues to impact urban environments and their residents, cities are increasingly adopting strategies to enhance resilience and inclusivity. However, a critical aspect of these strategies integrating gender perspectives often remains overlooked. To address these inequities, urban climate adaptation strategies must prioritize gender inclusivity as a core component. This section outlines actionable measures to incorporate gender-responsive approaches into urban planning and climate adaptation efforts.

*Collection of Gender-Disaggregated Data:* By prioritizing the collection of gender-disaggregated data, cities can better understand the unique vulnerabilities faced by women and marginalized groups. This data should inform both the design and monitoring of urban climate adaptation plans, ensuring these initiatives are tailored to meet the specific needs of women.

*Tracking Women's Participation and Benefits:* Gender data should track women's participation in climate projects and quantify the benefits they receive, enabling more inclusive and equitable urban adaptation strategies.

*Expanding Women's Roles in Urban Planning:* Expanding women's roles in urban planning can integrate gender perspectives into climate adaptation strategies, making them more inclusive, responsive, and sustainable. Currently, in many Indian cities, women's involvement in climate resilience planning remains peripheral, with limited input into projects that directly affect their lives.

*Training for Gender Responsiveness:* Training urban planners, policymakers, and other stakeholders in gender responsiveness is essential to inculcate a sense of inclusivity in project development and implementation.

*Gender Audits of Climate Projects:* Conducting gender audits of climate projects can help identify areas where women's needs have been overlooked, ensuring projects are more equitable and effective in addressing gender-specific concerns.

*Dedicated Financial Resources:* Gender-responsive strategies must be supported by dedicated financial resources. Smart cities should adopt gender-responsive budgeting to allocate adequate funds to initiatives specifically addressing women's needs, such as safe public spaces, sanitation facilities, and employment opportunities in green industries.

*Model Gender-Inclusive Projects:* Cities like Coimbatore and Bhopal offer successful models of community-driven, gender-inclusive climate projects. Expanding these initiatives citywide and ensuring women lead these projects will be crucial for scaling the impact of gender-responsive climate adaptation.

*Community-Driven Climate Adaptation:* Climate adaptation efforts should be community-driven, focusing on empowering women and marginalized groups to take on leadership roles, thereby promoting inclusivity and resilience at the grassroots level.

Addressing the gap in gender inclusivity necessitates institutionalizing gender mainstreaming, enhancing women's participation in decision-making processes, and securing adequate funding for climate resilience projects. These measures are crucial for fostering urban environments that are not only resilient to the impacts of climate change but also promote gender equity and social justice, ensuring holistic and inclusive urban development.

## References

Ahvenniemi H, Huovila A, Pinto-Seppa I, Airaksinen M (2017) What are the differences between sustainable and smart cities? Cities 60:234–245

Amit SNKB, Aoki Y (2017) Disaster detection from aerial imagery with a convolutional neural network. International electronics symposium on knowledge creation and intelligent computing (IES-KCIC). IEEE, pp 239–245

Angelidou M (2015) Smart cities: a conjuncture of four forces. Cities 47:95–106

Bansal N, Mukherjee M, Gairola A (2017) Smart cities and disaster resilience. In: Seta F et al. (ed) From poverty, inequality to smart city. Springer Transactions in Civil and Environmental Engineering

Benali N, Abdelkafi I, Feki R (2018) Natural-disaster shocks and government's behaviour: evidence from middle-income countries. Int J Disaster Risk Reduction 27:1–6. https://doi.org/10.1016/j.ijdrr.2016.12.014

Calzada I, Cobo C (2015) Unplugging: Deconstructing the Smart City. J Urban Technol 22(1):23–43

Caprotti F, Cowley R (2017) Interrogating urban experiments. Urban Geogr 38(9):1441–1450

Caragliu A, Bo CD, Nijkamp P (2014) Smart Cities in Europe. In: Deakin M (ed) Smart cities: governing, modelling and analyzing the transition. Routledge, London, pp 173–195

Cardullo P, Kitchin R (2018) Being a 'citizen' in the smart city: up and down the scaffold of smart citizen participation. The Programmable City, Working Paper 30, Maynooth University, Maynooth

Cardullo P, Kitchin R (2018) Smart urbanism and smart citizenship: the Neoliberal Logic of 'Citizen-focused' Smart Cities in Europe. The Programmable City, Working Paper 39, Maynooth University, Maynooth
Chang DL, Sabatine-Marques J, da Costa EM, Selig PM, Yigitcanlar T (2018) Knowledge-based, smart and sustainable cities: a provocation for a conceptual framework. J Open Innov Technol Market Complex 4(5):1–17
Cheng MY, Chiu KC, Hsieh YM, Yang IT, Chou JS, Wu YW (2017) BIM integrated smart monitoring technique for building fire prevention and disaster relief. Autom Constr 84:14–30
Cowley R, Joss S, Dayot Y (2017) The smart city and its publics: insights from across six UK cities. Urb Res Pract 11(1):53–77
Crivello S (2015) Urban policy mobilities: the case of turin as a smart city. Eur Plan Stud 23(5):909–921
Datta A (2015) New urban utopias of postcolonial India: 'entrepreneurial urbanization' in Dholera Smart City. Gujarat Dialog Hum Geogr 5(1):3–22
Dodman D et al (2022) Climate Change 2022: impacts, adaptation and vulnerability. In: Pörtner H et al (ed) Cambridge University Press, pp 907–1040
Erdelj M, Król M, Natalizio E (2017a) Wireless sensor networks and multi-UAV systems for natural disaster management. Comput Netw 124:72–86
Erdelj M, Natalizio E, Chowdhury KR, Akyildiz IF (2017b) Help from the sky: leveraging UAVs for disaster management. IEEE Pervasive Comput 16(1):24–32
Evans J, Karvonen A, Raven R (eds) (2016) The experimental city. Routledge, Abingdon
Food and Agriculture Organization (FAO) (2011) The state of food and agriculture 2010–11. FAO, Rome
Gabrys G (2014) Programming environments: environmentality and citizen sensing in the smart city. Environ Plan Soc Sp 32(1):30–48
Garschagen M, Porter L, Satterthwaite D, Fraser A, Horne R, Nolan M, Solecki W, Friedman E, Dellas E, Schreiber F (2018) The new urban agenda: from vision to policy and action. Plan Theor Pract 19(1)
Gencer EA, UNISDR (2017) Local government powers for disaster risk reduction: a study on local-level authority and capacity for resilience. UNISDR
Govt. of India (2015) Smart cities mission guidelines. Ministry of Housing and Urban Affairs, New Delhi. https://smartcities.gov.in/themes/habikon/files/SmartCityGuidelines.pdf
Govt. of India (2020) Climate smart cities assessment framework 2.0. Ministry of Housing and Urban Affairs, New Delhi
Haarstad H (2017) Constructing the sustainable city: examining the role of sustainability in the 'smart city' discourse. J Environ Planning Policy Manage 19(4):423–437
Habiba M, Akhter S (2013) A cloud-based natural disaster management system. Int Conf Grid Pervasive Comput 152–161
Habtezion S (2016) Gender and climate change: gender, adaptation and disaster risk reduction. Policy Brief 2, United Nations Development Programme
Hari V, Dharmasthala S, Koppa A, Karmakar S, Kumar R (2021) Climate hazards are threatening vulnerable migrants in Indian megacities. Nat Clim Chang 11:636–638
Hollands R (2015) Critical Interventions into the Corporate Smart City. Camb J Reg Econ Soc 8(1):61–77
Ikeda K (1995) Gender differences in human loss and vulnerability in natural disasters: a case study from Bangladesh. Indian J Gend Stud 2(2):171–193
Intergovernmental Panel on Climate Change (IPCC) (2012) Special report on managing the risks of extreme events and disasters to advance climate change adaptation. In: Field CB et al. (ed). Cambridge University Press, Cambridge
Ivić M (2019) Artificial Intelligence and geospatial analysis in disaster management. ISPRS Int Arch Photo Grammetry Rem Sens Spat Inform Sci 161–166
Joss S, Cook M, Dayot Y (2018) Smart cities: towards a new citizenship regime? A discourse analysis of the British smart city standard. J Urban Technol 24(4):29–49

Joss S, Sengers F, Schraven D, Caprotti F, Dayot Y (2019) The smart city as global discourse: storylines and critical junctures across 27 cities. J Urban Technol 26(1):3–34
Kaku K (2018) Satellite remote sensing for disaster management support: a holistic and staged approach based on case studies in Sentinel Asia. Int J Disaster Risk Reduction 33:417–432
Karvonen A, Cugurullo F, Caprotti F (2018) Inside smart cities: place, politics and urban innovation. Routledge, Abingdon
Kaur M, Kaur PD, Sood SK (2022) ICT in disaster management context: a descriptive and critical review. Environ Sci Pollut Res 29:86796–86814. https://doi.org/10.1007/s11356-022-21475-5
Khan MTI, Anwar S, Sarkodie SA, Yaseen MR, Nadeem AM, Ali Q (2023) Natural disasters, resilience building, and risk: achieving sustainable cities and human settlements. Nat Hazards. https://doi.org/10.1007/s11069-023-06021-x
Khatoun R, Zeadally S (2016) Smart cities: concepts, architectures, research opportunities. Commun ACM 59(8):46–57. https://doi.org/10.1145/2858789
Kitchin R (2015) Making sense of smart cities: addressing present shortcomings. Cambridge Reg Econ Soc 8(1):131–136
Klopp J, Petretta D (2017) The urban sustainable development goal: indicators, complexity and the politics of measuring cities. Cities 63:92–97. https://doi.org/10.1016/j.cities.2016.12.019
Kudva S, Ye X (2017) Smart cities, big data, and sustainability union. Big Data Cogn Comput 1(4):1–13
Lützkendorf T, Balouktsi M (2017) Assessing a sustainable urban development: a typology of indicators and sources of information. Procedia Environ Sci 38:546–553
Lytras MD, Visvizi A, Sarirete A (2019) Clustering smart city services: perceptions, expectations, responses. Sustainability 11(1669)
Ministry of Housing and Urban Affairs (2020) climate smart cities assessment framework 2.0
Ministry of Housing and Urban Affairs (2023) Climate change and resilient cities: Part B of SAAR
Murray V, Maini R, Clarke L, Eltinay N (2017) Coherence between the Sendai framework, the SDGs, the climate agreement, new urban agenda and world humanitarian summit, and the role of science in their implementation
National Institute of Urban Affairs (NIUA) (2023) Smart cities and Academia towards action and research: SAAR Part B—Climate Change and Resilient Cities
Neumayer E, Plümper T (2007) The gendered nature of natural disasters: the impact of catastrophic events on the gender gap in life expectancy, 1981–2002. Ann Assoc Am Geogr 97(3)
Oke TR, Mills G, Christen A, Voogt JA (2017) Urban climates. Cambridge Univ, Press
Oxfam (2005) The Tsunami's impact on women. Briefing note
Peters K, Langston L, Tanner T, Bahadur A (2016) Resilience across the post-2015 frameworks: towards coherence? Overseas Development Institute
Przeybilovicz E, Cunha MA, Macaya JFM, Albuquerque JPD (2018) A tale of two 'smart cities': investigating the echoes of new public management and governance discourses in smart city projects in Brazil. In: Proceedings of the 51st Hawaii international conference on system sciences. Theme Smart Cities and Governance.
Rahman MS, Di L, Esraz-Ul-Zannat M (2017) The role of big data in disaster management. In: Proceedings international conference on disaster risk mitigation, pp 23–24
Raven R, Sengers F, Spaeth P, Xie L, Cheshmehzangi A, de Jong M (2017) Urban experimentation and institutional arrangements. European Planning Studies, pp 1–25. https://doi.org/10.1080/09654313.2017.1393047
Ray PP, Mukherjee M, Shu L (2017) Internet of things for disaster management: state-of-the-art and prospects. IEEE 5:18818–18835
Sakhardande P, Hanagal S, Kulkarni S (2016) Design of disaster management system using IoT-based interconnected network with smart city monitoring. In: International conference on internet of things and applications (IOTA), pp 185–190
Saunders T, Baeck P (2015) Rethinking smart cities from the ground up. Smart City Expo World Congress (SCEWC) NESTA, London. http://www.smartcityexpo.com/en/

Scholl C, Kemp M (2016) City labs as vehicles for innovation in urban planning processes. Urb Plan 1(4):89–102
Sethi SS, Vinoj V (2024) Urbanization and regional climate change-linked warming of Indian cities. Nat Cit 1:402–405. https://doi.org/10.1038/s44284-024-00074-0
Sharifi A (2020) Urban resilience assessment: mapping knowledge structure and trends [Review]. Sustainability (Switzerland) 12(15):5918. https://doi.org/10.3390/SU12155918
Sharifi A, Yamagata Y (2022) Smart cities and climate-resilient urban planning. EPB: Urb Anal City Sci 49(5):1347–1353. https://doi.org/10.1177/23998083221102400
Singh C, Solomon D, Rao N (2021) How does climate change adaptation policy in India consider gender? An analysis of 28 state action plans. Clim Policy 21(7):958–975. https://doi.org/10.1080/14693062.2021.1953434
Söderström O, Paasche T, Klauser F (2014) Smart cities as corporate storytelling. City 18(3):307–320
Townsend AM (2013) Smart cities: big data, civic hackers, and the quest for a new Utopia. WW Norton, New York
Trindade EP, Hinnig MPF, Moreirade Costa E, Marques JS, Bastos RC, Yigitcanlar T (2017) Sustainable development of smart cities: a systematic review of the literature. J Open Innov Technol Mark Complex 3(11):1–14. https://doi.org/10.1186/s40852-017-0063-2
UNFCCC (2016) FCCC/CP/2015/10/Add.1
UN-Habitat (2023) Smart cities mission, India: localizing sustainable development goals. Ministry of Housing and Urban Affairs, New Delhi
United Nations (2015) About sustainable development goals (SDGs). United Nations (UN). https://www.un.org/sustainabledevelopment/sustainable-development-goals/
United Nations Development Programme (UNDP) (2011) Human Development Report (2011) Sustainability and equity: a better future for all. UNDP, New York
United Nations General Assembly (2015b) A/RES/69/313
United Nations General Assembly (2016a) A/70/709. https://undocs.org/A/70/709
United Nations General Assembly (2017b) A/RES/71/256*
United Nations International Strategy for Disaster Reduction (UNISDR) (2002) Natural disasters and sustainable development: understanding the links between development. Natural Disasters and Sustainable Development, Johannesburg
United Nations Office for Disaster Risk Reduction (2015) Sendai framework for disaster risk reduction 2015–2030. United Nations
United Nations Office for Disaster Risk Reduction (2019) Global assessment report on disaster risk reduction. United Nations Office for Disaster Risk Reduction (UNDRR), Geneva
UN-Habitat (2016) New urban agenda. In: United nations conference on housing and sustainable urban development (Habitat III) in Quito, Ecuador. United Nations
UN-Habitat and Ministry of Housing and Urban Affairs (2023) Smart cities mission: localizing sustainable development goals. UN-Habitat Report
Vanolo A (2014) Smartmentality: the smart city as disciplinary strategy. Urb Stud 51(5):883–898
Valero SD, Kaul S (2023) Why climate change matters for women? Research Highlight—Environment UN Women
Visvizi A, Lytras MD (2019) Smart cities research and debate: what is in there? Smart cities: issues and challenges. https://doi.org/10.1016/B978-0-12-816639-0.00001-6
Watson V (2015) The Allure of 'smart city' rhetoric: India and Africa. Dialog Hum Geogr 5(1):36–39
Westen CJ (2013) Remote sensing and GIS for natural hazards assessment and disaster risk management. Treatise Geomorphol 3:259–298
World Bank (2010) Cities and climate change: an urgent agenda. Urban Development Series Knowledge Papers
World Bank (2024) Gearing up for India's rapid urban transformation. Opinion Piece. World Bank Group. https://www.worldbank.org/en/news/opinion/2024/01/30/gearing-up-for-india-s-rapid-urban-transformation

World Health Organization (2011) Gender, climate change and health. www.who.int/globalchange/GenderClimateChangeHealthfinal.pdf

World Summit on Sustainable Development (WSSD) (2002) The Johannesburg Plan of Implementation (JPOI) adopted at WSSD

World Urbanization Prospects (2019) The 2018 revision. United Nations Department of Economic and Social Affairs

Yun SD, Kim A (2021) Economic impact of natural disasters: a myth or mismeasurement? Appl Econ Lett. https://doi.org/10.1080/13504851.2021.1896667

# Author Index

A
Abdullah, 377
Abhishek Kumar, 247
Amrita Sen, 105
Ananda Kumar Biswas, 123
Anindita Sarkar, 3
Aparajita Banerjee, 227
Aparna P. Krishnan, 141
Atia Sanjida Sushoma, 123

B
Badrun Susantyo, 377

C
Carlos A. Nobre, 65
Chandani Tiwari, 141

D
Debarchana Biswas, 105
Debora Pignatari Drucker, 65
Dhanya Punnoli, 183
Dibakar Sahoo, 21
Djaenudin, 377
Douglas Nyathi, 45

E
Euis Setiawati, 377

F
Farhat Naz, 269, 317

G
Gabriela Azevedo Couto, 65

H
Habibullah, 377
Hari Harjanto Setiawan, 377

I
Iffat Ara Jasmin, 123
Imran Hosen, 123
Ineu Maryani, 377

J
Jayarajan Kunnampalli, 183
Joram Ndlovu, 45
Junaid ul Shafi, 161

K
Karno, 377
Katyayini Sood, 247
Kritika Mishra, 269
Kunal Avishek Gounder, 87

L
Luthfi Setiadiwibawa, 377

M
Mahdi Ibrahim Tanjung, 377
Mamta Mehar, 339


A. Sarkar and N. Bandyopadhyay (eds.), *Gender-Transformative Approaches for Climate Change Adaptation*, Springer Climate,
https://doi.org/10.1007/978-3-031-90236-9

Maurício Trentin, 65
Mery Ganti, 377
Miguel Angel Trejo-Rangel, 65
Mumtaz Alam, 87

**N**
Nairwita Bandyopadhyay, 3
Neelam Singh, 21
Nkosilamandla Kunene, 45
Nyi Raden Irmayani, 377

**P**
Parul Gurjar, 291
Prasanta Moharaj, 21
Pratyay Raha, 105

**R**
Rituparna Bhattacharya, 397
Rizal Akbar Aldyan, 377
Ruslan Halifu, 377

**S**
Sakshi Bahuguna, 141
Sramana Mukherjee, 397
Sreeraj Punnoli, 183
Stanley Maphosa, 45
Suhandy Siswoyo, 377
Surabhi Singh, 317
Suranjana Karmakar, 201

**T**
Tanusree Paul, 201
Tarun Katoch, 291
Thembelihle Nyathi, 45
Twinkle Jaspal, 339

**V**
Vipasha Negi, 247

**Z**
Zelina Sultana, 355

# Subject Index


**A**

Adaptation, 3–9, 11–15, 21, 23, 25, 31–33, 38, 46, 49–51, 56, 57, 65, 71, 77, 87–89, 94, 96–98, 100, 101, 105–108, 114, 118, 119, 123, 138, 142–144, 154, 156–158, 175, 185, 195, 198, 202, 203, 205, 212, 219, 227–230, 237, 242, 243, 247, 248, 251, 252, 256–264, 269, 271, 274, 277, 278, 282, 283, 285–287, 292, 293, 295, 297, 318, 331, 339, 340, 344–347, 349, 350, 359, 365, 372, 379–382, 384, 386, 388–393, 400, 402, 406, 413, 417–420

Adaptation interventions, 6, 9, 197, 212

Agricultural practices, 32, 33, 38, 97, 143, 144, 147, 156–158, 227, 231, 248, 257, 263, 264, 296, 298, 299, 320, 329, 331, 342, 345, 346, 348

Agriculture, 6, 7, 12, 21–24, 27–29, 31, 33, 35, 36, 38, 47, 49, 51, 56, 91, 94, 97, 98, 100, 107, 108, 112, 115, 141, 142, 144–146, 149, 153–158, 164, 166, 184, 186, 189, 195, 196, 198, 202, 204, 212, 215, 216, 227, 228, 230, 231, 234–239, 241, 243, 250, 251, 253, 255, 257–260, 264, 269, 270, 287, 291–298, 302, 305, 308, 309, 312, 317–321, 323–325, 327–333, 339–341, 343, 345–347, 349–351, 356, 367

**B**

Bangladesh, 13, 89, 93, 107, 123–126, 128, 129, 135, 166, 169, 347, 356–360, 362–372

Bharathapuzha Basin, 187, 190

**C**

Climate action, 13, 21, 46, 47, 50, 57, 58, 69, 87, 90, 91, 95, 98, 100, 107, 138, 154, 155, 196, 212, 251, 258–261, 263, 264, 269, 271–277, 279, 281, 282, 285–287, 340, 358

Climate change, 3–5, 7, 8, 11–15, 21–25, 28, 29, 31–34, 36–39, 45–59, 71, 77, 79, 87–97, 99–101, 105, 107–110, 118, 119, 123–125, 127, 130, 138, 141–145, 147, 154–156, 161–167, 169, 171–175, 183–185, 190, 191, 194, 195, 197, 198, 202–204, 206, 207, 210, 212, 219, 220, 228, 229, 235, 241–243, 247–260, 262–264, 269–272, 274–278, 281, 282, 284–288, 291–294, 296–298, 302, 304, 309, 312, 317–321, 324, 327–333, 340, 341, 343–345, 347, 349, 351, 355–359, 361–372, 377–393, 397, 398, 400, 402–404, 406, 409, 419, 420

Climate change adaptation, 3–5, 8–10, 12, 14, 15, 23, 33, 49–51, 68, 69, 96, 119, 138, 143, 147, 148, 154, 202, 229, 230, 237, 243, 250, 251, 257, 271, 275–277, 283, 291, 345, 378, 379, 381, 384, 389, 391–393






A. Sarkar and N. Bandyopadhyay (eds.), *Gender-Transformative Approaches for Climate Change Adaptation*, Springer Climate,
https://doi.org/10.1007/978-3-031-90236-9

Climate disasters, 108, 111, 112, 115–118, 143, 360
Climate-related disasters, 13, 47, 56, 65, 66, 125, 163, 166, 258, 328, 366, 400, 408
Climate-resilient agriculture, 258, 302, 312, 339, 340, 343, 351
Climate-smart agriculture, 6, 12, 21, 27–29, 96, 155, 157, 184, 235, 237, 243, 257, 291, 293, 296, 297, 345, 350
Coastal communities, 105, 107, 377, 383, 384, 386, 388–391
Constructivism, 123, 125–127
Crisis, 45, 50, 55, 56, 66, 101, 108, 118, 126, 130, 133, 135, 142, 155, 166, 169–171, 191, 202, 211, 216, 220, 221, 228, 254, 270, 328, 329, 355–358, 360–362, 369–372, 379, 392, 418
Culture, 5, 15, 51, 73, 89, 108, 124, 145, 228, 243, 260, 288, 362, 365, 381, 402

**D**
Disaggregated data analysis, 66
Disaster management, 13, 105, 109, 110, 112, 117, 331, 359, 402
Displacement, 13, 45, 47–50, 54–57, 136, 161–165, 167–170, 172–175, 285, 292, 355–366, 368–372

**E**
Emotional geography, 202, 203, 213, 220, 221
Empowerment, 13–15, 21, 46, 55, 94, 96, 97, 100, 114, 138, 144, 145, 147, 149, 151, 154, 155, 172, 195, 230, 269, 271–277, 279, 281–287, 299, 311, 312, 317, 331, 339, 343, 344, 346–348, 351, 358, 366, 367, 369, 378, 386, 388, 392, 393, 403
Extreme climate, 32, 93, 96–99, 157, 164, 192, 197
Extreme events, 12, 67, 77, 88, 118, 123–126, 138, 185, 201, 356, 370

**F**
Feminist standpoint, 317, 320
Financial inclusion, 297, 303, 304, 312
Food security, 8, 13, 21–25, 27–29, 32–34, 37, 39, 50, 91–94, 115, 132, 141–148, 151, 153, 154, 156, 157, 165, 184, 248, 250, 255, 258–260, 287, 292, 317, 318, 327, 331, 346, 384, 391–393

**G**
Gender, 3–15, 21–29, 31–34, 36–39, 45–59, 65–69, 71–74, 76–78, 87–91, 94–101, 105–107, 109, 112–114, 117–119, 123–128, 130–132, 135, 138, 141, 143–145, 147–149, 151–158, 161–163, 165–175, 189, 196, 197, 201–203, 206, 213, 214, 227–230, 237–239, 241–243, 247, 250, 251, 259–261, 269–272, 274–279, 286, 287, 292–297, 304, 310–312, 318, 320, 321, 325, 326, 328, 330, 331, 339–351, 357–363, 365, 366, 369–372, 379, 384, 386, 392, 393, 397, 398, 402–408, 412–420
Gender-based violence, 45–50, 53–58, 89, 168, 169, 173, 175, 363, 364
Gender equality, 4, 5, 11, 14, 21, 25, 46, 50, 57–59, 71, 88, 89, 97, 98, 119, 124, 138, 151, 153, 155, 171, 172, 195, 196, 229, 242, 260, 269, 271, 274, 279, 287, 326, 330, 341, 343, 347, 348, 351, 365, 370, 378, 379, 384, 386, 403, 415
Gender-inclusive, 310, 351, 403
Gender inequalities, 4, 6, 10–12, 24, 48, 50, 51, 53, 66, 68, 71, 87–89, 97, 100, 130, 143, 144, 147–149, 151, 153, 161, 171, 184, 227, 229, 241, 260, 269, 275, 278, 292, 317–320, 323, 327, 332, 333, 342, 345–347, 355, 363, 371, 372, 386, 392
Gender integration, 339, 340, 343–345, 347, 349
Gender roles, 7, 9, 11, 33, 36, 49, 96, 109, 114, 123, 125–127, 131–133, 135, 138, 144, 151, 166, 201, 203, 240–243, 250, 264, 274, 288, 318–320, 324, 326, 330, 332, 342, 344, 348, 349, 378, 385, 389, 393, 403
Gender transformative approaches, 3, 124, 293, 342, 348, 349
Gender transformative climate change adaptation, 124, 349
Gender-transformative research methodologies, 4, 5, 10

Gender vulnerability, 6, 66, 124, 162
Global south, 4, 5, 7, 12, 21–25, 28–30, 37, 38, 56, 70, 78, 87–92, 94–101, 162, 171, 227, 228, 247, 250, 259, 262, 269, 271, 291, 356
Global warming, 118, 124, 210, 212, 228, 249, 377, 380

**H**
Himachal Pradesh, 247, 249, 258–261

**I**
India, 13, 32, 35–37, 91, 94, 95, 105, 111, 167, 183–185, 187, 201–203, 205, 227, 230, 231, 247, 250, 259, 260, 269–273, 278–282, 287, 292–296, 298, 304, 312, 317, 318, 320, 321, 326, 327, 329, 348, 364, 397, 403–409, 412, 413
Inequality, 4–7, 9–11, 14, 21, 32, 33, 37, 50, 51, 57, 58, 65–69, 71, 74, 77, 88, 90, 96, 109, 123–125, 127, 138, 147–149, 156, 165, 169, 171, 172, 174, 175, 228, 229, 242, 250, 254, 269, 270, 274, 284, 340, 342, 344, 346–350, 357, 358, 371, 389, 418, 419
Intersectionality, 9–11, 14, 47, 48, 66, 69, 71, 77, 78, 138, 169, 342, 356

**L**
Leadership, 98
Legal, 7, 9, 47, 52, 56, 66, 72, 91, 112, 154, 155, 158, 161–163, 165, 168–175, 236, 250, 293, 318, 326, 331, 332, 355, 357–359, 363, 365, 370–372
Legal reform, 175

**M**
Mal Pahariyas, 13, 227, 230, 231, 233–235, 237, 238, 241–243
Marginalized communities, 12, 13, 32, 47, 48, 237, 243, 269, 270, 272, 277, 284, 311, 404, 412, 417
Migration, 8, 12, 32, 36, 45, 47, 51, 56, 108, 123, 133, 134, 137, 161–172, 174, 175, 247, 248, 250, 251, 255, 285, 292, 317, 328, 359, 360, 362, 370, 379

**N**
Nutrition security, 227, 229, 230, 236, 237, 241, 320

**P**
Paddy farmers, 187
Policy intervention, 5, 10, 88, 90, 98, 144
Protection, 13, 46, 52, 54, 56, 58, 88, 96, 97, 108, 110, 161, 162, 169–175, 261, 277, 355–359, 365, 370–372, 384, 393

**R**
Refugees, 13, 45, 47, 76, 161, 162, 165–168, 170–175, 355, 358, 370, 414
Rice farming, 193, 202

**S**
Smart cities, 13, 397, 401–418, 420
Sobrevidas©, 65, 72–74, 77, 78
Sundarbans, 13, 105, 107–111, 113, 114, 116–119, 167, 368
Sustainable Development Goals (SDG), 4, 12, 21, 50, 97, 98, 171, 251, 269–272, 274–279, 282, 287, 288, 355, 357, 370–372, 406
Systematic literature review, 23, 29, 38

**T**
Transformation, 3–5, 8–15, 89, 101, 105, 108, 109, 138, 228, 243, 244, 263, 269, 270, 283, 312, 329, 342, 348, 351, 404

**U**
Urban planning, 397, 398, 400, 404, 407, 409, 413, 414, 418, 419

**V**
Vulnerabilities, 4, 6–9, 12–15, 21, 27, 36, 47–49, 51–55, 57, 65–69, 79, 87–90, 94, 96–98, 100, 105, 107–109, 111, 117, 118, 123, 125–127, 132–134, 136–138, 142–144, 156–158, 161, 162, 165, 166, 168–171, 173 175, 184, 228–230, 235, 236, 243, 247–250, 253–256, 275, 286, 291–293, 296, 298, 317–321, 328,

329, 333, 346, 349, 355–360, 362, 365–367, 369–372, 377–380, 385–387, 389, 392, 393, 397–400, 402–404, 408–411, 413, 417–419

**W**

Water, 8, 13, 23, 45, 47, 48, 50–53, 56, 67, 75, 87, 90–95, 97, 100, 107, 108, 110, 113, 115–118, 124, 131–137, 146, 156, 157, 163, 186, 191, 193–195, 197, 201–210, 212–221, 234–236, 240, 241, 247–249, 251, 255, 257, 258, 261, 264, 270, 272, 273, 276, 277, 281–283, 285, 287, 292, 296, 308, 309, 320, 321, 323, 328–332, 341, 356, 360, 362, 363, 369, 371, 378, 379, 387, 390, 392, 400, 403, 404, 409–412, 415–417

Weather warnings, 183

Woman resilience, 13, 21, 24, 34, 38, 77, 78, 87, 96–100, 105, 107, 114, 126, 127, 155, 165, 198, 256, 259–264, 281, 285, 287, 295, 296, 301, 312, 331, 333, 349, 379, 381, 382, 385, 390, 392, 393, 397, 404, 415, 419, 420

Women, 3–15, 21–25, 27, 31–39, 45–59, 65–68, 71, 74–77, 79, 87–101, 105–109, 111–114, 116–119, 123–128, 130–138, 141–149, 151–158, 161–163, 165–175, 183–185, 194–198, 201–203, 205, 208, 211–221, 227–231, 233–242, 247–264, 269–271, 273–276, 278–288, 291–299, 302–306, 308, 310–312, 317–321, 324–333, 339–351, 355–372, 377–382, 384–393, 397, 403, 404, 406, 411–420

Women capacities, 7, 66, 67, 77–79, 107, 144, 228, 238, 333, 349, 351, 385, 403

Women empowerment, 13–15, 21, 46, 55, 96, 97, 114, 138, 144, 145, 154, 155, 172, 195, 230, 269, 271, 273, 275, 282, 284, 285, 287, 311, 317, 331, 339, 346, 366, 367, 386, 388, 392, 393

Women farmers, 6, 13, 33, 37, 38, 56, 154, 156, 158, 183, 185, 193–198, 209, 221, 236, 258, 259, 291–294, 297–299, 301–304, 306–312, 317, 320, 324, 328, 333

Women farmers' drudgery, 183, 198

Women leaders, 9, 260, 309

The manufacturer's authorised representative in the EU is Springer Nature Customer Service Centre GmbH, Europaplatz 3, 69115 Heidelberg, Germany. If you have any concerns regarding our products, please contact ProductSafety@springernature.com

Printed and bound by CPI Group (UK) Ltd, Croydon, CR0 4YY
01/10/2026
02225871-0016